ADVANCES IN CHEMICAL PHYSICS

VOLUME 129

EDITORIAL BOARD

Advances in
CHEMICAL PHYSICS

Edited by

STUART A. RICE

Department of Chemistry
and
The James Franck Institute
The University of Chicago
Chicago, Illinois

VOLUME 129

AN INTERSCIENCE PUBLICATION
JOHN WILEY & SONS, INC.

For general information on our other products and services please contact our Customer Care Department within the U.S. at 877-762-2974, outside the U.S. at 317-572-3993 or fax 317-572-4002.

Wiley also publishes its books in a variety of electronic formats. Some content that appears in print, however, may not be available in electronic format.

Library of Congress Catalog Number: 58:9935

ISBN 0-471-44527-4

Printed in the United States of America

10   9   8   7   6   5   4   3   2   1

CONTRIBUTORS

A. I. BURSHTEIN, Department of Chemical Physics, The Weizmann Institute of Science, Rehovot 76100, Israel

MYUNG S. JHON, Department of Chemical Engineering and Data Storage Systems Center, Carnegie Mellon University, Pittsburgh, PA 15213-3890, U.S.A.

UDAYAN MOHANTY, Eugene F Merkert Chemistry Center, Department of Chemistry, Boston College, Chestnut Hill, MA 02467, U.S.A.

YURIY L. RAIKHER, Institute of Continuous Media Mechanics, Ural Branch of RAS, 1 Korolyev Street, Perm 614013, Russia

VICTOR I. STEPANOV, Institute of Continuous Media Mechanics, Ural Branch of RAS, 1 Korolyev Street, Perm 614013, Russia

INTRODUCTION

Few of us can any longer keep up with the flood of scientific literature, even in specialized subfields. Any attempt to do more and be broadly educated with respect to a large domain of science has the appearance of tilting at windmills. Yet the synthesis of ideas drawn from different subjects into new, powerful, general concepts is as valuable as ever, and the desire to remain educated persists in all scientists. This series, *Advances in Chemical Physics*, is devoted to helping the reader obtain general information about a wide variety of topics in chemical physics, a field that we interpret very broadly. Our intent is to have experts present comprehensive analyses of subjects of interest and to encourage the expression of individual points of view. We hope that this approach to the presentation of an overview of a subject will both stimulate new research and serve as a personalized learning text for beginners in a field.

STUART A. RICE

CONTENTS

PHYSICOCHEMICAL PROPERTIES OF NANOSTRUCTURED PERFLUOROPOLYETHER FILMS

MYUNG S. JHON

Department of Chemical Engineering and Data Storage Systems Center, Carnegie Mellon University, Pittsburgh, Pennsylvania, U.S.A.

CONTENTS

This chapter presents fundamental scientific tools as well as potential applications relevant to the emerging field of nanotechnology. In particular, understanding the behavior of molecularly thin lubricant films is essential for achieving durability and reliability in nanoscale devices, and the experimentation and theory for the physicochemical properties of ultrathin perfluoropolyether (PFPE) films are reviewed. A method for extracting spreading properties from the scanning microellipsometry (SME) for various PFPE/solid surface pairs and the rheological characterization of PFPEs are examined at length. The

Advances in Chemical Physics, Volume 129, edited by Stuart A. Rice
ISBN 0-471-44527-4   Copyright © 2004 John Wiley & Sons, Inc.

interrelationships among SME spreading profiles, surface energy, rheology, and tribology, are discussed as well. Phenomenological theories, including stability analysis and microscale mass transfer, are introduced to interpret ultrathin PFPE film nanostructures qualitatively. In addition, rigorous simulation tools, including a lattice-based simple reactive sphere model, the off-lattice bead-spring Monte Carlo method, and molecular dynamics method, are examined. These tools may accurately describe the static and dynamic behaviors of PFPE films consistent with experimental findings and thus will be suitable for describing the fundamental mechanisms of film dewetting and rupture due to instability arising from nanoscale temperature and pressure inhomogeneities. Nanotribological applications, such as finding an optimal disk lubricant based on a molecule-level interaction of the lubricant with solid surfaces, will be explored.

I. INTRODUCTION

Nanoscale confined polymers are important for their potential industrial applications. The functionalities of polymer chain and solid surfaces are key control factors in determining the material designs for these applications. A fluid confined in a nanoscale system will dramatically alter its structural and dynamic properties. Because of broad technological interest, numerous studies on nanoscale confined fluids have been investigated, both theoretically and experimentally by scientists and engineers from a variety of backgrounds, including data storage, synthetic catalysis, polymer synthesis and physics, tribology, robotics, and medicine [1]. The behavior of materials having constituents with dimensions on the nanometer scale is remarkably different from the behavior in bulk state, which has led to a new paradigm that we now refer to as *nanotechnology.*

Molecularly thin lubricant film is an important application of nanoscale confined polymeric fluids, and is the focus of this chapter. Ultrathin lubricant films are necessary in high-density data storage to increase the reliability and performance of hard-disk drive (HDD) systems [2–4]. Spinoff and intermittent contact between the slider (or head) and the lubricated disk [ultrathin perfluoropolyether (PFPE) films are applied to the disk's carbon-overcoated surface, as shown in Fig. 1.1] cause loss and reflow of the lubricant film. The relevant HDD technology is summarized briefly in the end-of-chapter Appendix Section A.I, which provides an overview of how certain information technology devices are controlled by nanoscale chemistry.

The lubricant dynamics can alter the nanoscale aerodynamics of the slider. Conversely, the lubricant morphology and dynamics may be altered because of the presence of the slider. For these types of applications, a molecule-level understanding of the lubricant interaction with nanoscale airbearing and solid surfaces is critical. The HDD industry must cope with problems of lubricant film uniformity, roughness [5], durability [6], and stability [7] in order to achieve its goal of increasing areal density.

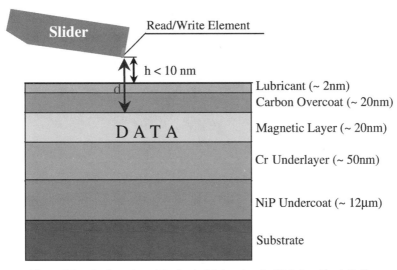

Figure 1.1. Configuration of the head–disk interface in HDD [see Fig. 1.4(*a*)].

The commercially available PFPE Z and Zdol (Montedison Co. [8] products) are random copolymers with the linear backbone chain structure

$$X-[(OCF_2-CF_2)_p-(OCF_2)_q]-X \quad (p/q \cong \tfrac{2}{3})$$

where X (endgroup) is CF_3 in PFPE Z and CF_2CH_2OH in PFPE Zdol. Note that Zdol has hydroxyl groups at both chain ends, which exhibit moderate interactions with solid surfaces, e.g., silica and carbon.

In addition, we examined PFPE ZdolTX [9] with "bulky" endgroups as a potential lubricant, which is shown below:

$$X = CF_2CH_2-(OCH_2CH_2)_{1.5}-OH$$

The structures of PFPE Z, Zdol, and ZdolTX are shown in Figure 1.2. Other PFPEs that have been investigated [10] include Ztetraol and AM2001; Xs are as follows:

$$\text{Ztetraol: } X = CF_2CH_2OCH_2CHCH_2-OH$$
$$\underset{OH}{|}$$

$$\text{AM2001: } X = CF_2CH_2OCH_2 \; \text{(benzodioxole ring with } CH_2 \text{)}$$

The use of additives, such as X1-P, may enhance the reliability of an HDD [11].

(a)

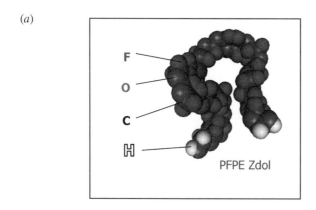

(b)

	Molecular Structure	PFPE/Solid Interaction	Backbone $(OCF_2\text{—}CF_2)_p\text{—}(OCF_2)_q$
			End Group
Z		weak	CF_3
Zdol		moderate	CF_2CH_2OH
ZdolTX		strong	$CF_2CH_2-(OCH_2CH_2)_{15}-OH$

Figure 1.2. (a) Molecular structure of PFPEs; (b) simplified view of (a); the larger circles indicate stronger endgroup interactions with the solid surface.

Scientists and engineers working in the information storage industries (e.g., Seagate, Hitachi, and Maxtor) have conducted numerous studies on PFPEs and their thin-film properties. Several academic institutions in the United States [University of California (Berkeley, Computer Mechanics Laboratory; San Diego, Center for Magnetic Recording Research), Ohio State University (Computer Microtribology and Contamination Laboratory), and Carnegie Mellon University (Data Storage Systems Center)] have been actively investigating the role of lubricants and their applications to data storage

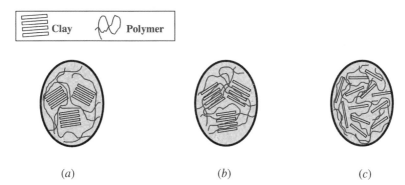

Figure 1.3. The microstructure of clay dispersed in the polymer matrix: (*a*) conventional composite; (*b*) extended polymer chains intercalated between the silicate layers, resulting in a well-ordered multilayer with alternation polymer/inorganic layers, and (*c*) silicate layers (1 nm thickness) exfoliated and dispersed in a continuous polymer matrix.

systems. The Data Storage Institute in Singapore, the largest such institute outside the United States, is involved in nanotribology research relevant to data storage as well. However, despite the plethora of research topics, we will focus only on fundamental scientific issues and our own findings regarding to data storage applications in this review. We will discuss the details of the PFPE experiments, qualitative analysis, and full-scale simulation. Although we will concentrate on selected highlights from our research, other topics will be touched on briefly.

Although not discussed in detail in this chapter, the scientific tools we explored and developed may be applicable to other areas of nanotechnology, such as nanocomposites. The nanocomposites have emerged as a new class of materials during the 1990s. For example, confined polymers solidify at temperatures well above the glass transition temperature as the intercalation rate slows down due to the increased affinity between the polymers and inorganic plate surface. It has been suggested [12] that the timescale of the intercalation process is relatively insensitive to the molecular weight of the polymer when compared to the diffusion coefficient in the confined slit. Various types of nanocomposites have been synthesized and characterized by our research group [13–20]. In addition to the well-known conventional composite (blend system), these nanocomposites can be classified into two distinct structures: intercalated and exfoliated (Fig. 1.3). Specific interactions through modified functional groups on polymer chains, tethered surface modifiers, and silicate surfaces play important roles in determining the molecular architecture for nanocomposites.

Using a variety of commercial and well-characterized polymers, intercalation has led to a wide variety of nanophase hybrids, many of which had not been previously synthesized using the traditional intercalation approaches. The

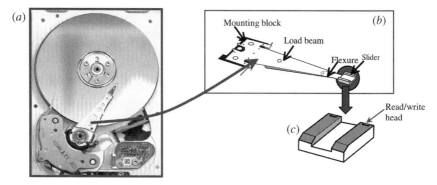

Figure 1.4. (*a*) HDD system; (*b*) schematic of suspension–slider assembly; (*c*) IBM 3370 slider.

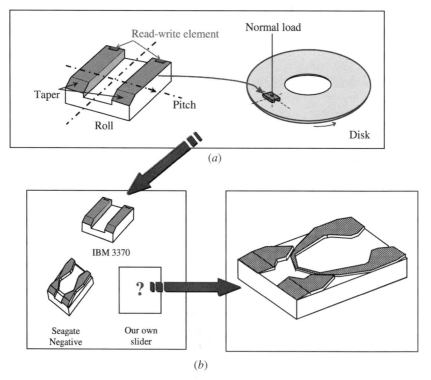

Figure 1.5. (*a*) Attitude and (*b*) shape of the slider.

intercalated nanocomposites are composed of macromolecules that are confined on a lengthscale that is small in comparison to the unperturbed molecular dimensions in a bulk state [21]. The energetic cost of perturbing chains is compensated by the topological "entanglement" constraints imposed on the chains.

Thermal and mechanical properties have been drastically improved by nanocomposites [12–33] dispersed with inorganic clays in a polymer matrix, which is characterized by nanometer lengthscale domains. These nanocomposite systems can be similarly examined by the methodology reported in this chapter.

Section A.1 of the Appendix discusses the background for the hard-disk drive (see also Figs. 1.4 and 1.5).

II. EXPERIMENTATION AND QUALITATIVE ANALYSIS

In this section, we will study spreading profiles measured from scanning microellipsometry (SME) and its relationship to various experiments, especially rheology, surface energy, and tribology. Phenomenological theory will be developed to interpret the various experiments as well as interrelation among experiments. Rheology of polymer melts and solutions and stability analysis based on thermodynamics will be studied at length as well.

Section A.2 of the Appendix discusses calculation of the film-thickness-dependent diffusion coefficient from a hydrodynamic model.

A. Scanning Microellipsometry

Figure 1.2 illustrates the various PFPEs. To demonstrate the essence of the spreading phenomena, we examined the "thought experiments" for different lubricant–surface and lubricant–lubricant interactions. This is sketched in Figure 1.6. O'Connor et al. [10,47,48] were the first to systematically carry out spreading experiments using SME to monitor monodisperse PFPE films on silica as they spread with time. Later, Ma et al. [49–52] reproduced and carefully examined these results for various carbon surfaces. Novotny [53], in actuality, pioneered the investigation of PFPEs, although his work mainly dealt with polydisperse PFPE samples. Our coworkers [10,47–52] carefully examined and extended the earlier work of Novotny by investigating the spreading of monodisperse PFPEs ($M_w/M_n < 1.1$, where, M_n and M_w are number average and weight average molecular weight, respectively) on silica and various carbon surfaces in a controlled temperature and humidity environment.

In the experiments of O'Connor et al. [10,47,48], monodisperse fractions of Z and Zdol, which were fractionated via supercritical fluid extraction in CO_2, were dip-coated onto the surface of silicon wafers. Film thickness was controlled

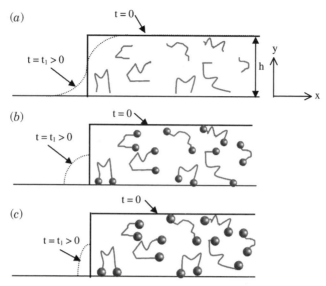

Figure 1.6. Sketch of the spreading profiles from SME as time progresses ($t = 0$ to t_1): (a) Z, (b) Zdol, and (c) ZdolTX. The coordinate system is drawn in (a).

by altering the PFPE concentration and draw rate as shown in Figure 1.7. For example, a draw rate of 6 mm/s with a PFPE concentration of 2 g/L generated a 50-Å thin film. An SME apparatus (its components are shown in Fig. 1.8) was used to measure the thickness of the film as it spread with time. The coated wafer was placed on a pedestal-like plate, housed in an environmental chamber with slits for passage of the incident and reflected beams. The chamber was mounted on a stage, which translated the sample across the beam area, and the thickness profile measurements were performed under a controlled temperature and humidity environment. The spreading profile obtained from SME is highly dependent on the PFPE–surface interactions, and provides a fingerprint for each pairing.

Typical SME thickness profiles for monodisperse Z and Zdol, as analyzed by O'Connor et al., are shown in Figures 1.9(a) and 1.9(b). As the film spreads with time, the spreading front travels along the surface of the silicon wafers. A sharp "spike" for the Z profile in Figure 1.9(a) was observed to decay over time. The spreading of Zdol exhibits a characteristic shoulder with a height on the order of the radius of gyration for the PFPE molecules, and the PFPE molecules separate from the initial film layer at a sharp boundary. In addition, Zdol appears to experience partial "dewetting" as indicated by the rough appearance of the SME scan beyond 2 mm [Fig. 1.9(b)]. This dewetting phenomenon (investigated

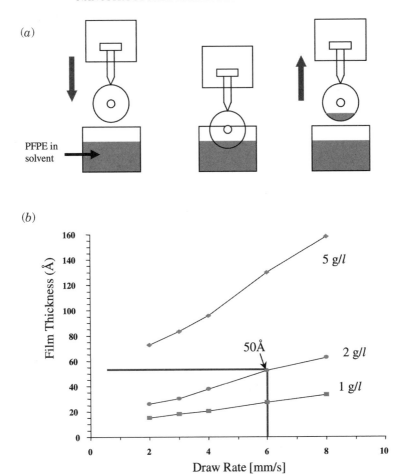

Figure 1.7. (*a*) Partial dip coating of a disk in a PFPE solution; (*b*) film thickness was determined from the PFPE concentration and draw rate.

for "mogul" dynamics [54,55] of nanoscale thin lubricant films) could become very important for airbearing design [40–44,56–59].

Similarly, Figures 1.10(*a*) and 1.10(*b*) represent Z and Zdol spreading experiments on amorphous carbon disks performed by Ma et al. [49–52] a few years after the observations made by O'Connor et al. Again, the Z front (unlike O'Connor et al. [10,47,48], they used polydisperse sample) traveled along the surface, gradually decreasing in height, and the Zdol profile showed a characteristic shoulder or layering structure. However, mass buildup for Z was

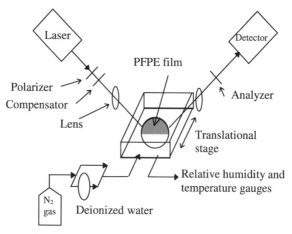

Figure 1.8. Schematic of the SME apparatus. The PFPE film is indicated in green.

not observed, and dewetting characteristics for Zdol were somewhat suppressed. Although the main features are similar, the major differences between the profiles in Figures 1.9 and 1.10 are the accumulation of mass at the leading edge of the profile shown in Figure 1.9(*a*) and also the severe dewetting characteristic shown in Figure 1.9(*b*). These phenomena clearly demonstrate the relevance of surface-energy-driven flow effects at nanometer-scale spreading processes, and could become important in establishing the design criteria for nanodevices or sensors. Figure 1.11 shows experimentation performed by Ma et al. [52,60] for various carbon surfaces (amorphous/hydrogenated/nitrogenated), illustrating the fingerprint for the various surface/PFPE molecular coupling. Note that Cazabat et al. [61–68] initially found a layered structure for polydimethylsiloxane (PDMS, without polar endgroups) in non-HDD applications and the spreading behavior of aqueous media containing a variety of nonionic surfactants has been investigated [69,70].

The spontaneous spreading of liquid films on solid surfaces described above has gained a considerable attention [71,72]. We examined the thickness-dependent diffusion coefficient $D(h)$, extracted from the experimental SME via two methods. The first method utilizes the spreading data at constant height (isoheight), while the second method utilizes the entire spreading profile. The first method does not assume spreading is a diffusion process, whereas the second method does. The length L, along which the leading edge of the advancing lubricant front traveled, is defined as the difference between the initial position of the leading edge and its position in each of the subsequent profiles. To determine the spreading rate, L and the corresponding times t obtained from

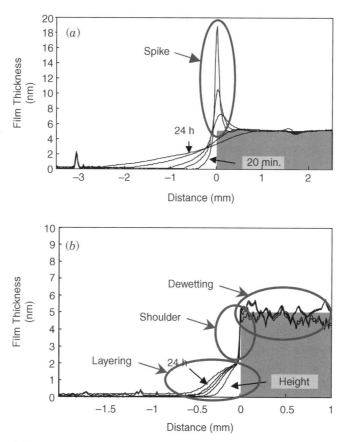

Figure 1.9. SME spreading profiles for PFPE on silicon wafers performed at IBM: (a) Z ($M_w = 13,800$ g/mol) and (b) Zdol ($M_w = 3,100$ g/mol). Both have initial thickness 5 nm for times of 20 min, 1.5 h, 3 h, 12 h, and 24 h at 26°C [47,48].

the profiles [Fig. 1.9] are plotted in Figure 1.12. The two piecewise straight lines in the log–log plot fit the data for the entire range of t (i.e., $L \propto t^{\alpha}$); that is, $\alpha \cong 1$ for short times (regime I), and after a gradual transition, $\alpha \cong \frac{1}{2}$ (regime II; diffusion region). Therefore, the surface diffusion coefficient D is estimated from the data in regime II ($L \propto t^{1/2}$) using the relationship $D = L^2/t$. We developed a second method called the *Boltzmann–Matano interface method* [50], which is more accurate for the purely diffusive process. Ma et al. [49–52] successfully employed this method as the spreading profiles they measured exhibit neither a severe mass buildup at the front nor dewetting. Because the distance of the migration of the film front is very small compared to the length

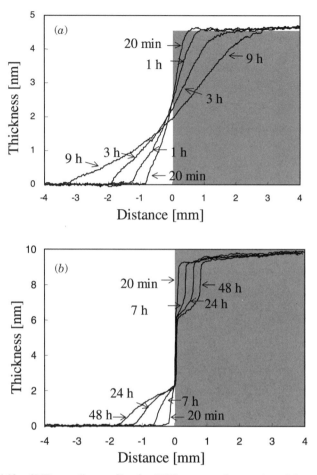

Figure 1.10. SME spreading profiles for PFPE on amorphous carbon disks performed at Seagate: (a) Z ($M_w = 2{,}500$ g/mol) with initial thickness 4.6 nm, and (b) Zdol ($M_w = 2{,}500$ g/mol) with initial thickness 10 nm at 26°C for times of (a) 20 min, 1 h, 3 h, and 9 h, and (b) 20 min, 7 h, 24 h, and 48 h [5,7].

of the boundary, the spreading can be described by the standard non-Fickian diffusion equation

$$\frac{\partial h(x,t)}{\partial t} = \nabla \cdot [D(h)\,\nabla h(x,t)] \tag{1.1}$$

where ∇ is the "del" (nabla) operator and $h(x,t)$ is the film thickness.

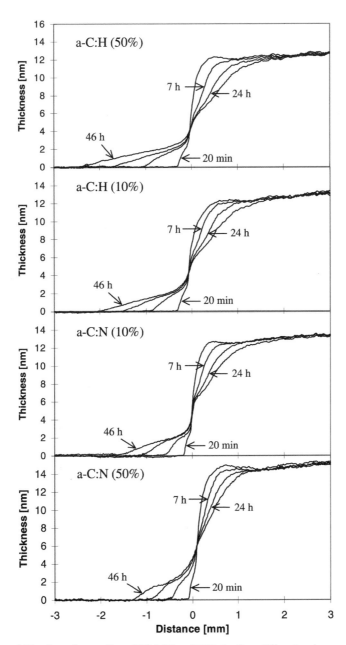

Figure 1.11. Spreading profiles of Zdol ($M_n = 2{,}500\,\text{g/mol}$) on different carbon overcoats at 20 min, 7 h, 24 h, and 46 h after coating the films [60].

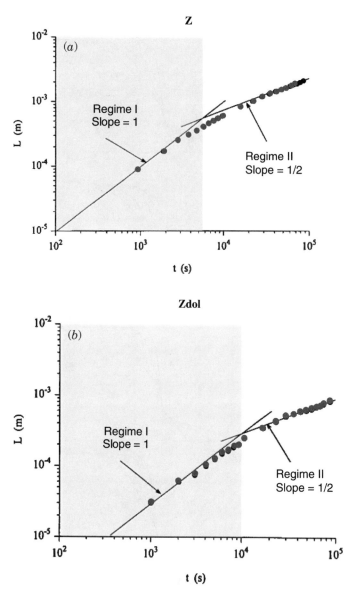

Figure 1.12. Travel length (L) as a function of time (t) obtained from Figure 1.9 for monodisperse (*a*) Z and (*b*) Zdol; $T = 26°C$ and the relative humidity of 0%.

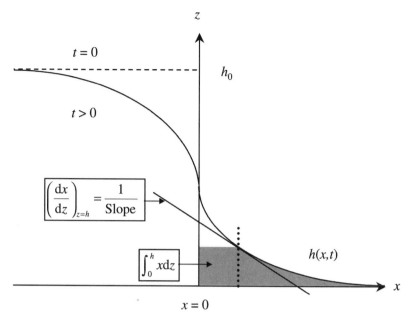

Figure 1.13. Components of the Boltzmann–Matano method in Eq. (1.2) applicable to an experimental spreading profile.

One can extract the diffusion coefficient from the Boltzmann–Matano technique (1D analysis) [50,52,73]:

$$D(h) = -\frac{1}{2t}\left(\frac{dx}{dz}\right)_{z=h}\int_0^h x\,dz, \text{ with the constraint } \int_0^{h_0} x\,dz = 0 \quad (1.2)$$

See Figure 1.13 for a schematic representation of h and the coordinates x and z.

B. Interpretation of L–t plot and D from Phenomenological Transport Model

Since the theories developed by Novotny [53] (conventional diffusion model) and Mate [74] (hydrodynamic model) fail to describe the $L \propto t$ behavior exhibited in regime I, the construction of a simple mesoscopic model (prior to investigating the molecular simulation; see Section III) has been suggested as described below to explain overall L–t behavior.

By introducing a time lag denoted as τ (which may be related to the relaxation time of the PFPEs), the flux vector **j**, and the concentration (or proportional to film thickness, h) gradient ∇h guarantee causality. Inspired by the microscale

heat transfer theory or Cattaneo equation in heat transfer theory [75–81], the
Fick's constitutive equation can be modified as [82,83]

$$\mathbf{j}(\mathbf{r}, t + \tau) = e^{\tau(\partial/\partial t)}\mathbf{j}(\mathbf{r}, t) = -D(h)\nabla h(\mathbf{r}, t) \tag{1.3}$$

Further, the mass conservation equation gives

$$\frac{\partial h(\mathbf{r}, t)}{\partial t} + \nabla \cdot \mathbf{j}(\mathbf{r}, t) = 0 \tag{1.4}$$

By combining Eqs. (1.3) and (1.4), and assuming $\tau\, \partial/\partial t$ to be small, we
obtained the following modified diffusion equation:

$$\tau \frac{\partial^2 h(\mathbf{r}, t)}{\partial t^2} + \frac{\partial h(\mathbf{r}, t)}{\partial t} = \nabla \cdot [D(h)\nabla h(\mathbf{r}, t)] \tag{1.5}$$

where τ quantifies the crossover behavior between regimes I and II. Note that Eq.
(1.5) reduces to Eq. (1.1) when $\tau = 0$. We found that the transition between these
two regimes depends on the value of τ. Increasing τ shifts the transition to a later
time. By setting $\tau = 10^3$ s in Eq. (1.5), we attained an excellent fit for the
experimental L–t data (Fig. 1.14).

From dimensional analysis arguments alone, we can qualitatively develop an
analysis of the L–t plot from Eq. (1.5), as follows. In regime II, the $\tau\partial^2 h/\partial t^2$
term can be neglected, and $h/t\ [=]\ Dh/L^2$ or $L^2/t\ [=]\ D$ (where L is the

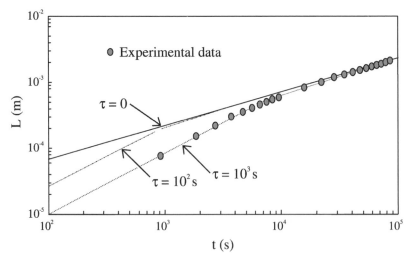

Figure 1.14. $L{-}t$ plot results using the modified diffusion equation [Eq. (1.5)] for several
values of τ.

characteristic length, t is the characteristic time, and [=] is read as "has the dimensions of"), implying $L^2 \propto t$. However, in regime I, the $\partial h/\partial t$ term becomes negligible and $\tau h/t^2$ [=] Dh/L^2 or $(t/L)^2[=]\tau/D$, implying $L \propto t$.

Waltman et al. [84] used the measured experimental disjoining pressure Π to qualitatively describe the spreading profile. Here, we will adopt an "inverse approach" to predict Π from the diffusion coefficient data. Note that Π can be calculated from the molecular simulation.

We developed the relationship between position-dependent viscosity $\eta(z)$, Π, and $D(h)$ via generalization of the hydrodynamic model (Reynolds equation):

$$D(h) = \int_0^h \frac{(h-z)^2}{\eta(z)} \left(\frac{\partial \Pi}{\partial h}\right) dz \tag{1.6a}$$

Or by assuming constant viscosity [i.e., $\eta(z) = \mu$ in Eq. (1.6a)], we obtain Eq. (1.6b) developed by Mate [74]:

$$D(h) = -\frac{h^3}{3\mu} \frac{\partial \Pi}{\partial h} \tag{1.6b}$$

A detailed derivation of Eq. (1.6a) using the no-slip boundary condition is provided in Section A.2 of the Appendix. If we were to generalize the analysis above with the partial-slip boundary condition, that is, $\partial v/\partial z = \beta v$ ($\beta \equiv$ slip parameter) instead of the no-slip condition in Eq. (A.2) at the lubricant/solid boundary with $\eta(z) = \mu$, we would obtain

$$D(h) = -\frac{1}{\mu} \left(\frac{h^2}{\beta} + \frac{h^3}{3}\right) \frac{\partial \Pi}{\partial h} \tag{1.7}$$

To procure a full-scale hydrodynamic model, we may need microrheological data [85–89] (especially for PFPEs with polar endgroups, i.e., Zdol). When the microrheological data are not available, one could use a simplified form of $\eta(z) = \mu_B f(z)$ to develop the improved hydrodynamic model. Here μ_B is the bulk viscosity and f is a function of z to be determined experimentally. A partial justification for the abovementioned functional form can be drawn from the temperature dependence of the surface diffusion coefficient and the bulk viscosity [10], or the fly stiction correlation with the bulk viscosity [9]. We examine the rheological properties of PFPE separately in Section II.C.

An alternative approach is based on the jump diffusion model adopted by Karis and Tyndall [90,91].

$$D(h) = -mh \frac{\partial \Pi}{\partial h} \tag{1.8}$$

where m is the mobility.

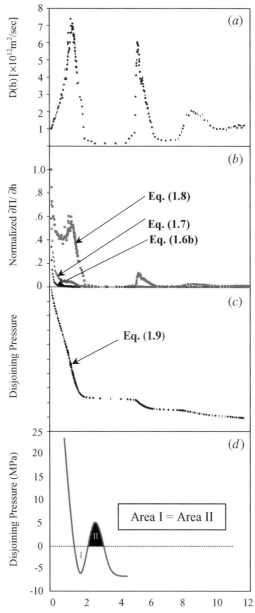

Figure 1.15. (a) $D(h)$ versus h [Fig. 3(a) of Ref. 49]; (b) normalized $\partial\Pi/\partial h$ calculated from Eqs. (1.6b)–(1.8); (c) calculation of Π from Eq. (1.9), $M_n = 1,860$ g/mol for (a)–(c); (d) Π versus h (Fig. 6 of Ref. 92), $M_n = 1,600$ g/mol.

Equations (1.6a,b) or (1.7) and (1.8) are inconsistent and should be carefully examined by correlating the SME profiles with surface energy measurements. The hydrodynamic model, Eq. (1.6a), describes the spreading characteristics of PFPE Z very well. The diffusion model, Eq. (1.8), also describes Zdol spreading qualitatively. From the experimental plot of $D(h)$ for Zdol ($M_n = 1,860$ g/mol) shown in Figure 1.15(a) [91], we plotted $\partial\Pi/\partial h$ against the initial film thickness h_0 from Eqs. (1.6)–(1.8). The results are normalized and shown in Figure 1.15(b), where the diffusion model depicted maxima, but the no-slip hydrodynamic model suppressed any maxima that appeared in the diffusion model. In other words, $\partial\Pi/\partial h$ appears to be decreasing monotonically on the normalized scale for the hydrodynamic model. When the initial film thickness is greater than 0.5 nm, the normalized disjoining pressure gradient is approximately zero. If we introduce a partial slip boundary condition [$\beta = 1000$ in Eq. (1.7)], the decay is extended to 2 nm. This implies that the diffusion model is more suitable for comparing spreading data to surface energy or disjoining pressure. We can also calculate the disjoining pressure from Eq. (1.8) as

$$\Pi(h) - \Pi(h_1) = -\int_{h_1}^{h} \frac{D(h)}{mh} dh \qquad (1.9)$$

Here $D(h)$ is plotted in Figure 1.15(a), and h_1 is the first data point. The disjoining pressure is shown in Figure 1.15(c) with arbitrary units; this is similar to the data reported in Figure 1.6 of Tyndall et al. [92] for Zdol with $M_w = 1600$, and this figure is reproduced as Figure 1.15(d). The difference is that Figure 1.15(c) shows $\partial\Pi/\partial h$ always negative, while Figure 1.15(d) shows some thickness range where $\partial\Pi/\partial h$ is positive.

If we were to utilize Maxwell's equal-area construction, since the Π-h plot in Figure 1.15(d) resembles a van der Waals isotherm in liquid–vapor equilibria [93], then Figures 1.15(c) and 1.15(d) would be qualitatively similar. Related issues can be further investigated via a stability analysis as described in Section II.D.

C. Rheological Measurement

In this section, we examine the bulk rheological properties of fractionated PFPEs with different molecular weights and chain-end functionalities (PFPE Z and Zdol with different C/O endgroup ratios: ZdolTX has an endgroup C/O ratio of 1.7, while Zdol has an endgroup C/O ratio of 1). As heat assisted magnetic recording technology [35] becomes increasingly important, the thermal stability [7,94–97] and temperature effects of PFPE become noteworthy topics. Therefore, in our rheological study we will focus on temperature effect with less attention on the confined geometric effects. It is interesting to note, however, that because of these subtle geometric effects, the viscosities in ultrathin fluid film may be a few order of magnitudes greater than the bulk [89] value. This finding was used in a novel VISqUS technology development [43,98]. Although we will not address micro/nano rheology in this section, readers who are interested in this subject can consult Refs. 86–89. Some of the results of bulk rheology reported here, however, may become very useful in HDD technology or micro-electro-mechanical system through the hybridization of wall effects examined in microrheology.

The bulk rheological properties of the PFPEs, including the melt viscosity (μ), storage modulus (G'), and loss modulus (G''), were measured at several different temperatures via steady shear and dynamic oscillation tests. Note that we denoted μ as melt viscosity and η as solution viscosity. An excellent description of the rheology is available in Ferry [99].

From the temperature dependence of viscosity, which yields an Arrhenius form, the activation energy for flow E_μ^* and the hydrodynamic volume μ_0 were determined as

$$\mu = \mu_0 \exp \frac{E_\mu^*}{RT} \qquad (1.10)$$

where R is the gas constant, T is absolute temperature, and $\mu_0 = N_A h / V$ (N_A is Avogadro's number, h is Planck's constant, and V is the molar volume). When $\ln \mu$ is plotted against $1/T$, it yields straight lines with a positive slope, which increases slightly with molecular weight for PFPE Z, Zdol, and ZdolTX, which are obtained as shown in Figure 1.16 [9,100]. This activation energy is an

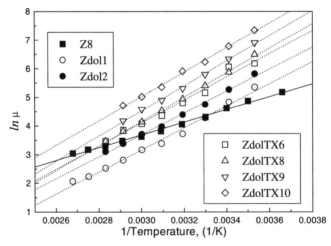

Figure 1.16. The melt viscosity dependence of PFPE Zdol and ZdolTX on temperature and molecular weight.

important molecular parameter for the structural PFPE properties and is interpreted in a manner similar to that of an elementary flow process. In these experiments, the activation energy increases with molecular weight and C/O ratio in the endgroups. For Z, the activation energy is lower than those of Zdol or ZdolTX, implying that the activation energy is dominated by the functional endgroup, not by the molecular weight. Temperature dependence in viscosity can be compared with SME measurements [10,47–52]. The activation energy for surface diffusion E_d^* calculated from the SME data using the Arrhenius equation $D = D_0 \exp(-E_d^*/RT)$ was carefully compared to the flow activation energies E_μ^* defined in Eq. (1.10). The lubricant's interaction with a surface was observable through the relationship between D and μ. It appears that $D\mu^R \cong$ constant (weak function of temperature), where $R = E_d^*/E_\mu^*$ depends on the lubricant–surface interaction via chain-end functionality. We observed that $R \cong 1$ for PFPE Z, while $R \cong 1.5$ for PFPE Zdol [10].

To quantify the abnormal viscoelastic properties of PFPE due to the polar endgroup, we conducted oscillatory experiments at different temperatures. We found that G'' is strongly dependent on temperature, while G' is weakly dependent on temperature. Figure 1.17 shows the behavior of the dynamic moduli at different temperatures.

From the oscillatory data, we can estimate the relaxation time of PFPEs as a measure of the time required for elastic recoil after stress removal. Polymer

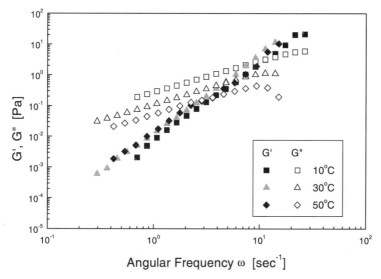

Figure 1.17. G' and G'' versus ω for Zdol1 for three different temperatures.

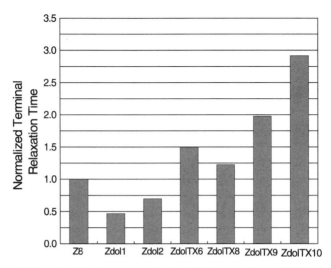

Figure 1.18. Relaxation times (normalized to Z8) for monodisperse PFPEs at 20°C.

relaxation times (τ_p) can be obtained from the Rouse theory [99] shown as follows

$$G' = G_0 \sum \frac{(\omega\tau_p)^2}{1 + (\omega\tau_p)^2}, \quad \text{and} \quad G'' = G_0 \sum \frac{\omega\tau_p}{1 + (\omega\tau_p)^2} \qquad (1.11)$$

with $G_0 \equiv \rho RT/M$, where ρ is the specific gravity, M is molecular weight, and ω is the angular frequency. In the terminal zone of the frequency scale, the properties are dominated by the terminal relaxation time $\tau_1 \equiv \tau_p(p = 1)$: $\tau_1 = 6\mu_0 M/\pi^2\rho RT$, where τ_1 is weakly dependent on chain length and strongly dependent on endgroup functionality. We found that the PFPE relaxation times were approximately 5 times greater for ZdolTX than Zdol [9,100]. Figure 1.18 shows the endgroup effect on τ_1 for PFPEs at 20°C [see Eq. (1.11)]. The Zdol relaxation times are slightly smaller than for Z, while those of ZdolTX are much larger than for Z. Figure 1.19 shows a modified Cole–Cole plot (correlation between the elastic and viscous properties) [101] of ZdolTX9 that exhibits high sensitivity to variation in temperature (the original Cole–Cole plot [102] deals with a complex dielectric constant). Because all the PFPE data in the modified Cole–Cole plot shown in Figure 1.19 do not collapse into a single line, we deduce the presence of temperature-dependent microstructures in the PFPE samples. To examine the essence of the modified Cole–Cole plot, we present a 3D plot

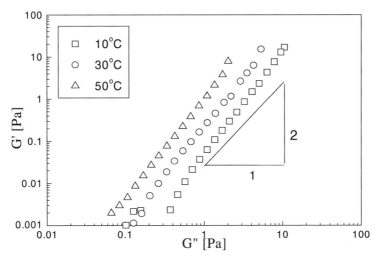

Figure 1.19. Modified Cole–Cole plot for PFPE Z8.

(G'–G''–ω relationship) in Figure 1.20 for Zdol1 at 10°C with the following remarks. The simple Rouse theory in Eq. (1.11) implies that

$$G'_R \equiv \frac{G'}{G_0} = \mathscr{F}_1(\omega\tau_1) \quad \text{and} \quad G''_R \equiv \frac{G''}{G_0} = \mathscr{F}_2(\omega\tau_1) \qquad (1.12)$$

By eliminating $\omega\tau_1$ from Eq. (1.12), we obtain

$$G'_R = \mathscr{F}_1(\omega\tau_1) = \mathscr{F}_1(\mathscr{F}_2^{-1}(G''_R)) = \mathscr{F}(G''_R) \qquad (1.13)$$

where $\mathscr{F}_1$, $\mathscr{F}_2$, and $\mathscr{F}$ are arbitrary functions. Equation (1.13) implies that the relationship between G'_R and G''_R is independent of temperature. Therefore, the G'–G'' plot (a modified Cole–Cole plot) will depend on temperature via G_0, which may strongly depend on temperature for PFPE systems.

In Figure 1.21, experimental observations show that the viscosity of polymer melts is dependent on molecular weight (M) as follows:

1. For M less than the critical value, M_C, the viscosity increases linearly with M or $\sqrt{M}$.

2. Above M_C, the viscosity increases far more rapidly with M.

Linear polymers with high molecular weight generally follow the well-known relation [99,103]

$$\mu = KM^{3.4} \qquad (1.14)$$

where K is constant (independent of M).

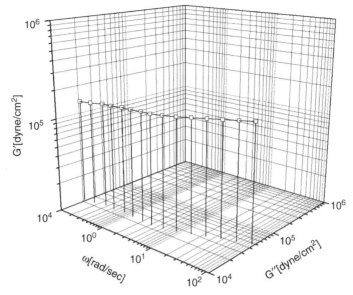

Figure 1.20. G'–G''–ω plot for Zdol1 at 10°C.

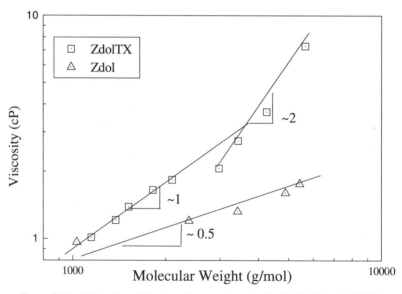

Figure 1.21.   Molecular weight dependence on viscosity for PFPE Zdol and ZdolTX.

The critical molecular weight M_C has been interpreted as the molecular weight at which entanglement coupling occurs. De Gennes proposed the "reptation" model for tagged polymer molecule in bulk polymeric systems [110], and stated that the polymer motion is much like that of a snake moving in a contorted "tunnel" formed by the surrounding polymer molecules. For PFPE systems, we obtained $\eta \propto M^{0.5}$ for Zdol for all M, whereas $\eta \propto M$ for $M < M_C$ and $\eta \propto M^2$ for $M > M_C$ for ZdolTX. An M_C on the order of 3600 g/mol was observed for ZdolTX. The apparent viscosity was found to increase with an increasing endgroup C/O ratio. Note that Ajrodi et al. [104] observed that the iso-free-volume viscosities of Zdol were linearly proportional to M_W above their glass transition temperatures. In contrast, we observed a crossover behavior for ZdolTX possessing bulky endgroups. Crossover behavior has been reported for the molecular weight dependence on viscosity for PFPEs with bulky endgroups [105,106].

The existence of an M_C in ZdolTX data is similar to the reptation processes normally found in high-molecular-weight polymers, which may arise from strong endgroup interactions by forming a "temporal tube"-like confinement. An illustrative description of this scenario is provided in Figure 1.22.

The static friction force after extended flying was also measured to evaluate HDD tribological performance. Fly stiction is believed to be caused by the collection of liquid lubricant on the head. The lubricant is transferred to the head surface while the head rests on the disk, resulting in high stiction on subsequent head takeoff. Because of its complex nature, fly stiction remains challenging to describe with complete scientific confidence. Nonetheless, we attempted to establish a correlation between fly sitction (tribology) and viscosity (rheology), through the shear dominant motion of the head prior to take-off.

The critical molecular weight M_C obtained from viscosity data exhibits a strong correlation with fly stiction, as demonstrated in Figure 1.23, implying that there exists an underlying similarity between the rheological properties and fly stiction for PFPEs. Karis and Jhon [107] were the first to demonstrate the relationship between the rheological properties and tribological performance of lubricants. We believe that rheological measurements will provide complementary information to the mechanical/tribological performance of HDD lubricants.

Shape effect of PFPE molecules or magnetic particles in suspension, including agglomeration phenomena at low concentration, interaction among these particles, and effects of flocs can be examined via solution viscosity (η) measurement. For a very dilute polymer solution [108], there is no interaction among polymer molecules, and the solution viscosity results from the contribution of the solvent plus the contribution of the individual polymer molecules. The intrinsic viscosity, therefore, is a measure of the hydrodynamic volume of a polymer molecule as well as the particle aspect ratio. Figure 1.24 shows the determination of the intrinsic viscosity for Zdol4000 in three different solvents.

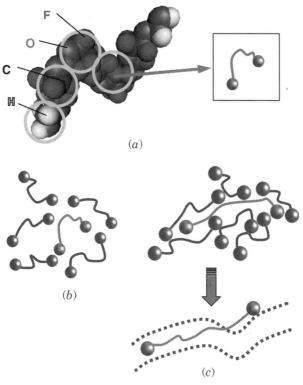

Figure 1.22. (*a*) Simple representation of PFPE molecules to illustrate "temporal tube"–like confinement. Relaxation processes and tube leakage of tagged PFPE molecules for (*b*) M < Mc and (*c*) M > Mc.

Since the density of a very dilute solution is approximately the same as the solvent itself, the relative viscosity ($\eta_r = \eta/\eta_s$) can be taken as the ratio of the time required for a given volume of fluid to flow through an Ubbelhode viscometer. The specific viscosity (η_{sp}) is then given by

$$\eta_{sp} = \frac{\eta - \eta_s}{\eta_s} = \eta_r - 1 \approx \frac{t}{t_s} - 1 \qquad (1.15)$$

where the subscript *s* stands for solvent.

The reduced viscosity (η_{red}) is obtained by dividing the specific viscosity by the PFPE concentration (c) in solution. The expression for η_{red} becomes [109]

$$\eta_{red} \equiv \frac{\eta_{sp}}{c} = [\eta] + k_H[\eta]^2 c + k_3 c^2 + \cdots \qquad (1.16)$$

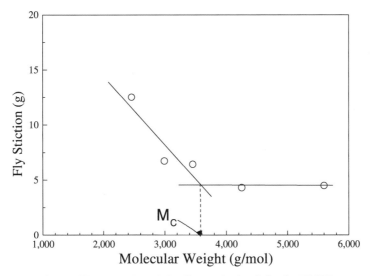

Figure 1.23. Molecular weight effect on the fly stiction for ZdolTX.

Here, the intrinsic viscosity [η] is defined as the infinite dilution limit of the reduced viscosity and k_H is Huggins' coefficient. This result is valid up to the second order in concentration. The PFPE solvent interaction is related to $k_H[\eta]^2$. For flexible polymer chains in a theta solvent, k_H has been found to vary from 0.4 to 1.0 [110]. From the intercept and slope, we obtain [η] and k_H. These values

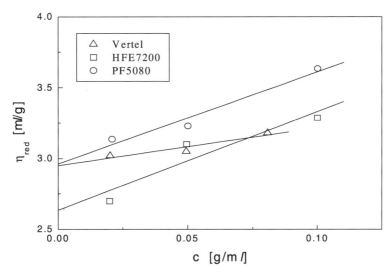

Figure 1.24. Reduced viscosities for Zdol for three different solvents.

TABLE I
$[\eta]$ and k_H for Three Different Solvent Systems

	Vertrel	HFE-7200	PF-5080
$[\eta]$ (Ml/g)	2.95	2.64	2.96
R_H (nm)	1.23	1.19	1.23
k_H	0.31	0.74	1.00

and the hydrodynamic radii R_H [111] of Zdol4000 in three different solvents are summarized in Table I.

According to the Einstein theory, the intrinsic viscosity of a spherical particle suspension is 2.5. However, for a colloidal suspension of nonspherical particles, $[\eta] > 2.5$. Jeffery [112] obtained the viscosity of an ellipsoidal particle suspension under shear. Incorporating Jeffery's results of velocity fields around the particle, Simha [113] obtained expressions for two explicit limiting cases of ellipsoids. Kuhn and Kuhn [114] also obtained an expression for intrinsic viscosity for the full range of particle aspect ratio (p) by taking an approach similar to Simha's method.

The $[\eta]$ for HFF-7200 is smallest and close to a 2.5 ($p = 1.49$ or 0.645). However, $[\eta]$ becomes substantially larger than 2.5 for Vertrel ($p = 2.07$ or 0.459) and PF-5080 ($p = 2.08$ or 0.455), implying that the apparent shape of the PFPE is "nonspherical" and suggesting the existence of cluster formations. Note that $[\eta]$ versus p is plotted in Figure 1.25. k_H values are strongly dependent on the solvent quality, indicating significant differences between the interaction among the "clusters" of PFPE molecules. This cluster formation in solution (or changes in aspect ratio) is related to the dispersion quality of the PFPE (see Fig. 1.26) and may result in media noise during manufacturing processes [the dip-coating process for the PFPE films, see Fig. 1.7 (a)]. This cluster formation of PFPE is similar to the quasiparticle formation in magnetic particle suspension [115], where $[\eta]$ or k_H can be used as the fingerprint analysis for the dispersion quality (or media noise).

Additionally, interesting investigations dealing with thermal stability of PFPE, as well as the effects of humidity and hydrocarbon contamination, have been examined in other studies [7,94–97,116]. Other topics of interest for PFPEs include chemical analyses [117–119], bonding mechanism of endgroups [84,120], molecular degradation [117–119,121–123], physical properties [105,124], and adsorption [125,126].

D. Thermodynamic and Qualitative Description

In this section, we will perform the stability analysis for PFPE Zdol and Ztetraol films via the Gibbs free-energy change (ΔG) for the PFPE system [7] to obtain criteria for uniform, stable thin films.

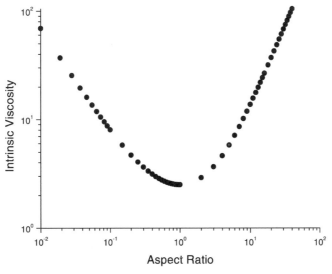

Figure 1.25. Intrinsic viscosity ($[\eta]$) versus aspect ratio (p) using Kuhn and Kuhn's method [114].

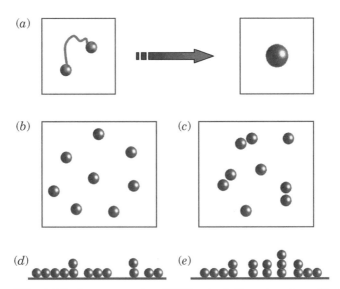

Figure 1.26. (*a*) Simplest representation of PFPE molecule illustrating solvent effect; (*b*) well-dispersed case ($p \cong 1$) and (*c*) poorly dispersed situation ($p \neq 1$). Dip coating after solvent evaporation to make film; (*d*) well-dispersed solution and (*e*) poorly dispersed solution (resulting in flying noise and corrosion).

Free-energy change ΔG of the film spreading is expressed as a function of thickness h, which is the sum over the dispersive (d) and nondispersive (nd) components.

$$\Delta G(h) = \Delta G^d(h) + \Delta G^{nd}(h) \qquad (1.17)$$

Following Sharma [127–129], the dispersive component is represented by

$$\Delta G^d(h) = -S^d \frac{z_0^2}{h^2} \qquad (1.18)$$

where S^d is the dispersive component of the spreading coefficient defined as $S^d = \gamma_S^d - \gamma_L^d - \gamma_{SL}^d$ and z_0 is the equilibrium separation distance between two phases in contact [128,130]. The interfacial energy γ_{SL}^d is estimated using Fowkes' method:

$$\gamma_{SL}^d = \left[\gamma_S^d \gamma_L^d \right]^{1/2} \qquad (1.19)$$

Similarly, the polar component becomes

$$\Delta G^{nd}(h) = -S^{nd} \exp\left(\frac{z_0 - h}{\ell} \right) \qquad (1.20)$$

where $S^{nd} = \gamma_S^p - \gamma_{L0}^p - \gamma_{SL}^p$ is the nd component of the spreading coefficient and ℓ is the characteristic length, which ranges between 0.2 and 1 nm [131]. Note that only $S^d > 0$ is required to have a stable film at any thickness for nonfunctional fluids. However, for nd fluids such as PFPE Z, the film is stable at any thickness for both $S^d > 0$ and $S^{nd} > 0$. The fact that the spreading coefficients satisfy these conditions does not guarantee that the stable film is on the order of several molecular layers thickness in PFPE films. Tyndall et al. [132] reported that the surface energy of PFPE Zdol films varied with a thickness between 0 and 4 nm, which is considered to be the thickness of the layered structure of PFPE films near the surface with chain ends facing each other or the solid surface. We incorporated the variation of surface energy into our stability analysis by assuming that the surface energy function is sinusoidal with an exponential decay envelope of

$$\gamma_L^{nd} = \gamma_{L0}^{nd} + \Delta\gamma_L^{nd} \cos\left(\frac{\pi h}{h_m} + \alpha \right) \exp\left(\frac{z_0 - h}{\ell} \right) \qquad (1.21)$$

where h_m is monolayer thickness, γ_{L0}^{nd} is the surface energy of bulk PFPE, $\Delta\gamma_L^{nd}$ is the amplitude, and α is the phase shift. By utilizing the nonlinear least-squares method, we obtained the optimal parameters ($h_m = 12$ Å, $\gamma_{L0}^{nd} = 0.01$ N/m, $\Delta\gamma_L^{nd} = 5.3 \times 10^{-3}$ N/m, $z_0 = 1.6$ Å, and $\ell = 10$ Å) to fit the data of Tyndall et al. [132].

From Eqs. (1.17)–(1.21), the ΔG as a function of thickness h becomes

$$\Delta G(h) = \Delta G^d(h) + \Delta G^{nd}(h)$$
$$= -S^d \frac{z_0^2}{h^2} - \left[S^{nd} - \Delta\gamma_L^{nd} \cos\left(\frac{\pi h}{h_m} + \alpha\right) \right] \exp\left(\frac{z_0 - h}{\ell}\right) \qquad (1.22)$$

Molecular weight (M) and chain-end functionality of PFPEs are important design parameters for the optimal performance of lubricants. The nd surface energy of PFPEs is approximately proportional to the areal density of polar endgroups, which may be inversely proportional to molecular weight:

$$\Delta\gamma_L^{nd} \propto M^{-1} \qquad (1.23)$$

Also, assuming Gaussian chain statistics for PFPE conformation in the surface normal direction, we can conjecture that the monolayer thickness is proportional to the average molecular size normal to the surface:

$$h_m \propto M^{0.5} \qquad (1.24)$$

Note that the disjoining pressure (Π) is defined as the first derivative of Gibbs free energy G, $\Pi = \partial G / \partial h$, and the slope of disjoining pressure, as a function of film thickness, provides the stability criteria. Thus, for the stable film, we have

$$\frac{d^2 \Delta G(h)}{dh^2} < 0 \qquad (1.25)$$

We constructed stability diagrams from Eq. (1.25) using our data for the surface energies of various PFPEs and carbon overcoats, which are shown in Figure 1.27 [7]. These stability diagrams provide molecular criteria for the uniformity of films (or dewetting).

Our results indicate that PFPE films become more stable with higher molecular weights, weaker polar interaction of PFPEs, and stronger carbon overcoats polar energy. Carbon overcoat derivatives cause PFPE films to stabilize because of higher polar surface energies. We also found that it is difficult to make a stable film below the thickness of 1.5 nm no matter what type of carbon overcoat is selected. This concept, which is strictly based on the thermodynamic argument alone, could provide important design criteria for nanoscale film coatings and useful guidelines for molecular simulations.

(a)   (b)

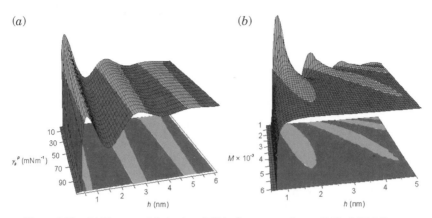

Figure 1.27. (*a*) The second derivative of Gibbs free-energy change (ΔG) of Zdol film versus thickness (*h*) and polar surface energy of carbon overcoat (γ_S^p) and (*b*) the second derivative of ΔG for Zdol film versus *h* and molecular weight (*M*).

III. SIMULATION

The immediate goal of our simulation is to construct reliable tools that accurately describe the dynamic behavior of ultrathin PFPE films, including the fundamental mechanisms of film dewetting and rupture due to instability arising from nanoscale temperature and pressure inhomogeneity as well as modeling thermally induced desorption or evaporation resulting in PFPE lubricant loss. Our ultimate goal is to construct a reliable numerical software to be used in an integrated simulation of the lubricant, airbearing, and nanoscale heat transfer in head/media (see Figs. 1.53 and 1.54 in Section IV).

A multitude of references deal with the general methodology behind molecular simulations [135]. For example, it has been reported that the simulation of nonreactive systems with simple internal structures [136–140] exhibited confinement effects on density profiles [137,140], conformations [136–140], and molecular relaxation [139,140]. Further, significant amount of work has been carried out on the surface diffusion of adsorbates on metal surfaces using the Monte Carlo (MC) method within non-HDD areas [141–147]. Since the molecular layering in the spreading of polydimethylsiloxane (PDMS) liquid on silica surfaces was reported, investigation of the spreading of liquid films on solids at the microscopic scale has gained considerable interest, and many simulations have been performed to examine molecular layering using MC and molecular dynamics (MD) methods [10,148–156]. To extract diffusion data from the SME profile in the HDD area, we constructed *L–t* plots in

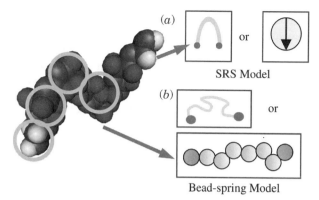

Figure 1.28. Molecular model of PFPE Zdol: (*a*) SRS model [157,158] and (*b*) bead-spring model [159], where blue spheres represent the backbone and red spheres represent the polar endgroups.

Section II (see Fig. 1.12). Simulation results of L–t plots for non-HDD systems are also available. For a simple liquid [152,153], the movement of the liquid front demonstrates a $(\log t)^{1/2}$ dependence, which is inconsistent with experimental observations. Bekink et al. [151] examined the spreading of simple liquids on solid surfaces, and reported a $(\log t)^{1/2}$ dependence. When the nearest-neighbor distance is reduced, $t^{1/2}$ dependence can be obtained. The spreading of polymeric liquids has also been studied by many investigators [10,81,154–156], and $t^{1/2}$ dependence has been reported as well. Nieminen et al. [154] reported an initial linear growth in time (L vs. t), crossing over at some characteristic time to the $t^{1/2}$ diffusion law, which agrees with the experimental observations reported in Section II by O'Connor et al. [10,47,48]. Figure 1.28 illustrates the two representative molecular models for PFPE to be examined in this chapter: the simple reactive sphere (SRS) [157,158] and the bead-spring models [159].

A. SRS Model

Here, we adopted a spin analogy/lattice gas model, or SRS model, as shown in Fig. 1.28(*a*), which represents an oversimplified molecular structure yet still captures the essence of the molecule–surface interactions for describing SME profiles. Similar techniques using the Ising model to study other physical systems have been investigated [148,149,160]; however, none of the literature deals with the simulation of PFPE lubricant dynamics described here.

MC simulations based on the SRS model were pioneered by Ma et al. [158] to explain the peculiar spreading profiles of PFPEs on amorphous carbon surfaces shown in Section II, via explicit adoption of four different interactions:

molecule–molecule, molecule–surface, endgroup–endgroup, and endgroup–surface (a molecule is denoted as a backbone in the absence of polar endgroups). Molecules are approximated as reactive spheres, where a spin $S = 1$ is assigned to an occupied lattice site (for a vacant site, $S = 0$). Excluding all endgroup interactions, simulations qualitatively describe the spreading properties of PFPE Z.

Endgroup interaction terms are required to explain the spreading behavior of molecules with polar endgroups (e.g., Zdol, Ztetraol) and are implemented through the spin parameter S^Z, which identifies the orientation of a polar endgroup ($S^Z = \pm 1$, where a positive value denotes an upward-pointing endgroup). As a result, the system Hamiltonian H in a simple cubic lattice can be represented as

$$
H = -\frac{J}{2} \sum_{i,j,k} S_{i,j,k} S_{i\pm 1, j\pm 1, k\pm 1} - A \sum_{i,j,k} \frac{S_{i,j,k}}{k^3}
$$
$$
+ \sum_{i,j,k} \frac{K(k)}{4} \left(S^Z_{i,j,k} + 1 \right) \left(S^Z_{i,j,k+1} - 1 \right) + \sum_{i,j,k} \frac{W(k)}{2} \left(S^Z_{i,j,k} - 1 \right)
$$

(1.26)

where k is a counting index normal to the surface. The first two terms of this equation describe molecules with nonpolar endgroups adequately. J is the nearest-neighbor, molecule–molecule coupling constant and A is the molecule–surface coupling constant, which is related to the Hamaker constant originating from van der Waals interactions. The third and fourth terms in Eq. (1.26) represent endgroup characteristics, which represent necessary for molecules with polar endgroups only; $K(k)$ and $W(k)$ are the endgroup–endgroup and the endgroup–surface couplings, respectively. Therefore, by setting $K(k) = W(k) = 0$, we recover the Hamiltonian for PFPEs with nonpolar endgroups, such as PFPE Z. $W(k)$ can be set as

$$
W(k) = W_0 \delta_{1,k}
$$

(1.27)

where $\delta_{1,k}$ is Kronecker delta. $K(k)$ requires additional explanation, which will be given in Eq. (1.29).

The spreading of PFPE Z can be easily simulated as follows. A 3D lattice with dimensions of $L \times M \times N$ is generated, where M is in the direction of the periodic boundary condition. Then, the initial film is formed on the substrate, as shown schematically in Figure 1.29. After the film is prepared, the spreading begins.

Simulations of PFPE Zdol include three steps: (1) the generation of film, (2) relaxation of the spin system, and (3) the spreading process. These steps are described carefully by Ma [52].

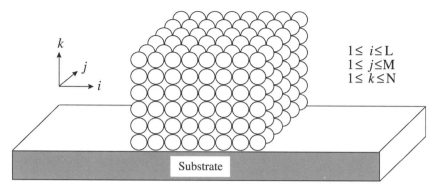

$1 \leq i \leq L$
$1 \leq j \leq M$
$1 \leq k \leq N$

Figure 1.29. A schematic of the simulation lattice.

1. *Generation of Film.* Similar to the PFPE Z case, a three-dimensional $(L \times M \times N)$ lattice is generated, where M is the direction of the periodic boundary condition in the simulation of spreading process. Then, a film of several layers with random orientation of spins is deposited on the substrate. This is shown schematically in Figure 1.30(*a*).

2. *Relaxation of the Spin System.* After the film is deposited, the interactions are turned on between the spins as well as between the spin and the substrate. For each MC cycle, the molecules are interrogated randomly as to whether their spins flip. After reaching equilibrium, the "ideal" final state [shown in Figure 1.30(*b*)] should be achieved; that is, all the spins in the first layer are pinned down to the surface, and the spins of molecules in the upper layer are antiparallel to those of molecules in the adjacent layers.

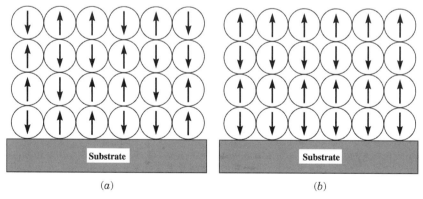

(*a*)   (*b*)

Figure 1.30. Schematic diagrams of the spin orientation of molecules on the surface in (*a*) initial state and (*b*) final state.

3. *The Spreading Process.* After the system is relaxed, spreading commences. The procedure is similar to that in the simulation of PFPE Z molecules; the only difference is that the interactions among endgroups and between endgroups and substrate are included. Each movement of a molecule includes a change of position as well as a flip of spin, which depends on whether it is energetically favorable.

Each MC step randomly interrogates every molecule exactly one time. The movement of an interrogated molecule depends on whether the projected site is empty and whether a favorable system energy change is associated with the movement. The probability of movement P into a vacant site is

$$P = \left(1 + \exp\left(\frac{\Delta H}{k_B T}\right)\right)^{-1} \tag{1.28}$$

where $\Delta H = H^f - H^i$ (H^i and H^f are the initial and final Hamiltonian of the system for a movement). Here, k_B and T represent the Boltzmann constant and absolute temperature, respectively.

The original simulations presented by Ma et al. [158] used $K(k) = K_0$ ($\equiv$ constant) to explain experimental spreading results over an amorphous carbon surface obtained from SME qualitatively. However, to describe spreading on hydrogenated and nitrogenated carbon surfaces [52,60], we introduced the concept of a screening length d for the endgroup/endgroup coupling, given by [160]

$$K(k) = K_0 \exp\left[-\frac{k}{d}\right] \tag{1.29}$$

thus inducing a transition from surface behavior to bulk behavior. In Eqs. (1.27) and (1.29), $K(k) > 0$ and $W(k) > 0$ imply favorable endgroup interactions.

A comparison between simulation results for molecules with nonpolar (PFPE Z) and polar (e.g., PFPE Zdol) endgroups is illustrated in Figure 1.31. A drastic difference is apparent between Figures 1.31(*a*) and 1.31(*b*)—the Z profile is relatively smooth and spreads more rapidly, while the Zdol profile posses a complex layering structure. These results are qualitatively consistent with the experimental observations by given in Figures 1.9 and 1.10, which show that molecules with polar endgroups demonstrate a first layer that is one molecule thick (the thickness of the first layer is on the order of the PFPE diameter of gyration in the bulk state), with subsequent layers approximately twice the thickness of the first layer. This is depicted in Figure 1.30(*b*). Also, Z with nonpolar endgroups spreads more rapidly than does Zdol with its polar endgroups and layered profile.

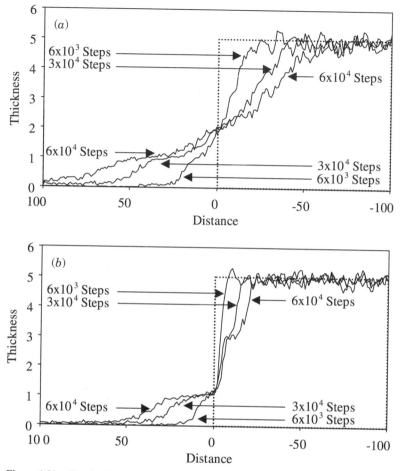

Figure 1.31. Simulated spreading profiles for (*a*) PFPE Z and (*b*) Zdol ($K_0/k_B T = 4$, $W_0/k_B T = 10$, $d \rightarrow \infty$, $J/k_B T = 1$ and $A/k_B T = 10$.)

    The spreading profile shows a strong dependence on the screening length *d*. A simulation result with a screening length $d = 4$ is presented in Figure 1.32. Comparing Figure 1.31(*b*) [where $d \rightarrow \infty$, implying constant K from Eq. (1.29)] and Figure 1.32, we note that the complex layered structure relaxed with a decreasing screening length, which is indicated by the decrease in steepness of the layering in the shoulder area. This behavior is similar qualitatively to experimental spreading data on hydrogenated and nitrogenated carbon surfaces [52,60]; it appears that hydrogen or nitrogen content decreases the screening length *d*. Figure 1.32 resembles SME spreading profiles obtained from

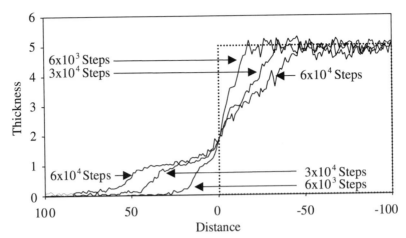

Figure 1.32. Simulated spreading profiles for Zdol with a finite screening length ($J/k_BT = 1$, $A/k_BT = 10$, $K_0/k_BT = 4$, $W_0/k_BT = 10$, $d = 4$).

hydrogenated carbon surfaces with a hydrogen-to-carbon ratio of 50%, as seen in Figure 1.11.

An alternative way to present the simulated spreading behavior is shown in Figure 1.33. The L–t plot was constructed by plotting the distance that an isoheight in the spreading profile travels from the initial sharp boundary (a position of constant thickness that changes with time) versus the number of

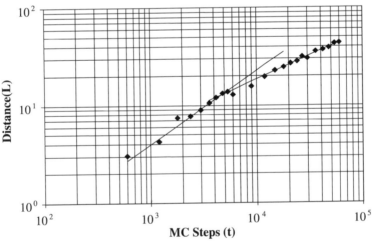

Figure 1.33. L–t plot for a 5% isoheight ($J/k_BT = 1$, $A/k_BT = 10$, $K_0/k_BT = 4$, $W_0/k_BT = 10$).

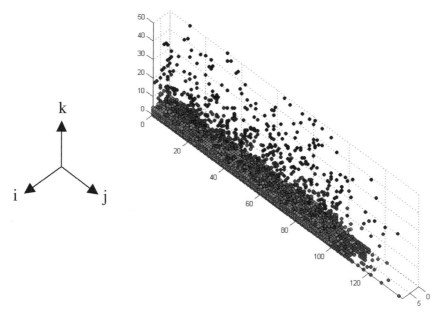

Figure 1.34. A schematic of particle evaporation.

MC steps. The isoheight was chosen as 5% of the initial five-molecule-thick profile, and the simulated L–t response shows a distinct transition between short and long times. This is qualitatively similar to the experimental L–t behavior shown in Figure 1.12. The long-term behavior exhibits $L \propto t^{1/2}$, thus meeting the criteria for the surface diffusion assumption [47,50].

Although qualitatively similar to the experimental SME results, our previous simulation results had difficulties with particle evaporation [160]. Figure 1.34 illustrates particle evaporation. To overcome evaporation, the model parameters have been adjusted. 3D visualization techniques were also used to monitor the issue, and the full-blown 3D capabilities [157] allow for a detailed, nanostructural analysis of the PFPE lubricant films.

The 3D plots for PFPE Z (Figs. 1.35 and 1.36) show that the front formation for $J/k_B T = 2$ is inhibited, which leads to a more abrupt profile with molecular layering and a small amount of fingering. The more pronounced front fingering (present in Fig. 1.35) exhibited fingerwidths of approximately 10 lattice units. Simulation widths are chosen to be larger than this finger size.

The 3D results for PFPE Zdol, (Figs. 1.37 and 1.38), provide more insight into the layer coupling. For the weaker endgroup interactions (Fig. 1.37), the 3D picture appears similar to Figure 1.35 (PFPE Z), except for the slower spreading dynamics. However, a difference manifests in the spacing of the individual layer

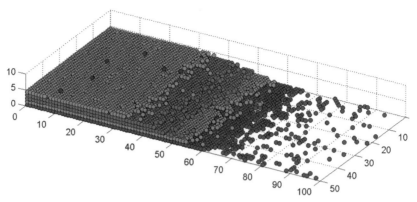

Figure 1.35. PFPE Z 3D simulation results: $J/k_BT = 1$, $A/k_BT = 1000$, 10^4 lattice steps.

fronts: the fronts for layers 2 and 3 and layers 4 and 5 in Figure 1.37 are closer together than are the evenly spaced fronts in Figure 1.35. Although the SME-type plot for PFPE Zdol with strong interactions has the appearance of a single, large shoulder, Figure 1.36 still shows independent couplings between layers 2 and 3 and layers 4 and 5. The averaged profile loses these finer details.

We extend this analysis to the multilayer situation to describe particle activation energies for 1D diffusion on a 3D cubic lattice. The constant source row boundary condition is extended to a thickness of up to five layers for this analysis. The diffusion front for each layer is defined using the "seawater" methodology of Sapoval et al. [161].

For front particles at the average position of the diffusion front, the mass M of the front particles scales with radius r as

$$M \propto r^{D_f} \tag{1.30}$$

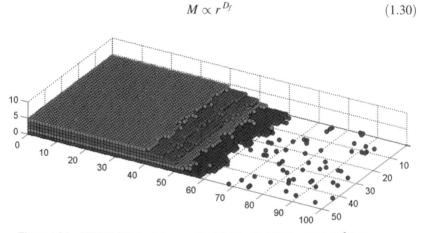

Figure 1.36. PFPE Z 3D simulation results: $J/k_BT = 2$, $A/k_BT = 1000$, 10^5 lattice steps.

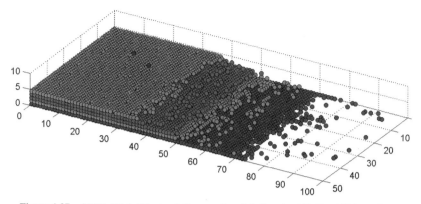

Figure 1.37. PFPE Zdol 3D simulation results: $J/k_BT = 1$, $A/k_BT = 1000$, $K_0/k_BT = 4$, $W_0/k_BT = 4$, $d \to \infty$, 10^5 lattice steps.

Here, D_f is the fractal dimension, which differs for the monolayer and multilayer situations, implying that D_f is determined not only by lateral movement (virtually independent of layer thickness) but also by a cascade effect from overlying layers. By quantifying D_f as a function of the molecule/surface coupling, as well as the number of monolayers, D_f can be used as a fingerprint to analyze lubricant/surface pairs, and possibly become a critical parameter for the selection of optimal lubricants.

In an effort to determine physically relevant model parameters, the D_f for each layer contour in the simulated film was analyzed [162]. D_f for each layer could be comparable to those extracted from the preliminary experimental data based on Brewster angle SME [163].

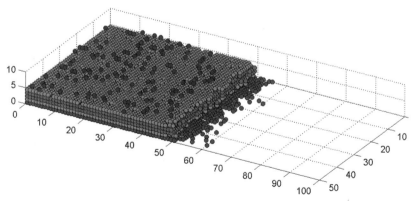

Figure 1.38. PFPE Zdol 3D simulation results: $J/k_BT = 1$, $A/k_BT = 1000$, $K_0/k_BT = 8$, $W_0/k_BT = 8$, $d \to \infty$, 10^6 lattice steps.

The fractal nature of monolayer diffusion fronts has been investigated both computationally [161] and analytically [164,165]. Findings for the case of a monolayer diffusing from a constant source row indicate a D_f of exactly 1.75 for a self-avoiding random walk ($H \equiv 0$) [165]. Sapoval et al. [161] found that D_f decreases below 1.75 when the equivalent of $J/k_B T$ in Eq. (1.26) exceeds a critical value of 1.76. Surely the minimum for D_f is 1.0 as $J/k_B T \rightarrow \infty$.

Because of the simplicity of this SRS model with lattice sites, an occupancy spin, and an endgroup orientation spin (toward or away from the surface) [157], we can qualitatively investigate the 3D, steady-state surface structure of PFPE films, describe its dewetting phenomenon and complex surface morphology, and examine its properties as a function of interaction parameters, surface loading, and initial surface structure. Surface structure evolution is monitored to better understand the mechanism for dewetting or "mogul" dynamics [54,55]. To examine the dewetting phenomenon as a function of endgroup–endgroup and endgroup–surface interaction energies, simulations are conducted on a 3D lattice with periodic dimensions of 512×512 parallel to the surface. For an initial condition, we randomly distributed a layer with 90%, 100%, and 110% loading onto the surface to represent a very rough, highly damaged layer as shown for the 100% initial loading condition in Figure 1.39. If the endgroup–endgroup interactions are too strong, the layer does not relax. The amount that a layer will relax is determined by the endgroup interaction strength.

To better examine the driving forces behind the nanoscale spreading process, we are currently developing a continuous spin-state SRS model for an off-lattice scenario. The endgroups in this case are represented as smaller, internal spheres at the molecular backbone sphere surface, as shown in Figure 1.40. However, the endgroup orientation is no longer limited in its direction toward or away from the surface. Examining the disjoining pressure for the film is a natural extension of this work with the goal of studying the anisotropic pressure tensor behavior for a thin-film situation [166].

B. Monte Carlo Simulation with Bead-Spring Model

To represent the molecular structure with reasonable accuracy as well as to reduce computational time, the coarse-grained, bead-spring model [Fig. 1.28(b)] was employed to approximate a PFPE molecule. This simplifies the detailed atomistic information while preserving the essence of the molecular internal structure [167]. The off-lattice MC technique with the bead-spring model was used to examine nanoscale PFPE lubricant film structures and stability with internal degrees of freedom [168].

In our model, a PFPE molecule is composed of a finite number of beads with different physical or chemical properties [Fig. 1.41]. For simplicity, we assume that all the beads, including the endbeads, have the same radius. Lennard-Jones

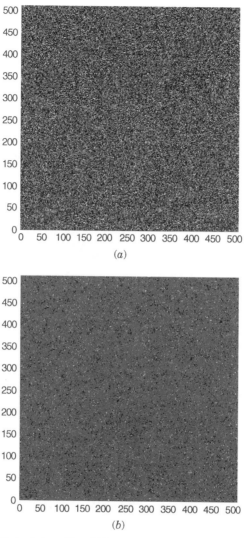

Figure 1.39. 3-D simulation with a SRS model: (*a*) initial condition for 100% layer loading and (*b*) relaxed film with a rough, dewetted surface. Black indicates a bare surface; white indicates peaks.

(LJ1) and van der Waals (LJ2) potentials were used for nonpolar bead–bead and bead–wall interactions, respectively. For polar interactions, exponential potential functions (EXP1,2) were added to both bead–bead and bead–wall cases. For the bonding potential between adjacent beads in the chain, a finitely extensible nonlinear elastic (FENE) model was used. For example, PFPE Zdol

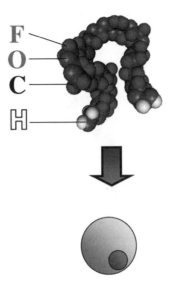

Figure 1.40. A schematic of a SRS model with a continuous spin state [166].

can be characterized differently from PFPE Z by assigning to the endbead a polarity originating from the hydroxyl group in the chain end.

Neighboring beads are connected via a FENE spring as follows

$$U_{\text{FENE}}(l) = \begin{cases} -\dfrac{1}{2} k R_0^2 \ln \left[1 - \left(\dfrac{l - l_0}{R_0} \right)^2 \right] & l_0 - R_0 \leq l \leq l_0 + R_0 \\ \infty & \text{otherwise} \end{cases} \quad (1.31)$$

where l is the bondlength, $l_0 = 1.3\,\sigma$ is the equilibrium bondlength, $R_0 = 0.3\,\sigma$ is the maximum extensible range of the spring, and $k = 40\,\varepsilon/\sigma^2$ is the spring constant that quantifies the rigidity of the bond. Here, σ is the bead diameter.

All beads, including the endbeads in PFPE Zdol, interact with each other by a pairwise, dispersive, truncated Lennard-Jones (LJ) potential as follows

$$U_{\text{LJ}}(r) = \begin{cases} 4\varepsilon \left[\left(\dfrac{\sigma}{r} \right)^{12} - \left(\dfrac{\sigma}{2} \right)^6 \right] & r < 2.5\sigma \\ 0 & r \geq 2.5\sigma \end{cases} \quad (1.32)$$

where r is the distance between beads and ε represents the well depth of LJ interaction.

For the dispersive bead (db)–wall interaction, we used the attractive potential, $U_{\text{db-wall}}(z)$:

$$U_{\text{db-wall}}(z) = \begin{cases} -\dfrac{\varepsilon_w \sigma^3}{(z+\sigma)^3} & z \geq 0 \\ \infty & z \leq 0 \end{cases} \tag{1.33}$$

where z is the distance from the wall and $\varepsilon_w = 4\varepsilon$ is the potential depth.

The mechanism of the endgroup interaction in PFPEs has not yet been clarified. Previous studies have indicated that the coupling between endgroups and their binding to the solid surface is more like a hydrogen-bonding interaction [7] from an ab initio calculation [169]. While other studies considered the possibilities of hydrogen transfer [170] and esterification as a result of annealing [132], in this study, we aimed to demonstrate the short-range interaction potential functions without losing the essence of the problem.

In the PFPE Zdol model, due to the polarity of endgroups induced by the hydroxyl group, the atomistic interaction is different from that in backbone beads. Here, the polarity interaction is assumed to occur within a short range, and is modeled as an exponential decay function. The potential function among endbeads is

$$U_{\text{eb-eb}}(r) = -\varepsilon^p \exp\left(-\frac{r-r_c}{d}\right) \tag{1.34}$$

where ε^p represents the interaction strength between endbeads, $r_c = 1.0\,\sigma$ is the critical distance between a pair of endbeads, which is defined by the diameter of an endbead, and $d = 0.3\sigma$ is the decay length in the exponential function, which must be small to meet the criteria of the short-range interaction. Further, endbeads interact with the solid wall as

$$U_{\text{eb-wall}}(z) = -\varepsilon^p_w \exp\left(-\frac{z-z_c}{d}\right) \tag{1.35}$$

where $z_c = \sigma$ is the critical distance of the interaction. Note that larger the ε^p_w, the stronger the interaction.

Figure 1.41 is a schematic of the potential functions given in Eqs. (1.32)–(1.35).

We assumed that all polymer chains in the system have the same number of beads, that is, that they are monodisperse. The number of beads N_p in each polymer chain is chosen as $N_p = 6$, 10, or 16. PFPEs have rigid fluorocarbon backbone units connected via ether bonds, which give flexibility to the chain while keeping PFPEs stable in a liquid state at room temperature. The beads in our model reproduce the rigid units and are connected only via their

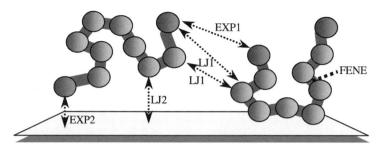

Figure 1.41. Potential energies for the bead-spring model: LJ1—Lennard-Jones potential; LJ2—van der Waals potential; EXP1, EXP2—short-range polar potential; FENE—finitely extensible nonlinear elastic potential.

stretching potential (i.e., without torsional or bending potential) to demonstrate the flexible ether bonds. Sixteen repeating units of a typical PFPE backbone unit C_2F_4O correspond to about a molecular weight of 2,000 g/mol. Although the chains with $N = 6$, 10 are slightly shorter than they would be for typical application, they are useful for investigating the endgroup functionality effect because they are expected to have stronger endgroup dependences. The wall is a square with $L = 10\sigma$ for each side. Polymer chains are positioned at lattice sites and packed loosely in the initial structure with no beads initially in contact with the wall. Each chain generates a random walk around the initial bead.

A single bead is picked randomly for the MC trial movement during each attempt. The displacement, $\Delta\mathbf{r} = \Delta x\mathbf{i} + \Delta y\mathbf{j} + \Delta z\mathbf{k}$ with $-r_{max}/2 < \Delta x, \Delta y,$ $\Delta z < r_{max}/2$, is set randomly in a cube centered by the selected bead.

The value of ΔU is calculated from the potential energy functions given above and used for the Metropolis acceptance probability W for the trial movement,

$$W = \min\left[\exp\left(-\frac{\Delta U}{k_B T}\right), 1\right] \qquad (1.36)$$

where k_B and T are the Boltzmann constant and the absolute temperature, respectively.

When W exceeds a random number, the trial movement is accepted. In our simulations, the acceptance ratio ranged between 0.15 and 0.25. After 300,000 MC steps for each bead, a stable state was reached where ΔU became statistically constant. We generated five initial configurations for each condition and collected 10 datasets from each run. The data sets are separated by more than 10,000 MC steps for each bead. Therefore, 50 runs were carried out for each system to collect statistical data. The simulation cell was divided into sublattices to speed up the potential energy calculations and reduce computation time.

We examined the molecular conformations of the confined film via the anisotropic radius of gyration ($R_g^2 = R_\perp^2 + 2R_\parallel^2$) in transverse ($\parallel$) and longitudinal ($\perp$) components:

$$R_\parallel^2 = \frac{1}{2N_P} \sum_{i=1}^{N_P} \left[(x_i - x_g)^2 + (y_i - y_g)^2 \right]$$

$$R_\perp^2 = \frac{1}{N_P} \sum_{i=1}^{N_P} (z_i - z_g)^2$$

(1.37)

where x_g, y_g, z_g are the locations of the chain's center of mass and N_p is the number of monomers. The anisotropic radius of gyration as a function of distance from the wall z (Fig. 1.42) demonstrates that the chain conformation becomes more oblate as the wall distance decreases. The chains in the film extend more in a transverse direction than in a longitudinal direction, resulting in a layered structure. However, the endbead interaction does not significantly affect the conformation of both $\varepsilon_b^p = \varepsilon_w^p \equiv \varepsilon$ and $\varepsilon_b^p = \varepsilon_w^p \equiv 2\varepsilon$, where $R_\perp$ oscillates around that of chains with $\varepsilon_b^p = \varepsilon_w^p = 0$ (denoted as $R_\perp^\circ$). The difference $\Delta R_\perp \equiv R_\perp - R_\perp^\circ$ plotted in Figure 1.43 shows that $\Delta R_\perp$ maximizes near $z = 3$ and $z = 8$ and minimizes near $z = 5$ and $z = 12$. This conformational fluctuation is due to the coupling and orientation of the functional endbeads. Figure 1.44 shows the density profile as a function of the distance from the wall. The polymer chains ($N_p = 10$) with functional endbeads ($\varepsilon_b^p = \varepsilon_w^p = \varepsilon$ and $\varepsilon_b^p = \varepsilon_w^p = 2\varepsilon$) exhibit pronounced peaks, while the nonfunctional endbeads ($\varepsilon_b^p = \varepsilon_w^p = 0$) are monotonic. The endbeads are believed to couple near the peak. Because the endbeads in the first layer attach to the wall as a result of the attractive functional interaction, the orientation of endbeads is induced in the second layer, and likewise, the orientation propagates to subsequent layers. In the case of the chains located within the endbeads' density peak, the functional endbeads also stay within the peak to couple with other endbeads resulting in a flatter conformation. The chains located out of the peak unfold themselves so that their endbeads can reach other endbeads. The characteristic oscillation in the molecular conformation (Fig. 1.43) and the density profile (Fig. 1.44) is induced by functional endbeads.

To establish a qualitative relationship between orientation and the layer structure, we examined the number of layers at z and the endbead density as plotted in Figure 1.45 for $\varepsilon_b^p = \varepsilon_w^p = 2\varepsilon$. The bead density profile has three peaks: at the wall, between the second and third layers, and between the fourth and fifth layers. The adsorption of functional endbeads results in an alternate ordering in the subsequent layers: up in the second layer, down in the third, and

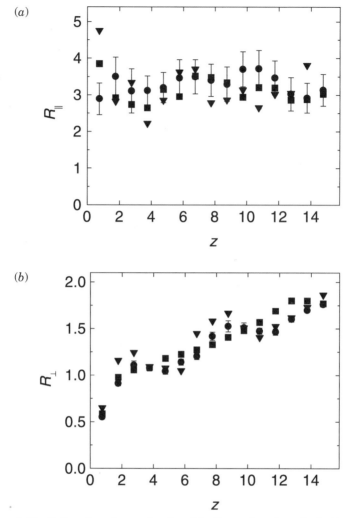

Figure 1.42. Radius of gyration as a function of the distance from the wall z for $N_p = 10$: (a) transverse component $R_\parallel$ and (b) longitudinal component $R_\perp$ for (■) $\varepsilon_b^p = \varepsilon_w^p = 0$, (●) $\varepsilon_b^p = \varepsilon_w^p = \varepsilon$, and (▼) $\varepsilon_b^p = \varepsilon_w^p = 2\varepsilon$. Error bars indicate one half of the standard deviation and are shown only for $\varepsilon_b^p = \varepsilon_w^p = \varepsilon$.

so on. Our result provides a direct interpretation of experimental surface energy data for PFPE films with functional endgroups [169]. The nondispersive component of surface energy exhibited an oscillatory pattern with increasing film thickness and was shown to be approximately proportional to endgroup density, as demonstrated in our previous study [7,159]. Our simulation results

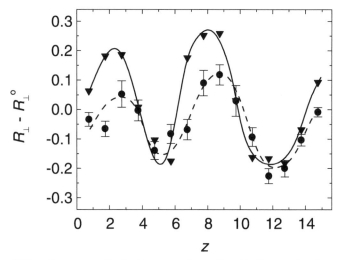

Figure 1.43. $\Delta R_\perp \equiv R_\perp - R_\perp^\circ$ for the chains ($N_p = 10$) with functional endbeads for (●) $\varepsilon_b^p = \varepsilon_w^p = \varepsilon$ and (▼) $\varepsilon_b^p = \varepsilon_w^p = 2\varepsilon$. Error bars indicate one-half of the standard deviation and are shown only for $\varepsilon_b^p = \varepsilon_w^p = \varepsilon$.

also suggest that the density variation of the endgroups is related to the characteristic behavior of the surface energy of PFPE films.

Furthermore, functional endgroup density variation could qualitatively explain the spreading profiles of PFPE films. The surface diffusion coefficient

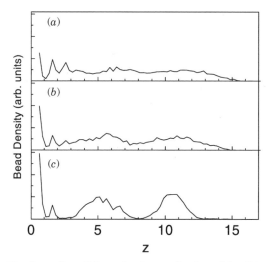

Figure 1.44. Density profiles of the endbeads as a function of the distance from the wall: (a) $\varepsilon_b^p = \varepsilon_w^p = 0$; (b) $\varepsilon_b^p = \varepsilon_w^p = \varepsilon$; (c) $\varepsilon_b^p = \varepsilon_w^p = 2\varepsilon$.

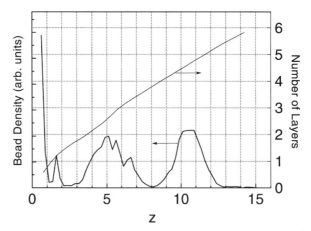

Figure 1.45. Density profiles of endbeads and number of layers ($N_p = 10$, and $\varepsilon_b^p = \varepsilon_w^p = 2\varepsilon$).

obtained from the experimental spreading profile depends strongly on the layer positions via endgroup orientation [160]. In other words, the diffusion coefficient between the second and third layers is smaller than that between the first and second layers because of the endbead coupling, which is consistent with the observation made by Ma et al. [49], who also provided a qualitative explanation for the anomalous peaks in thickness-dependent diffusion coefficients.

As temperature increases, the layer expands and the orientation of endbeads smears. This is shown in Figure 1.46. The observed expansion of layer thickness is attributed mainly to the temperature dependence of the intermolecular interaction and excluded volume effect not due to bond stretching. Temperature dependence, which is pertinent to the annealing process of thin polymeric lubricant films, has been carefully examined by Hsia et al. [171].

The steady-state nanoscopic properties were examined, and the result agrees qualitatively with the simulation results obtained via the SRS model discussed in Section III.A. Our results provide a direct interpretation of experimental surface energy data for PFPE films with functional endgroups [7,92,132,159]. Stable films did not experience dewetting or film rupture. However, a rougher surface morphology was observed for lower molecular weights [Fig. 1.47(b)] and strong endbead functionality [Fig. 1.47(c)]. Visualizations of surface morphology could provide a powerful tool for describing airbearing stability.

Another noteworthy observation is the segregation, or localization, of chain ends at the film surface, which was found in both nonfunctional and functional chain ends [172]. The endbead density obtained from the MC simulations is

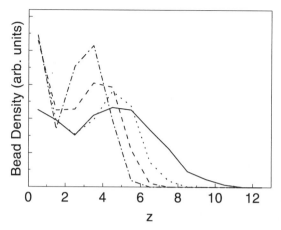

Figure 1.46. Temperature effect on the density profiles of endbeads $\left(\varepsilon_w^p = 2\varepsilon_b^p = 2\varepsilon\right)$. $T = 0.5\varepsilon/k_B$ (dotted-dashed line); $T = 0.8\varepsilon/k_B$ (dashed line); $T = \varepsilon/k_B$ (dotted line); $T = 1.5\varepsilon/k_B$ (solid line).

normalized with respect to the total number of beads and plotted in Figure 1.48 as a function of endbead functionality. The segregation of the nonfunctional chain ends at the surface $(\varepsilon^p = \varepsilon_w^p = 0)$ ensures that the driving force is entirely entropic. The chains can avoid the decrease in entropy by having the nonfunctional ends localized at the film surface where the number of configurations is reduced as a result of chain folding. The mildly functional endbeads $(\varepsilon^p = \varepsilon_w^p = \varepsilon)$ still localize at the film surface, which suggests that the entropic contribution is more dominant and, as a result, the free energy of the system is minimized by having the endbeads localized. The strongest functional endbeads, on the other hand, are totally depleted from the surface $(\varepsilon^p = \varepsilon_w^p = 3\varepsilon)$. The contribution of surface enthalpy to the free energy is too large to be compensated by the entropic component. In this case, therefore, all functional endbeads minimize the energy of their functional interaction by coupling with other endbeads inside the film. In the same manner, the segregation of low-molecular-weight fraction at interfaces were observed from our binary MC simulations [172].

C. Molecular Dynamics Simulation

So far, we have demonstrated that the MC simulation (lattice-based SRS model and off-lattice bead-spring model) results are in qualitative agreement with the experiments. A complementary approach is molecular dynamics (MD) simulation using the bead-spring model. Since MD study for PFPE is still the infant stage, we will discuss it only briefly. The equation of motion can be expressed in

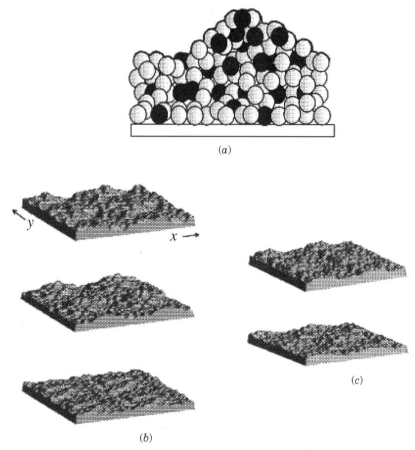

Figure 1.47. Morphology of PFPE films from MC simulations: (*a*) schematic of simulated surface roughness for (*b*) molecular weight dependence [$N_p = 10$ (upper), 15 (middle), 20 (bottom)] and (*c*) endbead functionality dependence [$\varepsilon_w^p = \varepsilon^p = 1$ (upper) and $\varepsilon_w^p = \varepsilon^p = 4$ (bottom)].

the following Langevin equation form [140,173,174]:

$$m\frac{d^2 r_{\alpha i}}{dt^2} = -\frac{\partial U}{\partial r_{\alpha i}} - \zeta \cdot \frac{dr_{\alpha i}}{dt} + f_{\alpha i}(t) \qquad (1.38)$$

Here, i spans $1, 2, \ldots, N$ and α spans $1, 2, \ldots, N_p$, where N and N_p imply the number of PFPE molecules and monomers, respectively; m is the mass of the beads; and U is the potential energy of the system expressed as some or all potential energy given in Eqs. (1.39) and (1.40); [Note that Eqs. (1.32) and

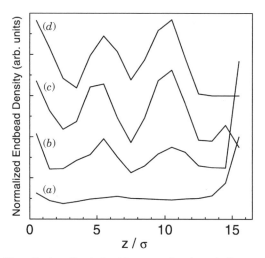

Figure 1.48. Normalized endbead densities as a function of distance from the surface $(N = 10)$ for (a) $\varepsilon^p = \varepsilon_w^p = 0$, (b) $\varepsilon^p = \varepsilon_w^p = \varepsilon$, (c) $\varepsilon^p = \varepsilon_w^p = 2\varepsilon$, and (d) $\varepsilon^p = \varepsilon_w^p = 3\varepsilon$.

(1.33) are replaced by Eqs. (1.39) and (1.40), respectively]; specifically, $U = U_{\text{bead}} + U_{\text{endbead}}$:

$$U_{\text{LJ}}(r_{bb}) = \begin{cases} 4\varepsilon\left[\left(\dfrac{\sigma}{r_{bb}}\right)^{12} - \left(\dfrac{\sigma}{r_{bb}}\right)^{6}\right] - 4\varepsilon\left[\left(\dfrac{\sigma}{r_{\text{cutoff}}}\right)^{12} - \left(\dfrac{\sigma}{r_{\text{cutoff}}}\right)^{6}\right] & r_{bb} \leq r_{\text{cutoff}} \\ 0 & r_{bb} > r_{\text{cutoff}} \end{cases}$$

$$(1.39)$$

$$U_{\text{db-surface}}(z) = \varepsilon_w\left[\frac{1}{2}\left(\frac{\sigma}{z}\right)^{9} - \frac{3}{2}\left(\frac{\sigma}{z}\right)^{3}\right] \qquad (1.40)$$

where U_{bead} represents the total potential energy of the nonpolar PFPE molecules (i.e., PFPE Z) including LJ1, LJ2, and FENE potentials, while U_{endbead} contains extra contributions on potential energy due to the endbeads, including the EXP1 and EXP2 potentials.

The term ζ is the frictional tensor. We assume it to be isotropic: $\zeta = \Gamma\mathbf{I}$ ($\mathbf{I}$ is unit tensor). $\mathbf{f}_{\alpha i}$ is assumed to be Gaussian white noise that is generated according to the fluctuation–dissipation theorem [173,174]

$$\langle \mathbf{f}_{\alpha i}(t)\mathbf{f}_{\beta j}(t')\rangle = 2k_B T\,\Gamma\delta_{\alpha\beta}\delta_{ij}\delta(t - t')\mathbf{I} \qquad (1.41)$$

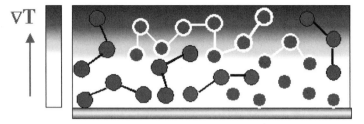

Figure 1.49. A schematic of a tagged (red) molecule in the self-diffusion process.

Here, the angular brackets denote an ensemble average, which is the same as time average from the ergodic hypothesis. $\delta_{\alpha\beta}$ and δ_{ij} are Kronecker delta, and $\delta(t - t')$ is Dirac's delta function.

MD simulation is advantageous for obtaining dynamic properties directly, since the MD technique provides not only particle positions but also particle velocities that enable us to utilize the response theory (e.g., the Kubo formula [175,176]) to calculate the transport coefficients from time-dependent correlation functions. For example, we will examine the self-diffusion process of a tagged PFPE molecular center of mass (Fig. 1.49) from the simulation to gain insight into the excitation of translational motion, specifically, spreading and replenishment. The squared displacement of the center mass of a molecule or a bead is used as a measure of translational movement. The self-diffusion coefficient D can be represented as a velocity autocorrelation function

$$D = \lim_{t\to\infty} \frac{1}{6t} \langle |\mathbf{r}(t) - \mathbf{r}(0)|^2 \rangle = \int_0^\infty \langle \dot{\mathbf{r}}(t) \cdot \dot{\mathbf{r}}(0) \rangle dt$$
$$= \int_0^\infty \langle \dot{x}(t)\dot{x}(0) \rangle dt + \int_0^\infty \langle \dot{y}(t)\dot{y}(0) \rangle dt + \int_0^\infty \langle \dot{z}(t)\dot{z}(0) \rangle dt$$
$$\equiv D_x + D_y + D_z = 2D_\parallel + D_\perp \tag{1.42}$$

where $\dot{\mathbf{r}}(t) \equiv \dot{x}(t)\mathbf{i} + \dot{y}(t)\mathbf{j} + \dot{z}(t)\mathbf{k}$ denotes the velocity vector of the center of mass for a PFPE molecule. Figure 1.50 illustrates the results for the $D_\parallel$ and $D_\perp$ (the components of the diffusion coefficients in the directions parallel and perpendicular to the solid surface) obtained from our preliminary MD simulations.

The molecular relaxation process has been studied by the autocorrelation function of normal modes for a linear polymer chain [177]. The relaxation spectrum can be analyzed by the Kohlrausch–Williams–Watts function [177,178]:

$$C(t) = C(0)\exp\left[-\left(\frac{t}{\tau^*}\right)^\beta \right] \tag{1.43}$$

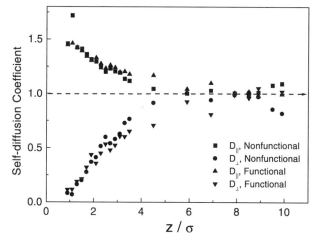

Figure 1.50. Self-diffusion coefficients ($N_p = 6$, $T = 0.5$) as a function of distance from the surface for nonfunctional ($\varepsilon_w^p = \varepsilon^p = 0$, $\varepsilon_w = 5$) and functional chain ends ($\varepsilon_w^p = \varepsilon^p = 2$, $\varepsilon_w = 5$).

Here, τ^* represents the elementary relaxation time and the stretching parameter β is descriptive for the nonexponential decay process (or distribution of relaxation times), or it physically quantifies the confinement effect; that is, the larger β is, the stronger the confinement becomes. Figure 1.51 demonstrates the

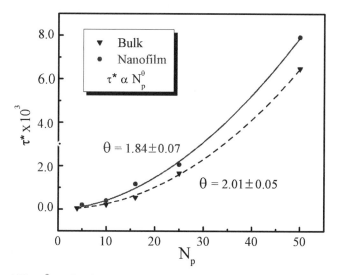

Figure 1.51. τ^* as a function of N_p for the PFPE ($\varepsilon^p = \varepsilon_w^p = 2\varepsilon$) in the bulk state ($\varepsilon_w = 0$) and in the nanofilm ($\varepsilon_w = 4\varepsilon$) with $T = 1.0\varepsilon/k_B$.

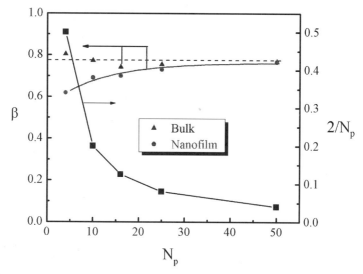

Figure 1.52. The stretching parameter β and endgroup ratio $2/N_p$ as a function of N_p for the PFPE ($\varepsilon^p = \varepsilon_w^p = 2\varepsilon$) in the bulk state ($\varepsilon_w = 0$) and in the nanofilm ($\varepsilon_w = 4\varepsilon$) with $T = 1.0\,\varepsilon/k_B$.

contribution of the solid surface to the relaxation process for end-functional PFPEs ($\varepsilon^p = \varepsilon_w^p = 2\varepsilon$) by comparing the bulk ($\varepsilon_w = 0$) with nanofilm ($\varepsilon_w = 4\varepsilon$). A longer τ^* is observed for the nanofilm than the PFPE bulk, but the MW dependence, [i.e., $\tau^* \propto (N_p)^\theta$] is weakened. For instance, $\theta = 2.01 \pm 0.05$ for the bulk as the Rouse model predicts [103], but it decreases to 1.84 ± 0.07 for the nanofilm. In addition, β for the nanofilm increased slightly (about 0.15) in MW as seen from the solid line in Figure 1.52. Note that the endgroup ratio $2/N_p$ decreases for higher MW and that the functional endgroups interact strongly with the solid surface. Consequently, the higher the endgroup ratio in a PFPE chain is, the stronger the PFPE–solid surface interaction that results. Compared with lower-MW PFPEs, the effective interaction with the surface reduces for higher-MW PFPEs.

We are currently calculating the density profile and surface morphology of PFPE as well as other time-dependent correlation functions to examine the nanoscale transport coefficients for PFPE in a confined geometry.

IV. CONCLUSION

In this section we present fundamental scientific tools as well as potential applications relevant to the emerging field of nanotechnology. For instance, understanding the behavior of molecularly thin lubricant films via experimental and theoretical study of the physicochemical properties of ultrathin perfluoropolyether (PFPE) films is essential for achieving durability and reliability in nanoscale devices.

A method for extracting spreading properties from scanning microellipsometry (SME) for various PFPE/solid surface pairs and the rheological characterization of PFPEs are examined at length. The interrelationships among SME spreading profiles, surface energy, rheology, and tribology, have been discussed as well. Phenomenological theories, including stability analysis and microscale mass transfer, have been introduced to qualitatively interpret ultrathin PFPE film nanostructures.

Rigorous simulation tools were examined as well. We employed lattice-based, simple reactive sphere Monte Carlo (MC) techniques to examine the fundamentals of PFPE dynamics. Unique features included introduction of the fractal dimension for possible use in a fingerprint analysis of lubricant/surface pairs as well as calculation of the disjoining pressure. An off-lattice bead-spring MC model was also introduced to capture the detailed PFPE nanofilm structure, and molecular dynamics (MD) was implemented for full-scale nanostructural analyses of PFPE thin films. These simulations will accurately capture the static and dynamic behaviors of PFPE films consistent with experimental findings and are thus suitable for describing the fundamental mechanisms of complex physical phenomena including film dewetting and rupture due to instability arising from nanoscale temperature and pressure nonhomogeneities.

Using MC simulations, we examined the nanostructure in confined PFPE films, including anisotropic molecular conformations, layering structures, segregations in the film, and surface morphology. For the first time, MD simulation with an off-lattice, bead-spring model was employed in PFPE systems. The mean-square displacement was calculated to examine PFPE dynamics, such as the self-diffusion coefficient. We also examined the transport coefficients by calculating the suitable autocorrelation functions. Nonexponential relaxation spectra (Kohlrausch–Williams–Watts function) were used as well to examine the nanoscale transport processes enabling the establishment of fingerprint analyses for lubricant–solid surface coupling.

One of our goal is to develop an integrated simulation tool capable of designing a heat-assisted magnetic recording (HAMR) slider (Fig. 1.53) [35], as well as to examine lubricant film structures from a monolayer to multilayers in the presence of harsh external conditions (e.g., pressure and temperature nonhomogeneity). To achieve these goals, multiscale modeling with hybridization of MD (lubricants), the Boltzmann transport equation (descriptive for heat transfer [179,180] and airbearing in nanoscale [40,45]), and fluid mechanics (airbearing) are desirable. The strategy for the integrated HDI design tool development from the fundamental scientific basis is summarized in Figure 1.54. The research reviewed here will generate important nanoscale transport process

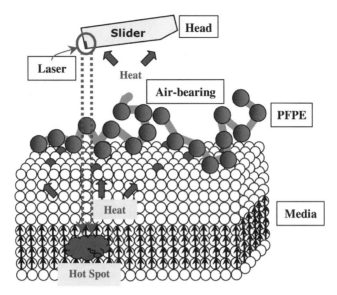

Figure 1.53. The bead-spring model may be used to simulate PFPE molecules in head–disk interface (HDI), which couple with slider dynamics via steep pressure and temperature gradients, which appears in HAMR technology [35].

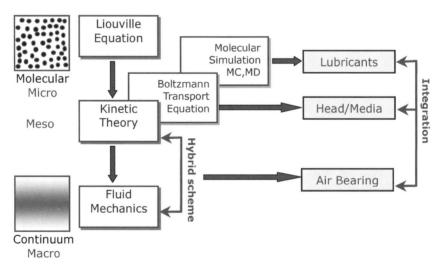

Figure 1.54. Multiscale, integrated modeling of HDI descriptive for Figure 1.53.

fundamentals, design tools for nanotechnology, and novel computational tools for nano/information devices and sensors.

APPENDIX

A.1. Background for Hard-Disk Drive (HDD)

(This section of the Appendix pertains to Section I of the text.) Information storage devices are critical vehicles on the road toward the multimedia era. Since the HDD was first introduced in 1957 as a component of IBM's RAMAC 350 [3], it has become the most prolific information storage device because of its fast access time, high data transfer rate, low cost, and high volumetric storage capacity. The volume of information generated by corporations doubles every year, such that the need for several-terabyte-sized databases has become a norm. Companies have begun to store much of their data on these storage systems and online, where the information is readily accessible, necessitating continued progresses in data storage technology. Additionally, server-based applications, such as e-commerce, medicine, and libraries, require modest access times but very large storage capacities and significant data transfer rates. Today, disk capacities double every 9 months, and quickly outpace advances in computer chips, which still obey Moore's law (doubling only every 18 months) [34]. These improvements are driven by the industry's astonishing ability to improve the areal density of HDD storage. We focus solely on HDDs in this section as similar technological concerns and stumbling blocks exist in other technologies, such as hybrid and heat assisted recording systems [35].

The "hard disk" gets its name from the part of the system that actually stores information: a rigid disk called a "platter," which is rotated by a spindle motor at a rotational speed of 10,000 rpm (revolutions per minute) in current state-of-the-art HDDs [Fig. 1.4(a)]. Disk size is generally 3.5 in. in diameter; however, 2.5-in. and 1.5-in. products are becoming popular in mobile computing devices, such as notebook and handheld computers. To increase storage capacity, most HDDs feature two or more platters. Information is written to and read from a platter by the read/write head, a tiny electronic circuit attached to the end tip of the slider (or head) [Fig. 1.4(c)]. These read/write heads are positioned on both sides of a platter and travel across the "data zone" during read/write operation. Sliders are located in the head gimbal assembly [Fig. 1.4(b)], and an actuator arm holds this assembly in place. When the drive is powered off, this device locks the read/write head into a safe "landing zone" to prevent data loss. This zone on the platter is typically laser-textured [36–39] to protect the slider from stiction failure and to reduce wear during the contact–start–stop operation.

It is remarkable that sliders are designed to fly less than 10 nm above the surface of the disk—this spacing is several orders of magnitude smaller than a smoke particle or the HIV virus. The rapid spinning of the disk creates a tiny cushion of air above each platter, permitting the slider to float, or "fly," just above the surface, such that the sliders never touch the surface of the platters while the disk is spinning. When this cushion of air between the head and the disk is compressed, a lift force is created and the disk operates like a "nanoscale pump." This cushion of air existing at the head–disk interface (HDI) surface is called a *nanoscale airbearing* [40–44]. Note that to describe nanoscale air bearing, we must use molecular or kinetic theory instead of the continuum-based Navier–Stokes equation [40,45]. If contamination or severe shock causes the head to touch the surface, or "crash," the head or data surface can be damaged, data can be lost, and in the most extreme cases, the drive may be rendered inoperable. Readers may notice why tribology (a subject dealing with lubrication, friction, and wear) is important in HDI reliability [3,6,46].

A hard disk typically consists of an Al-Mg alloy (or glass substrates), a NiP undercoat, magnetic layer, carbon overcoat, and a very thin layer of lubricant, as illustrated in Fig. 1.1. Since the Al-Mg alloy is mechanically soft, a hard undercoat is applied to provide adequate impact resistance to the head–disk interactions. The magnetic layer where the information is stored is typically sputtered directly onto a bare NiP/Al-Mg disk substrate. A carbon overcoat is sputtered to enhance wear and/or corrosion resistance. Finally, a molecularly thin layer of PFPEs (the subject of this chapter) along with airbearing is added to further reduce both the wear of the overcoat and stiction between the head and disk.

The sliders are free to move vertically around their load point (i.e., point of attachment) and can also rotate about the pitch and roll axes, as illustrated in Figure 1.5(*a*). As the HDI spacing is reduced to a nanometer scale, novel slider shapes, as shown in Figure 1.5(*b*), are necessary to maintain HDD reliability. This will remain a major design problem for sliders having ultrahigh recording density capabilities.

A.2.   Calculation of $D(h)$ from Hydrodynamic Model

(This section of the Appendix pertains to Section II.B of the text.) If the liquid in the film can be treated as a Newtonian fluid (an incorrect assumption, however), the Navier–Stokes equation with position dependent viscosity can be used to describe the multilayer film with the form [140,141]

$$-\frac{\partial}{\partial z}\left(\eta(z)\frac{\partial v}{\partial z}\right) = \frac{\partial \Pi}{\partial x} \tag{A.1}$$

where z is the direction normal to the surface, x is parallel to the surface and normal to the spreading, $\eta(z)$ is the position-dependent viscosity of the liquid film, and v is the velocity in the spreading direction. The no-slip boundary condition is imposed at the solid–film interface to simplify the analysis:

$$v = 0 \quad \text{at} \quad z = 0. \tag{A.2}$$

Also at the liquid film–air interface $(z = h)$, the shear stress-free boundary condition is generally applicable:

$$\frac{\partial v}{\partial z} = 0 \quad \text{at} \quad z = h. \tag{A.3}$$

The velocity profile satisfying Eqs. (A.1)–(A.3) is

$$v(z) = \int_0^z \frac{h - z'}{\eta(z')} \left(\frac{\partial \Pi}{\partial x} \right) dz'. \tag{A.4}$$

By integrating the continuity equation for the entire film, we obtain

$$\frac{\partial h}{\partial t} = -\frac{\partial q}{\partial x}. \tag{A.5}$$

From Eqs. (1.1) and (A.5), we obtain the Reynolds equation for incompressible fluids:

$$q = -D(h) \frac{\partial h}{\partial x}. \tag{A.6}$$

The value of q can be obtained via the integration over the entire film thickness and gives the following expression for the volumetric flow rate per unit width q:

$$q = \int_0^h v(z) dz. \tag{A.7}$$

From Eqs. (A.4), (A.6), and (A.7), we obtain

$$D(h) = \int_0^h dz \int_0^z \frac{h - z'}{\eta(z')} \left(\frac{\partial \Pi}{\partial h} \right) dz' = \int_0^h f(z) dz. \tag{A.8}$$

With integraton by parts, $D(h)$ is expressed in a more compact form as shown in Eq. (1.6a):

$$D(h) = zf(z)]_{z=0}^{z=h} - \int_0^h \frac{z(h - z)}{\eta(z)} \left(\frac{\partial \Pi}{\partial h} \right) dz = \int_0^h \frac{(h - z)^2}{\eta(z)} \left(\frac{\partial \Pi}{\partial h} \right) dz. \tag{1.6a}$$

Acknowledgments

This research was supported, in part, by the National Science Foundation under grant INT-9910429. I would like to thank Drs. T. Karis, S-C. Kang, and B. Marchon (Hitachi), Drs. X. Ma, R. Crone, and Y-T. Hsia (Seagate), and Prof. H.J. Choi (Inha University, Korea) for numerous discussions and suggestions on this subject. I also acknowledge my current graduate students, S. Izumisawa, Q. Guo, H. Chen, S. Ghai, and research associates, J.W. Kim, H.J. Kim for their valuable assistant.

References

1. M.C. Roco and W.S. Bainbridge, Eds. *Converging Technologies for Improving Human Performance.* Kluwer Academic Publisher, Norwell, MA, 2003.

2. B. Bhushan, Magnetic media tribology—state-of-the art and future challenges, *Wear* **136**, 169–197 (1990).

3. B. Bhushan, *Tribology and Mechanics of Magnetic Storage Devices,* Springer-Verlag, New York (1990).

4. M. S. Jhon and H. J. Choi, Lubricant in future data storage technology, *J. Ind. Eng. Chem.* **7**, 263–275 (2001).

5. C. M. Mate, B. K. Yen, D. C. Miller, M. F. Toney, M. Scarpulla, and J. E. Frommer, New methodologies for measuring film thickness, coverage, and topography, *IEEE Trans. Magn.* **36**, 110–114 (2000).

6. T. E. Karis, Chemistry of magnetic recording disk surfaces with intermittent asperity contacts, *IEEE Trans. Magn.* **37**, 924–928 (2001).

7. S. Izumisawa and M. S. Jhon, Stability analysis of ultra-thin lubricant films with chain-end functional groups, *Tribol. Lett.* **12**, 75–81 (2002).

8. Ausimont, http://www.ausimont.com/docs/fom_thin.html.

9. R.-N. Kono, M. S. Jhon, H. J. Choi, and C. A. Kim, Effect of reactive end groups on the rheology of disk lubricant systems, *IEEE Trans. Magn.* **35**, 2388–2390 (1999).

10. B. G. Min, J. W. Choi, H. R. Brown, D. Y. Yoon, T. M. O'Connor, and M. S. Jhon, Spreading characteristics of thin liquid films of perfluoropolyalkylethers on solid surfaces: Effects of chain-end functionality and humidity, *Tribol. Lett.* **1**, 225–232 (1995).

11. C. L. Jiaa and Y. M. Liu, Tribological evaluation and analysis of the head/disk interface with perfluoropolyether and X1-P phosphazene mixed lubricants, *Tribol. Lett.* **7**(1), 11–16 (1999).

12. J. Y. Lee, A. R. C. Baljon, R. F. Loring, and A. Z. Panagiotopoulos, Simulation of polymer melt intercalation in layered nanocomposites, *J. Chem. Phys.* **109**, 10321–10330 (1998).

13. J. Choi, J. W. Kim, M. H. Noh, D. C. Lee, M. S. Suh, M. J. Shin, and M. S. Jhon, SAN-Na^+-Montmorillonite nanocomposite for electrorheological material, *J. Mat. Sci. Lett.* **18**(18) 1505–1507 (1999).

14. J. W. Kim, S. G. Kim, H. J. Choi, M. S. Suh, M. J. Shin, and M. S. Jhon, Synthesis and electrorheological characterization of polyaniline and Na^+-montmorillonite clay nanocomposite, *Int. J. Mod. Phys. B* **15**, 657–664 (2001).

15. Y. H. Hyun, S. T. Lim, H. J. Choi, and M. S. Jhon, Rheology of poly(ethylene oxide)/organoclay nanocomposites, *Macromolecules* **34**(23), 8084–8093 (2001).

16. T. H. Kim, L. W. Jang, D. C. Lee, H. J. Choi, and M. S. Jhon Synthesis and rheology of intercalated polystyrene/Na^+-montmorillonite nanocomposite, *Macromol. Rapid Commun.* **23**(3), 191–195 (2002).

17. S. T. Lim, Y. H. Hyun, H. J. Choi, and M. S. Jhon, Synthetic biodegradable aliphatic polyester/ montmorillonite nanocomposites, *Chem. Mat.* **14**(4), 1839–1844 (2002).

18. C. H. Lee, S. T. Lim, Y. H. Hyun, H. J. Choi, and M. S. Jhon, Fabrication and viscoelastic properties of biodegradable polymer/organophilic clay nanocomposites, *J. Mat. Sci. Lett.* **22**(1), 53–55 (2003).

19. S. T. Lim, H. J. Choi, and M. S. Jhon, Dispersion quality and rheological property of polymer/ clay nanocomposites—ultrasonification effect, *J. Ind. Eng. Chem.* **9**(1), 51–57, (2003).

20. D. P. Park, J. H. Sung, S. T. Lim, H. J. Choi, and M. S. Jhon, Synthesis and characterization of organically soluble polypyrrole/clay nanocomposites, *J. Mat. Sci. Lett.* **22**(18), 1299–1302 (2003).

21. R. A. Vaia, B. B. Sauer, O. K. Tse, and E. P. Giannelis, Relaxations of confined chains in polymer nanocomposites: Glass transition properties of poly(ethylene oxide) intercalated in montmorillonite, *J. Polym. Sci. B: Polym. Phys.* **35**(1), 59–67 (1997).

22. J. Wu, and M. M. Lerner, Structural, thermal, and electrical characterization of layered nanocomposites derived from Na^+-montmorillonite and polyethers, *Chem. Mat.* **5**, 835–838 (1993).

23. Y. Kojima, A. Usuki, A. Kawdsumi, A. Okada, T. Kurauchi, O. Kamigaito, and K. Kaji, Novel preferred orientation in injection-moled Nylon 6-clay hybrid, *J. Polym. Sci. Part B: Polym. Phys.* **33**, 1039–1045 (1995).

24. X. Kornmann, R. Thomann, R. Mülhaupt, J. Finter, and L. Berglünd, Synthesis of amine-cured, epoxy-layered silicate nanocomposites: The influence of the silicate surface modification on the properties, *J. Appl. Polym. Sci.* **86**, 2643–2652 (2002).

25. Y. I. Tien and K. H. Wei, High-tensile-property layered silicates/polyurethane nanocomposites by using reactive silicates as pseudo chain extenders, *Macromolecules* **34**, 9045–9252 (2001).

26. S. S. Ray, K. Yamada, M. Okamoto, and K. Ueda, New polylactide-layered silicate nanocomposites. 2. Concurrent improvements of material properties, biodegradability and melt rheology, *Polymer* **44**, 857–866 (2003).

27. M. Pramanik, S. K. Srivastava, B. K. Samantaray, and A. K. Bhowmick, Rubber–clay nanocomposite by solution blending, *J. Appl. Polym. Sci.* **87**, 2216–2220 (2003).

28. V. Krikorian, M. Kurian, M. E. Galvin, A. P. Nowak, T. J. Deming, and D. J. Pochan, Polypeptide-based nanocomposite: Structure and properties of poly(L-lysine)/Na+-montmoril-lonite, *J. Polym. Sci. Part B: Polym. Phys.* **40**, 2579–2586 (2002).

29. R. A. Vaia, H. Ishii, and E. P. Giannelis, Synthesis and properties of two-dimensional nanostructures by direct intercalation of polymer melts in layered silicates, *Chem. Mat.* **5**, 1694–1696 (1993).

30. P. Aranda and E. Ruizhitzky, New polyelectrolyte materials based on smectite polyoxyethylene intercalation compounds, *Acta Polym.* **45**(2), 59–67 (1994).

31. R. A. Vaia and E. P. Giannelis, Lattice model of polymer melt intercalation in organically-modified layered silicates, *Macromolecules* **30**, 7990–7999 (1997).

32. M. Dijkstra, J.-P. Hansen, and P. Madden, Statistical model for the structure and gelation of smectite clay suspensions, *Phys. Rev.* **E55**, 3044–3053 (1997).

33. V. V. Ginzburg and A. C. Balazs, Calculating phase diagrams of polymer-platelet mixtures using density functional theory: Implications for polymer/clay composites, *Macromolecules* **32**, 5681–5688 (1999).

34. G. E. Moore, Cramming more components onto integrated circuits, *Electronics* **38**(8), 114–117 (1965).

35. J. W. Toigo, The Technology of computer hard drives is fast approaching a physical barrier imposed by the superparamagnetic effect. Overcoming it will require tricky innovations, *Sci. Am.* **282**(5), 58–74 (May 3, 2000); T.W. McDaniel, W.A. Challener, and K. Sendur, "Issues in heat-assisted perpendicular recording," *IEEE Trans. Magn.* **39**, 1972–1979 (2003).

36. R. Ranjan, D. N. Lambeth, M. Tromel, P. Goglia, and Y. Li, Laser texturing for low flying-height media, *J. Appl. Phys.* **69**, 5745–5747 (1991).

37. X. Zhao and B. Bhushan, Lubrication studies of head-disk interfaces in a controlled environment Part 1: effects of disk texture lubricant thermal treatment and lubricant additive on the durability of the head-disk interface, *P. I. Mech. Eng.* **214**(J6), 535–546 (2000).

38. C. D. Lin, J. S. Lin, and H. W. Huang, Lube build-up and migration behavior on laser texture media after long term parking, *IEEE Trans. Magn.* **36** (5) 2711–2713 (2000).

39. J. J. Liu, W. J. Li, and K. E. Johnson, Current and future approaches for laser texturing of thin film media, *IEEE. Trans. Magn.* **36**(1), 125–132 (2000).

40. S.-C. Kang, R. M. Crone, and M. S. Jhon, A new molecular gas lubrication theory suitable for head-disk interface modeling, *J. Appl. Phys.* **85**(8), 5594–5596 (1999).

41. R. M. Crone, M. S. Jhon, B. Bhushan, and T. E. Karis, Modeling the flying characteristics of a rough magnetic head over a rough rigid-disk surface, *J. Tribol.* **113**, 739–749 (1991).

42. R. M. Crone, M. S. Jhon, T. E. Karis, and B. Bhushan, The behavior of a magnetic sliders over a rigid-disk surface: A comparison of several approximations of the modified Reynolds equations, *Adv. Inform. Storage Syst.* **4**, 105–121 (1992).

43. P. R. Peck, M. S. Jhon, R. F. Simmons, and T. J. Janstrom, Mathematical modeling of lubrication for the head-disk interface using incompressible fluids, *J. Appl. Phys.* **75**(10), 5747–5749 (1994).

44. P. R. Peck, K.-O. Park, M. S. Jhon, S.-C. Kang, and T. I.-P. Shih, Transient response of ultra-low flying sliders over contaminated and textured surfaces, *J. Appl. Phys.* **79**(8), 5785–5787 (1996).

45. S. Fukui and R. Kaneko, A database for interpolation of Poiseuille flow rates for high knudsen number lubrication problems, *ASME J. Tribol.* **112**, 78–83 (1990).

46. T. E. Karis, G. W. Tyndall, and R. J. Waltman, Lubricant bonding effects on thin film disk tribology, *Tribol. Trans.* **44**(2), 249–255 (2001).

47. T. M. O'Connor, M. S. Jhon, C. L. Bauer, B. G. Min, D. Y. Yoon, and T. E. Karis, Surface diffusion and flow activation energies of perfluoropolyalkylether, *Tribol. Lett.* **1**, 219–223 (1995).

48. T. M. O'Connor, *Spreading of Ultra-Thin Perfluoropolyalkylether Films,* Ph.D. thesis, Carnegie Mellon Univ., Pittsburgh, PA, 1995.

49. X. Ma, J. Gui, L. Smoliar, K. Grannen, B. Marchon, M. S. Jhon, and C. L. Bauer, Spreading of perfluoropolyalkylether films on amorphous carbon surfaces, *J. Chem. Phys.* **110**(6), 3129–3137 (1999).

50. X. Ma, J. Gui, L. Smoliar, K. Grannen, B. Marchon, C. L. Bauer, and M. S. Jhon, Complex terraced spreading of perfluoropolyalkylether films on carbon surfaces, *Phys. Rev. E.* **59**(1), 722–727 (1999).

51. X. Ma, J. Gui, B. Marchon, M. S. Jhon, and C. L. Bauer, Lubricant replenishment on carbon coated disks, *IEEE Trans. Magn.* **35**, 2454–2456 (1999).

52. X. Ma, *Spreading of Perfluoropolyalkylether Films on Amorphous Carbon Surfaces,* Ph.D. Thesis, Carnegie Mellon Univ., Pittsburgh, PA, 1998.

53. V. J. Novotny, Migration of liquid polymers on solid surfaces, *J. Chem. Phys.* **92**, 3189–3196 (1990).

54. R. Pit, B. Marchon, S. Meek, and V. Velindandla, Formation of lubricant "moguls" at the head/disk interface, *Tribol. Lett.* **10**, 133–142 (2001).

55. L. Xu, D. F. Ogletree, M. Salmeron, H. A. Tang, J. Gui, and B. Marchon, De-wetting of K. E. lubricants on hard disks, *J. Chem. Phys.* **112**(6), 2952–2957 (2000).

56. M. H. Azarian, C. L. Bauer, T. M. O'Connor, R. M. Crone, and M. S. Jhon, Critical operating conditions for a novel flying tester, *Tribol. Trans.* **36**, 525–534 (1993).

57. R. M. Crone and M. S. Jhon, Mathematical modeling of the head-disk interface, *J. Appl. Phys.* **73**, 6016 (1993).

58. R. M. Crone, P. R. Peck, M. S. Jhon, and T. E. Karis, Scaling criteria for slider miniaturization using the generalized Reynolds equation, *J. Tribol.* **115**, 566–572 (1993).

59. M. S. Jhon, P. R. Peck, R. F. Simmons, and T. J. Janstrom, Behavior of the head-disk interface in future disk drives, *IEEE Trans. Magn.* **30**, 410–416 (1994).

60. X. Ma, J. Gui, K. J. Grannen, L. A. Smoliar, B. Marchon, M. S. Jhon, and C. L. Bauer, Spreading of PFPE lubricants on carbon surfaces: Effect of hydrogen and nitrogen content, *Tribol. Lett.* **6**, 9–14 (1999).

61. A. M. Cazabat, N. Fraysse, F. Heslot, and P. Carles, Spreading at the microscopic scale, *J. Phys. Chem.* **94**, 7581–7585 (1990).

62. A. M. Cazabat, N. Fraysse, and F. Heslot, Thin wetting films, *Colloids Surf.* **52**, 1–8 (1991).

63. F. Heslot, N. Fraysee, and A. M. Cazabat, Molecular layering in the spreading of wetting liquid drops, *Nature* **338**, 640–642 (1989).

64. F. Heslot, A. M. Cazabat, and P. Levinson, Dynamics of wetting of tiny drops: Ellipsometric study of the late stages of spreading, *Phys. Rev. Lett.* **62**, 1286–1289 (1989).

65. M. P. Valignat, N. Fraysee, A. M. Cazabat, and F. Heslot, Molecular networks in the spreading of microdroplets, *Langmuir* **9**, 601–603 (1993).

66. S. Gerdes, A. M. Cazabat, G. Strom, and F. Tiberg, Effect of surface structure on the spreading of a PDMS droplets, *Langmuir* **14**, 7052–7057 (1998).

67. M. Voue, M. P., Valignat, G. Oshanin, A. M. Cazabat, and J. De. Coninck, Dynamics of spreading of liquid microdroplets on substrates of increasing surface energies, *Langmuir* **14**, 5951–5958 (1998).

68. F. Heslot, A. M. Cazabat, P. Levinson, and N. Fraysse, Experiments on wetting on the scale of nanometers: Influence of the surface energy, *Phys. Rev. Lett.* **65**, 599–602 (1990).

69. T. Stobe, Z. Lin, R. M. Hill, M. D. Ward, and H. T. Davis, Enhanced spreading of aqueous films containing ethoxylated alcohol surfactants on solid substrates, *Langmuir* **13**, 7270–7275 (1997).

70. T. Stobe, R. M. Hill, M. D. Ward, and H. T. Davis, Enhanced spreading of aqueous films containing ionic surfactants on solid substrates, *Langmuir* **13**, 7276–7281 (1997).

71. W. Adamson, *Physical Chemistry of Surfaces*, 4th ed., Wiley, New York, 1982.

72. P. G. de Gennes, Wetting: Statics and dynamics, *Rev. Mod. Phys.* **57**, 828–863 (1985).

73. J. Crank, *The Mathematics of Diffusion*, Oxford Univ. Press, New York, 1975.

74. C. M. Mate, Application of disjoining and capillary pressure to liquid lubricant films in magnetic recording, *J. Appl. Phys.* **72**, 3084–3090 (1992).

75. J. I. Frankel and B. Vick, General formulation and analysis of hyperbolic heat conduction in composite media, *Int. J. Heat Mass Transf.* **30**, 1293–1305 (1987).

76. K. E. Goodson, *Thermal Conduction in Microelectronic Circuits*, Ph.D. thesis, Massachusetts Inst. Technol. (MIT), Feb. 1993.

77. M. N. Özisik and D. Y. Tzou, On the wave theory in heat conduction, *ASME J. Heat Transf.* **116**, 526–535 (1994).

78. A. Barletta and E. Zanchini, Hyperbolic heat conduction and thermal resonances in a cylindrical solid carrying a steady-periodic electric field, *Int. J. Heat Mass Transf.* **39**, 1307–1315 (1996).

79. D. Y. Tzou, A unified field approach for heat conduction from macro to micro-scales, *ASME J. Heat Transf.* **117**, 8–16 (1995).

80. A. Haji-Sheikh, W. J. Minkowycz, and E. M. Sparrow, Certain anomalies in the analysis of hyperbolic heat conduction, *ASME J. Heat Transf.* **124**, 307–319 (2002).

81. Y. Bayazitoglu and G. P. Peterson, eds., *Fundamental Issues in Small Scale Heat Transfer*, ASME HTD-227, 1992.

82. T. M. O'Connor, Y.-R. Back, M. S. Jhon, B. G. Min, D. Y. Yoon, and T. E. Karis, Surface diffusion of thin perfluoropolyalkylether films, *J. Appl. Phys.* **79**(8), 5788–5790 (1996).

83. M. N. Özisik and D. Y. Tzou, On the Wave Theory in Heat Conduction, *Fundamental Issues in Small Scale Heat Transfer*, ASME HTD-227, p. 13, 1992.

84. R. J. Waltman, G. W. Tyndall, and J. Pacansky, Computer-modeling study of the interactions of Zdol with amorphous carbon surfaces, *Langmuir* **15**, 6470–6483 (1999).

85. M. Ruths and S. Granick, Tribology of confined Fomblin-Z perfluoropolyalkylethers: Molecular weight dependence and comparison between unfunctionalized and telechelic chains, *Tribol. Lett.* **7**(4), 161–172 (1999).

86. M. Ruths and S. Granick, Tribology of confined Fomblin-Z perfluoropolyalkylethers: Role of chain-end chemical functionality, *J. Phys. Chem. B* **103**(41), 8711–8721 (1999).

87. X. Y. Zhang and S. Granick, Nanorheology of aqueous polyethylene glycol PEG, *Macromolecules* **35**(10), 4017–4022 (2002).

88. G. Reiter, A. I. Demirel, and S. Granick, From static to kinetic friction in confined liquid-films, *Science* **263**(5154), 1741–1744 (1994).

89. S. Granick, Motions and relaxations of confined liquids, *Science* **253**, 1374–1379 (1991).

90. T.E. Karis and G.W. Tyndall, Calculation of spreading profiles for molecularly-thin films from surface energy gradients, *J. Non-Newtonian Fluid Mechanics* **82**, 287–302 (1999).

91. M. S. Jhon, D. Phillips, S. J. Vinay, and C. Messer, The dynamic behavior of thin-film lubricants, *IEEE Trans. Magn.* **35**(5), 2334–2337 (1999).

92. G. W. Tyndall, T. E. Karis, and M. S. Jhon, Spreading profiles of molecularly thin perfluoropolyether films, *Tribol. Trans.* **42**, 463–470 (1999).

93. M.S. Jhon, J.S. Dahler, and R.C. Desai, A theory of the liquid-vapor interface, Advances in Chemical Physics, I. Prigogine and S.A. Rice, Wiley, New York, Vol. 46, 279–362 (1981).

94. R. J. Lei, A. J. Gellman, and P. Jones, Thermal Stability of Fomblin Z and Fomblin Zdol thin films on amorphous hydrogenated carbon, *Tribol. Lett.* **11**, 1–5 (2001).

95. J. L. Brand and S. M. George, Effects of laser-pulse characteristics and thermal desorption parameters on laser-induced thermal desorption, *Surf. Sci.* **167**, 341–362 (1986).

96. R. Z. Lei, K. R. Paserba, A. J. Gellman, N. Shukla, and L. M. Cornaglia, PFPE lubricant bonding to carbon overcoats *Proc. ASME/STLE Symp. Nanotribology and Nanotechnology for 1 Tbit/in²* A. A. Polycarpou and C. S. Bhatia, eds., 2001, pp. 45–53.

97. R. Z. Lei, *Fundamental Studies of Molecular Interactions between Lubricants and Contaminants on Amorphous Carbon Overcoats*, Ph.D. thesis, Carnegie Mellon Univ., Pittsburgh, PA, 2002.

98. J. U. Lemke and W. W. French, US Patent 5097368 (1992).

99. J. D. Ferry, *Viscoelastic Properties of Polymers*, 3rd ed., J Wiley, New York, pp. 241–254, 1980.

100. R.-N. Kono, S. Izumisawa, M. S. Jhon, C. A. Kim, and H. J. Choi, Rheology of perfluoro-polyether lubricants, *IEEE Trans. Magn.* **37**(4), 1827–1829 (2001).

101. C. D. Han and M. S. Jhon, "Correlation of the first normal stress difference with shear stress and of the storage modulus with loss modulus for homopolymers, *J. Appl. Polym. Sci.* **32**(3), 3809–3840 (1986).

102. K. S. Cole and R. H. Cole, Dispersion and absorption in dielectrics, *J. Chem. Phys.* **9**, 341–351 (1941).

103. P. de Gennes, *Scaling Concepts in Polymer Physics*, Cornell Univ. Press, Ithaca, NY, 1979.

104. G. Ajrodi, G. Marchionni, and G. Pezzin, The viscosity-molecular weight relationships for diolic perfluoropolyethers, *Polymer* **40**, 4163–4164 (1999).

105. G. Marchionni, G. Ajroldi, P. Cinquina, E. Tampellini, and G. Pezzin, Physical properties of perfluoropolyethers: Dependence on composition and molecular weight, *Polym. Eng. Sci.* **30**, 829–834 (1990).

106. D. Sianesi, V. Zamboni, R. Fontanelli, and M. Binaghi, Perfluoropolyethers: Their physical properties and behavior at high and low temperatures, *Wear* **18**, 85–100 (1971).

107. T. E. Karis and M. S. Jhon, The relationship between PFPE molecular rheology and tribology, *Tribol. Lett.* **5**, 283–286 (1998).

108. M. S. Jhon, S. Izumisawa, and H. J. Choi, Solution viscosity of polar perfluoropolyether lubricants, *J. Ind. Eng. Chem.* **9**(5), 508–512 (2003).

109. M. Huggins, The viscosity of dilute solutions of long-chain molecules. IV. Dependence on concentration, *J. Am. Chem. Soc.* **64**, 2716–2718 (1942).

110. T. Sakai, Huggins constant k for flexible chain polymers, *J. Polym. Sci. Part A-2* **6**, 1535–1549 (1968).

111. A. Sanguineti, P. A. Guarda, G. Marchionni, and G. Ajroldi, Solution properties of perfluoro-polyether polymers, *Polymer* **36**, 3697–3703 (1995).

112. G. B. Jeffery, The motion of ellipsoidal particles immersed in a viscous fluid, *Proc. Roy. Soc. A* **102**, 161–179 (1922).

113. R. Simha, The influence of brownian movement on the viscosity of solutions, *J. Phys. Chem.* **44**, 25–34 (1940).

114. W. Kuhn and H. Kuhn, Die abhängigkeit der viskositä vom strömungsgefälle bei hoch verdünnten suspensionen und lösungen, *Helv. Chim. Acta* **28**, 97–127 (1945).

115. T. M. Kwon, M. S. Jhon, and H. J. Choi, Viscosity of magnetic particle suspensions, *J. Mol. Liq.* **75**, 115–126 (1998).

116. R. Z. Lei and A. J. Gellman, Humidity effects on PFPE lubricant bonding to a-CHx overcoats, *Langmuir* **16**(16), 6628–6635 (2000).

117. D. E. Fowler, R. D. Johnson, D. van Leyen, and A. Benninghoven, Determination of molecular weight and composition of a perfluorinated polymer from fragment intensities in time-of-flight secondary ion mass spectrometry, *Anal. Chem.* **62**, 2088–2092 (1990).

118. D. E. Fowler, R. D. Johnson, D. van Leyen, and A. Benninghoven, Quantitative time-of-flight secondary ion mass spectrometry of a perfluorinated polyether, *Surf. Interf. Anal.* **17**, 125–136 (1991).

119. I. A. Bletsos, D. Fowler, D. van Leyen, A. Benninghoven, and D. M. Hercules, Time-of-flight secondary ion mass spectrometry of perfluorinated polyethers, *Anal. Chem.* **62**, 1275–1284 (1990).

120. P. H. Kasai and A. M. Spool, Z-DOL and carbon overcoat: Bonding mechanism, *IEEE Trans. Magn.* **37** 929–933 (2001).

121. P. H. Kasai, Degradation of perfluoropolyethers catalyzed by Lewis acids, *Adv. Inform. Storage Syst.* **4**, 291 (1991).

122. P. H. Kasai, Perfluoropolyethers: Intramolecular disproportionation, *Macromolecules* **25**, 6791–6799 (1992).

123. T. E. Karis, R. D. Johnson, V. J. Novotny, and M. S. Jhon, Ultrasonic scission of perfluoropolyethers, *J. Magn. Soc. Japan* **18** (S1), 509–512 (1994).

124. P. M. Cotts, Solution properties of a group of perfluoropolyethers: comparison of unperturbed dimensions, *Macromolecules* **27**, 6487–6491 (1994).

125. T. E. Karis, Water adsorption on thin film media, *J. Colloid Interf. Sci.* **225**(1), 196–203 (2000).

126. N. Shukla, J. Gui, and A. J. Gellman, Adsorption of fluorinated ethers and alcohols on graphite, *Langmuir* **17**(8), 2395–2401 (2001).

127. A. Sharma, Relationship of thin-film stability and morphology to macroscopic parameters of wetting in the apolar and polar systems, *Langmuir* **9**, 861–869 (1993).

128. A. Sharma and R. Khanna, Pattern formation in unstable thin liquid films, *Phys. Rev. Lett.* **81**, 3463–3466 (1998).

129. G. Reiter, A. Sharma, A. Casoli, M.-O. David, R. Khanna, and P. Auroy, Thin film instability induced by long-range forces, *Langmuir* **15**, 2551–2558 (1999).

130. J. Van Oss, M. K. Chaudhury, and R. J. Good, Interfacial Lifshitz-van der Waals and polar interactions in macroscopic systems, *Chem. Rev.* **88**, 927–941 (1988).

131. J. Van Oss, Surface free energy contribution to cell interactions, in *Biophysics of the Cell Surface*, R. Glaser, and D. Gingell, eds. (Springer-Verlag, Berlin, Germany, 1990), pp. 131–152.

132. G. W. Tyndall, R. J. Waltman, and D. J. Pocker, Concerning the interactions between zdol perfluoropolyether lubricant and an amorphous-nitrogenated carbon surface, *Langmuir* **14**, 7527–7536 (1998).

133. M. C. Kim, D. M. Phillips, X. Ma, and M. S. Jhon, Microscopic spreading characteristics of nonpolar perfluoropolyether films, *Chem. Eng. Commun.* **191**, 1–16 (2004).

134. M. C. Kim, D. M. Phillips, X. Ma, and M. S. Jhon, The molecular spreading of non-polar perfluoropolyether films on amorphous carbon surfaces, *J. Colloid Interf. Sci.* **228**(2), 405–409 (2000).

135. D. Frenkel and B. Smit, *Understanding Molecular Simulation: From Algortihm to Applications*, 2nd ed., Academic Press, New York, 2002.

136. M. Bishop, D. Ceperley, H. L. Frisch, and M. H. Kalos, Investigation of static properties of model bulk polymer fluids, *J. Chem. Phys.* **75**, 3228–3235 (1980).

137. A. Milchev, W. Paul, and K. Binder, Polymer-chains confined into tubes with attractive walls—a Monte-Carlo simulation, *Macromol. Theory Simul.* **3**, 305–323 (1994).

138. A. Milchev and K. Binder, Static and dynamic properties of adsorbed chains at surfaces: Monte Carlo simulation of a bead-spring model, *Macromolecules* **29**, 343–354 (1996).

139. K. Binder, J. Baschnagel, C. Bennemann, and W. Paul, Monte Carlo and molecular dynamics simulation of the glass transition of polymers, *J. Phys. Condens. Matt.* **11**, A47–A55 (1999).

140. T. Aoyagi, J. Takimoto, and M. Doi, Molecular dynamics study of polymer melt confined between walls, *J. Chem. Phys.* **115**, 552–559 (2001).

141. M. Bowker and D. A. King, Adsorbate diffusion on single crystal surfaces: I. The influence of lateral interactions, *Surf. Sci.* **71**, 583–598 (1978).

142. D. A. Reed and G. Ehrlich, Surface diffusion, atomic jump rates and thermo-dynamics, *Surf. Sci.* **102**, 588–609 (1981).

143. D. A. Reed and G. Ehrlich, Surface diffusivity and the time correlation of concentration fluctuations, *Surf. Sci.* **105**, 603–628 (1981).

144. M. Tringides and R. Gomer, A Monte Carlo study of oxygen diffusion on the (110) plane of tungsten, *Surf. Sci.* **145**, 121–144 (1984).

145. M. Tringides and R. Gomer, Models of surface diffusion: I. Anisotropy in activated diffusion, *Surf. Sci.* **166**, 419–439 (1986).

146. A. Sadiq and K. Binder, Diffusion of absorbed atoms in ordered and disordered monolayers at surfaces, *Surf. Sci.* **128**, 350–382 (1983).

147. C. H. Mak, H. C. Anderson, and S. M. George, Monte Carlo studies of diffusion on inhomogeneous surfaces, *J. Chem. Phys.* **88**, 4052–4069 (1988).

148. A. Lukkarinen, K. Kaski, and D. B. Abraham, Mechanisms of fluid spreading: Ising model simulations, *Phys. Rev. E* **51**, 2199–2202 (1995).

149. H. Hinridusen and E. Domany, Damage spreading in the Ising model, *Phys. Rev. E* **56**, 94–98 (1997).

150. J. D. de Coninck, S. Hoorelbecke, M. P. Valignat, and A. M. Cazabat, Effective microscopic model for the dynamics of spreading, *Phys. Rev. E* **48**, 4549–4555 (1993).

151. S. Bekink, S. Karaborni, G. Verbist, and K. Esselink, Simulating the spreading of a drop in the terraced wetting regime, *Phys. Rev. Lett.* **76**, 3766–3769 (1996).

152. J. Yang, J. Koplik, and J. R. Banavar, Molecular dynamics of drop spreading on a solid surface, *Phys. Rev. Lett.* **67**, 3539–3542 (1991).

153. J. Yang, J. Koplik, and J. R. Banavar, Terraced spreading of simple liquids on solid surfaces, *Phys. Rev. A* **46**, 7738–7749 (1992).

154. J. A. Nieminen and T. Ala-Nissila, Spreading dynamics of polymer microdroplets: A molecular-dynamics study, *Phys. Rev. E* **49**, 4228–4236 (1994).

155. L. Wagner, Molecular dynamics simulations of polymer droplets, *Phys. Rev. E* **51**, 499–503 (1995).

156. U. D'Ortona, J. D. Coninck, J. Koplik, and J. R. Banavar, Terraced spreading mechanisms for chain molecules, *Phys. Rev. E* **53**, 562–569 (1996).

157. D. M. Phillips and M. S. Jhon, Dynamic simulation of nanoscale lubricant films, *J. Appl. Phys.* **91**(10), 7577–7579 (2002).

158. X. Ma, M. S. Jhon, C. L. Bauer, J. Gui, and B. Marchon, Monte Carlo simulations of liquid spreading on a solid surface, *Phys. Rev. E.* **60**, 5795–5801 (1999).

159. S. Izumisawa and M. S. Jhon, Stability analysis and computer simulation of nanoscale lubricant films with chain-end functional groups, *J. Appl. Phys.* **91**(10), 7583–7585 (2002).

160. S. J. Vinay, D. M. Phillips, Y. S. Lee, C. M. Schroeder, X. Ma, M. C. Kim, and M. S. Jhon, Simulation of ultra-thin lubricant films spreading over various carbon overcoats, *J. Appl. Phys.* **87**(9), 6164–6166 (2000).

161. B. Sapoval, M. Rosso, and J. F. Gouyet, The fractal nature of a diffusion front and the relation to percolation, *J. Phys. Lett.* **46**(4), L149–L156 (1985).

162. D. M. Phillips, A. S. Khair, and M. S. Jhon, Mathematical simulation of ultra-thin polymeric film spreading dynamics, *IEEE Trans. Magn.* **37**(4), 1866–1868 (2001).

163. G. W. Tyndall (private discussion)

164. Bunde and J. F. Gouyet, On scaling relations in growth-models for percolating clusters and diffusion fronts, *J. Phys. A—Math Gen.* **18**(6), L285–L287 (1985).

165. H. Saleur and B. Duplantier, Exact determination of the percolation hull in 2 dimensions, *Phys. Rev. Lett.* **58**(22), 2325–2328 (1987).

166. D. Phillips, *Nanoscale Modeling for Ultra-Thin Liquid Films: Spreading and Coupled Layering*, Ph.D. thesis, Carnegie Mellon Univ., Pittsburgh, PA, 2003.

167. M. S. Jhon, S. Izumisawa, Q. Guo, D. M. Phillips, and Y. T. Hsia, Simulation of nanostructured lubricant films, *IEEE Trans. Magn.* **39**(2), 754–758 (2003).

168. S. Izumisawa and M. S. Jhon, Molecular simulation of thin polymer films with functional endgroups, *J. Chem. Phys.* **117**(8), 3972–3977 (2002).

169. R. J. Waltman, G. W. Tyndall, and J. Pacansky, Computer-modeling study of the interactions of Zdol with amorphous carbon surfaces, *Langmuir* **15**, 6470–6483 (1999).

170. P. H. Kasai and A. M. Spool, Z-DOL and carbon overcoat: Bonding mechanism, *IEEE Trans. Magn.* **37**, 929–933 (2001).

171. Y. T. Hsia, Q. Guo, S. Izumisawa, and M. S. Jhon, The dynamic behavior of ultrathin lubricant films, *Microsyst. Technol.* (in press).

172. S. Izumisawa, Q. Guo, and M. S. Jhon, The structural nonuniformity and molecular motion of molecularly thin polymeric film, (manuscript in preparation).

173. M. S. Jhon, G. Sekhon, and R. Armstrong, The response of polymer molecules in a flow, in *Advances in Chemical Physics*, I. Prigogine and S. A. Rice, eds., Wiley, New York, Vol. 66, 153–211 (1987).

174. Q. Guo, S. Izumisawa, D. M. Phillips, and M. S. Jhon, Surface morphology and molecular conformations for molecularly-thin lubricant films with functional chain endgroups, *J. Appl. Phys.* **93**(10), 8707–8709 (2003).

175. K. Kremer and G. S. Grest, Dynamics of entangled linear polymer melts—a molecular-dynamics simulation, *J. Chem. Phys.* **92**, 5057–5086 (1990).

176. R. Kubo, Brownian-motion and nonequilibrium statistical-mechanics, *Science* **233**, 330–334 (1986).

177. K. Binder, *Monte Carlo and Molecular Dynamics Simulations in Polymer Science*, Oxford Univ. Press, Oxford, UK, 1995.

178. Q. Guo, S. Izumisawa, M. S. Jhon, and Y. T. Hsia, Transport properties of nanoscale perfluoropolyether films, *IEEE Trans. Magh.* (in press).

179. G. Chen, Ballistic-diffusive heat-conduction equation, *Phys. Rev. Lett.* **86**(11), 2297–2300 (2001).

180. R.A. Escobar, S.S. Ghai, M.S. Jhon, and C. Amon, Time-dependent simulations of sub-continuum heat generation effects in electronic devices using the lattice-Boltzmann method, *Micro-Electro-Mechanical Systems* **5**, 603–612 (2003).

FRAGILITY METRICS IN GLASS-FORMING LIQUIDS

UDAYAN MOHANTY

Department of Chemistry, Boston College[*]
Chestnut Hill, Massachusetts

CONTENTS

[*] Current address: Department of Physics, Stanford University, Stanford, California.

Advances in Chemical Physics, Volume 129, edited by Stuart A. Rice
ISBN 0-471-44527-4   Copyright © 2004 John Wiley & Sons, Inc.

I. FRAGILITY

A. General Considerations

Molecular, ionic and metallic liquids can often be made to bypass crystallization on cooling below their melting point. On further cooling, these liquids exhibit a dramatic slowing down of their dynamics by as much as ten to twelve orders of magnitude, so that ultimately with lowering of temperature the liquids structure is frozen in on the timescale of laboratory experiments. A supercooled liquid falls out of equilibrium at a characteristic temperature called the *glass transition temperature* T_g, the temperature at which the shear viscosity is 10^{13} poise or the enthalpy relaxation reaches 10^2 s. For temperatures lower than T_g, one obtains a glass or an amorphous state of matter.

There have been renewed experimental and theoretical efforts in elucidating the nature of supercooled and glassy states of matter [1–26]. A wealth of experimental data spanning a wide range of glass-forming liquids indicates that they have dynamical attributes that display semiuniversal properties [1–26]. This includes the fact that glassy state is intricately nonlinear and that the temperature variation of the slow α relaxation time is non-Arrhenius and is often described by the Vogel–Tamman–Fulcher (VTF) equation [1–6,17,18]. Another distinct feature of these liquids is the "stretched exponential" or Kohlrausch–Williams–Watts form of time dependence of response function near T_g [1–6, 8–10]. There is experimental evidence that this slowdown in highly viscous liquids is due to spatial hetereogeneity [11,12,27,28]. In particular, the dynamics in a particular region of glass forming liquid could be orders of magnitude faster or slower than another region located several nanometers apart [11,12,27,28]. These hetereogeneities are found to strongly influence the relaxation of a nonequilibrium glass towards equilibrium [11,12,27,28].

The existence of fast β relaxation is unique to glassy materials and has been attributed to the nature of the packing statistics of the molecules and to intramolecular reorientation of the molecules [20,21,29]. The fast beta relaxation exists below T_g, and its temperature dependence is Arrhenius [20,21,28]. At high temperatures, β and slow α relaxations often merge near a temperature termed the crossover temperature T_x, while at lower temperatures they bifurcate [7,20,21,29,30,31]. A recently discovered attribute of supercooled liquids is the observation that the natural logarithm of the relaxation time at or the vicinity of the crossover temperature, $-\log\tau(T_x)$, is -7 ± 1 [32–34].

For temperatures below $1.2\,T_g$, molecules on the average translate considerably further than that based on viscosity through the Stokes–Einstein relation [12,35,36]. In fact, current experimental evidence suggests that on the average, for every rotation, a given molecule translates further as temperature is lowered [12,35,36]. The decoupling between viscosity and translational diffusion as well as that between translational diffusion and rotational diffusion is a universal feature in glass formers that span a wide range of kinetic behavior [12,35,36].

There is considerable evidence experimentally and theoretically that the kinetics and the thermodynamic behavior of deeply supercooled liquids, which are far away from equilibrium, are intricately linked [1–7,9,13,14,16–18,23,25, 26,32–34,37–105]. Although precise theoretical description of this connection is a major unsolved theoretical challenge, three key experimental and theoretical advances have provided novel insights into the decisive role played by thermodynamic factors in glass-forming liquids [37,38,52,54–56]:

1. Inherent structure analysis of diffusion via molecular dynamics of a deeply supercooled binary Lennard-Jones fluid have provided renewed impetus to the decisive role played by thermodynamic factors [52,53]. The location of the mode crossover temperature and the onset of super-Arrhenius behavior were related to the static structure of the liquid via the potential energy hypersurface [52,53].

2. Ito et al. observed a remarkable similarity between kinetic fragility plot and normalized entropy data, namely, Kauzmann plot, exhibited in a scaled-T_g form sheds considerable light into the role of excess entropy [37]. As will be discussed in more detail later, the kinetic fragility can be correlated with excess entropy for a wide class of glass-forming liquids [38]. The excess entropy has contributions from both vibrational and configurational degrees of freedom.

3. An explicit expression relating kinetic fragility to thermodynamic behavior of supercooled liquids was accomplished for the first time by Mohanty and coworkers [55,56] and independently by Speedy [54]. These authors derived an expression for the steepness parameter, a measure of kinetic fragility, from the temperature variation of the relation time or viscosity, with the ratio of excess entropy and heat capacity changes at the glass transition temperature [54–56]. A detailed description of this work will be provided later in the review chapter.

This review chapter is focussed on surveying current developments on a topic of considerable current interest, namely, the concept and the corresponding phenomenonology of steepness or fragility index as a quantitative measure of equilibrium (thermodynamic) and dynamical properties of glass-forming liquids. Because of the scope of the review, various themes in supercooled liquids and glassy solids, such as mode coupling theory and colloidal glasses, will not be discussed.

B. Historical Background

In a paper that appeared in 1957, Oldekop investigated the temperature dependence of viscosity η of supercooled liquid by judiciously plotting $\ln \eta$ as a function of reduced inverse temperature T_g/T, where T_g is the glass transition temperature [66]. The viscous flow was interpreted in terms of the potential energy barrier picture [66]. This idea was then picked up by Nemilov and coworkers, who attempted to correlate the slope of $\ln \eta$ with respect to T_g/T for various glass-forming liquids with changes in the corresponding heat capacity at the glass transition temperature [66].

The pioneering work of Gibbs–DiMarzio and Adam–Gibbs configurational entropy theories of glass-forming liquids introduced the notion and the importance of a new characteristic thermodynamic temperature T_2, a temperature at which the configurational entropy vanishes [59–62]. Half a decade later, Adam and Gibbs argued based on the idea of a cooperative rearranging region as part of the system that can rearrange into a configuration independent of its environment, that the ratio of glass transition temperature to T_2, was universal [41,42]. A notable correlation between the Vogel–Tamman–Fulcher temperature T_0 and transport properties of various liquids was soon thereafter observed by Angell [104].

The concept that melting temperature could be fruitful as a correspondence states parameter for viscous flow has its origin in the work of Greet and Magill and Laughlin and Uhlmann [67,68]. These authors noted that a weakness of utilizing melting temperature to correlated dynamic data is that the characteristic of the melting point is governed by both the liquid and the crystalline phases [67,68].

For a series of organic liquids, such as salol, α-phenyl-o-cresol, o-terphenyl, and tri-α-naphthylbenzene, Laughlin and Uhlmann investigated in detail the possibility and the identity of the glass transition temperature, defined as the value of temperature where the viscosity reaches a characteristic value of 10^{15} P (poise), as a normalization measure of the viscosity data [67]. Corresponding state analysis of the viscosity data on KNO_3-$Ca(NO_3)_2$ melt and Au-Si-Ge metal alloy lead to a $\log \eta$ versus T_g/T plot that is remarkably similar to that of organic glass formers [67]. In contrast, oxide glass formers such as SiO_2, GeO_2, B_2O_3, sodium disilicate, potassium disilicate, and NBS standard soda lime silicate, display shallow $\log \eta$ versus T_g/T behavior compared with organic liquids [67]. Laughlin and Uhlmann concluded that the corresponding state parameter T_g is a useful methodology to characterize glass-forming liquids since they can be divided into approximately three groups with distinct viscous flow characteristics [67].

The modern era on the classification of glass formers as "strong" and "fragile" liquids and the concept of fragility began in the groundbreaking work on T_g-scaled Arrhenius viscosity plots presented by Angell at the Blackburg Workshop on Relaxation in Disordered Systems in 1983 [22]. In this work, log η viscosity was analyzed as function of normalized inverse temperature, T_g/T, as well as the corresponding liquid to crystal heat capacity under constant pressure as a function of reduced temperature T/T_g, for 17 supercooled liquids [22]. In this work, the glass transition was defined as the temperature corresponding to where the viscosity is 10^{13} P [22].

Glass formers that exhibit stability in short- and intermediate-range order, often have a small change in heat capacity at T_g and are classified as "strong" [1–6,22]. The temperature dependence of the average relation time of these supercooled liquids is almost Arrhenius. In contrast, fragile liquids have a large change in heat capacity at T_g, and the temperature dependence of the average relation time with temperature is strongly non-Arrhenius [1–6,8,9,22]. In this context the term "fragile" refers to the fact that on increasing the temperature above the glass transition temperature, these glass formers would have a breakdown of their configurational structure [1–6,8,9,22]. Fragile liquids often exhibit a broader distribution of relaxation times than do strong liquids.

The concept of fragility is a qualitative one and is related to deviations of the relaxation time of a liquid from Arrhenius-like behavior and to the topology of the potential energy landscape of the system. The classification of liquids into strong and fragile thus provides a fundamental framework for quantitatively describing equilibrium and dynamical properties of supercooled liquids and glassy states of matter [1–6,8,9,22,37,38,52,54–56,88–91,103].

II. MEASURES OF FRAGILITY

A. Steepness Parameter

The steepness index or the fragility index m is defined as follows [105]:

$$m = \frac{d \log \tau}{d(T_g/T)} \bigg|_{T=T_g} \qquad (2.1)$$

This quantity is related to the effective activation enthalpy at the glass transition temperature

$$m = \frac{\Delta h_{\text{eff}}(T_g)}{(\ln 10)RT_g} \qquad (2.2)$$

where R is the gas constant.

If the relaxation time of the supercooled liquid is described by the Vogel–Tammann–Fulcher (VTF) equation

$$\tau(T) = \tau_0 \exp \frac{B}{T - T_0} \qquad (2.3)$$

where B, T_0, and τ_0 are the VTF constants with respect to temperature, the temperature dependence of the effective enthalpy is then

$$\frac{\Delta h_{\text{eff}}}{R} \equiv \frac{d \ln \tau}{d(1/T)}$$
$$= \frac{B}{(1 - T_0/T)^2} \qquad (2.4)$$

Thus, the steepness parameter can be expressed in terms of experimentally determined VTF constants B and T_0 as [18]

$$m = \frac{BT_g}{\ln(10)(T_g - T_0)^2} \qquad (2.5)$$

B.   Strength Parameter

The strength parameter D is defined through the VTF relation [23]

$$\tau(T) = \tau_0 \exp \frac{DT_0}{T - T_0} \qquad (2.6)$$

A comparison of Eqs. (2.6) and (2.3) shows that D is related to the VFT parameters as

$$B = DT_0 \qquad (2.7)$$

Several comments are in order regarding the strength parameter: (1) tabulated values of VTF equation usually refer to $\log \tau$—consequently, the strength parameter is related to VTF constants via $B = DT_0/\ln(10)$ [23]; (2) the data fit to VTF equation is done by taking τ_0 to characteristic of phonon vibration lifetimes, $\approx 10^{-14}$ s or by assuming that T_0 is approximated by the Kauzmann temperature [23,43]; and (3) Eq. (2.6) can be derived from the Adam–Gibbs relation for the relaxation time $\tau_{AD}(T)$ together with an assumption on the temperature variation of the heat capacity [17,43]

$$\Delta C_P(T) \approx \Delta C_P(T_g) \frac{T_g}{T} \qquad (2.8)$$

The configurational entropy at T_g is

$$S(T_g) = \int_{T_0}^{T_g} \Delta C_P(T_g) \frac{T_0}{T} \frac{dT}{T}$$

$$= \Delta C_P(T_g) \frac{T_g - T_0}{T_0} \qquad (2.9)$$

Substituting Eq. (2.9) in the Adam–Gibbs relation for relaxation time $\tau_{AD}(T)$ yields an expression for the strength parameter [18,41]

$$D = \frac{N_A \Delta \mu s^*}{k_B T_g \Delta C_p(T_g)} \qquad (2.10)$$

where N_A is the Avogadro number, $\Delta \mu$ is the change in chemical potential per mole of segments, s^* is the entropy of the smallest rearranging region, and k_B is the Boltzmann constant. Consequently, the strength parameter is inversely proportional to heat capacity change at T_g.

C. Kauzmann Plot

Consider the ratio of the difference of entropy between the liquid and the crystal, $\Delta S(T)$, to the corresponding entropy difference ΔS_m at the melting temperature T_m. A plot of this ratio as a function of normalized temperature T/T_m is referred to as the *Kauzmann plot* [25]. The apparent ground state of the liquid is at the Kauzmann temperature T_K where $\Delta S(T_K)/\Delta S_m$ vanishes [25,26,41,42,59–62].

The Kauzmann plot can be exploited to introduce thermodynamic fragility, which bears a close resemblance to the kinetic fragilty plot discussed earlier: log (viscosity) versus T_g/T. Martinez and Angell have accomplished this goal by plotting, instead, the ratio $\Delta S(T_g)/\Delta S(T)$ versus inverse normalized inverse temperature T_g/T for temperature above T_g [38]. Several important results follow from such an analysis: (1) the important scaling role played by excess entropy $S_{exc}^{(T)} = \Delta S(T)$ [37,38], (2) ordering of the various liquids was qualitatively the same in both thermodynamic and kinetic metrics [37,38], (3) thermodynamic and kinetic ground state temperatures are almost identical [37,38], and (4) correlation between thermodynamic fragility and kinetic fragility [37,38].

D. Scaled VTF and Crossover Temperatures

To capture the relaxation times change from molecular vibration times to values prototypical near the glass transition temperature, other metrics of fragility have been used that exploits the shape of the normalized excess entropy data in a reduced-T_g form. This includes the ratio F of VFT temperature, where the relaxation time diverges, to the glass transition temperature, T_0/T_g [18,107] or the ratio of the glass transition temperature to the crossover temperature; the later is obtained by appropriate scaling of the relaxation time or diffusion data [43,108]. The value of the crossover temperature is independent of the fitting function used to fit the relaxation data over the entire range of temperatures [43,108].

An ambiguity of the F metric or the choice of the ratio T_g/T_K, where T_K is the Kauzmann temperature, is that experimental data have to be extrapolated to obtain the VTF and the Kauzmann temperatures [37,43].

E. $F_{1/2}$ Metric

To measure the departure from an Arrhenius-like behavior and to decrease the ambiguity in the use of fragility as a quantitative probe of the liquid state, the so-called $F_{1/2}$ metric has been introduced. It is defined as the value of T_g/T at the midway of the relaxation time on a log scale, specifically, between the high-temperature phonon vibration lifetimes 10^{-14} s and the relaxation time at T_g, namely, $\tau(T_g)$, which is generally taken to be 10^2 s [37]. An advantage of this definition is that the midway values for the relaxation time are readily and accurately accessible by viscosimetric and by dielectric measurements [37,43].

Let $T_{1/2}$ be temperature at which $\tau = 10^{-6}$ s. Now define a quantity $F_{1/2}$ as follows [37,43]:

$$F_{1/2} = 2\frac{T_g}{T_{1/2}} - 1 \qquad (2.11)$$

Note that the magnitude of $F_{1/2}$ lies between zero and unity. The quantity $F_{1/2}$ is a measure of the fragility of the supercooled liquid. The $F_{1/2}$ fragility metric has two notable virtues: (1) it is independent of the fitting function employed to analyze the experimental data [37,43] and (2), because of the definition of the midway temperature, the relaxation time at this characteristic temperature is accurately accessible by various experimental techniques [37,43].

F. $F_{1/2;thermo}$ and $F_{3/4;thermo}$ Metrics

As mentioned in Section I, the melting temperature has been probed as a correspondence states parameter for viscous flow [68]. This idea has been pushed further in defining a thermodynamic fragility as [37,88]

$$F_{1/2;thermo} = \frac{T_{1/2}}{T_m} \qquad (2.12)$$

Here $T_{1/2}/T_m$ is the value of T/T_m at which half of the entropy of fusion of the liquid is lost on supercooling, where T_m is the melting temperature. The success of this type of scaling is due to the fact for a number of substances the "$\frac{2}{3}$ rule" of glass formers apply, specifically, the glass transition is approximately two-thirds of the melting temperature [37,88].

An alternative and perhaps more accurate definition of thermodynamic fragility that can be determined experimentally for almost all liquids would be to determine the fraction of entropy of fusion lost by $T/T_m = 0.8$ [37,88]. In both the definitions, the property of the crystal phase enters; this disadvantage is missing in the kinetic measures of fragility [37,88].

In parallel with Eq. (2.12), the $F_{3/4;\text{thermo}}$ metric is defined as the value of (T_g/T) for which $S_{\text{exc}}(T_g)/S_{\text{exc}}(T) \approx \frac{3}{4}$. With this definition, a correlation was observed between kinetic $F_{1/2}$ and thermodynamic $F_{3/2}$; thermo fragilities [38].

G. $\Delta C/\Delta S$ Metric

Mohanty and coworkers [55,56] and independently Speedy [54] made the provocative suggestion that the potential energy landscape that governed dynamics of viscous was to a large extent themodynamic in nature. Mohanty et al. showed that the fragility index is expressible in terms of experimentally measurable thermodynamic properties of a supercooled liquid such as the excess entropy and heat capacity changes frozen in at the glass transition temperature [54–56]

$$m/m_{\text{min}} = 1 + \frac{\Delta C_p(T_g)}{S(T_g)} \qquad (2.13)$$

where, by definition, $m_{\text{min}} = 16$. Note that the significance of the ratio of the excess heat capacity to entropy changes to dynamics of supercooled liquid has been suggested by Goldstein [106], but the association of this ratio to kinetic fragility was established by Mohanty and coworkers [55,56].

Although, we will investigate Eq. (2.13) in detail in the next section, let us point out the importance of Eq. (2.13): (1) the expression for the fragility index indicates the importance of the role of excess entropy and heat capacity; (2) as will be shown later, Eq. (2.13) has quantitative predictive powers—in fact, over a dozen or so strong and fragile liquids accurately satisfies this relation; (3) the minimum fragility m_{min} is not universal for polymeric glass formers; and (4) Eq. (2.13) is the closest one can come in expressing fragility, a dynamical property, in terms of measurable thermodynamic quantities.

III. RELATIONSHIP BETWEEN FRAGILITY INDEX AND CHARACTERISTICS OF GLASS-FORMING LIQUIDS

A. Thermodynamic Model

As mentioned in Section II, the steepness or the fragility index m is defined as [105]

$$m = \frac{d \log \tau}{d(T_g/T)}\bigg|_{T=T_g} \qquad (2.14)$$

which we rewrite as

$$m = -\frac{T_g}{\ln(10)}\frac{d\ln\tau}{d(T)}\bigg|_{T=T_g} \tag{2.15}$$

To proceed further, we make use of the Adam–Gibbs model for the temperature dependence of relaxation time $\tau(T)$ of cooperative rearranging regions in glass-forming liquids [41]

$$\tau(T) = \tau_0 \exp\frac{C}{TS_c} \tag{2.16}$$

where S_c is the configurational entropy, $C = \Delta\mu s^* N_A/k_B$; $\Delta\mu$ is the change in potential energy per monomer segment; k_B is the Boltzmann constant; s^* is the critical entropy of the smallest rearranging region; and N_A is Avogadro's number. Making use of the identity

$$\frac{d}{dT}\frac{1}{TS_c} = -\frac{1}{T^2 S_c}\left(1 + \frac{C_P}{S_c}\right) \tag{2.17}$$

the expression for the fragility index simplifies to

$$m = \frac{C}{T_g S_c(T_g)\ln(10)}\left(1 + \frac{\Delta C_P(T_g)}{S_c(T_g)}\right) \tag{2.18}$$

We introduce a quantity called the minimum fragility $m_{\min}$ defined as [18]

$$m_{\min} = \log\left(\frac{\tau}{\tau_0}\right)\bigg|_{T=T_g} \tag{2.19}$$

From the Adam–Gibbs model, $m_{\min}$ can be written as

$$m_{\min} = \frac{C}{T_g S_c(T_g)\ln(10)} \tag{2.20}$$

A comparison of Eqs. (2.20) and (2.18) leads to a remarkable expression that relates the fragility index, a kinetic property, with measurable thermodynamic quantities

$$\frac{m}{m_{\min}} = 1 + \frac{\Delta C_P(T_g)}{S_c(T_g)} \tag{2.21}$$

This equation was first derived by Mohanty et al. [55].

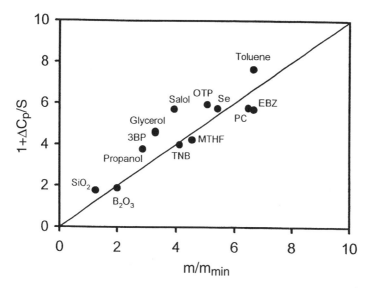

Figure 2.1. A plot of the ratio of the change in heat capacity to the entropy (excess or configuration) at the glass transition temperature versus scaled fragility index, m/m_{min} for glass-forming liquids that cover a wide range of dynamical behavior (adapted from Ref. 55). The solid line denotes the points for which the y-axis and the x-axis are equal, is a guide to the eye. The predicted correlation depicts a deep connection between thermodynamic and kinetic characteristics of organic and inorganic glass-forming liquids [55].

If configurational entropy is proportional to excess entropy, then experimental data on excess entropy and heat capacity can be used to test the predicted fragility with experiments. Since the relaxation time at T_g is 10^2 s and τ_0 is associated with the high-temperature relaxation time, specifically, 10^{-14} s, $m_{min} = 16$. Using this value for minimum fragility, the prediction of Eq. (2.21) for inorganic and organic glass formers that span a wide range of kinetic behavior is shown in Figure 2.1 [55]. The results indicate that fragility index can indeed be correlated with measurable thermodynamic properties of supercooled liquids, namely, the ratio of heat capacity and entropy changes at the glass transition temperature [55].

A comment is in order regarding the derived relation between fragility index and thermodynamics. A generalization of Eq. (2.21) can be formulated that bypasses one key assumption of Adam–Gibbs, namely, the relation between critical size z^* and configurational entropy S_c, namely, $z^* = N_A s^* / S_c$, where N_A is the Avogadro number. If no such assumption is made, then arguments similar to that given above leads to the following relation for the steepness index

$$\frac{m}{m_{\min}} = 1 - \frac{T_g}{z^*}\left(\frac{\partial z^*}{\partial T}\right)_{P;T=T_g} \tag{2.22}$$

In this relation, $m_{\min}$ is as defined by Eq. (2.19).

For polymer glass formers, there is experimental evidence that the value of 16 for the minimum fragility is less than adequate [89,109]. Mohanty and coworkers have directly evaluated the fragility index based on Eq. (2.18) of several polymer systems such as PVC, PMMA, PS, PVAc, and PIB based on Gibbs–DiMarzio (GD) and DiMarzio–Dowell (DD) configurational entropy model of glasses [59–62]. In the DiMarzio–Dowell model, lattice vibrations are incorporated into the Gibbs–DiMarzio model. The predictions of the fragility index based on Eq. (2.18) but within the framework of GD, DD, and AG models are in good agreement with experimental data [56]. The result underscores the importance of including vibration contributions to configurational heat capacity and entropy changes at the glass transition temperature for a quantitative prediction of the steepness index (see also Fig. 2.2) [56].

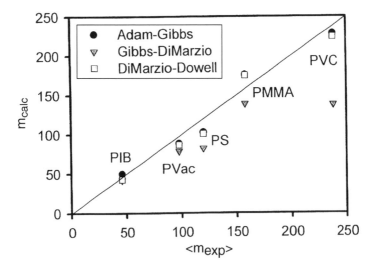

Figure 2.2. A plot of the predicted fragility of several glass-forming polymeric liquids versus average experimental fragility index (adapted from Ref. 56). The solid line denotes the points for which the y-axis and the x-axis are equal, is a guide to the eye. The predicted correlation indicates an intricate connection between thermodynamic and kinetics of supercooled polymeric liquids [56].

B. VTF Route

The temperature dependence of relaxation in glass-forming liquids is often described by the Vogel–Tammann–Fulcher (VTF) equation

$$\tau(T) = \tau_0 \exp\frac{B}{T - T_0} \tag{2.23}$$

where B, T_0, and τ_0 are constants. The fragility index and the minimum value of fragility is then, respectively, given by [18,48]

$$m = \frac{B}{T_g \ln(10)[1 - T_0/T_g]^2} \tag{2.24}$$

$$m_{min} = \frac{B}{T_g \ln(10)[1 - T_0/T_g]} \tag{2.25}$$

Elimination of T_0/T_g between Eqs. (2.24) and (2.25) yields [18]

$$B = \frac{T_0 \ln(10)m_{min}^2}{m - m_{min}} \tag{2.26}$$

Similarly, eliminating B between Eqs. (2.24) and (2.25) leads to

$$\frac{1 - m_{min}}{m} = \frac{T_g}{T_0} \tag{2.27}$$

Consequently, we have a relationship between fragility index and the VTF constant B [18,48]:

$$m = m_{min} + \frac{T_0 \ln(10)m_{min}^2}{B} \tag{2.28a}$$

$$= m_{min} + \frac{\ln(10)m_{min}^2}{D} \tag{2.28b}$$

The quantity D, which is defined as the product of VTF parameters B and T_0 [see Eq. (2.7)], is referred to as the "strength" parameter [48]. Since $m_{min} \approx 16$, we have an expression due to Bohmer et al. that relates fragility to strength parameter [48,18,105]:

$$m = 16 + \frac{589.5}{D} \tag{2.29}$$

As discussed in Section II, for monomeric organic and inorganic and glass-forming liquids, the temperature dependence of heat capacity change is often described by hyperbolic temperature dependence [17,43]:

$$\Delta C_p(T) = \frac{\Delta C_p(T_g)}{T/T_g} \tag{2.30}$$

Assuming the validity of Adam–Gibbs equation for relaxation dynamics and the hyperbolic temperature dependence of heat capacity, the strength parameter is found to be inversely proportional to the change in heat capacity [see Eq. (2.10)] at the glass transition temperature [48,105].

C.   Random First-Order Transition Model

There has been a wealth of activity based on the idea that glassy dynamics is due to some underlying thermodynamic transition [1–25]. If a glass former shows a jump in some an appropriate order parameter without the evolution of latent heat, then such a system is said to exhibit a random first-order transition [94,95]. Models of this kind, which include the p-spin glasses [110], and the random energy model [111], do not have symmetry between states but do have quenched random long-range interactions and exhibit the so-called Kauzmann entropy crisis.

Another approach to liquid glass transition is the self-consistent phonon theory or density functional theory applied to aperiodic structures [112–114]. These theories predict the Lindemann stability criterion for the emergence of a density wave of a given symmetry. Although the finite Lindemann ratio implies a first-order phase transition, the absence of latent heat in glassy systems suggests the presence of an exponentially large number of aperiodic structures that are frozen in at T_g [94,95,110,111].

The free energy of a "entropic droplet" of a radius r obtained within the framework of density functional arguments is [39]

$$F(r) = -\frac{4\pi Ts_c r^3}{3} + 4\pi\sigma r^2 \tag{2.31}$$

where s_c is the configurational entropy density. By finding r that maximizes $F(r)$ and substituting this value back into Eq. (2.31), one obtains the free-energy barrier [39]

$$\Delta F^\# = \frac{16\pi\sigma^3}{3(Ts_c)^2} \tag{2.32}$$

Random first-order transition suggests that, due to universality of the Lindemann ratio, the surface tension σ is universal for temperatures much lower than the model coupling temperature T_c [39].

In contrast, the barrier for cooperative rearrangements in the Adam–Gibbs model is inversely proportional to configurational entropy density [41]

$$\Delta F^{\#} \propto \frac{s_c^* \Delta\mu}{s_c} \qquad (2.33)$$

One way to improve the Adam–Gibbs model is to include details of the structure of the interface between the various aperiodic minima [39]. Near the Kauzmann temperature, the interface broadens, and correct scaling laws are obtained by wetting the droplet surface [39]. In this case, the surface tension of the "entropic droplet" is a function of its radius and can be obtained by renormalization group arguments. Analysis reveals that the activation barrier to configuration rearrangement is [39]

$$\Delta F^{\#} = \frac{3\pi\sigma_0^2 r_0}{Ts_c} \qquad (2.34)$$

which can be rewritten in the form of a Vogel–Fulcher form [39]

$$\Delta F^{\#} = k_B TD \frac{T_K}{T - T_K} \qquad (2.35)$$

The coefficient D is the steepness index of the supercooled and within the framework of the random first-order transition it is given by [39]

$$D = \frac{27}{16}\pi \frac{nk_B}{\Delta\tilde{C}_p} \ln^2 \frac{\alpha_L r_0^2}{\pi e} \qquad (2.36)$$

Here, $\alpha_L^{-1/2}$ is the root-mean-square displacement, $\Delta\tilde{C}_p$ is the heat capacity jump per unit volume, n is the density of particles, and r_0 is the average lattice or interparticle spacing. Xia and Wolynes observed a correlation between D and $1/\Delta C_p$ for several glass-forming liquids [39].

D. Narayanaswamy–Moynihan Nonlinear Parameter

It is well established that structural relaxation within the glass transition region is nonlinear. Can the degree of nonlinearity be connected with fragility? If so, the nonlinearlity of structural relaxation would have a thermodynamic foundation.

During annealing, the temperature dependence of relaxation time of macroscopic quantities is often described by the Narayanaswamy–Moynihan equation

$$\tau(T, T_f) = A \exp\left(\frac{x\Delta h^*}{RT} + \frac{(1 - x)\Delta h^*}{RT_f}\right) \tag{2.37}$$

where R is the gas constant, Δh^* is an effective activation energy at the glass transition temperature, A is a parameter, and x is a measure of degree of nonlinearity. T_f is the fictive temperature, the temperature at which a nonequilibrium measurable property would be in equilibrium. Thus, for temperatures above the glass transition temperature, the glass-forming liquid is in an equilibrium state and one has $T_f = T$; in the nonequilibrium glassy state, the fictive temperature is constant.

For the equilibrium liquid, we have

$$
\begin{aligned}
\left.\frac{d\log\tau}{d(1/T)}\right|_{\text{liquid}} &= -\frac{T^2}{\ln(10)}\left.\left(\frac{d\ln\tau(T)}{dT}\right)\right|_{\text{liquid}} \\
&= \frac{1}{\ln(10)}\left[\frac{\Delta\mu z^*}{k_B} - \frac{\Delta\mu T}{k_B}\left(\frac{\partial z^*}{\partial T}\right)_P\right] \\
&= \frac{1}{\ln(10)}\left(\frac{\Delta h^*}{R}\right)
\end{aligned}
\tag{2.38}
$$

Similarly, in the glassy state the variation of relaxation time with inverse temperature is

$$
\begin{aligned}
\left.\frac{d\log\tau}{d(1/T)}\right|_{\text{glass}} &= \frac{1}{\ln(10)}\left(\frac{z^*\Delta\mu}{k_B}\right) \\
&= \frac{1}{\ln(10)}\left(\frac{x\Delta h^*}{R}\right)
\end{aligned}
\tag{2.39}
$$

The ratio of Eqs. (2.38) to (2.39) evaluated at the glass transition temperature yields the following expression for the Narayanaswamy–Moynihan nonlinear parameter x [115]:

$$x = \frac{1}{1 - \frac{T_g}{z^*}\left(\frac{\partial z^*}{\partial T}\right)_{P;T=T_g}} \tag{2.40}$$

A comparison of this equation with Eq. (2.22), yields the desired relationship between fragility index and Narayanaswamy–Moynihan nonlinear parameter x [115]

$$\frac{m}{m_{\text{min}}} = \frac{1}{x} \tag{2.41}$$

To put it differently, Eqs. (2.41), (2.40), and (2.21) establish that nonlinearity of structural relaxation in a glass-forming liquid does have an intricate link with thermodynamics of the system. On combining Eqs. (2.41) and (2.21), one observes that if entropy change at the glass transition is greater than the heat capacity jump, then the nonlinear parameter x is high. Hodge and O'Reilly [18,64] have deduced approximate cases of Eq. (2.41).

E. Width of Glass Transition Regime

1. Inorganic Glasses

A remarkable correlation between activation enthalpy Δh_η for shear viscosity in 17 inorganic glasses, such as NBS710, ZBLAN20, NaS, KS, ZBLA, BSC, and B_2O_3, and the corresponding width of the glass transition region (100), has been observed. To be precise, for inorganic glasses, Moynihan argued the universality of the relation (100)

$$\frac{\Delta h_\eta}{R}\left(\frac{1}{T_g} - \frac{1}{T_g'}\right) = \Gamma \tag{2.42}$$

where Γ is a constant (4.8 ± 0.4), T_g is the glass transition temperature, and T_g' is the glass transition temperature corresponding to the upper end of the glass transition region obtained in a graph of heat capacity versus temperature. Note that the width of the glass transition is defined to be $\Delta T_g = (T_g' - T_g)$, while the activation energy is proportional to the fragility index

$$m = \frac{1}{T_g \ln(10)}\frac{\Delta h}{R} \tag{2.43}$$

Consequently, Eq. (2.42) expresses a metric between fragility index and the width of the glass transition region.

The derivation of Eq. (2.42) starts with the observation that the heating or cooling rate governs the location of the glass transition temperature. Indeed, the dependence of cooling rate q_c with glass transition temperature T_g is defined by [100,101]

$$\frac{d\ln|q_c|}{d(1/T_g)} = -\frac{\Delta h}{R} \tag{2.44}$$

The structural relaxation time defines the activation enthalpy Δh for small deviation from equilibrium

$$\frac{d\ln\tau(T)}{d(1/T)}\bigg|_{T_f=T} = \frac{\Delta h}{R} \tag{2.45}$$

where T_f is the fictive temperature. Consequently, the width of the glass transition regime depends on the variation of the structural relaxation time with temperature [100,101].

Cooper and Gupta have argued that during coding the upper and lower ends of the region that define the glass transition region could be accurately described by $(dT_f/dT) = 1 - \delta$ and $(dT_f/dT) = \delta$, respectively, where δ is a small positive number [101]. These authors showed empirically that for a range of behavior of the temperature dependence of structural relaxation time, $q_c(d\tau/dT) = \delta$, when $T = T_u$, and $q_c(d\tau/dT) = 5\delta$, when $T = T_l$ [101]. The transition to glassy state occurs in the range of temperature where the Lillie number Li, defined as $Li = q(d\tau/dT)$, is of order unity [101].

During heating or cooling, the breadth of the glass transition region within the framework of Cooper–Gupta model satisfies the constraint [100,101]

$$\frac{\left(\frac{d\tau}{dT}\right)\big|_{T_l}}{\left(\frac{d\tau}{dT}\right)\big|_{T_u}} = \text{const} \qquad (2.46)$$

Using Eq. (2.45), Eq. (2.46) can be reexpressed as [100]

$$\frac{\tau(T_l)/T_l^2}{\tau(T_u)/T_u^2} = \text{const} \qquad (2.47)$$

Since the ratio T_l/T_u is near unity, Eq. (2.45) implies that $\tau(T_l)/\tau(T_u) = \text{constant}$; the latter relation can be recast in a form due to Moyhihan [100]

$$\frac{\Delta h}{R}\left(\frac{1}{T_l} - \frac{1}{T_u}\right) = \ln(\text{const}) \qquad (2.48a)$$

As mentioned earlier, the proximity of the rise of heat capacity $C_p(T)$ versus temperature has been utilized to identify the lower end of the glass transition region: $T_l = T_g$. Similarly, the upper end of the glass transition region, $T_u = T_g'$, is obtained from suitable analysis of the overshoot in heat capacity data. The term in brackets is proportional to $\Delta T_g/T_g T_g'$. But the activation energy is proportional to the fragility index, $\Delta h/R = m T_g \ln(10)$. Consequently, the width of the glass transition region is inversely proportional to steepness index:

$$m = \frac{\ln(\text{const})}{\ln(10)\{\Delta T_g/T_u\}} \qquad (2.48b)$$

Equation (2.48) has been extensively tested for wide range of high-T_g inorganic glasses. An important result is that the right-hand side is found to be virtually constant: 4.8 ± 0.4 for 17 inorganic glassy systems [100].

2. Molecular Glasses

Ito et al. have generalized the correlation given above for inorganic glasses to include generalized to molecular glasses such as butyronitrile, propylene glycol, OTP, toluene, and sorbitol [37]. Let $\Delta T_g / T_g$ denote normalized or reduced glass transition width. Ito et al. derive the following relation between the reduced glass transition width and $F_{1/2}$ metric [37]:

$$\frac{\Delta T_g}{T_g} = 0.151 \frac{1 - F_{1/2}}{1 + F_{1/2}} \tag{2.49}$$

The ordering parameter $F_{1/2}$ is defined as $F_{1/2} = 2T_g/T_{1/2} - 1$, where $T_{1/2}$ is such that the relaxation time $\tau(T_{1/2}) = 10^{-6}$ s. In deriving Eq. (2.49), VTF temperature dependence for relaxation time $\tau(T)$ or shear viscosity $\eta(T)$ was assumed, and it was further assumed that between lower and upper ends of the glass transition region, the relaxation time and shear viscosity changed by 2.3 orders of magnitude [37].

F. Kohlrausch Exponent

An intimate connects exists between the shape of the relaxation function $\phi(t)$ and steepness index [3,5,48,89,116,117]. Strong liquids have less broad relaxation functions compared with fragile glass formers. The degree of nonexponentiality is reflected in the Kohlrausch exponent β

$$\phi(t) = \phi_0 \exp - \left(\frac{t}{\tau}\right)^{\beta} \tag{2.50}$$

where τ is the relaxation time. For a large class of glass-forming liquids one has the empirical observation [48]

$$m = 250(\pm 30) - 320\,\beta \tag{2.51}$$

A theoretical justification of Eq. (2.51) has not yet been formulated.

G. Crossover Temperature

It has been revealed that the relaxation time at the crossover temperature, i.e., $-\log \tau(T_x)$ is universal, namely, -7 ± 1 for glass-forming liquids that spans a wide range of kinetic behavior [118].

The universality of the relaxation time at the crossover temperature can be elucidated by introducing the concept of minimum fragility at the cross-over temperature T_x as discussed in Eq. (2.19). Within the framework of the configuration entropy model [41], $m_{\min}(T_x)$ can be expressed in terms of its value at the glass transition temperature T_g [119]

$$m_{\min}(T_x) = m_{\min}(T_g) \frac{(T_g/T_x)}{z(T_g)/z(T_x)} f(\Delta\mu) \tag{2.52}$$

where $\Delta\mu$ is the potential energy barrier per particle, $f(\Delta\mu) = [\Delta\mu(T_x)]/[\Delta\mu(T_g)]$, and $z(T)$ is the number of molecules in a cooperative rearranging region at temperature T [41].

On expanding $z(T_g)$ in a Taylor series around $T = T_x$ to second order, one finds [119]

$$m_{\min}(T_x) \approx m_{\min}(T_g) \frac{(T_g/T_x)}{1 + (C_P(T_x)/S(T_x)) - (T_g/T_x)(C_P(T_x)/S(T_x)) + D} \tag{2.53}$$

where D is

$$D = \frac{(1 - T_k/T_x)^2}{2} \left(\frac{2(C_P(T_x))}{S(T_x)} \right)^2 + \frac{C_P(T_x)}{S(T_x)}$$
$$- \frac{C_P(T_x)}{S(T_x)} \frac{T_x}{C_P(T_x)} \left(\frac{dC_P}{dT} \right)_{T=T_x} \tag{2.54}$$

In deriving Eq. (2.54), it has been assumed that $\Delta\mu$, the potential energy barrier per particle, is weakly temperature-dependent compared with configurational or excess entropy [41,119].

For 3-bromopentane, 2-methyltetra-hydrofuran, salol, o-terphenyl and n-propanol, the minimum fragility at the crossover temperature is calculated from Eqs. (2.53) and (2.54) by using as input the experimental heat capacity and entropy data [43]. One finds that $m_{\min}(T_x)$ is 6.95, 7.78, 8.35, 8.50, and 9.53 for 3-bromoprntane, 2-methyltetra-hydrofuran, salol, o-terphenyl, and n-propanol, respectively [119]. Consequently, the model predicts that the average value of $m_{\min}(T_x)$ for these glass-forming liquids is 8.2, implying that the relaxation time $\tau(T_x)$ is $10^{-5.8}$ s [119].

The universality of the relaxation time near the crossover temperature also originates in the dynamic nature of supercooled liquids. The idea here is that supercooled liquids have collective excitations. These "elementary excitations" have characteristics of phonons [119–122]. Furthermore, there is a unique temperature at which the lifetime for the elementary excitation becomes comparable to the lifetime of hopping dynamics on the potential energy surface [119]. Analysis indicates that the value of crossover relaxation time at this characteristic temperature is $\phi \times 10^{-7.5}$ s, where ϕ varies between 1 and $\frac{1}{10}$ [119].

H. Low-Temperature Properties

1. Boson Peak

The low-frequency vibration properties of amorphous solids probed by Raman scattering, infrared absorption or inelastic scattering is found to be anomalous [123–130]. For a wide range of supercooled liquids that form glasses at low temperatures, this anomalous behavior is observed as an excess heat capacity relative to Debye-like behavior of crystalline solids [123–130]. Although the origin of this Boson peak is still not well understood [131–133], the observation of an intensity maximum, the so-called Boson peak, in low-frequency Raman scattering, for example, is an universal feature of glasses [123–130]. For some glasses, the Boson peak tends to persist into the supercooled state [125,123,130].

Hassan et al. has showed that the low-frequency vibration features of glasses are correlated with fragility index [123]. These authors introduced a quantity called the dynamical correlation length l_c defined as the ratio of the velocity of sound to the frequency at which the Boson peak is at its maximum [123]. l_c is in the range of 0.5–5 nm for glassy systems [123]. For molecular, ionic, and network oxide glasses, a correlation was observed between fragility index and l_c. In fact, l_c increases as fragility index decreases [123].

Sokolov and coworkers have introduced a quantity $R_1(T)$ that is a measure of the relaxational to the vibrational contribution in low-frequency Raman scattering [124]. If $R_1(T)$ is defined as the ratio of intensity minimum to intensity maximum at the Boson peak at a temperature T, then a correlation is observed between $R_1(T_g)$ and the fragility index for glassy systems such as SiO_2, GeO_2, B_2O_3, PS, salol, PB, PVC, and o-terphenyl [124].

At temperatures below 1 K, the ratio of the constant pressure heat capacity C_p of the glass relative to the Debye heat capacity C_{Debye} of a solid increases [124]. Consequently, we introduce a quantity R_2 defined as the ratio of $C_p(glass)$ to that of the Debye heat capacity $C_{Debye}(solid)$ [124]. Specifically, if R_2 is normalized as $R_2 = [(C_p/C_{Debye})_{min}]/[(C_p/C_{Debye})_{max}]$, then a correlation is observed between this quantity and the fragility index [124].

A mechanism for the emergence of Boson peak in an amorphous solid has been proposed as a consequence of coupling of two factors: (1) coupling between the transverse sound modes and slowly varying density fluctuations and (2) slow decay of defect density [134,135]. Let δ denote the slowly varying defect density. By connecting the defect density δ with the emergence of the Boson peak, Das argues of a qualitative correlation between fragility index and defect density [135].

2. Two-Level Tunneling States

For amorphous materials, it is widely regarded that "tunneling two-level system" (TLS) is capable of explaining a wealth of experimental data for low-temperature

($T < 1$K) glasses such as linear specific heat and the $T^{1.8}$ temperature dependence of thermal conductivity [133,136–138].

Although the microscopic origin of the two-level system is far from well understood [131–133], Zhu found a correlation between the density of two-level tunneling states, the latter obtained form the low-temperature characteristics of specific-heat data, and the fragility index [137,138]. Specifically, glasses with higher density of tunneling states are found to be more fragile [137,138].

3.   Glassy Phase

In this section we briefly address the issue of whether there is a connection between fragility index and the characteristics of glass-forming liquids below the glass transition, that is, in the glassy state.

Detailed high-frequency (terahertz) dynamical studies of glasses have been probed by inelastic X-ray scattering (IXS) [139]. The advantage of this technique is that with reliable measurements it allows determination of the so-called nonergodicity parameter $f(q, T)$ as a function of wavevector q; this quantity is defined by the long time limit of the density–density correlation function $F(q, t)$ divided by the static structure factor [15].

The density–density correlation function $F(q, t)$ can be expressed as [139]

$$F(q,t) = e^{-W(q)}[S_{IS}(q) + F_{inel}(q,t)] \tag{2.55}$$

The quantity $e^{-W(q)}$ is the Debye–Waller factor, while $S_{IS}(q)$ is the structure factor due to packing statistics or the inherent structure. The contributions from the anharmonic vibrational degrees of freedom about the inherent structure are denoted by $F_{inel}(q, t)$. In the longtime limit, the nonergodicity parameter becomes [139]

$$
\begin{aligned}
f(q,T) &= \lim_{t \to \infty} \Gamma(q,t) \\
&= \frac{1}{1 + S_{inel}(q)/S(q)}
\end{aligned}
\tag{2.56}
$$

Here, $\Gamma(q, t)$ is the normalized density–density correlation function, namely, $\Gamma(q, t) = F(q, t)/S(q)$, where $S(q)$ is the static structure factor of the system. In Eq. (2.56), $S_{inel}(q)$ is approximated by the $t = 0$ limit of the density–density correlation function $F(q, t)$ [139].

What is accessible in an IXS experiment is the Fourier transform of $F(q, t)$, which is called the *dynamical structure factor*, $S(q, \omega)$. Thus, the nonergodicty parameter is obtained from the dynamical structure factor by taking the ratio of the elastic to inelastic contribution [139]. In the harmonic approximation for vibration motions, the nonergodicity parameter is given by [139]

$$f(q,T) = \left[1 + \frac{k_B T q^2}{M S_{IS}(q)} \sum_i \frac{\left| \sum_j \vec{q} \cdot \vec{e}_i(j) e^{iqr_j} \right|^2}{N \omega_i^2} \right]^{-1} \qquad (2.57)$$

where N is the number of particles, the sum j is over 1 through N, k_B is the Boltzmann constant, M is the molar mass, and the sum i is over the normal modes, whose eigenvalues and eigenvectors are respectively ω_i and $\vec{e}_j$. In the low-wavevector limit, the nonergodicity factor reduces to [139]

$$f(q \to 0, T) = \frac{1}{1 + \alpha \frac{T}{T_g}} \qquad (2.58)$$

where the dimensionless quantity α has embedded within it the molecular details of the glass-forming liquid.

For glycerol one finds $\alpha = 0.32$, while $\alpha = 0.58$ and 0.19 for OTP and silica, respectively [139]. One finds a strong correlation between fragility index and the parameter α. Larger α implies larger values for the fragility index. The results give strong support to the model by Mohanty et al., which predicts an intricate connection between the curvature of the potential energy minima and molecular motions in the supercooled state [140].

I. Replica Symmetry Breaking

At low temperatures, the configuration space of glass-forming liquids partitions into exponentially large number of disjoint regions [82,83,156–160]. The Boltzmann weight for such a region is very small. Hence, it is difficult to formulate an order-parameter that labels the metastable states in a quantitative fashion.

A way out of this difficulty, first recognized by Parisi, is to introduce an order-parameter that gauges the similarity between the disjoint regions [156–160]. To illustrate this idea, consider two configurations or replicas labeled z and y, where the symbols z and y refer to the coordinates of all the particles. The overlap between the two replicas is defined by an order parameter $q(z,y)$. Thus, if z and y have configurations that are similar, q will have large values [156–160]. On the otherhand, if the configurations are not as similar, then q will acquire low values [156–160].

Let the configuration y be weighted by the Boltzmann–Gibbs distribution. Assume that configuration y does not influence z. Configuration z, however, feels the influence of configuration y. It is governed by a restricted Boltzmann–Gibbs distribution proportional to $\exp(-H/k_B T) \times \delta(q(z,y) - q)$, where $\delta(z)$ is the Dirac delta function and H is the Hamiltonian [156–160]. Thus,

Boltzmann weight is nonzero for those configurations which are at a specified distance from configuration y [156–160]. To put it differently, the low temperature glassy is viewed as being in either a confined or a deconfined "phase" [156–160].

Computer simulations have pointed toward the fact that for temperatures below the Kauzmann temperature T_K, replica symmetry is broken [157,158, 160]. The variable that describes this phenomenon, $m_{\text{replica}}(T)$, is proportional to temperature [157,158,160]:

$$m_{\text{replica}}(T) = \frac{T}{T_K} \qquad (2.59)$$

To state this differently, at low temperatures the diverse minima in configuration space are as dissimilar as imaginable [157,158,160].

By a synthesis of the partition function for a supercooled liquid at some fictive temperature in the inherent structure formalism [1,4] with the configurational entropy obtained by restricting m_{replica} replicas to be in the same state [161], Mohanty has uncovered a relationship between Parisi's replica symmetry-breaking parameter $m_{\text{replica}}(T)$, and the Narayanaswamy–Moynihan non-linear parameter x, a parameter that provides a metric on the deviation of the glass-forming system from equilibrium [162]:

$$m_{\text{replica}}(T) \approx 1.28 \left(\frac{1}{x} - 1 \right) \frac{T}{T_K} - \frac{C_{\text{vib}}(T)}{S_{\text{vib}}(T)} \qquad (2.60)$$

Here, $C_{\text{vib}}(T)$ and $S_{\text{vib}}(T)$ are the vibrational contributions to the heat capacity and the entropy, respectively. Note that the slope of the replica symmetry-breaking parameter with respect to temperature is not unity as predicted by one-step replica symmetry breaking. Rather, the slope is governed by three factors: the Narayanaswamy–Moynihan nonlinearity parameter x, the Kauzmann temperature, and the *ratio* of the Kauzmann temperature to the glass transition temperature.

J. Potential Energy Landscape and Inherent Structures

Lennard-Jones binary mixture of particles is a prototypical model that describes glass-forming liquids [52,53,158,162–165]. The temperature and the density dependence of diffusivity $D(T, \rho)$ have been obtained by computer simulations for the Lennard-Jones binary mixture in the supercooled state. To relate fragility of binary Lennard-Jones mixture to thermodynamic properties necessitates determination of the configurational entropy $S_c(T, \rho)$ as well as the vibration entropy $S_{\text{vib}}(T, \rho)$ at a given temperature and density.

The inherent structure formalism allows one to obtain both the configurational and the vibrational entropies. In this formalism, one separates the configurational space into packing statistics and anharmonic vibrational motions about the packing statistics, namely, the local potential energy minima [1,4,9].

Thus, the configurational entropy $S_c(T, \rho)$ expresses the number of local potential energy minima and hence can be evaluated by potential energy landscape and thermodynamic integration methodology [53]. The vibration entropy was calculated within the framework of a harmonic approximation to each basin [165,166], an approximation that is valid at low temperatures [53].

Three notable results were obtained by Sastry [53]:

1. Analysis of the so-called fragility plot, in which $-\log[D(T, \rho)]$ is depicted as a function of reduced temperature T_r/T, where the reference temperature T_r is taken to the temperature at which the diffusivity reaches some small but fixed value, indicates that with an increase in density, the fragility of the liquid increases [53].

2. The Adam–Gibbs relation, namely, $\log[D(T, \rho)]$ versus $(TS_c)^{-1}$, was found to be valid at each density [53].

3. The characteristics of the potential energy landscape was linked to fragility via the following steps [53]:

    a. The configurational entropy density $S_c(\Phi)$ was calculated from knowledge of the probability distribution $P(\Phi, T)$ of an inherent structure of energy Φ at temperature T. $P(\Phi, T)$ was deduced from the inherent structures number density at energy Φ, $\Omega(\Phi)$, the vibrational free energy of the basin, and the partition function $Q_N(T, \rho) = \int d\Phi P(\Phi, T)$ for the N-particle system.

    b. Since the distribution of $\Omega(\Phi)$ is expected to be Gaussian [167], the entropy density was fitted to an inverted parabola [53]

$$\frac{S_c(\Phi)}{Nk_B} = \alpha - \frac{(\Phi - \Phi_c)2}{\sigma^2} \qquad (2.61)$$

    where Φ_c is the mean of the distribution, σ is the variance of the distribution, and α is the height of the parabola.

    c. Simulation shows that $\Delta S_{\text{vib}}(\Phi, T) = S_{\text{vib}}(\Phi, T) - S_{\text{vib}}(\Phi_c, T)$ is linear in Φ; this suggests the functional form for $\Delta S_{\text{vib}}(\Phi, T)$, where $\Delta S_{\text{vib}}(\Phi, T) = \delta S(\Phi - \Phi_c)$, where δS is the change in entropy between the highest and the lowest energy inherent structures visited by the glass-forming liquid [53].

    d. Assume a Gaussian approximation to the number of inherent structures with energy Φ. The partition function is readily evaluated to yield the average energy [53]

$$\langle \Phi \rangle (T) = \Phi_c^{\text{eff}} - \frac{\sigma^2}{2Nk_BT} \qquad (2.62)$$

where $\Phi_c^{\text{eff}} = \Phi_c + \sigma^2 \delta S / 2Nk_B$.

On combining steps 3a–d, the temperature dependence of the configurational entropy is found to be [53]

$$TS_c(T) = K_{\text{Adam-Gibbs}}(T)\left(\frac{T}{T_K} - 1\right) \qquad (2.63a)$$

$$K_{\text{Adam-Gibbs}}(T) = \left(\frac{\sigma\sqrt{\alpha}}{2} + \frac{\sigma^2\delta S}{4Nk_B}\right)\left(1 + \frac{T_K}{T}\right) - \frac{\sigma^2\delta S}{2Nk_B} \qquad (2.63b)$$

where $T_K = \sigma(2Nk_B\sqrt{(\alpha + \sigma\delta S)^{-1}})$ is the so-called Kauzmann temperature, that is, where $S_c(T_K)$ vanishes. $K_{\text{Adam-Gibbs}}(T)$ is the thermodynamic fragility within the framework of the Adam–Gibbs model and the potential energy landscape picture [53]. Equation (2.63b) indicates that thermodynamic fragility is governed by (1) the total number of configurational states as measured by the parameter α, (2) the spread of these states through the variance σ, and (3) the change in the entropy δS defined by the highest and the lowest energy inherent structures visited by the glass-forming liquid at a fixed density ρ and temperature T [53].

In the inherent structure (IS) description of a many body system, the transition between the metabasins represents the α process. In contrast the transitions between the IS of a metabasin denote the β process. On the basis of this idea, Chakrabarti and Bagchi have proposed a model for cooperative relaxation that sheds light into the fragility of glass-forming liquids [96]. The assumptions of the model are [96]: (1) A cooperative rearranging region has embedded within it N_β two-level noninteracting degrees of freedom—thus, N_β denotes the number of beta process; (2) a two-level system is separated by some characteristic energy scale; (3) for relaxation of the alpha (α) process, excitation of a critical number N_c out of the total number of beta process must occur; (4) a Poisson waiting time distribution between two level systems is assumed; (5) a time-dependent order parameter $Q(t) = \sum_{j=1}^{N_\beta} \xi_j(t)$ is introduced, where the variable $\xi_j(t)$ is either unity or zero (if the jth two-level system is not occupied at time t, it is assigned the value unity, and zero otherwise); and (6) relaxation occurs when $Q(t)$ reaches N_c. The probability that the random variable Q takes on a value p at time t satisfies a master equation [96].

The master equation is solved numerically from which the mean-first passage time is extracted. Analysis of the mean-first passage time indicates that even a moderate increase in the critical number N_c of the beta process leads to entropic slowdown of the dynamics [96]. Furthermore, the fragility of the system is controlled by the ratio of the critical number N_c to total number N_β of beta process [96].

K. Tables of Fragility Index

1. Polymer Glass Formers

Polymer	Fragility Index	Ref(s).
Polyhexylmethyacrylate	34	141
Poly(tetramethylene oxide)	35	142
Polyethylene	46	117
Polyisobutylene	46	48
Polybutylmethacrylate	56	141
Poly(propylene oxide)	74	48,88
Polyisoprene	76	143,144
Polyvinylacetate	95	48
Poly(dimethyl siloxane)	100	48
Poly(methyl acrylate)	102	48
1,4-Polybutadiene	107	145
Polyvinylethylene	135	145,146
Polypropylene	137	147
Polystyrene	143	90
Polymethylmethacrylate	145	48,99
Poly(ethylene terephthalate)	156	148
Polyvinylchloride	191	48

2. Inorganic Glass Formers

Inorganic Molecules	Fragility Index	Ref.
GeO_2	33	124
B_2O_3	32	43
SiO_2	20	43
$4SiO_2-Na_2O$	44	124
$3SiO_2-Na_2O$	44	124
$B_2O_3-xNa_2O$ ($x = 0.15$)	41	149
$B_2O_3-xNa_2O$ ($x = 0.25$)	56	149
$B_2O_3-xNa_2O$ ($x = 0.35$)	65	149
$B_2O_3-xLi_2O$ ($x = 0.15$)	45	149
$B_2O_3-xLi_2O$ ($x = 0.25$)	62	149
$Ca(NO_3)_2-8H_2O$	93	150
$Ca(NO_3)_2-KNO_3$	93	48
Selenium	86.4	2

3.   Organic Glass Formers

Inorganic Molecules	Fragility Index	Ref(s).
n-Proponal	35	43,151
PropGlyc	52	43,48
3-MePent	36	43,152
3-Bromopentane	53	43
Glycerol	53	43,48
Phenyl salicylate (Salol)	63	43
2-Methyltetra-hydrofuran (MTHF)	65	43
o-Terphenyl (OTP)	81	43,48
Propylene carbonate	104	43,48
Triphenylphosphite (TPP)	160	43,2,153
Tolune	107.2	48
Metatricresylphosphate	87	48
Metatoludine	77	154
Butylbenzene	53	155,139

It has been recently realized that the activation enthalpy of the Johari-Goldstein beta (β) relation is cross-correlated with the glass transition temperature of the alpha (α) relaxation of various glass formers [168,169]. This indicates a deep link between the fragility index of the α and the β relaxation [170].

Acknowledgments

The author gratefully acknowledges the support of this work by the National Science Foundation. I thank Steven Chu for his hospitality at Stanford University and to H. C. Andersen for penetrating discussions.

References

1. U. Mohanty, *Adv. Chem. Phys.* **89**, 89 (1995).
2. C. A. Angell, *J. Res. NIST* **102**, 171 (1997).
3. C. A. Angell, *Science* **267**, 1924 (1995).
4. F. H. Stillinger, *Science* **267**, 1935 (1995).
5. C. A. Angell, *Proc. Natl. Acad. Sci. USA* **92**, 6675 (1995).
6. C. A. Angell, *J. Phys. Condens. Matt.* **12**, 6463 (1999).
7. U. Mohanty, G. Dietzmann, and I. Oppenheim, *J. Phys. Condens. Matt.* **12**, 6431 (1999).
8. C. A. Angell, *Nature* **393**, 521 (1998).
9. P. G. Debenedetti and F. H. Stillinger, *Nature* **410**, 259 (2001).
10. P. G. Debenedetti, *Metastable Liquids, Concepts and Principles* (Princeton Univ. Press, Princeton, NJ, 1996).

11. H. Sillescu, *J. Non-Cryst. Solids* **243**, 81 (1999).

12. (a) M. D. Ediger, *Annu. Rev. Phys. Chem.* **51**, 99 (2000); (b) M. D. Ediger, C. A. Angell, and S. Nagel, *J. Phys. Chem.* **100**, 13200 (1996).

13. E. Donth and E. J. Donth, *The Glass Transition. Relaxation Dynamics on Liquids and Disordered Materials*, Springer Series in Material Science, Vol. 48, 2001.

14. J. T. Fourkas, D. Kivelson, U. Mohanty, and K. Nelson, in *Supercooled Liquids: Advances and Novel Applications*, J. T. Fourkas et. al., eds., ACS Symposium Series, 1997.

15. (a) W. Gotze and L. Sjorgen, *Rep. Prog. Phys.* **55**, 55 (1992); (b) W. Gotze and L. Sjorgen, *Trans. Theor. Stat. Phys.* **24**, 801 (1995).

16. S. Nemilov, *Thermodynamics and Kinetic Aspects of the Vitreous State*, CRC Press, Boca Raton, FL, 1995.

17. I. M. Hodge, *J. Non-Cryst. Solids* **169**, 211 (1994).

18. I. M. Hodge, *J. Non-Cryst. Solids* **202**, 164 (1996).

19. (a) G. Harrison, *The Dynamical Properties of Supercooled Liquids*, Academic Press, New York, 1976; (b) S. Brawer, *Relaxation in Viscous Liquids and Glasses* (*Am. Ceram. Soc.*, Columbus, OH, 1985).

20. G. P. Johari, in *Plastic Deformation of Amorphous and Semicrystalline Materials*, 1982; Les Houches Lectures, Les Editions de Physique, France, 1982, pp. 109–141.

21. G. P. Johari, Lecture Notes in Physics, Vol. 277, Springer, Berlin, 1987.

22. C. A. Angell, in *Relaxation in Complex Systems*, K. Ngai and G. B. Wright, eds., Natl. Technol. Inform Ser., U.S. Dept. Commerce, Springfield, VA, 1985, p. 1.

23. C. A. Angell, *J. Non-Cryst. Solids* **131–133**, 13 (1991).

24. R. V. Chamberlin, in *Supercooled Liquids: Advances and Novel Applications*, J. T. Fourkas et al., eds., ACS Symposium Series, 1997.

25. W. Kauzmann, *Chem. Rev.* **43**, 219 (1948).

26. P. G. Wolynes, *J. Res. NIST* **102**, 187 (1997).

27. E. V. Russell and N. E. Israeloff, *Nature* **408**, 695 (2000).

28. R. Bohmer, G. Hinze, G. Dietzemann, G. Geil, and H. H. Sillescu, *Europhys. Lett.* **36**, 55 (1996).

29. G. P. Johari and M. Goldstein, *J. Chem. Phys.* **53**, 2872 (1970).

30. G. Diezmann, U. Mohanty, and I. Oppenheim, *Phys. Rev. E* **59**, 2067 (1999).

31. E. Rossler, U. Warschewske, P. Eiermann, A. P. Sokolov, and D. Quitmann, *J. Non-Cryst. Solids* **172–174**, 113 (1994).

32. V. N. Novikov and A. P. Sokolov, *Phys. Rev. E* **67**, 031507 (2003).

33. U. Mohanty, G. Dietzemann, and J. T. Fourkas, *J. Chem. Phys.* **113**, 3719 (2000).

34. G. Gerardin, S. Mohanty, and U. Mohanty, *J. Chem. Phys.* (in press).

35. F. Fujara, B, Geil, H. H. Silescu, and G. Fleischer, *Z. Phys. B. Condens. Matt.* **88**, 195 (1992).

36. M. T. Cicerone and M. D. Ediger, *J. Chem. Phys.* **103**, 5684 (1995).

37. K. Ito, C. T. Moynihan, and C. A. Angell, *Nature* **398**, 492 (1999).

38. L. M. Martinez and C. A. Angell, *Nature* **410**, 663, (2001).

39. Xia, X. and P. G. Wolynes, *Phys. Rev. Lett.* **86**, 5526–5529 (2001).

40. K. L. Ngai and O. Yamamuro, *J. Chem. Phys.* **111,** 10403 (1999).

41. G. Adam and J. H. Gibbs, *J. Chem. Phys.* **43**, 139 (1965).

42. U. Mohanty, *Physica A* **177**, 345 (1991).

43. R. Richert and C. A. Angell, *J. Chem. Phys.* **108**, 9016 (1998).

44. M. Goldstein, *Ann. NY Acad. Sci.* **484**, 291 (1986).

45. H. Fujimori and M. Oguni, *J. Chem. Therm.* **26**, 367 (1994).

46. O. Yamamuro, I. Tsukushi, A. Lindquist, S. Takahara, M. Ischikawa, and T. Matsuo, *J. Phys. Chem. B* **102**, 1605 (1998).

47. E. Hempel, G. Hempel, A. Hensel, C. Schick, and E. Donth, *J. Phys. Chem. B* **104**, 2460 (2000).

48. R. Bohmer, K. L. Ngai, C. A. Angell, and D. J. Plazek, *J. Chem. Phys.* **99**, 4201 (1993).

49. G. P. Johari, *J. Chem. Phys.* **112**, 8958 (2000).

50. S. S. N. Murthy, *J. Mol. Liq.* **44**, 119 (1990).

51. A. Ha, I. Cohen, X. Zhao, M. Lee, and D. Kivelson, *J. Chem. Phys.* **100**, 1 (1996).

52. S. Sastry, P. G. Debenedetti, and F. H. Stillinger, *Nature* **393**, 354 (1998).

53. S. Sastry, *Nature* **409**, 164 (2001).

54. R. J. Speedy, *J. Phys. Chem. B* **103**, 4060 (1999).

55. U. Mohanty, N. Craig, and J. T. Fourkas, *J. Chem. Phys.* **114**, 10577 (2001).

56. (a) U. Mohanty, N. Craig, and J. T. Fourkas, *Phys. Rev. E* **64**, 10501 (2001); (b) U. Mohanty and J. T. Fourkas (unpublished).

57. G. P. Johari, *J. Phys. Chem. B* **105**, 3600 (2001).

58. G. P. Johari, *J. Chem. Phys. Chem.* **112**, 7518 (2000).

59. J. H. Gibbs and E. A. DiMarzio, *J. Chem. Phys.* **28**, 373 (1958).

60. E. A. DiMarzio and A. J. M. Yang, *J. Res. Natl. Stand. Technol.* **102**, 135 (1997).

61. E. A. DiMarzio and F. Dowell, *J. Appl. Phys.* **50**, 6061 (1979).

62. M. R. Montero, U. Mohanty, and B. Brey, *J. Chem. Phys.* **99**, 9979 (1993).

63. P. A. O'Connell and G. B. McKenna, *J. Chem. Phys.* **110**, 11054 (1999).

64. I. M. Hodge and J. M. O'Reilly. *J. Phys. Chem. B* **103**, 4171 (1999).

65. I. M. Hodge, *Macromolecules* **20**, 2897 (1996).

66. W. Oldekop, *Glastech. Ber.* **30**, 8 (1957).

67. R. J. Greet and J. H. Magill, *J. Phys. Chem.* **71**, 1746 (1967).

68. W. T. Laughlin andd D. R. Uhlmann, *J. Phys. Chem.* **76**, 2317 (1972).

69. U. Mohanty, *J. Chem. Phys.* **100**, 5905 (1994).

70. C. T. Moynihan and J. Schroder, *J. Non-Cryst. Solids* **160**, 52 (1994).

71. M. Goldstein, *J. Chem. Phys.* **51**, 3728 (1969).

72. T. Keyes, *Phys. Rev. E* **59**, 3207 (1999).

73. L. Angelani, R. DiLeonardo, G. Ruocco, A. Scala, and F. Sciortino, *Phys. Rev. Lett.* **85**, 5356 (2000).

74. F. H. Stillinger, P. G. Debenedetti, and S. Sastry, *J. Chem. Phys.* **109**, 3983 (1998).

75. F. H. Stillinger and P. G. Debenedetti, *J. Phys. Chem. B* **103**, 4052 (1999).

76. F. Sciortino, W. Kob, and P. Tartaglia, *Phys. Rev. Lett.* **83**, 3214 (1999).

77. S. Buchner and A. Heuer, *Phys. Rev. E* **60**, 6507 (1999).

78. A. Scala, F. W. Starr, E. La Nave, F. Sciortino, and H. E. Stanley, *Nature* **406**, 166 (2000).

79. D. Kivelson, S. A. Kivelson, Z. Zhao, Z. Nussinov, and G. Tarjus, *Physica* A **219**, 27 (1995).

80. D. Kivelson and G. Tarjus, *J. Non-Cryst. Solids* **235–237**, 86 (1998).

81. D. Kivelson and G. Tarjus, *J. Chem. Phys.* **109**, 5481 (1998).

82. T. R. Kirkpatrick and P. G. Wolynes, *Phys. Rev. B* **36**, 8552 (1987).

83. T. R. Kirkpatrick, D. Thirumalai, and P. G. Wolynes, *Phys. Rev. A* **40**, 1045 (1989).

84. M. Mezard and G. Parisi, *Phys. Rev. Lett.* **82**, 747 (1999).

85. F. H. Stillinger, *J. Chem. Phys.* **88**, 7818 (1988).

86. F. H. Stillinger, *Phys. Rev. E* **59**, 48 (1999).

87. F. H. Stillinger, *J. Phys. Chem. B* **102**, 2807 (1998).

88. C. M. Roland, P. G. Santangelo, and K. L. Ngai, *J. Chem. Phys.* **111**, 5593 (1999).

89. D. J. Plazek and K. L. Ngai, *Macromolecules* **25**, 1222 (1991).

90. P. G. Santangelo and C. M. Roland, *Macromolecules* **31**, 4581 (1998).

91. P. G. Santangelo and C. M. Roland, *Phys. Rev. B* **58**, 14121 (1998).

92. Y. Bottinga, *Phys. Rev. B* **49**, 95 (1994).

93. R. Casalini, S. Capaccioli, M. Lucchesi, P. A. Rolla, and S. Corezzi, *Phys. Rev. E* **63**, 031207 (2001).

94. T. R. Kirkpatrick and P. G. Wolynes, *Phys. Rev. A* **35**, 3072 (1987).

95. T. R. Kirkpatrick, D. Thirumalai, and P. G. Wolynes, *Phys. Rev. A* **40**, 1045 (1989).

96. D. Chakrabarti and B. Bagchi, *J. Stat. Phys.* (in press; preprint 2003).

97. S. Bhattacharyya and B. Bagchi, *Phys. Rev. Lett.* **89**, 25004 (2002).

98. N. Giovambattista, S. V. Buldyrev, F. W. Starr, and H. E. Stanley. *Phys. Rev. Lett*, **90**, 85506 (2003).

99. E. Hempel, G. Hempel, A. Hensel, S. Schick, and E. Donth, *J. Phys. Chem. B* **104**, 2460 (2000).

100. C. T. Moynihan, *J. Am. Ceram. Soc.* **76**, 1081 (1993).

101. A. R. Cooper, Jr. and P. K. Gupta, *Phys. Chem. Glasses* **23**, 44 (1982).

102. V. Velikov, S. Borick and C. A. Angell, *J. Phys. Chem. B* **106**, 1069 (2002).

103. D. M. Zhu, *Phys. Rev B* **54**, 6287 (1996).

104. C. A. Angell, *J. Phys. Chem.* **70**, 2793 (1966).

105. R. Bohmer and C. A. Angell, *Phys. Rev. B* **45**, 10091 (1992).

106. M. Goldstein, *Ann. NY Acad. Sci.* **484**, 291 (1986).

107. E. Donth, *J. Non-Cryst. Solids* **53**, 325 (1982).

108. E. Rossler and A. P. Sokolov, *Chem. Geol.* **128**, 143 (1996).

109. C. A. Angell, in *Structure and Properties of Glassy Polymers*, M. R. Tant and A. J. Hill, eds., ACS Symposium Series Vol. 710, 1998.

110. E. Gardner, *Nucl. Phys. B* **257**, 747 (1985).

111. M. Mezad, G. Parisi, and M. A. Virasoro, *Spin Glass Theory and Beyond* (World Scientific, Singapore, 1987).

112. J. P. Stoessel and P. G. Wolynes, *J. Chem. Phys.* **80**, 4502 (1984).

113. Y. Singh, J. P. Stoessel, and P. G. Wolynes, *Phys. Rev. Lett.* **54**, 1059 (1985).

114. M. Jaric and U. Mohanty, *Phys. Rev. B* **38**, 9434 (1988).

115. U. Mohanty (preprint; unpublished, 2001).

116. C. M. Roland and K. L. Ngai, *Macromolecules* **25**, 5764 (1992).

117. K. L. Ngai and C. M. Roland, *Macromolecules* **26**, 6824 (1993).

118. V. N. Novikov and A. P. Sokolov, *Phys. Rev. E* **67**, 031507 (2003).

119. J. Gerardine, S. Mohanty, and U. Mohanty, *J. Chem. Phys.* (2003).

120. N. Ailawadi, *Physica* **49**, 345–359 (1970).

121. R. Nossal and R. Zwanzig, *Phys. Rev.* **157**, 120–126 (1967).

122. R. Zwanzig, *J. Chem. Phys.* **79**, 4507 (1983).

123. A. K. Hassan, L. Borjesson and L. M. Torell, *J. Non-Cryst. Solids* **172–174**, 154–160 (1994).

124. (a) A. P. Sokolov et al., *J. Non-Cryst. Solids* **172–174**, 138–174 (1994); (b) A. P. Sokolov, E. Rossler, A. Kisliuk, and D. Quittman, *Phys. Rev. Lett.* **71**, 2062 (1993); (c) see also S. Yannopoulos and G. Papatheodorou, *Phys. Rev. E* **62**, 3728 (2000).

125. R. Shuker and R. W. Gammon, *Phys. Rev. Lett.* **25**, 222 (1970).

126. K. Malinovsky and A. P. Sokolov, *Solid State Commun.* **57**, 757 (1986).

127. R. J. Nemanich, *Phys. Rev.* **B16**, 1665 (1977).

128. U. Strom and P. C. Taylor, *Phys. Rev.* **B16**, 5512 (1977).

129. U. Buchenau, M. Prager, N. Nucker, A. J. Dianoux, and W. A. Phillips, *Phys. Rev.* **B34**, 5665 (1986).

130. L. Borjesson, A. K. Hassan, J. Swenson, and L. M. Torell, *Phys. Rev. Lett.* **70**, 127 (1993).

131. V. Lubchenko and P. G. Wolynes, preprint 2003.

132. V. Lubchenko and P. G. Wolynes, *Phys. Rev. Lett.* **87**, 195901-195901-4.

133. A. J. Leggett, *Physica* **B169**, 322–327 (1991).

134. P. D. Fleming III, C. Cohen, and J. H. Gibbs, *Phys. Rev.* **B13**, 500 (1976).

135. S. P. Das, *Phys. Rev. E* **59**, 3870–3875 (1999).

136. P. W. Andersen, B. I. Halperin, and C. M. Varma, *Philos. Mag.* **25**, 1 (1972).

137. D. M. Zhu, *Prog. Theor. Phys. Supplement* **126**, 97–101 (1997).

138. D. M. Zhu, *Phys. Rev.* **B54**, 6287–6291 (1996).

139. T. Scopigno, G. Ruocco, F. Sette, and G. Monaco, *Science* **302**, 849–852 (2003).

140. U. Mohanty, I. Oppenheim, and C. H. Taubes, *Science* **266**, 425–427 (1994).

141. E. Hempel, M. Beiner, T. Renner, and E. Donth, *Acta. Polym.* **47**, 525 (1997).

142. P. G. Santangelo, K. L. Ngai, and C. M. Roland, *Macromolecules* **26**, 2682 (1983).

143. C. M. Roland and K. L. Ngai, *Macromolecules* **24**, 5315 (1991); **25**, 1844 (1992).

144. P. G. Santangelo and C. M. Roland, *Macromolecules* **31**, 3715 (1998).

145. R. Zora, G. B. McKenna, L. Winter, and D. Richter, *Macromolecules* **28**, 8552 (1995).

146. C. M. Roland, *Macromolecules* **27**, 4242 (1994).

147. P. G. Santangelo, K. L. Ngai, and C. M. Roland, *Macromolecules* **29**, 3651 (1996).

148. K. L. Ngai and C. M. Roland, *Macromolecules* **26**, 2688 (1993).

149. G. D. Chryssikos, J. A. Duffy, J. M. Hutchinson, M. D. Ingram, E. I. Kamitos, and A. J. Pappin, *J. Non-Cryst. Solids* **172–174**, 378 (1994).

150. J. H. Ambrus, G. T. Moynihan, and P. B. Macedo, *J. Electrochem. Soc.* **119**, 192 (1972).

151. C. Hansen, F. Stickel, T. Berger, R. Richert, and E. W. Fischer, *J. Chem. Phys.* **107**, 1086 (1997).

152. M. Baranek, M. Breslin, and J. G. Berberian, *J. Non-Cryst. Solids* **172–174**, 223 (1994).

153. B. Schiener, A. Loidl, R. V. Chamberlin, and R. Bohmer, *J. Mol. Liq.* **69**, 243 (1996).

154. B. M. Erwin and R. H. Colby, *J. Non-Cryst. Solids* **307–310**, 225 (2002).

155. C. Hansen, F. Stickel, T. Berger, R. Richert, and E. W. Fischer, *J. Chem. Phys.* **108**, 6408 (1998).

156. G. Parisi, in *Supercooled Liquids: Advances and Novel Applications*, J. Fourkas, D. Kivelson, U. Mohanty, and K. A. Nelson, eds., ACS Symposium Series Vol. 676, pp. 110–121, 1997.

157. (a) S. Franz and G. Parisi, *cond-mat/9701033*; (b) S. Franz and G. Parisi, *cond-mat/9711215*.

158. B. Coluzzi, G. Parisi, and P. Verrocchio, *J. Chem. Phys.* **112**, 2933 (2000).

159. M. Mezard and G. Parisi, *cond-mat/9812180*.

160. M. Cardenas, S. Franz, and G. Parisi, preprint, Dec. 1997.

161. R. Monasson, *Phys. Rev. Lett.* **75**, 2847 (1995).

162. U. Mohanty, in *Liquid Dynamics: Experiment, Simulation, and Theory*, J. Fourkas, ed., ACS Symposium Series Vol. 820, 2002, p. 249.

163. F. Sciortini, W. Kob, and P. Tartaglia, *Phys. Rev. Lett.* **83**, 3214 (1999).

164. W. Kob and H. C. Andersen, *Phys. Rev. E* **51**, 4626 (1995).

165. (a) S. Sastry, *Phys. Rev. Lett.* **85**, 590 (2000); (b) R. K. Muraka and B. Bagchi, *Phys. Rev. E* **67**, 41501 (2003).

166. S. Buechner and A. Heuer, *Phys. Rev. E* **60**, 6507 (1999).

167. A. Heuer and S. Buechner, *J. Phys. Condens. Matt.* **12**, 6535 (2000).

168. K. L. Ngai and S. Capaccioli, *Phys. Rev. E 69* (2004; in press).

169. A. Kudlik, C. Tschirwitz, S. Benkhof, T. Blochowicz, and E. Rössler, *Europhys. Lett.* **40**, 649 (1997).

170. U. Mohanty (unpublished, 2004).

NON-MARKOVIAN THEORIES OF TRANSFER REACTIONS IN LUMINESCENCE AND CHEMILUMINESCENCE AND PHOTO- AND ELECTROCHEMISTRY

A. I. BURSHTEIN

Department of Chemical Physics
The Weizmann Institute of Science
Rehovot Israel

CONTENTS

Advances in Chemical Physics, Volume 129, edited by Stuart A. Rice
ISBN 0-471-44527-4   Copyright © 2004 John Wiley & Sons, Inc.

I. INTRODUCTION

The first non-Markovian approach to chemical reactions in solutions, developed by Smoluchowski [1], was designed for contact irreversible reactions controlled by diffusion. Contrary to conventional (Markovian) chemical kinetics in the Smoluchowskii theory, the reaction "constant" of the bimolecular reaction, $k(t)$, becomes a time-dependent quantity instead of being truly constant. This feature was preserved in the Collins–Kimball extension of the contact theory, valid not only for diffusional but for kinetic reactions as well [2].

The remote reactions of energy or electron transfer proceeding at rate $W(r)$ at any given distance r were also subjected to the non-Markovian rate description with the so-called differential encounter theory (DET). Developed independently but intuitively in a few Russian works [3–5], DET gave $k(t)$ a more general definition in terms of a position-dependent transfer rate, $W(r)$, and encounter diffusion D, that randomly modulates the inter-reactant distance $r(t)$. DET was shown to be a binary theory [6], not breaking with the rate concept but takes into account the space and free-energy dependence of W, as well as the internal quantum states of the reactants [7]. It had a tremendous success when applied to the irreversible bimolecular quenching of fluorescence [8–13]. Not only the quenching kinetics but also the diffusional dependence of the effective quenching radius were specified by DET [7,14,15] and inspected a few times experimentally using solvents of different viscosities [16–18].

The theory of geminate recombination experienced a similar evolution: from primitive "exponential model" and contact approximation [19,20], to distant recombination carried out by backward electron transfer [21]. However, all these theories have an arbitrary parameter: initial separation of reactants in a pair, r_0. This uncertainty was eliminated by unified theory (UT) proposed in two articles published almost simultaneously [22,23]. UT considers jointly the forward bimolecular electron transfer and subsequent geminate recombination of charged products carried out by backward electron or proton transfer. The forward transfer creates the initial condition for the backward one. This is the distribution of initial separations in the geminate ion pair $f(r_0)$, closely analyzed theoretically [24,25] and inspected experimentally [26,27]. It was used to specify the geminate recombination kinetics accompanied by spin conversion and exciplex formation [28–31]. These and other applications of UT have been covered in a review published in 2000 [32].

However, the very first attempt to justify DET starting from the general multiparticle approach to the problem led to a surprising result: it revealed the integral kinetic equation rather than differential one [33]. This equation constitutes the basis of the so-called integral encounter theory (IET), which is a kind of memory function formalism often applied to transport phenomena [34] or spectroscopy [35], but never before to chemical kinetics. The memory functions that are the kernels of the integral equation were given an exact microscopic definition through the transfer rates $W(r)$ and pair distribution affected by the reaction and diffusion [36]. This is the main advantage of IET over many integral analogues, using arbitrary or model expressions for memory functions. However, this and other advantages of IET were not immediately recognized because the main goal of the pioneering works was to provide proof for DET. The latter as well as UT can be obtained from IET in some approximation reducing integral equations to differential ones [37,38]. At one time both DET and IET were considered as equivalent theories, and the choice of a proper one for any particular application was usually made for the sake of convenience [37].

The actual relationship between IET and DET is more complex. The original derivation of IET used concentration expansion, preserving only the lowest-order terms. Not even all contributions of binary origin were collected. As a result, some of the multiparticle effects accounted for by DET were lost in IET. To minimize the difference between the theories, IET was revised and transformed to the modified encounter theory (MET) [39–42], which takes into account the nonlinear concentration corrections to the IET quenching rate and corrects the long-time kinetics of quenching distorted in IET. MET developed in the matrix form is well suited for studying reversible and multistage reactions [43–45].

However, MET is not a unique theory accounting for multiparticle effects. There are some others competing between themselves, but they all can be reduced to the integral equations of IET distinctive only by their kernels. Depending in a different way on the concentration of quenchers c, the kernels of all contact theories of irreversible quenching coincide with that of IET in the low concentration limit ($c \rightarrow 0$) [46]. IET of the reversible dissociation of exciplexes is also the common limit for all multiparticle theories of this reaction, approached at $c = 0$ [47]. This universality and relative simplicity of IET makes it an irreplaceable tool for kinetic analysis in dilute solutions.

There are some general advantages of integral theories over their rate alternatives. IET demonstrates these features in full measure. The first precedent was established by the IET study of reversible energy transfer between excited reactants having different lifetimes [48]. In this case DET fails to compete with IET because the rate constant of either the forward or backward energy transfer inevitably diverges in DET [49,50]. Two other theories [51,52] were subjected to essential revision to reproduce the quenching kinetics and quantum yields of fluorescence obtained with IET.

Another advantage of IET (and MET) is the matrix formulation of the theory making it applicable to reactions of almost arbitrary complexity. A subject of special attention here will be photochemical reactions composed from sequential geminate and bimolecular stages and accompanied by spin conversion, thermal decay, and light saturation of the excited reactants. The quantum yields of fluorescence as well as the yields of charged and excited products of photoionization are extensively studied experimentally. In liquid solutions their dependence on the free energy of forward and backward transfer is known to deviate fundamentally from the free-energy-gap (FEG) law, established for the rate of transfer between immobile reactants $W(r)$. The origin of this effect, first discovered by Rehm and Weller [53], will be discussed several times, in Sections II, VII, and XI. Another important feature of the phenomena under study is the diffusional dependence of the experimentally measured rates and yields. There is not as much experimental data describing the viscosity effects, but all of them are collected and analyzed here. We demonstrate that the nontrivial free-energy and viscosity dependencies are well fitted with IET, using noncontact, distance-dependent transfer rates whose parameters are subjected to specification.

IET serves as a theoretical basis not only for fluorescence and photochemistry but also for photoconductivity and for electrochemiluminescence initiated by charge injection from electrodes. These and other related phenomena are considered. The kinetics of luminescence induced by pulse and stationary excitation is elucidated as well as the light intensity dependence of the fluorescence and photocurrent. The variety and complexity of applications proves that IET is a universal key for multichannel reactions in solutions, most of which are inaccessible to conventional (Markovian) chemical kinetics.

This chapter is organized as follows. In Section II the non-Markovian rate theories of irreversible reactions will be presented: classical contact theory and DET. In Sections III and IV the simplest IET will be outlined and applied to reversible bimolecular energy transfer, both inter- and intramolecular. In Section V the model description of geminate recombination will be replaced by a more reasonable contact approximation (CA) especially appropriate for proton transfer. In Section VI, IET and UT will be applied to irreversible but remote electron transfer to specify the kinetics of ion accumulation, distribution, and geminate recombination. The free energy and diffusional dependence of the recombination efficiency will be analyzed in Section VII. The reversible photoionization, delayed fluorescence, and electrochemiluminescence will be treated in Section VIII mainly in the contact approximation. The response of the nonlinear reacting system to either instantaneous or stationary excitation of moderate strength is considered in Section IX, while in Section X only the stationary regime is analyzed but for arbitrary strong light. The spin and magnetic field effects in photochemistry and chemiluminescence are described in Section XI.

II. DIFFERENTIAL ENCOUNTER THEORIES

The simplest irreversible bimolecular reaction in solution is usually represented as $A + Q \Longrightarrow B + Q$. To be more specific, let us consider Q as a quencher and A and B as the excited and ground states of active molecule D. If in the absence of quenchers the excitation has the finite lifetime τ, then the reaction scheme can be rewritten as follows:

$$D^* + Q \overset{k(t)}{\Longrightarrow} D + Q \tag{3.1}$$
$$\downarrow \tau$$

The time-dependent (non-Markovian) rate constant $k(t)$ determines the rate of energy quenching in the differential kinetic equation that constitutes the basis of this theory:

$$\dot{N}^* = -k(t)\, c\, N^*(t) - \frac{N^*}{\tau} \tag{3.2}$$

Here $N^*(t) = [D^*]$ is the survival probability of excitation, provided $N^*(0) = 1$ and $c = [Q]$ is the concentration of quenchers. The latter is conserved since the quenchers are not expendable in this reaction. Equation (3.2) turns to its Markovian analogue only in the long time limit, when $k(t)$ approaching its asymptotic value becomes truly a constant

$$k = \lim_{t \to \infty} k(t) = \lim_{s \to 0} s\tilde{k}(s) \tag{3.3}$$

where $\tilde{k}(s) = \int_0^\infty k(t)e^{-st}dt$.

The usual subject of experimental study with time-resolved methods is the nonexponential quenching kinetics, which is system response to δ-pulse excitation:

$$R(t) = \frac{N^*(t)}{N^*(0)} = e^{-t/\tau - c\int_0^t k(t')dt'} \tag{3.4}$$

Alternatively the excitation can be carried out continuously with the rate $I(t)$, which may be considered as a density of δ pulses in a timescale. The total number of excitations is the sum of the responses to those preceding the observation time t

$$N^*(t) = N \int_{-\infty}^t I(t')R(t - t')dt' \tag{3.5}$$

where N is the number of nonexcited particles D that are assumed to be present in great excess: $N^* \ll N \approx const$ (the light saturation will be accounted for in Section X).

The physical interpretation of the convolution recipe (3.5) is as follows. Let us consider the system response to an arbitrary pulse as a sequence of responses resulting from absorption of separate photons. If the time evolution after each excitation is the same due to the linearity of the problem, one can sum them up to get the total response to the original pulse. In non-Markovian differential theory, this is the only way to calculate the number of light-induced excitations. The insertion of $I(t)N$ in the rhs of Eq. (3.2) is justified only in the Markovian approximation, when $k(t) \equiv k = \text{const}$. In general, the convolution recipe (3.5) should be used instead to avoid a fallacy.

The permanent illumination switched on through the stepwise function $f(t)$ is represented by

$$I(t) = I_0 f(t) \qquad \text{where} \quad f(t) = \begin{cases} 0 & \text{at} \quad t < 0 \\ 1 & \text{at} \quad t > 0 \end{cases}$$

The stationary concentration of excitation reached as $t \to \infty$ is

$$N_s^* = I_0 N \int_{-\infty}^{\infty} f(t') R(t') dt' = I_0 N \int_0^{\infty} R(t') dt' \qquad (3.6)$$

If the rate of spontaneous light emission is A, then the stationary intensity of fluorescence at this level of excitation is

$$F(c) = A N_s^*(c) = A I_0 N \int_0^{\infty} R(t) dt \qquad (3.7)$$

It is less than the fluorescence in the absence of quenchers: $F(0) = A I_0 N \tau$. The ratio $F(c)/F(0)$ is known as a *relative fluorescence quantum yield*:

$$\eta = \frac{N_s^*(c)}{N_s^*(0)} = \frac{N_s^*(c)}{I_0 N \tau} \qquad (3.8)$$

In reality the quantum yield is often calculated by integrating the short time kinetics of system response with and without quenchers [54,55]:

$$\eta_0 = \frac{\int_0^{\infty} N^*(c, t) dt}{\int_0^{\infty} N^*(0, t) dt} = \frac{1}{\tau} \int_0^{\infty} R(t) dt \qquad (3.9)$$

This is in fact the yield of fluorescence from only the initial geminate part of quenching. If the transfer is reversible, the stationary detection of fluorescence also includes, besides this part, the contribution from the bulk recombination of transfer products back to the excited state (see Sections VIII and XI). However,

for irreversible quenching this contribution is zero and the result following from
Eqs. (3.8) and (3.6) is exactly the same as in Eq. (3.9)

$$\eta = \frac{1}{\tau}\int_0^\infty R(t)dt = \frac{\tilde{R}(0)}{\tau} \tag{3.10}$$

where $\tilde{R}(s) = \int_0^\infty R(t)\exp(-st)dt$.

The fluorescence quantum yield is usually expected to obey the conventional
Stern–Volmer law [11,32,56]

$$\eta = \frac{1}{1+c\kappa\tau} \tag{3.11}$$

where the Stern–Vomer constant κ is in general a concentration dependent
parameter. However, in the lowest-order concentration expansion of $R(t)$
following from (3.4), we obtain

$$\eta = \int_0^\infty \frac{e^{-t/\tau}}{\tau}\left[1 - c\int_0^t k(t')dt'\right]dt = 1 - c\kappa_0\tau \tag{3.12}$$

where κ_0 is really a constant (ideal Stern–Volmer constant):

$$\kappa_0 = \frac{1}{\tau}\int_0^\infty e^{-t/\tau}k(t)\,dt = \frac{\tilde{k}(1/\tau)}{\tau} \tag{3.13}$$

As follows by comparing Eqs. (3.3) and (3.13), the stationary (Markovian)
constant

$$k = \lim_{\tau\to\infty}\kappa_0 \tag{3.14}$$

but in general these constants do not coincide and are worthy of separate
investigation.

A. Irreversible Contact Reaction

In the contact Smoluchowskii model refined by Collins and Kimball, the reaction
takes place only at the closest approach distance σ and the non-Markovian rate
constant is a product of the contact rate constant k_c and the pair density of
reactants $n(r,t)$ at $r = \sigma$ [2]:

$$k(t) = k_c\,n(\sigma,t) \tag{3.15}$$

To specify $k(t)$, one should find $n(r,t)$ that obeys the auxiliary diffusional equation for the reactants whose diffusional motion toward each other is governed by the operator $\hat{L}$:

$$\dot{n} = \hat{L}n = \frac{D}{r^2}\frac{\partial}{\partial r}r^2 e^{-V(r)/T}\frac{\partial}{\partial r}e^{V(r)/T}n \qquad (3.16)$$

Here $D = D_D + D_Q$ is the coefficient of encounter diffusion and $V(r)$ is the interparticle potential (from this point on we took the Boltzmann constant $k_B = 1$). The corresponding initial and boundary conditions are

$$n(r,0) = \exp\frac{-V(r)}{T} \qquad \text{and} \qquad 4\pi\sigma^2 D e^{-V(r)/T}\frac{\partial}{\partial r}e^{V(r)/T}n\Big|_{r=\sigma} = k_c\, n(\sigma,t)$$

$$(3.17)$$

The former accounts for the initial correlation in the relative distribution of reactants in a given potential (see Section VII.C). The latter accounts for the reaction that proceeds only at the contact distance with the rate constant k_c.

When diffusion is very fast, the reaction is under kinetic control that is $n(r,t) = n(r,0)$ and the "kinetic" rate constant

$$k_0 \approx k_c e^{-V(\sigma)/T} \qquad (3.18)$$

does not depend on time. In the opposite, slow diffusion limit, the reaction is under diffusional control and the time dependence of the diffusional "constant" $k(t)$ can be specified by the solution to Eqs. (3.16) and (3.17). Even for the simplest $V(r)$, it is a rather complex problem that was attacked a number of times with different approximate methods, which are discussed in detail in a few reviews [55,56,63]. For illustration of the principal points, it suffices to consider the simplest case.

1. Diffusional Quenching in a Free Space

If there is no inter-particle interactions out of contact ($V = 0$), then the auxiliary equation (3.16) takes the simplest form

$$\dot{n} = D\Delta n \qquad (3.19)$$

as well as the initial and boundary conditions (3.17):

$$n(r,0) = 1 \qquad \text{and} \qquad 4\pi\sigma^2 D\frac{\partial n}{\partial r}\Big|_{r=\sigma} = k_0\, n(\sigma,t) \qquad (3.20)$$

The kinetic rate constant is $k_0 \equiv k_c$ in this particular case. From the solution to this problem inserted into definition (3.15) the following time-dependent rate constant appears [2,56]:

$$k(t) = k\left[1 + \frac{k_0}{k_D}e^{\alpha^2 t}\text{erfc}\left(\alpha\sqrt{t}\right)\right] = \begin{cases} k_0 & \text{at} \quad t = 0 \\ k & \text{at} \quad t = \infty \end{cases} \tag{3.21}$$

Here $k_D = 4\pi\sigma D$ is the diffusional rate constant, $\alpha = \sqrt{(D/\sigma^2)}[1 + (k_0/k_D)]$ and $k = k_0 k_D/(k_0 + k_D)$ is the stationary (Markovian) rate constant for the contact reactions. The latter is usually presented as

$$\frac{1}{k} = \frac{1}{k_0} + \frac{1}{k_D} \tag{3.22}$$

In the kinetic limit ($k_D \gg k_0$) there is no difference between the initial, k_0, and final, $k \approx k_0$, values of the time dependent rate constant; that is, $k(t) \approx$ const and the theory becomes Markovian. In the opposite limit ($k_D \ll k_0$) the reaction is controlled by diffusion and the final value of the rate constant $k = k_D$ is less than the initial one. Hence, the time dependence of $k(t)$ is well pronounced and at long times takes the following well-known form [56]:

$$k(t) = 4\pi\sigma D\left(1 + \sqrt{\frac{\sigma^2}{\pi D t}}\right) \tag{3.23}$$

The last term represents the non-stationary (transient) quenching, which precedes the stationary (steady-state) one. Thus, the diffusional reactions are essentially non-Markovian.

2. Proton Transfer

The transient effect was confirmed experimentally by monitoring the proton transfer between excited photoacid (2-naphthanol) and anionic bases [57,58]:

$$\text{ROH}^* + \text{B}^- \overset{k(t)}{\Longrightarrow} \text{RO}^{-*} + \text{BH} \tag{3.24}$$

Since the transfer of protons is truly contact, this reaction is best suited for comparison with the contact and spinless theory given above. However, the authors who first monitored it in the time domain tried to fit the fluorescence signal as a biexponential one [59]. The similar reaction but of a more reactive photoacid (2-naphthol-6-sulfonate) with an acetate anion has been studied, and its kinetics, which is neither exponential nor biexponential, was fitted to the true theory of contact quenching [60]. It is especially important that the fluorescence

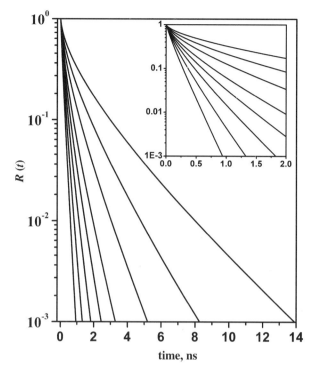

Figure 3.1. Calculated survival probability $R(t)$ of 2-naphthol-6-sulfonate in its reaction with 2 M acetate anion for various solvent compositions (bottom left to upper right): 0, 10, 20, 30, 40, 50, 60, and 70% (vol) of glycerol. (From Ref. 60.)

of the photoacid was measured in a water–glycerol mixture that allows variation of the solvent viscosity coming close to the diffusional limit. As the viscosity increases from 1 to 40 cP (centipoise), the fluorescence decay becomes increasingly nonexponential (Fig. 3.1). The theoretical $R(t)$ was obtained from Eq. (3.4) with $k(t)$ calculated from Eqs. (3.15)–(3.17) accounting for the Debye–Hückel screened Coulomb potential $V(r)$. Then $R(t)$ was convoluted with the independently measured instrument response function, and the result fitted to the experimental data for eight solvents of different viscosities was depicted in Figure 1. Assuming that the contact distance is known, the fitting procedure was aimed at finding two other parameters of the problem: k_c and D. The former was found to be the same for all studied solvents while the diffusion varies significantly:

$$k_c = 9.4 \times 10^9 \, \mathrm{M^{-1} s^{-1}}, \qquad D = 0.97 \div 0.023 \times 10^{-5} \, \mathrm{cm^2/s} \quad \text{at} \quad \sigma = 7\mathring{\mathrm{A}}$$

$$(3.25)$$

The values obtained for D are in good agreement with those calculated from the measured viscosities by the Stokes law.

A similar fitting procedure was also used in Ref. [61], but for determination of the contact radius from the purely diffusional quenching of N-acetyltrypto-phanamide by iodine ions whose quenching efficiency was known to be 1 ($k_c = \infty$) [62]. These ions catalyze the intersystem crossing according to the general scheme (3.1), where Q is I^-. Since electrostatic interaction are negligible in water ($V = 0$), the simplest Smoluchowski expression (3.23) was used in Eq. (3.4) to find the best fit:

$$\sigma = 3.40 \pm 0.12\,\text{Å}, \qquad D = (1.12 \pm 0.05) \times 10^{-5}\,\text{cm}^2/\text{s}, \qquad \text{at} \quad k_c = \infty$$

$$(3.26)$$

The authors were satisfied correctly that the resulting reaction radius $3.4\,\text{Å}$ "agrees with the literature where it is usually assumed that I^- is a contact quencher and must approach the fluorophore very closely [62]. The identification of the contact and reaction radii justifies a posteriori the application of contact theory to this kind of the reaction. As will be shown below, for the remote (energy and electron) transfer, these are two different quantities.

3.   Relative Quantum Yield

The Laplace transformation of the general expression (3.21) has the following form [63]:

$$s\tilde{k}(s) = \frac{k_0 k_D[1 + \sqrt{s\tau_d}]}{k_0 + k_D[1 + \sqrt{s\tau_d}]}$$

By substituting this expression into Eq. (3.13), we obtain the ideal Stern–Volmer constant for contact quenching [25,64].

$$\frac{1}{\kappa_0} = \frac{1}{k_0} + \frac{1}{k_D\left(1 + \sqrt{\tau_d/\tau}\right)} \tag{3.27}$$

where

$$\tau_d = \frac{\sigma^2}{D} \tag{3.28}$$

is the encounter time (the separation time of the contact pair). As follows from comparing Eqs. (3.22) and (3.27), the difference between the Markovian and Stern–Volmer constants is pronounced in the diffusional limit especially at short

τ, but completely disappears in kinetic control limit. Moreover, in the latter case ($k_D \to \infty$, $\alpha \to 0$) it follows from (3.21) and (3.4) that $k(t) = k_0$ at any t and there is no nonstationary quenching at all: $R = \exp(-t/\tau - ck_0 t)$. Using this result in Eq. (3.10), we see that

$$\eta = \frac{1}{1 + ck_0\tau} \qquad \text{at} \quad D \to \infty \tag{3.29}$$

for all concentrations. In other words, for kinetic quenching $\kappa = \kappa_0$ at all concentrations.

On the contrary, under diffusion control the nonstationary quenching lasts almost the whole lifetime and its contribution into quantum yield is dominant at least for large concentrations. Some insights into the concentration dependence of the corresponding Stern–Volmer constant, $\kappa(c)$, one can obtain, using in Eqs. (3.10) and (3.4), the simplest Smoluchowski expression (3.23), which is valid at any time in a limit of diffusional control (when $k_0 \gg k_D$). For this limit the analytic solution was given in Ref. [54]

$$\eta = \frac{Y(x)}{1 + ck_D\tau} = \frac{1}{1 + c\kappa\tau} \qquad \text{at} \ D \to 0 \tag{3.30}$$

where

$$Y(x) = 1 - x\sqrt{\pi} \, \exp x^2 \, \text{erfc} \, x \tag{3.31}$$

and

$$x = 4\sigma^2 c \sqrt{\frac{\pi D}{1/\tau + ck_D}}$$

At weak quenching ($ck_D\tau \ll 1$) the expression (3.31) can be subjected to concentration expansion that brings the Stern–Volmer constant to the form

$$\kappa = \kappa_0 + \beta c + O(c^2) + \cdots \tag{3.32}$$

where

$$\kappa_0 = k_D\left(1 + \sqrt{\tau_d/\tau}\right) \qquad \text{and} \qquad \beta = 2k_D\tau\sigma^3\left[2\pi - 4 + \sqrt{\tau/\tau_d}\right] \tag{3.33}$$

The more general result extended to the kinetic limit is given by the Collins–Kimball model. The integral over t from the rate constant, Eq. (3.21), is well known [65–67]:

$$\int_0^t k(t')dt' = \frac{k_0 k_D}{k_0 + k_D}\left(t + \frac{k_0}{k_D\alpha^2}\left[e^{\alpha^2 t}\mathrm{erfc}(\alpha\sqrt{t}) + 2\alpha\sqrt{t/\pi} - 1\right]\right)$$

Using this expression in Eq. (3.4) for $R(t)$ and integrating the latter in Eq. (3.10), one can get the general contact η and the corresponding Stern–Volmer constant, which is an increasing function of quencher concentration c. There are also a number of competing contact theories that do the same but with slightly different results. They were compared in Ref. 46, reviewed in Section XII.

B.   Irreversible Remote Transfer

The remote transfer in condensed matter is characterized by the position-dependent rate $W(r)$, which is the input data for encounter theory. In its differential version (DET), the main kinetic equation (3.2) remains unchanged, but the rate constant acquires the definition relating it to $W(r)$:

$$k(t) = \int W(r)n(r,t)d^3r \qquad (3.34)$$

The auxiliary equation (3.16) for $n(r,t)$ is also changed accordingly:

$$\dot{n} = -W(r)n + \hat{L}n \qquad (3.35)$$

It takes into account that the transfer reaction occurs at any distance between reactants. If there is no intermolecular interaction, then $V(r) = 0$ and $\hat{L} = D\Delta$. In such a case the initial and the boundary conditions (3.17) reduce to the following ones:

$$n(r,0) = 1 \qquad \text{and} \qquad \left.\frac{\partial n}{\partial r}\right|_{r=\sigma} = 0 \qquad (3.36)$$

Using the initial condition from Eq. (3.36) in Eq. (3.34), we see that the time evolution of $k(t)$ starts from the kinetic rate constant, which acquires the following microscopic definition:

$$k(0) = \int_\sigma^\infty W(r)4\pi r^2 dr = k_0 \qquad (3.37)$$

In the long time limit $k(t)$ approaches its stationary (Markovian) value

$$k = \int W(r)n_s(r)d^3r \qquad (3.38)$$

where n_s obeys the stationary diffusional equation following from (3.35) at $\dot{n} = 0$:

$$\hat{L}n_s = W(r)n_s \tag{3.39}$$

If the position-dependent rate can be represented as

$$W(r) = \frac{k_0}{4\pi r^2}\delta(r - \sigma) \tag{3.40}$$

then the theory reduces to its contact precursor appropriate for proton and atom transfer. However, this is only a rough approximation for the long-range energy or electron transfer.

Substituting $W(r)n_s(r)$ from Eq. (3.39) into Eq. (3.38) and taking into account the reflecting boundary condition (3.36), we can obtain, after integration, the following result [7]:

$$k = \int_\sigma^\infty \hat{L}n_s(r)4\pi r^2 dr = 4\pi D \lim_{r\to\infty} r^2\frac{dn_s(r)}{dr} = 4\pi DR_Q \tag{3.41}$$

Here $R_Q(D)$ is the diffusion-dependent effective quenching radius, which plays in the liquid-state kinetics the same role as the energy-dependent reaction cross sections in gas-phase kinetics. This is the single parameter that determines the universal form of $n_s(r)$ at large r following from Eq. (3.41) and boundary condition $n_s(\infty, t) = 1$ [7]:

$$n_s(r) = 1 - \frac{R_Q}{r} \tag{3.42}$$

For each particular quenching rate, k or R_Q should be deduced from either Eq. (3.38) or (3.41) after insertion of $n_s(r)$ obtained as a solution of Eq. (3.39). The upper and lower bounds for $k = 4\pi R_Q D$ can be established for arbitrary $W(r)$ and $V(r)$ [68].

C. Energy Quenching

The best example of a noncontact reaction is the energy transfer governed by a multipole interparticle interaction. It proceeds according to the following scheme:

$$D^* + A \overset{k(t)}{\Longrightarrow} D + A^* \tag{3.43}$$

The dipole–dipole transfer of energy that was the first subject of DET [3–7] occurs with the rate

$$W(r) = \frac{C}{r^6} \tag{3.44}$$

Unlike contact quenching, which is impossible without diffusion, the remote quenching occurs even in solids or viscous liquids when $D = 0$ ("static quenching"). In this case we have from Eqs. (3.35) and (3.34) that $n = \exp(-Wt)$, $k(t) = -\int \dot{n} d^3 r$, and kinetics of static quenching obtained from Eq. (3.4) is given by the following general expression:

$$R_0(t) = e^{-t/\tau} \exp\left(-c \int_\sigma^\infty \left[1 - e^{-W(r)t}\right] 4\pi r^2 dr\right) \tag{3.45}$$

Initially this decay proceeds as usual: $e^{-t/\tau} \exp(-ck_0 t)$, but then becomes essentially nonexponential and noncontact. Setting $\sigma = 0$ and using (3.44) in Eq. (3.45), one obtains for the final stage the famous Förster kinetics of dipole-dipole quenching [11,69]:

$$R_0(t) = e^{-t/\tau} \exp(-c\sqrt{\mu t})$$

where $\mu = (16\pi^3/9)C$.

In liquids the static kinetics precedes the diffusion accelerated quenching, which ends by stationary quenching. The rate of the latter $k = 4\pi R_Q D$ has a few general properties. In the fast diffusion (kinetic control) limit $R_Q \to 0$ while $k \to k_0$. In the opposite diffusion control limit R_Q essentially exceeds σ and increases further with subsequent retardation of diffusion. As the major quenching in this limit occurs far from contact, the size of the molecules plays no role and can be set to zero. This is the popular point particle approximation ($\sigma = 0$), which simplifies the analytic investigation of diffusional quenching. For the dipole–dipole mechanism the result has been known for a very long time [70]:

$$R_Q = \frac{2\Gamma(3/4)}{\Gamma(1/4)} \left(\frac{C}{D}\right)^{1/4} \tag{3.46}$$

For the quenching of higher multiplicity, the necessary generalization is also well known [71,72].

D. Electron Transfer Rate

The electron transfer fills the gap between highly remote energy transfer and near-contact proton or atom transfer reactions. The length of the electron tunneling L is small but detectable, and the static electron transfer in solids was studied theoretically and experimentally long ago [73–75]. The rate of transfer between electron donor and acceptor at fixed distance r was the subject of longstanding quantum-mechanical investigations summarized in a few more recent work [76–78]. However, throughout this review we will use only the perturbation theory estimate of $W(r)$, assuming that the tunneling is weak almost everywhere. Only in the nearest vicinity of contact where the tunneling may become strong can the transfer be limited by diffusion along the energy or reaction coordinate. If the reaction is under kinetic control, the contribution of this narrow region into integral (3.37) should be small and negligible. It is even less important if the reaction is diffusional and occurs far away from the contact.

In the simplest model of a single-channel transfer (without excitation of reaction products), the rate obtained with the perturbation theory of R. Marcus is given by the following formula [32,79]:

$$W(r) = \frac{V_0^2}{\hbar} \exp\left(-\frac{2(r-\sigma)}{L}\right) \frac{\sqrt{\pi}}{\sqrt{\lambda T}} \exp\left(-\frac{(\Delta G + \lambda)^2}{4\lambda T}\right) \qquad (3.47)$$

It was assumed that the electron transfer assisted by the intramolecular mode and polar media requires the reorganization energy

$$\lambda(r) = \lambda_i + \lambda_s(r) = \lambda_i + \lambda_0\left[2 - \frac{\sigma}{r}\right] \qquad (3.48)$$

Here $\lambda_i = $ const is a contribution of an intramolecular mode, while the second term, $\lambda_s(r)$, accounts for the Coulomb interaction with the polarized solvent. The contact value of this term

$$\lambda_0 = \left(\frac{1}{\epsilon_0} - \frac{1}{\epsilon}\right)\frac{e^2}{\sigma} = \gamma\,\frac{e^2}{\sigma} \qquad (3.49)$$

is half as much as at an infinite separation of ions (ϵ and $\epsilon_0 = n^2$ are static and optical dielectric permittivities).

1. Free-Energy-Gap Law

According to the Marcus classification, there are two alternative regions: *normal*, where $-\Delta G < \lambda$, and *inverted*, where $-\Delta G > \lambda$. The transfer rate (3.47) considered as a function of ΔG in a semilogarithmic plot is represented by a symmetric parabola

$$\ln\frac{W}{W_0} = -\frac{(\Delta G + \lambda)^2}{4\lambda T} \qquad (3.50)$$

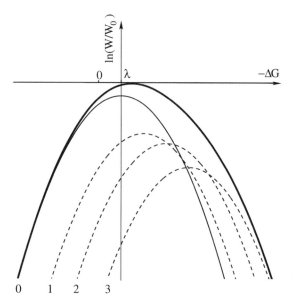

Figure 3.2. The free-energy-gap law for single-channel (thin line) and multichannel (thick line) electron transfer, decomposed into vibronic components (dashed lines).

with a maximum at $-\Delta G = \lambda$ (thin line in Fig. 3.2). This is the famous free-energy-gap (FEG) law of Marcus. In the case of the multichannel electron transfer, to and from the excited vibronic states of the reaction partners, the total rate takes the more complex form [32,80,81]

$$W = \sum_{-\infty}^{\infty} W_n \exp\left[-\frac{(\Delta G + \lambda + \hbar\omega n)^2}{4\lambda T}\right] \qquad (3.51)$$

where n is the number of vibronic quanta and $\hbar\omega$ is their energy. Each summand in (3.51) is represented by the same symmetric parabola shifted to the right (by $\hbar\omega n$) and down (because the multiphonon transitions, partially forbidden, have smaller W_n). The total rate is shown in Figure 3.2 by the thick curve bending all of them.

The FEG law was checked and confirmed experimentally but only for intramolecular electron transfer when the donor–acceptor separation r is fixed and what is measured is really $W(r)$ (Fig. 3.3). The results were reviewed and discussed in Ref. 83 and in other articles published in the same special issue of the *Journal of Physical Chemistry*. Similar results were obtained in solids where the reactants are also immobile.

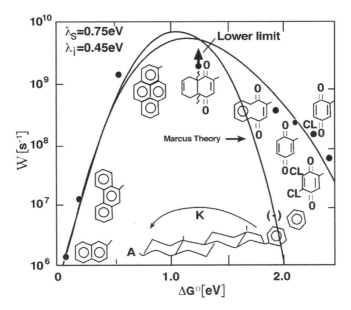

Figure 3.3. Intramolecular rate constants in a 2-methyloxacyclopentane solution at 296 K, as a function of the free-energy change. The electron transfer occurs from biphenylanion to the eight different acceptor moieties (shown adjacent to the data points), in the eight bifunctional molecules of the general structure shown in the center. (From Ref. 82.)

However, the situation in liquids is qualitatively different. Since r is modulated by the encounter diffusion, the measured parameter is the reaction constant of the electron transfer from donor D to acceptor A after excitation of either of them:

$$D^* + A \xrightarrow{W_I} D^+ + A^-, \qquad D + A^* \xrightarrow{W_I} D^+ + A^- \qquad (3.52)$$
$$\downarrow \tau_D \qquad\qquad\qquad \downarrow \tau_A$$

To make the theoretical predictions for these reactions one should first solve the kinetic equation (3.35) using $W(r)$ as input data. This is not a simple task even for a single-channel reaction. The sharp exponential decrease of the tunnelling competes in Eq. (3.47) with the space dependence of the Arrhenius factor, which follows from the r dependencies of ΔG and λ [32]. Even in highly polar media where the Coulomb attraction between ions is negligible and $\Delta G = \text{const}$, the Arrhenius factor is affected by $\lambda_s(r)$ from Eq. (3.48), which is doubled when r increases from the contact distance to infinity. In addition, the shape of $W(r)$ essentially depends on the exergonicity of the electron transfer ($\Delta G < 0$) that should be related to the contact reorganization energy, $\lambda_c = \lambda(\sigma)$.

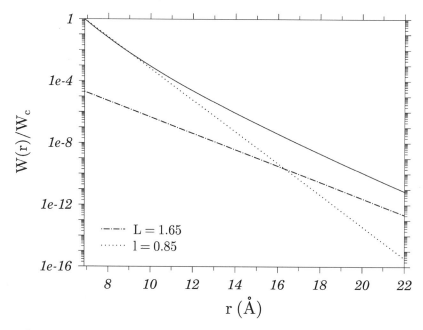

Figure 3.4. The distance dependence of the Marcus transfer rate (1.47) in the normal region ($\Delta G_i = -0.59$ eV) with $L = 1.65$ Å and $\lambda_c = 1.3$ eV (thick line) in comparison to its exponential approximations for short (dotted line) and long (dashed–dotted line) distances. (From Ref. 18.)

2. Model Rates

In the normal region ($-\Delta G < \lambda_c$), $W(r)$ decreases monotonously from contact and appears to be a biexponential dependence (Fig. 3.4). The long asymptote of tunneling is not distorted and vanishes with the true decrement $2/L$, while the near-contact part strongly affected by the Arrhenius factor shows a much sharper decrease. Since only this part is usually responsible for a detectable reaction, it is popular to represent the normal transfer as a monoexponential model for $W(r)$:

$$W(r) = W_c e^{-2(r-\sigma)/l} \qquad (3.53)$$

where $l < L$. In the specific example shown in Figure 3.4, the effective tunnelling length $l = 0.85$ Å is almost half as long as the true one, $L = 1.65$ Å.

With increasing exergonicity the rate of electron transfer is subjected to qualitative transformation [84]. From the monotonous function of r in the normal region, it turns in the inverted region ($-\Delta G > \lambda_c$) into a bell-shaped

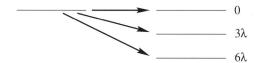

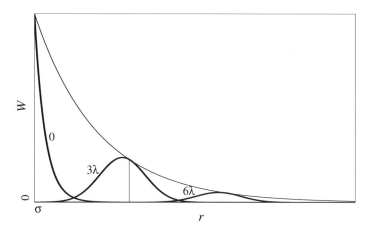

Figure 3.5. Transition from contact to remote transfer in a polar solvent with increasing contact reorganization energy $\lambda(\sigma) = \lambda_i + \lambda_0$. Above: the energy scheme of electron transfer at different exergonicity. Below: the r dependence of the corresponding rates (thick lines) in comparison to an exponential decrease of the normalized tunneling rate $V(r)$ (thin line).

function, shifting from the contact and becoming lower with a further increase of $-\Delta G$ (Fig. 3.5). The bell shape can be modeled as in Ref. 85:

$$W(r) = \frac{W_0}{\text{ch}^2\left(\frac{r-R}{\Delta}\right)} \qquad (3.54)$$

Even simpler, the rectangular model of a reaction layer (Fig. 3.6) was often used instead:

$$W(r) = \begin{cases} 0 & \sigma < r \leq R - \Delta \\ W_0 & R - \Delta < r < R + \Delta \\ 0 & R + \Delta \leq r \end{cases} \qquad (3.55)$$

The bell model (1.54) is better suited for an investigation of noncontact, diffusion-accelerated bimolecular reactions.

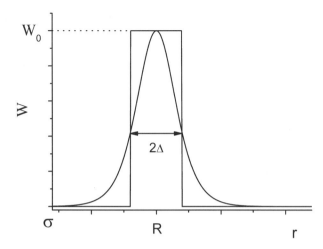

Figure 3.6.   The bell-shaped and rectangular models of the electron transfer rate $W(r)$.

E.   Transient Effects in Fluorescence Quenching

It was rigorously proved that quenching by remote transfer has exactly the same asymptotic behavior as the contact one [12,86–88]. Thus, the long-time asymptote of the time dependent rate constant $k(t)$ is given by the same equation [Eq. (3.23)], except that the effective quenching radius R_Q is substituted for σ:

$$k(t) = 4\pi R_Q D \left(1 + \sqrt{\frac{R_Q^2}{\pi D t}} \right) \qquad (3.56)$$

The last term in this expression represents the transient effect in the long-time asymptote of quenching, which makes it nonexponential.

If $R_Q > \sigma$, the quenching is diffusional and R_Q acquires a sense of a "black sphere radius." This is the radius of a reaction sphere outside which neither excitation is yet quenched while inside the sphere all of them are already deactivated. In reality, of course, the border between outskirts and interior of the sphere is not as sharp, but R_Q fixes this boundary and specifies the stationary (Markovian) rate of quenching (3.41).

Only at the fastest diffusion, when $R_Q \leq \sigma$, the contact theory is applicable to electron transfer. Under this condition the Collins–Kimball expression (3.21) integrated in Eq. (3.4) constitutes a reasonable approximation to $R(t)$ and for both long and short times. However, it was recognized long ago that for slower diffusion the Collins–Kimball model works better if σ is considered as a fitting parameter R that can be a bit larger than σ [66,67,89]. When such a generalized model was used for fitting the fluorescence decay in Ref. 89, the values of R were obtained "scattered between 5.4 and 13.8 Å" and "any systematic changes of this parameter with the change of the other parameters such as ΔG were not recognized." Moreover, studying the diffusion-influenced quenching reaction of rhodamine B and ferrocyanide, Fleming et al. came to the conclusion that even the generalized Collins–Kimball (GCK) model cannot consistently explain both the rapid initial decay (upconversion data) and the slower decay investigated with time-correlated single photon counting [13]. The fitting parameters, R and k_0, which are good for short times, are bad for long-time decay and vice versa. This deficiency is inherent in the contact approximation, which completely ignores the fast-static-quenching, preceding diffusional one. Finally, it was recognized that "as long as we adopt realistic values of diffusion coefficients, the experimentally obtained decay curves ... cannot be satisfactorily reproduced by the Collins and Kimball model, whatever values of the parameters are assumed" [90].

Calculations of this sort were also done with the rectangular [63] and exponential [91–93] models of the electron transfer rate. However, the rectangular model is practically equivalent to the Collins–Kimball model with varying parameters [94], while the exponential model is just a simplification of the true Marcus rate (3.47). The latter was also used by Tachiya et al. [90] for fitting the entire quenching kinetics, but having only two fitting parameters, V_0 and L, the authors failed to find the unique values for them in low-viscosity solvents. The convolution of $R(t)$ with the instrument response function was equally well fitted to the experimental curve when the fitting parameters varied between $V_0 = 30\,\mathrm{cm}^{-1}, 2/L = 0.78\,\mathrm{\mathring{A}}^{-1}$ and $V_0 = 500\,\mathrm{cm}^{-1}, 2/L = 3.68\,\mathrm{\mathring{A}}^{-1}$. Only for high-viscosity ethylene glycol were they able to fix reasonable values, but the choice of ethylene glycol was inappropriate for the reasons presented in Ref. 95 and confirmed later [18]. To reduce the number of parameters $L = 2\,\mathrm{\mathring{A}}$ was arbitrarily taken in Ref. 96 as "a value usually admitted in the literature." This choice allowed the authors to fit closely the transient quenching kinetics and get a reasonable $V_0 = 6 \div 7$ meV.

1. Effective Quenching Radius

The failure of the contact and other models to fit the entire quenching kinetics does not make the problem unsolvable. The situation reverses if one turns to the asymptotic expression (3.56), which is nonmodel and allows extracting the true value of R_Q. In fact, not the R_Q value itself but its dependence on diffusion and the parameters of electron transfer is really informative.

To find $R_Q(D)$ theoretically, one should calculate the Markovian (stationary) rate constant $k = 4\pi R_Q D$, by substitution the solution of Eq. (3.39) into

Eq. (3.38). For the exponential model (3.53), this was first done in Ref. 7, and the following diffusion dependent quenching radius was obtained

$$R_Q = \sigma + \frac{l}{2}\left[\ln(\gamma^2 \beta_m) + 2\theta\left(\beta_m, \frac{2\sigma}{l}\right)\right] \qquad (3.57)$$

where

$$\theta(x, y) = \frac{K_0(2\sqrt{x}) - y\sqrt{x}K_1(2\sqrt{x})}{I_0(2\sqrt{x}) + y\sqrt{x}I_1(2\sqrt{x})}, \qquad \beta_m = \frac{W_c l^2}{4D}, \qquad \gamma = e^C \approx 1.781$$

where C is the Euler constant and $K(x)$ and $I(x)$ are the modified Bessel functions. Unfortunately, the very first experimental verification of the expected diffusional dependence $R_Q(D)$ led to confusion. It was done by studying the quenching of excited pheophytin a by toluquinone in solvents of different viscosity [17]. The effective radii found from the measured values of k by Eq. (3.41) fell in the range between 3 and 15 Å. The best fit for this data was obtained at $W_c = 1.8 \times 10^{10}\,\text{s}^{-1}, \sigma = 4\,\text{Å}$, and $l = 5.4\,\text{Å}$. This value of l is abnormally large, not to mention L, which should be even larger.

The expected value of the tunneling length must not be more than 1 or 2 Å. Therefore, a few more attempts were undertaken to find the realistic fit to the same data. The last one was based on the bell-shaped model for $W(r)$ (3.54), for which it was found that [85]

$$R_Q = R + \Delta\left\{C + \xi(s+1) + \frac{\Delta Q_s(\alpha) + \sigma\sqrt{1-\alpha^2}Q_s^1(\alpha)}{\Delta P_s(\alpha) + \sigma\sqrt{1-\alpha^2}P_s^1(\alpha)}\right\} \qquad (3.58)$$

where

$$\alpha = \tanh\left(\frac{\sigma - R}{\Delta}\right), \qquad s = \frac{1}{2}\left[\sqrt{1 - \frac{4W_c\Delta^2}{D}} - 1\right]$$

where $\xi(x)$ is the logarithmic derivative of the Γ function and $P_s^M(\alpha)$ and $Q_s^M(\alpha)$ are the attached Legendre functions. Assuming $\sigma = 10\,\text{Å}, \Delta = 2\,\text{Å}$, a reasonable agreement with the experimental results was reached, which is better, the further the maximum of $W(r)$ is from the contact. The best agreement was reached when R was chosen to be 16 Å. Only deep in the inverted region the shift of the reaction layer may be as large as this. For systems under investigation, this is definitely not the case. Thus, the question as to how the reliable value of the tunneling length can be extracted from the experimental study of quenching kinetics remained open for a decade and was answered only recently (as of 2002–2003).

The progress of experimental techniques made possible much more accurate kinetic investigation of the transfer kinetics between rhodamine 3B (R3B) and N,N-dimethylaniline in the normal Marcus region in seven solvents of different viscosities [97]. If the fluorescence quenching kinetics $R(t)$ were considered as an exponential function, $\exp(-t/\tau - ckt)$, and used to yield the stationary rate constant $k = 4\pi R_Q D$ and the corresponding R_Q, then the latter, as well as the tunneling length found from its diffusional dependence, would be greatly overestimated. Using an analysis similar to that performed for pheophytin a, it was shown that $l \approx 4\,\text{Å}$. This is also too much. Such an overestimation of the tunneling length stimulated the joint critical inspection of the way in which R_Q is extracted from the experimentally studied kinetics [18]. It was shown that the systematic error was made assuming the quenching kinetics to be exponential. In fact, it remains nonstationary even at the very end of the available time interval.

2. Asymptotic Analysis of Nonexponential Quenching

To analyze the quenching kinetics in the pure state, one should deal with $P(t) = N^*(t,c)/N^*(t,0) = R(t)\exp(t/\tau)$ whose asymptotic behavior has the simple form following from Eqs. (3.56) and (3.4) [1,2]:

$$\ln P = -c\left[4\pi R_Q Dt + 8R_Q^2\sqrt{\pi Dt}\right] \tag{3.59}$$

In Figure 3.7 we demonstrate how R_Q can be found from the best fit of expression (3.59) to $\ln P(t)$ obtained experimentally [18]. The initial discrepancy between them is due to the convolution of the excitation pulse with the system response function. This procedure makes the latter smoother at the top of the signal, while the long-time asymptote (3.59) extrapolated into this region is much sharper then the true $\ln P(t)$. On the other hand, the time interval of fitting is also restricted from above by noise, which sharply increases with time. However, even in a limited time interval, the fitting, which accounts for the non-Markovian development of the quenching rate, is much better than the identification of the stationary rate with the tangent to the kinetic curve at any time, even at the end of the available time interval. This tangent greatly overestimates ck as well as R_Q calculated from this false k.

The proper procedure should be quite the opposite. R_Q has to be found by fitting the non-Markovian asymptote (3.59) to the experimental data, provided D has been measured or calculated from the Stokes equation. Only then can the true Markovian rate constant k be found as $4\pi R_Q D$. The pure exponential decay with this rate constant is shown by the dashed line in Figure 3.7 for comparison with the true nonstationary kinetics.

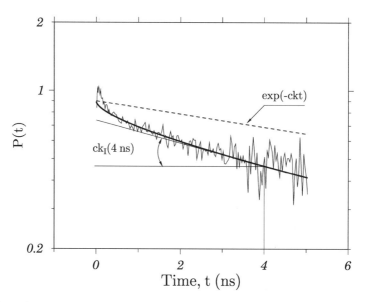

Figure 3.7. The fit of the non-Markovian electron transfer kinetics (thick curve) to the experimental data obtained in propylene glycol in Ref. 97. The thin line represents the tangent to the kinetic curve at the largest time within the available interval ($t = 4$ ns). Its slope is the time-dependent rate of ionization $ck_I(t)$, which differs essentially from that of the dashed straight line. The latter represents the purely exponential decay, with the stationary rate constant $k = 4\pi R_Q D$. (From Ref. 18.)

3. Fitting Diffusional Dependence of Transfer Rate

The GCK approximation includes the reaction layer adjacent to the contact in the reaction sphere and thus magnifies its external radius to the size of R considered as a fitting parameter. Using it instead of σ, we obtain $4\pi RD$ instead of k_D and transform Eq. (3.22) into the following expression:

$$\frac{1}{k} = \frac{1}{k_0} + \frac{1}{4\pi RD} \tag{3.60}$$

This approximation is good only for low viscosity (fast diffusion region) where the real $R_Q \lesssim R$. Having k for a number of solutions with different viscosities, one can fit to these data the linear dependence (3.60) and use it to obtain good estimates for k_0 and R. An example studied in Ref. 98 is perylene quenched by N,N'-dimethyl-aniline in a dimethylsulfoxide (DMSO)–glycerol mixture. The latter allows changing viscosity with composition not affecting other physical properties of the solvent. From the results shown in Figure 3.8(a), we obtained

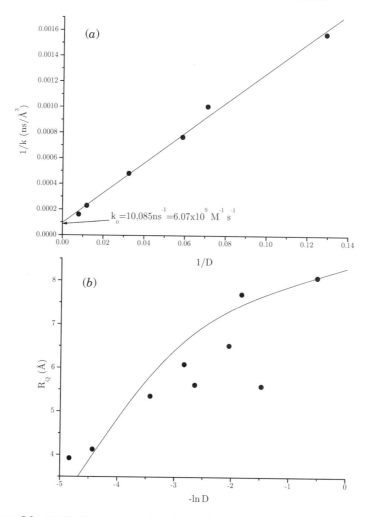

Figure 3.8. (*a*) The linear viscosity dependence of the inverse ionization rate in the reaction studied in Ref. 98. Bullets—experimental points; solid line—fit performed with the generalized Collins–Kimball model. (*b*) The effective quenching radius for the same reaction in the larger range of the viscosity variation. Bullets—experimental points; solid line—fit performed with the encounter theory for the exponential transfer rate. The diffusion coefficient D given in $\text{Å}^2/\text{ns}$ was calculated from the Stokes–Einstein relationship corrected by Spernol and Wirtz [100].

$R = 8.14\,\text{Å}$, which is slightly larger than the $\sigma = 7.2\,\text{Å}$ estimated from the known molecular radii of the reactants. In the same figure (bottom) the diffusional dependence of R_Q was compared with that predicted for the exponential transfer rate in Eq. (3.57). Since $k_0 = 2\pi W_c(\sigma^2 l + \sigma l^2 + l^3/2) = 6 \cdot 10^9\,\text{M}^{-1}\text{s}^{-1}$

is already known, the relationship between W_c and l is fixed, and by varying only l, it was found from the best fit:

$$l = 0.81 \, \mathring{A}, \qquad W_c = 29.12 \, \text{ns}^{-1}$$

A similar analysis performed even earlier for the reaction of excited rhodamine 3B with N,N-dimethylaniline led to almost the same result [18]. From the best fit of the exponential model to the experimental data, $l = 0.85 \, \mathring{A}$ was found, which corresponds to the true tunnelling length $L = 1.65 \, \mathring{A}$. The proper extraction of the reliable values for the tunneling lengths manifests the resolution of a long standing paradox. The overestimation of these parameters was overcome by taking proper account of the nonstationary quenching kinetics. Besides, it elucidates the difference between original and generalized Collins–Kimball models and encounter theory of remote transfer.

The diffusional dependence of R_Q in the GCK model is qualitatively the same as in the original contact approximation (except that R is a bit larger than σ):

$$R_Q = R \frac{k_0}{k_0 + 4\pi RD} \leq R \tag{3.61}$$

Thus, it approaches R instead of σ but then remains constant at higher viscosities while the effective quenching radii $R_Q(D)$ exceeds this value and tends to increase with a further decrease of D [see Fig. 3.9(a)]. However, the experimental data tend to exceed this value, indicating that the transfer furthermore should be considered as remote. For the exponential model of transfer (3.53), there should be a logarithmic asymptotic behavior of $R_Q(D)$ at high viscosities, following from Eq. (3.57):

$$R_Q = \sigma + \frac{l}{2} \ln \frac{\gamma^2 W_c l^2}{4D} \qquad \text{at} \qquad R_Q \gg \sigma \tag{3.62}$$

This is shown by the thick solid line in Figure 3.9(b).

However, it is important to remember that the largest R_Q results from the most remote electron transfer represented by the long-time asymptote of the Marcus transfer rate (3.47) (see Fig. 3.4). The latter also decays exponentially but with the true tunneling length L, which is twice as large as l in a given system. Therefore the asymptotic formula (3.62) gives way to a similar formula, but with L substituted for l. As a result, the final asymptotic behavior of $R_Q(\ln D)$ in Figure 3.9(b) is twice as steep as the heavy line calculated with the exponential model of the initial decay. However, experiments at such a high viscosity seem unattainable. Even the initial change in slope of the data is not

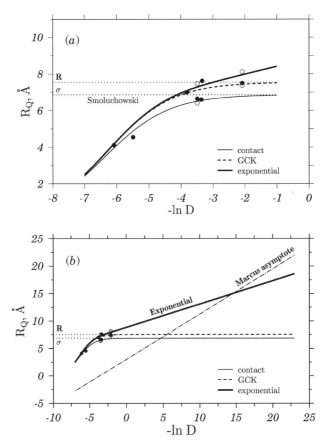

Figure 3.9. The dependence on diffusion of the effective electron transfer radius R_Q. The experimental data, indicated by circles, are approximated by thin and dashed lines, representing the contact and the generalized Collins–Kimball relationships, respectively. The thick line depicts the same dependence, but for the exponential transfer rate with $l = 0.85$ Å and $W_c = 180 \, \text{ns}^{-1}$. The dashed–dotted line at the bottom represents the high-viscosity asymptote of such dependence but for the Marcus transfer rate. (From. Ref. 18.)

definitive. This means that not only the kinetic, but also the diffusional, electron transfer remains near-contact in the available range of viscosities, at least for a given system.

4. Extracting Parameters from the Quantum Yield

The transient kinetics is not a unique source of information about the quenching parameters. Alternatively, the same parameters may be extracted from the stationary fluorescence data. In principle, the viscosity and concentration dependencies of the relative quantum yield or corresponding Stern–Volmer constant provide the same scope of information.

At the lowest concentrations the Stern–Volmer law can always be linearized, as shown in the following form, identical to Eq. (3.12):

$$\frac{1}{\eta} = 1 + c\kappa_0\tau \tag{3.63}$$

Inserting the Laplace transformation of Eq. (3.34) into Eq. (3.13), we get the general DET definition of ideal Stern–Volmer constant:

$$\kappa_0 = \frac{1}{\tau}\int W(r)\tilde{n}\left(r,\frac{1}{\tau}\right)d^3r \tag{3.64}$$

The contact estimate of this quantity obtained in Eq. (3.27) is different from the Markovian (stationary) constant (3.22). Both of them approach the same kinetic value k_0 when $D \to \infty$ (Fig. 3.10), but their asymptotic behavior at slow

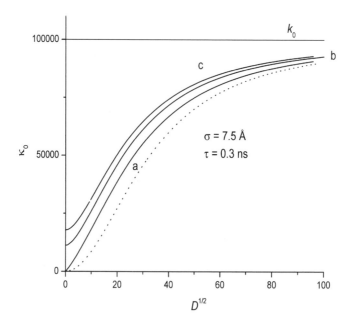

Figure 3.10. The ideal Stern–Volmer constants κ_0 (solid lines) as functions of diffusion in the contact approximation (a) and for the exponential transfer rate with different tunnelling lengths: $l = 1.6\text{Å}$ (b) and $l = 2.5\,\text{Å}$ (c) (From Ref. 46.) The contact stationary constant k (dashed line) is shown for comparison with contact κ_0 (a).

diffusion varies:

$$\kappa_0 \approx 4\pi\sigma D\left(1 + \sqrt{\frac{\sigma^2}{D\tau}}\right) \to 4\pi\sigma^3\sqrt{\frac{D}{\tau}}, \quad k \approx 4\pi\sigma D, \quad \text{at } D \to 0 \quad (3.65)$$

Unlike k, κ_0 is proportional to $\sqrt{D}$ because of nonstationary quenching at slow diffusion. To demonstrate this linearity, $\sqrt{D}$ is used as an abscissa in Figure 3.10.

For remote transfer the analytic calculation of κ_0 from (3.64) is hardly possible, but for the exponential quenching rate (3.53) and polar solvents ($\hat{L} = D\Delta$), at least $\tilde{n}(r,s)$ can be found from Eq. (3.35) [46]:

$$\tilde{n}(r,s) = \frac{1}{s}\left[1 + \frac{\Phi}{r}I_v(w) + \frac{\pi l}{2r\sin v\pi}F_{vv}(w)\right] \quad (3.66)$$

Here $w = w_0\exp\left[-(r-\sigma)/l\right]$, $w_0 = l\sqrt{W_c/D}$, $v = l\sqrt{s/D}$, and

$$F_{\mu v}(w) = I_{-\mu}(w)G_v(w) - I_\mu(w)G_{-v}(w), \quad G_v(w) = \int_0^w x\ln(x/\lambda)I_v(x)dx$$

where I_v is the modified first-kind Bessel function of the order v and

$$\Phi = -\frac{\pi l}{2\sin v\pi}\frac{[1 + \sqrt{s\tau_d}]F_{vv}(w_0) + \sqrt{W_c\tau_d}F_{v+1,v}(w_0)}{[1 + \sqrt{s\tau_d}]I_v(w_0) + \sqrt{W_c\tau_d}I_{v+1}(w_0)}$$

After the substitution of expression (3.66) into Eq. (3.64), the integral was taken numerically at any diffusion coefficient except for its very small values, where the quenching is quasistatic. However, for this very region there is an approximate expression derived in Ref. 99:

$$\kappa_0 = \int_\sigma^\infty \frac{W(r)4\pi r^2 dr}{1 + W(r)\tau} + D\int_\sigma^\infty \frac{(\tau dW/dr)^2 4\pi r^2 dr}{(1 + W(r)\tau)^4} \quad \text{at} \quad D \to 0 \quad (3.67)$$

For exponential $W(r)$ this serves as an alternative to the contact estimate of κ_0 at slow diffusion given in Eq. (3.65). The latter tends to zero as $\sqrt{D} \to 0$ while κ_0 from (3.67) approaches the lowest but finite static value, $\lim_{D\to 0}\kappa_0(D) \neq 0$. The Stern–Volmer constant increases monotonously with diffusion from this value up to the kinetic rate constant $k_0 = \lim_{D\to\infty}\kappa_0(D)$. As shows Figure 3.10, at the same k_0 the more efficient the remote transfer is, the greater the tunneling length l.

In the kinetic limit $\kappa_0 \equiv k_0$ and Eq. (3.63) coincides with Eq. (3.29), which is exact at any concentrations. Hence, no nonlinear terms can arise in Eq. (3.63) at fast diffusion. This is not the case in viscous solutions, where the transfer reaction is under diffusion control. Even in the contact approximation there is a pronounced transient effect that gives rise to essential modification of $\eta(c)$ dependence in Eq. (3.30) and brings to existence the higher-order concentration correction in Eq. (3.33), as $\beta \neq 0$ etc.

We need to determine how to account for the same effect when diffusional transfer is remote and occurs far from the contact. Even the generalized Smoluchowski approximation, substituting R for σ in contact formulae, is not enough. The fitting parameter R having the same physical meaning as R_Q is the radius of the "black" sphere around the acceptor. For the exponential transfer rate it is defined by the following condition:

$$\frac{W(R)l^2}{D} = \frac{W_c l^2}{D} \exp\left(-\frac{2(R-\sigma)}{l}\right) = 1 \qquad (3.68)$$

The R obtained from this condition is approximately equal to R_Q and follows the same logarithmic dependence on diffusion (3.62). All excitations crossing the reaction layer of width $l/2$ adjacent to the black sphere are quenched during their residence time there, $l^2/4D$. However, most of those that were initially inside this sphere are quenched where they were at the moment of excitation. These excitations disappear first, during static quenching, which is often considered as "instantaneous" compared to subsequent stages limited by diffusion, which delivers the excitations from outside into the black sphere.

The excitations that survive after static (instantaneous) quenching should have no surrounding acceptors, inside the reaction sphere of a certain volume v. If there are N acceptors in the volume V, then the probability of finding neither of them inside the selected spherical volume (in the limit of infinitely large sample) is

$$\lim_{N,V\to\infty} \left(1 - \frac{v}{V}\right)^N = e^{-cv} \qquad (3.69)$$

Such a contribution of the "instantaneous" (static) quenching into the fluorescence quantum yield was recognized as essential a long ago [101]. It is usually added to the yield of subsequent diffusional quenching by just multiplying Eq. (3.30) (with R substituted for σ in k_D and x) and Eq. (3.69) [16,63,102,103]:

$$\eta = \frac{Y(x)}{1 + 4\pi RDc\tau} e^{-cv} \qquad (3.70)$$

where

$$x = 4R^2 c \sqrt{\frac{\pi D}{1/\tau + 4\pi RDc}} \qquad (3.71)$$

The main question is "What is v?" The common answer is [13,16,102]:

$$v = \frac{4\pi(R^3 - \sigma^3)}{3} \tag{3.72}$$

which is the volume of the entire reaction sphere with the exception of the inaccessible volume. This answer turns R into a single fitting parameter of $\eta(c)$ dependence (3.70) making it possible to extract it from the experimental data if D and σ are known.

However, the values of R extracted in this way are less reliable than R_Q obtained above from asymptotic analysis, not only because this is a pseudocontact model but also because of uncertainty in v [104]. The rigorous alternative to this model is the straightforward calculation of η from Eq. (3.10), using in Eq. (3.4) an exact $k(t)$ obtained as a solution of the DET equations (3.34) and (3.35). The authentic concentration dependence of $\kappa(c)$ obtained with contact DET will be compared in Section XII.B with other estimates of the same quantity made with a number of competing theories and approximations.

5. Harnessing of Kinetic and Stationary Data

In order to fit the kinetic and stationary data together to get both R_Q and D in one stroke, they were proposed by Costa et al. [16]. The authors stated that the quenching kinetics that they detected is almost an ideal exponent:

$$R(t) = e^{-ut}, \qquad \text{where} \qquad u = \frac{1}{\tau} + ck \tag{3.73}$$

This is doubtful if the reaction is under diffusion control. In such a case (shown in Fig. 3.7), the exponential stationary quenching is unattainable in the available time interval and the employment of the exponential approximation at earlier times leads to an essential overestimation of k as well as R_Q [18].

Even if k is somehow measured, there are two alternative theoretical estimates for it, in the kinetic and diffusional limits:

$$k = 4\pi R_Q D = \begin{cases} k_0 & \text{at} \quad R_Q < \sigma \quad \text{(kinetic)} \\ 4\pi RD & \text{at} \quad R_Q \approx R \geq \sigma \quad \text{(diffusional)} \end{cases} \tag{3.74}$$

The quantum yields for kinetic quenching [Eq. (3.29)] and for the diffusional one [Eq. (3.70)] are also different and should be compared in pairs.

In the kinetic limit, both measurable quantities, $1/\eta - 1$ and $u\tau - 1$, are linear functions of c with the same slope, $k_0\tau$:

$$\frac{1}{\eta} - 1 = u\tau - 1 = ck_0\tau \qquad k_0 \ll k_D \tag{3.75}$$

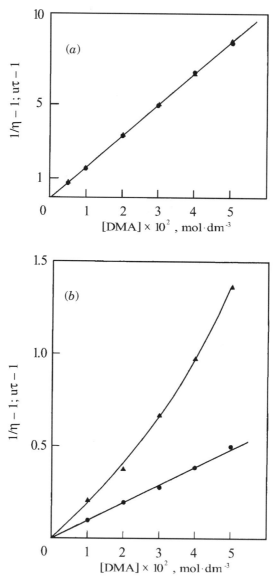

Figure 3.11. The fluorescence quenching of benzyl 1-pyrenoate by dimethylaniline (DMA) in (a) hexane and (b) paraffin/heptane (90 : 10): (▲) $1/\eta - 1 = ck_q\tau$, (●) $u\tau - 1 = ck\tau$ (From Ref. 16.)

In contrast, in the diffusional limit $1/\eta - 1$ becomes essentially nonlinear because of the concentration dependence of $Y(c)$ and $\exp(-cv)$ in Eq. (3.70). Only $u\tau - 1 = ck_D\tau$ remains a linear function of c. Setting them together we see that

$$\eta u\tau = Y(c)\exp(-cv), \qquad u\tau - 1 = 4\pi RD\tau\, c, \qquad k_D \ll k_0 \qquad (3.76)$$

From measured η and u the concentration dependence of the product $\eta u\tau$ can be specified experimentally and $Y(c)\exp(-cv)$ can be fitted to it with two varying parameters, R and D, whose known product $RD = k_D/4\pi$ is fixed. Starting an iteration procedure with a trial radius, the authors of Ref. 16 obtained from the best fit the reasonable values of both R and D, as well as the dependence $R(\ln D)$. The linearity of this dependence was expected to be seen in Figure 3.9(b), but was not reached experimentally.

The data shown in Figure 3.11(a) indicates that the reaction in a solution of the lowest viscosity (hexane) is under kinetic control; the functions $1/\eta - 1$ and $u\tau - 1$ are both linear and indistinguishable as in Eq. (3.75). In contrast, the data in Figure 3.11(b) prove that the reaction in viscous paraffin/heptane solvents are diffusion-controlled; in full agreement with Eq. (3.76), the concentration dependence of $1/\eta$ is strongly nonlinear, unlike that of $u\tau$.

However, Costa et al. considered all of them as diffusion-limited [Fig. 3.12(a)]. If the kinetic rate constant is large enough, it could be that the diffusion control of the transfer occurs at rather small and even moderate D. But it is doubtful that the reaction remains diffusional up to the largest D, when R_Q becomes smaller than the contact distance σ. This is particularly true for the last two points in the circles (for cyclohexane and hexane). They fall on the horizontal line $R = \sigma$ if only "one assumes that charge transfer in these solvents takes place at collisional distances" [16].

Accepting this assumption, the authors follow the theories developed in Refs. 14 and 15. Unlike the conventional DET outlined in Section II.B, these theories presume that remote quenching is diffusional and also does not give way to kinetic quenching even at the contact distance. They employ the absorbing Smoluchowski condition

$$n(\sigma, t) = 0 \qquad (3.77)$$

instead of the reflecting boundary condition (3.36). This makes quenching diffusional everywhere, guaranteeing that $\lim_{D\to\infty} R_Q = \sigma$, instead of being zero. If the authors were not so trusting they could get k as the slope of the line shown in Figure 3.11(a) and estimate $R_Q = k/4\pi D$, using their own D. The $R_Q = 3.5\,\text{Å}$ so obtained is the true ordinate of the last point in Figure 3.12(a) [105]. It is essentially less than σ and the corresponding point

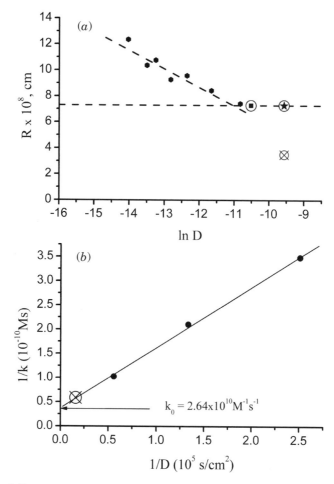

Figure 3.12. (a) Variation of R with ln D in the system benzyl 1-pyrenoate/DMA with several mixtures of paraffin/heptane (hexagons), cyclohexane (circled box), and hexane (circled star) [16]. The bending dashed line represents the high-viscosity asymptote (3.62), and the point ⊗ indicates the value of R_Q found from Figure 3.11(a). (b) The same data in the anamorphosis of Figure 3.8(a) specifying the kinetic rate constant k_0.

falls on the same line (3.60) as the other points [Fig. 3.12(b)]. The intersection of this line with the ordinate axis determines the true kinetic rate constant $k_0 = 2.64 \times 10^{10} \, M^{-1} \, s^{-1}$, which is quite reasonable.

Being different as $D \to \infty$, the theories with finite k_0 and $k_0 = \infty$ are identical as $D \to 0$ and R_Q essentially exceeds σ. When quenching is accomplished at the external border of the large black sphere, it is of no importance what is going on deeply inside it, at the contact region. Therefore, the logarithmic diffusional dependence of $R_Q(D)$, (3.62), is the same at all boundary conditions. As confirmed in Ref. 16, it was fitted with $l = 2.8\,\text{Å}$ (Fig. 3.12). Unfortunately, this value is almost certainly overestimated as always occurs when exponential simplification (3.73) is used for analysis of the quenching asymptote, instead of the expression (3.59), which accounts for the transient effect [18]. A similar mistake was made in Ref. 17, but it was obviated in subsequent publications [18,98].

F. The FEG Law for Electron Transfer in Liquids

At slow ionization and fast diffusion the electron transfer is expected to be under kinetic control, and its rate constant k_0 defined in Eq. (3.37) is diffusion-independent. Moreover, if a sharp exponential function (3.53) is a good model for $W(r)$, the kinetic rate constant may be approximately estimated as follows:

$$k_0(\Delta G_i) \approx W_c v \qquad (3.78)$$

The argument $\Delta G_i = \Delta G_I(\sigma)$ is the contact value of the ionization free energy $\Delta G_I(r)$ inherent in the reactions (3.52). Since the reaction volume $v \approx 2\pi(\sigma^2 l + \sigma l^2 + l^3/2)$ does not depend on the free energy, it was expected from the very beginning that the $k_0(\Delta G_i)$ dependence reproduces that of $W_c(\Delta G_i)$, which is the Marcus free-energy-gap law (3.50) for $W(\sigma)$. The same is true for the Stern–Volmer constant $\kappa_0 = k_0$ in the kinetic control limit ($D \to \infty$) as follows from Eq. (3.27).

In contrast to this expectation, the very first inspection of the FEG law in liquid solution made by Rehm and Weller [53] revealed that κ_0 does not depend on the free energy in the inverted region (Fig. 3.13). Note that in the original presentation of their data reproduced in this figure the authors used $\Delta G_i \equiv \Delta G_{23}$ for the abscissa instead of $-\Delta G_i$, as in Figures 3.2 and 3.3. Therefore, the inverted region for ionization in Figure 3.13 is to the left, while in Figure 3.14 as well as in Figures 3.2 and 3.3 it is to the right. The absence of the inverted branch in the $\kappa_0(\Delta G_i)$ dependence discovered by Rehm and Weller [53], casts doubt on the whole theory because it contradicts the expectation that $\kappa_0 \approx k_0$ reproduces the FEG law inherent in k_0.

This confusion initiated a lot of experimental inspections and theoretical revisions of the FEG law. There is no need to review all of them here because the most reliable explanation of the effect was given by Marcus and Siders [108]. In view of the fact that all transient effects are ignored in the Markovian theory, where $k(t) \equiv k$, it follows from Eq. (3.13) that the Markovian κ_0 coincides with k but not with k_0. Therefore, Marcus and Siders calculated k by means of DET and concluded that the fastest transfer (at the top of the FEG

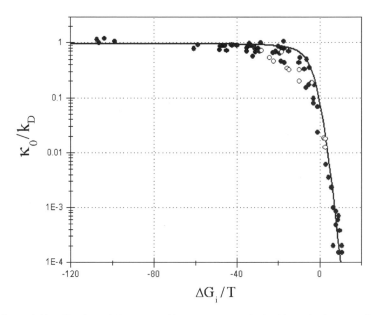

Figure 3.13. The Stern–Volmer quenching constant κ_0 obtained in Ref. 53 versus the free energy of ionization ΔG_i. The open circles are the data borrowed from another work [106]. The free-energy dependence shown by the solid line represents the theoretical expectation of Rehm and Weller deduced from their original approach to the simplest reaction mechanism of quenching [53]. (From Ref. 107.)

curve) is controlled by diffusion rather than by the reaction. Even from the contact estimate of k in (3.22), it is clear that what is measured in this region is the diffusional rate constant (k_D), which does not depend on the free energy.

Qualitatively the same is true also for the non-Markovian Stern–Vomer constant for the irreversible quenching represented by κ_0 from Eq. (3.27), but the plateau that cuts the FEG parabola is shifted up more, the shorter the excitation lifetime τ. The qualitative confirmation of the diffusional nature of the plateau was obtained experimentally in Ref. 89, where the true kinetic constant $k_0(\Delta G_i)$ was measured as the initial value of the time-dependent rate constant (3.37). The latter roughly follows the expected FEG dependence (Fig. 3.14), as it is essentially larger than κ_0 in the activationless region $(-\Delta G_i \approx \lambda_c)$, where

$$\kappa_0 \approx k_D\left(1 + \sqrt{\tau_d/\tau}\right) = \text{const} \tag{3.79}$$

This is the height of the plateau that cuts the FEG parabola. Nonetheless, there are still two problems related to the very same data.

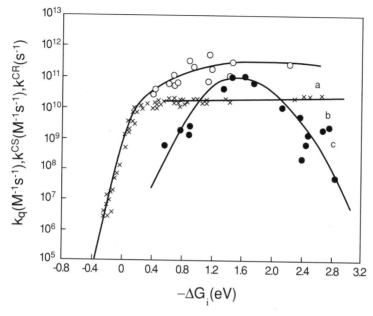

Figure 3.14. Experimental data for the free-energy dependence of the Stern–Volmer constant $\kappa_0 \approx k$ (×) obtained by Rehm and Weller [53], initial (kinetic) rate constant $k^{CS} = k_0$ (○), obtained in Ref. 89, and "charge recombination rate" $k^{CR} \propto W_R$ (●), obtained by Mataga et al. [109,115].

1. Rehm–Weller Paradox

The one most known is the extra large width of the diffusional plateau seen in Figure 3.14. The plateau is much longer than a single FEG parabola could be, as is obvious when comparing it with the rate of the backward electron transfer, which is never as wide. A number of causes were offered to explain why the ionization is so strong (and therefore controlled by diffusion) in such a wide region.

The transition from the contact to distant electron transfer causes the inverted branch of the FEG curve to slope more gently than normal [32,110,111]. This branch is extended to an even greater extent if the transfer is multichannel (see Fig. 3.2). In this case the inverted branch is composed from the tops of the partial FEG curves, where the transfer is activationless. Therefore, their sum is also temperature independent and smooth. But the best way of stretching the diffusional plateau is by taking into account the space dependence of $\lambda(r)$. Since maximum $W_I(r)$ moves away with increasing $-\Delta G_i$, the effective λ corresponding to the position of the maximum also increases with $|\Delta G_i|$, broadening the whole FEG curve. The extra-stretching of the diffusional plateau shown in Fig. 3.14 was reached, assuming that the $\lambda(r)$ dependence differs from the classical law (3.48). Starting from the Monte Carlo estimates [112] of this dependence, the authors of Refs. 113 and 114 concluded that λ changes from $\lambda_c = 0.2\,eV$ up to $\lambda(\infty) = 2.7\,eV$, that is, 14 times instead of 2.

It is questionable whether such a revision of Eq. (3.48) is well grounded, but without it the stretching is not as great. Although significant, it is not enough to extend the plateau up to a border group of points with the highest exergonicity (Fig. 3.15). All of them are related to donors quenched by TCNE (tetracyanoethylene). The charge transfer complexes (exciplexes) of TCNE with different donors have such a low ground-state energy that can be formed without any excitation, although with a rather small equilibrium constant [116]. The energy of the excited exciplex is also lower than that of the excited reactant. This makes reasonable the revival of an old idea that was discussed a few times [79,115,116], but then rejected in Ref. 117 for insufficient reasons. The formation of an excited exciplex can successfully compete with the normal

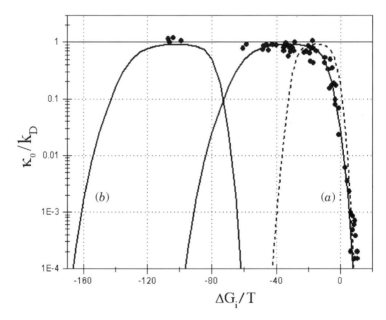

Figure 3.15. The Markovian FEG law for the Stern–Volmer constant of distant multichannel ionization (solid lines), producing stable ions (*a*) or electronically excited products (*b*) in comparison with the single-channel contact ionization (dashed curve). The filled circles are the experimental points of Rehm and Weller taken from Figure 3.13 (From Ref. 107.)

creation of the ground-state exciplex:

$$D^* + A \nearrow \begin{array}{ccc} [D^+A^-]^* & \to & (D^+)^* + A^- \\ \downarrow & & \downarrow \\ [D^+A^-] & \to & D^+ + A^- \end{array} \qquad (3.80)$$

The same is true regarding the straightforward generation of the solvent separated ions in the excited state bypassing the exciplex. Forward electron transfer into these states is much faster than into ground states, if the corresponding free energy $\delta G_i = \Delta G_i + \epsilon^*$ is essentially reduced by adding the energy of ion excitation ϵ^*. A simple shift of the stretched curve (a) to the left (for $\epsilon^* = 1.71\,\text{eV} \approx 70\,T$), makes possible the very reasonable explanation of the position of the TCNE points at the top of the resulting (b) curve (Fig. 3.15). Since the transfer is limited by diffusion in this region, all the points are located on the same plateau as the others. There is also the straightforward experimental evidence for the generation of electronically excited radical ions, in the highly exergonic fluorescence quenching studied in other systems: antracenecarbonitriles (D) and aromatic amines or aminobenzenes (A) [118,119].

2.   Transfer Reversibility

Another, much less known paradox lies in the fact that the good fitting in the quasi-resonance region ($\Delta G \approx 0$), seen in Figures 3.13 and 3.15, is an illusion. It was obtained by a theory of irreversible electron transfer, while at $\Delta G \geq 0$ the reverse electron transfer to the excited donor state becomes equally fast and even faster than the forward one. The reverse transfer has to reduce significantly the efficiency of fluorescence quenching near $\Delta G \approx 0$, violating good agreement between experiment and theory of irreversible transfer.

   This effect was ignored by later authors, but not by Rehm and Weller themselves. Using elementary Markovian theory, they scrutinized the quenching by contact but reversible electron transfer, resulting in formation of exciplex $[D^+A^-]$, which decays with the rate W_R:

$$D^* + A \leftrightarrow [D^* \cdots A] \underset{W_B}{\overset{k_f}{\rightleftharpoons}} [D^+A^-] \qquad (3.81)$$

$$\downarrow \tau_D \qquad \downarrow \tau_D \qquad \downarrow W_R$$

The overall rate constant of exciplex formation is identical to that in Eq. (3.22):

$$k = k_f \frac{k_D}{k_f + k_D} \qquad (3.82)$$

where the association constant $k_f = W_c v$ is actually the kinetic rate constant. The overall rate of the backward reaction, k', is expressed in the same way through the monomolecular dissociation constant W_B [53,107,120]:

$$k' = W_B \frac{k_D}{k_f + k_D} \tag{3.83}$$

This is because the dissociation products diffusing in the liquid cage can come in contact again and again, restoring the exciplex so that the rate of their final separation is given by Eq. (3.83). The ratio of k to k' as well as k_f to W_B fits the detailed balance principle for the reversible reaction (3.81):

$$K_{eq} = \frac{k}{k'} = \frac{k_f}{W_B} = v \exp\left(-\frac{\Delta G_i}{T}\right) \tag{3.84}$$

where K_{eq} is the equilibrium rate constant while ΔG_i is the free energy of ionization resulting in the exciplex formation.

Since the quantum yield of reversible quenching was found in Ref. 53 using Markovian chemical kinetics, the nonstationary quenching and the concentration dependence of the Stern–Volmer constant were lost. As a result, the linear Stern–Volmer law (3.63) was reproduced but with constant depending on both forward and reverse transfer rates:

$$\kappa_0 = \frac{k}{1 + k'/W_R} \tag{3.85}$$

By substituting k' from Eq. (3.84) into Eq. (3.85), we obtain

$$\kappa_0 = \frac{k}{1 + k/W_R K_{eq}} \tag{3.86}$$

The simplest Markovian result (3.86) was recently approved and subjected to a non-Markovian generalization by means of IET[107] [see Eq. (3.372) and Section XI.E]. IET is usually used instead of DET when the occasion requires accounting for the reversibility of the transfer reaction.

As follows from Eq. (3.13) in the Markovian limit ($\tau = \infty$), the Stern–Volmer constant of irreversible quenching κ_0 is identical to the stationary ionization constant k. This is not the case for the reversible quenching whose Stern–Volmer constant (3.85) can be essentially reduced, due to backward transfer to the excited state. As seen from Eq. (3.86), $\kappa_0 \approx k$ only at fast exciplex recombination, when

$$k \ll W_R K_{eq} \tag{3.87}$$

Under these conditions the reaction is practically irreversible. However, in the opposite case, when transfer is reversible, the quenching is controlled by recombination:

$$\kappa_0 \approx W_R K_{eq} \ll k \qquad (3.88)$$

As a result, the Stern–Volmer constant for reversible ionization $\kappa_0(\Delta G_i)$ goes down long before the increasing ΔG_i approaches the zero value.

To shift this border to the right, one should facilitate the exciplex recombination transforming the reversible transfer into irreversible transfer. Rehm and Weller chose this very way assuming that $W_R = $ const and is larger than k/K_{eq} everywhere. This is a very astonishing and unacceptable assumption since $W_R(\Delta G_r)$ should also follow FEG law, which is common for any transfer rate. Since the energy of the excited reactant

$$\mathcal{E} = -\Delta G_i - \Delta G_r \qquad (3.89)$$

is kept constant in all pairs studied, the change of the ionization free energy ΔG_i inevitably entails a corresponding change in the recombination free energy ΔG_r. Therefore, $W_R(\Delta G_r)$ is not a constant but is represented by the bell-shaped FEG curve similar to that shown in Figure 3.14. It is narrower than a plateau, and its maximum is shifted out of the resonance region. To make the recombination fast enough, one should involve some additional channels of product dissipation that are much more efficient. This can hardly be done without an enlargement of the reaction scheme. In fact, the reaction is rarely accomplished by exciplex formation as in the Rehm–Weller scheme (3.81). Even unseparable ion pairs can be subjected to reactions with solvent or solutes that exclude one or both partners from the further exchange of electron. This makes the whole process irreversible even though the forward electron transfer is endothermic [121,122].

Impressive evidence of such a reaction was provided in Ref. 121, where trivalent phosphorus compound Z_3P undergoes single-electron transfer to the photoexcited rhodamine 6C (Rho^{+*}), generating the corresponding trivalent phosphorus radical cation $Z_3\dot{P}^+$. However, this cation undergoes the ionic reaction with water in the solvent, which is so fast that electron transfer becomes irreversible:

$$Z_3P + Rho^{+*} \leftrightarrow [Z_3P \cdots Rho^{+*}] \underset{W_B}{\overset{k_f}{\rightleftharpoons}} [Z_3\dot{P}^+ \cdots \dot{R}ho] \qquad (3.90)$$

$$\downarrow \tau_A \qquad\qquad \downarrow \tau_A \qquad\qquad W_R \downarrow +ROH/-H^+$$

$$Z_3\dot{P} - OR$$

If W_R is really large, this reaction proceeds as an irreversible one up to $\Delta G_i = 0$. This border between exergonic and endergonic electron transfer is located at

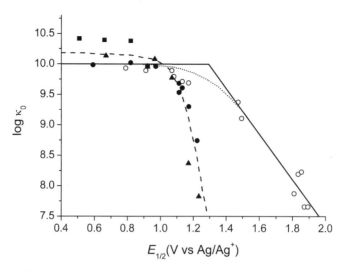

Figure 3.16. The Stern–Volmer constant as a function of the half-wave potential $E_{1/2}$ for the reactions with trivalent phosphorus compounds (○), aromatic amines (■), aliphatic amines (●), and alkoxybenzenes (▲). The solid broken line represents the diffusional (horizontal) and kinetic branches of κ_0 for the irreversible reaction, while the dotted line is k from Eq. (3.22). The dashed curve represents κ_0 for reversible reactions. (From Ref. 121.)

$E_{1/2} = 1.3$ in Figure 3.16. The reaction (3.90) remains diffusional up to this point and becomes kinetic only at larger $E_{1/2}$ (at positive ΔG_i). However, the reactions of the same rhodamine 6C but with other electron donors (amines and alkoxybenzenes) that are not followed by any subsequent ion transformation remain reversible. As a result, the corresponding Stern–Volmer constant reduced much earlier and becomes 100 times smaller than the diffusional one when ΔG_i approaches zero (Fig. 3.16).

Another opportunity appears when a singlet exciplex $[D^+A^-]$ reversibly dissociates into an ion pair, $[D^+ \cdots A^-]$, where the spin conversion can occur, opening the way for subsequent recombination through the triplet channel. Even if there is no spin conversion in the geminate pair, the latter can separate and enable free ions meet in the bulk with uncorrelated spins. The new pairs find themselves in the triplet state with the probability $\frac{3}{4}$. As a result of smaller exergonicity, the recombination of the triplet pair can proceed faster than that of the singlet one, making the total process irreversible. The same is true regarding reactions (3.52), where the ionization proceeds, bypassing the exciplex formation. This opportunity considered in Ref. 123 will be discussed in more detail in Section XI.

III. INTEGRAL ENCOUNTER THEORY

Any kinetic equations for the bimolecular reactions result from one or another reduction of the multiparticle problem to the pair problem. For encounter theory, this was done in the pioneering works of Sakun and Doktorov [33,36], but the first theory recognized as an integral one operated with a density matrix formalism [124]. It was applicable to quantum systems subjected to intramolecular relaxation and a bimolecular transfer reaction. If all these processes are incoherent, only the diagonal elements of the density matrices (populations of states) are of interest. In this case the theory was adopted to the rate (balance) equations and given matrix formulation, making determination of the integral equation kernels much easier and natural [48,49]. In Section X.C it will be reformulated as in Ref. 125 to specify the zero elements of the matrix kernel and thus reduce the rank of the problem from the very beginning. The modification of the integral theory known as MET will also be considered later.

We are starting with the simplest though matrix integral theory. It is outlined by the example of the quasi-resonant energy transfer, which is first considered in an immobile pair and only then in the course of encounter diffusion.

A. Reversible Energy Transfer in an Immobile Pair

The pair problem represented in DET by the auxiliary Eqs. (3.35), describes the evolution of the pair distribution, affected by both the reaction and diffusion. If there is no diffusion, Eq. (3.35) reduces to one that represents only intermolecular transfer between two reactants at a given (fixed) distance between them:

$$\dot{n} = -W(r)n \tag{3.91}$$

In the case of energy transfer, n is the survival probability of excitation generated by an instantaneous light pulse.

However, the adequate kinetic description of such a process is richer. Let us derive it from the rate consideration of the pair, subjected to reversible energy transfer accompanied by light excitation and radiationless decay. According to the simplest reaction scheme

$$[A^* \cdots B] \underset{W_B}{\overset{W_A}{\rightleftharpoons}} [A \cdots B^*] \tag{3.92}$$

$$I_0 \uparrow\downarrow \tau_A \qquad\qquad \downarrow \tau_B$$

there are four collective energy levels of such a pair, AB, AB^*, A^*B, and A^*B^* (Fig. 3.17). At stationary light pumping their populations, g, μ, ν, and f, obey the

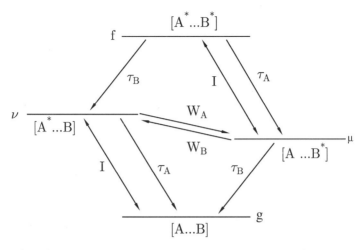

Figure 3.17. Four-level energy diagram for the light-induced transfer reaction.

following set of equations:

$$\dot{f} = -\left(\frac{1}{\tau_A} + \frac{1}{\tau_B} + I_0\right)f + I_0\mu \tag{3.93a}$$

$$\dot{v} = -\left(\frac{1}{\tau_A} + I_0 + W_A\right)v + W_B\mu + \frac{1}{\tau_B}f + I_0g \tag{3.93b}$$

$$\dot{\mu} = -\left(\frac{1}{\tau_B} + I_0 + W_B\right)\mu + W_Av + \left(\frac{1}{\tau_A} + I_0\right)f \tag{3.93c}$$

$$\dot{g} = -I_0g + \left(\frac{1}{\tau_A} + I_0\right)v + \frac{1}{\tau_B}\mu \tag{3.93d}$$

The rates of the forward and backward transfers with the free-energy excess ΔG relate to each other according to the detailed balance principle:

$$\frac{W_A}{W_B} = \exp\left(-\frac{\Delta G}{T}\right) = K \tag{3.94}$$

where ΔG is negative for the exergonic forward transfer.

The set (3.93) may be represented as a matrix equation for the column vector **F** composed (from bottom to top) from the abovementioned four densities, $\langle g, \mu, v, f \rangle$

$$\dot{\mathbf{F}} = (\hat{\mathbf{W}} + \hat{Q})\mathbf{F} \tag{3.95}$$

where

$$
\hat{W}(r) = \begin{pmatrix} 0 & 0 & 0 & 0 \\ 0 & -W_A & W_B & 0 \\ 0 & W_A & -W_B & 0 \\ 0 & 0 & 0 & 0 \end{pmatrix}, \quad
\hat{Q} = \begin{pmatrix} -\left(\frac{1}{\tau_A}+\frac{1}{\tau_B}+I_0\right) & 0 & I_0 & 0 \\ \frac{1}{\tau_B} & -\frac{1}{\tau_A}-I_0 & 0 & I_0 \\ \frac{1}{\tau_A}+I_0 & 0 & -\frac{1}{\tau_B}-I_0 & 0 \\ 0 & \frac{1}{\tau_A}+I_0 & \frac{1}{\tau_B} & -I_0 \end{pmatrix}
$$

$$(3.96)$$

The relaxation of the system after weak δ-pulse excitation at $t = 0$ is easier to describe because $f = I_0 = 0$ and the set (3.93) reduces to only two equations:

$$\dot{v} = -\left(\frac{1}{\tau_A} + W_A\right)v + W_B\mu \tag{3.97a}$$

$$\dot{\mu} = -\left(\frac{1}{\tau_B} + W_B\right)\mu + W_A v \tag{3.97b}$$

For large negative ΔG, the transfer becomes irreversible: $W_B \ll W_A$. Setting $W_B = 0$ and $W_A \equiv W(r)$, we obtain under this condition a single equation for the survival probability of the initially excited state:

$$\dot{v} = -\frac{v}{\tau_A} - W(r)v \tag{3.98}$$

This equation accounts for the decay of the excited state with the rate $1/\tau_A$ ignored by equation (3.91). The difference between these equations retains when they turn to the auxiliary equations for IET and DET by appending diffusional terms to the rhs of them. However, the usage of the auxiliary equations in these theories is also different; one of them is designed for the memory function of IET and another, for the time-dependent rate constant of DET. In spite of all these differences, the results of DET and IET were shown to be identical in the case of irreversible transfer [124].

The extension of DET to reversible energy transfer was intuitively made only once [7,64] and for the particular case of the resonance intramolecular transfer ($\Delta G = 0$), when $W_A = W_B = W$. The set of equations that substituted for Eq. (3.91) assumed the following form:

$$\dot{n} = -Wn + Wm \tag{3.99a}$$

$$\dot{m} = -\left(\frac{1}{\tau_B} + W\right)m + Wn \tag{3.99b}$$

This is identical to the set (3.97) only in the limit $\tau_A \to \infty$. At any finite τ_A the difference in the final results of DET and IET is noticeable and at $\tau_A < \tau_B$ becomes qualitative. The rate constant of transfer to the less stable particle diverges, rendering DET inapplicable [48]. Only IET compatible with all decay rates provides us with the right kinetic description of reversible energy quenching [49,50]. The memory functions of this theory are expressed through the auxiliary equations that naturally account for any relaxation in the reacting pair, like Eqs. (3.97) and (3.93). This makes possible the straightforward extension of the matrix IET to reactions of even higher complexity, unattainable by any alternative approaches.

B. Bimolecular Reaction after Instantaneous Excitation

The two-level particles are represented by the column vectors composed from the populations of the ground and excited states, N and N^*:

$$\mathbf{N}_A = \begin{pmatrix} N_A^* \\ N_A \end{pmatrix} \qquad \mathbf{N}_B = \begin{pmatrix} N_B^* \\ N_B \end{pmatrix} \tag{3.100}$$

In the absence of intermolecular reactions, the particles are subjected just to excited-state decay, represented by the following matrix equations

$$\dot{\mathbf{N}}_A = \hat{\mathbf{Q}}_A N_A \qquad \dot{\mathbf{N}}_B = \hat{\mathbf{Q}}_B N_B, \tag{3.101}$$

where

$$\hat{\mathbf{Q}}_A = \begin{bmatrix} -\frac{1}{\tau_A} & 0 \\ \frac{1}{\tau_A} & 0 \end{bmatrix} \qquad \hat{\mathbf{Q}}_B = \begin{bmatrix} -\frac{1}{\tau_B} & 0 \\ \frac{1}{\tau_B} & 0 \end{bmatrix} \tag{3.102}$$

The general IET equations for the reversible reaction (3.92) have a form typical of memory function formalism [49,50]:

$$\dot{\mathbf{N}}_A = Sp_B \int_0^t \mathbf{M}(t - \tau)\mathbf{N}_A(\tau) \otimes \mathbf{N}_B(\tau)d\tau + \hat{\mathbf{Q}}_A \mathbf{N}_A \tag{3.103a}$$

$$\dot{\mathbf{N}}_B = Sp_A \int_0^t \mathbf{M}(t - \tau)\mathbf{N}_A(\tau) \otimes \mathbf{N}_B(\tau)d\tau + \hat{\mathbf{Q}}_B \mathbf{N}_B \tag{3.103b}$$

Here $\mathbf{N}_A \otimes \mathbf{N}_B$ is a direct product on the basis of the collective states AB, AB*, A*B, A*B*, while Sp_A or Sp_B are the traces over the states of the corresponding particle. In our particular case

$$
\mathbf{N}_A \otimes \mathbf{N}_B = \begin{pmatrix} N_A^* N_B^* \\ N_A^* N_B \\ N_A N_B^* \\ N_A N_B \end{pmatrix}; \quad Sp_A \begin{pmatrix} 4 \\ 3 \\ 2 \\ 1 \end{pmatrix} = \begin{pmatrix} 2+4 \\ 1+3 \end{pmatrix}, \quad Sp_B \begin{pmatrix} 4 \\ 3 \\ 2 \\ 1 \end{pmatrix} = \begin{pmatrix} 3+4 \\ 1+2 \end{pmatrix}
$$

The most important part of the integral equations is their kernel or the matrix memory function $\mathbf{M}(t)$, which is an operator of rank 4 defined by its Laplace transformation,

$$
\tilde{\mathbf{M}}(s) = \int \hat{\mathbf{W}}(r)\tilde{\mathbf{F}}(r,s)[s\hat{\mathbf{I}} - \hat{\mathbf{Q}}]d^3r \tag{3.104}
$$

It is expressed via the transfer operator $\hat{\mathbf{W}}(r)$ and the matrix $\mathbf{F}$, composed of four vectors whose components are the same but differ in their initial values:

$$
\mathbf{F(t)} = \begin{pmatrix} f_1 & f_2 & f_3 & f_4 \\ v_1 & v_2 & v_3 & v_4 \\ \mu_1 & \mu_2 & \mu_3 & \mu_4 \\ g_1 & g_2 & g_3 & g_4 \end{pmatrix} \qquad \mathbf{F(0)} = \begin{pmatrix} 1 & 0 & 0 & 0 \\ 0 & 1 & 0 & 0 \\ 0 & 0 & 1 & 0 \\ 0 & 0 & 0 & 1 \end{pmatrix} \tag{3.105}
$$

This quantity obeys the auxiliary equation that describes the evolution of the mobile reaction pair during the encounter:

$$
\dot{\mathbf{F}} = \left(\hat{\mathbf{W}} + \hat{\mathbf{Q}}\right)\mathbf{F} + \hat{\mathbf{L}}F \tag{3.106}
$$

This is the same as Eq. (3.95) except that the last term with operator of encounter diffusion $\hat{\mathbf{L}}$ is added. This operator is diagonal, but its elements may differ because of different inter-particle interactions or diffusion coefficients for different pairs. Nothing like that is expected for the given example of energy transfer, so that

$$
\hat{\mathbf{L}} = \hat{L}\hat{\mathbf{I}} = D\Delta \begin{pmatrix} 1 & 0 & 0 & 0 \\ 0 & 1 & 0 & 0 \\ 0 & 0 & 1 & 0 \\ 0 & 0 & 0 & 1 \end{pmatrix} \tag{3.107}
$$

Using this operator together with $\hat{\mathbf{W}}(r)$ and $\hat{\mathbf{Q}}$ (at $I_0 = 0$) in matrix equation (3.106), we can represent the latter as the set of equations for its component:

$$\dot{f} = -\left(\frac{1}{\tau_A} + \frac{1}{\tau_B}\right)f + D\Delta f \tag{3.108a}$$

$$\dot{v} = -\left(\frac{1}{\tau_A} + W_A\right)v + W_B\mu + \frac{1}{\tau_B}f + D\Delta v \tag{3.108b}$$

$$\dot{\mu} = -\left(\frac{1}{\tau_B} + W_B\right)\mu + W_A v + \frac{1}{\tau_A}f + D\Delta\mu \tag{3.108c}$$

$$\dot{g} = \frac{1}{\tau_A}v + \frac{1}{\tau_B}\mu + D\Delta g \tag{3.108d}$$

With initial condition $\mathbf{F}(r,0) = \mathbf{I}$, many elements of $\mathbf{F}$ are zero, $f_1 = \exp(-t/\tau_A - t/\tau_B)$ and $g_4 = 1$:

$$\mathbf{F(t)} = \begin{pmatrix} f_1 & 0 & 0 & 0 \\ v_1 & v_2 & v_3 & 0 \\ \mu_1 & \mu_2 & \mu_3 & 0 \\ g_1 & g_2 & g_3 & 1 \end{pmatrix} \tag{3.109}$$

Using the Laplace transformation of this operator in Eq. (3.104), we obtain

$$\tilde{\mathbf{M}}(s) = \begin{bmatrix} 0 & 0 & 0 & 0 \\ -\langle\tilde{P}\rangle & -\langle\tilde{R}\rangle & \langle\tilde{S}\rangle & 0 \\ \langle\tilde{P}\rangle & \langle\tilde{R}\rangle & -\langle\tilde{S}\rangle & 0 \\ 0 & 0 & 0 & 0 \end{bmatrix} \tag{3.110}$$

where $\langle\cdots\rangle$ means the space averaging $\int\cdots d^3r$ of the following entities:

$$\tilde{S} = \left(s + \frac{1}{\tau_B}\right)(W_B\tilde{\mu}_3 - W_A\tilde{v}_3), \quad \tilde{R} = \left(s + \frac{1}{\tau_A}\right)(W_A\tilde{v}_2 - W_B\tilde{\mu}_2) \tag{3.111}$$

$$\tilde{P} = W_B\left[\frac{1}{\tau_A}(\tilde{\mu}_3 - \tilde{\mu}_1) + \frac{1}{\tau_B}(\tilde{\mu}_2 - \tilde{\mu}_1) - s\tilde{\mu}_1\right] - W_A\left[\frac{1}{\tau_A}(\tilde{v}_3 - \tilde{v}_1) + \frac{1}{\tau_B}(\tilde{v}_2 - \tilde{v}_1) - s\tilde{v}_1\right]$$

The last kernel accounts for the transfer resulting from the excitation decay during the encounter. At $(1/\tau_A) = (1/\tau_B) = 0$ this kernel is also zero, since in this case $v_1 = \mu_1 = 0$ as well.

Now we are ready to write the operator Equation (3.103) as a set of integrodifferential equations. Substituting Eqs. (3.110) and (3.102) into

Eq. (3.103) we obtain

$$\dot{N}_A^* = \int_0^t \langle S(t-\tau)\rangle N_A(\tau)N_B^*(\tau)d\tau - \int_0^t \langle R(t-\tau)\rangle N_A^*(\tau)N_B(\tau)d\tau$$

$$+ \int_0^t \langle P(t-\tau)\rangle N_A^*(\tau)N_B^*(\tau)d\tau - \frac{N_A^*}{\tau_A} = -\dot{N}_A, \qquad (3.112a)$$

$$\dot{N}_B^* = -\int_0^t \langle S(t-\tau)\rangle N_A(\tau)N_B^*(\tau)d\tau + \int_0^t \langle R(t-\tau)\rangle N_A^*(\tau)N_B(\tau)d\tau$$

$$- \int_0^t \langle P(t-\tau)\rangle N_A^*(\tau)N_B^*(\tau)d\tau - \frac{N_B^*}{\tau_B} = -\dot{N}_B. \qquad (3.112b)$$

If the excitation is weak, the particles in the ground state are present in great excess and their number remains practically unchanged:

$$N_A^* \ll N_A = \text{const}, \qquad N_B^* \ll N_B = \text{const} \qquad (3.113)$$

Under this condition the terms that are second-order in the excited-state concentrations are negligible, and the set (3.112) reduces to the following system of linearized equations [50]:

$$\dot{N}_A^* = N_A \int_0^t \langle S(t-\tau)\rangle N_B^*(\tau)d\tau - N_B \int_0^t \langle R(t-\tau)\rangle N_A^*(\tau)d\tau - \frac{N_A^*}{\tau_A} \qquad (3.114a)$$

$$\dot{N}_B^* = -N_A \int_0^t \langle S(t-\tau)\rangle N_B^*(\tau)d\tau + N_B \int_0^t \langle R(t-\tau)\rangle N_A^*(\tau)d\tau - \frac{N_B^*}{\tau_B} \qquad (3.114b)$$

These equations have to be solved with the initial conditions created by δ-pulse excitation at $t = 0$:

$$N_A^*(0) = N_0, \qquad N_B^*(0) = 0 \qquad (3.115)$$

The kernels of the integral equations (3.114) are defined by their Laplace transformations, equal to $\tilde{S}$ and $\tilde{R}$ from (3.111) averaged over space:

$$\langle \tilde{S}(s)\rangle = \left(s + \frac{1}{\tau_B}\right) \int [W_B(r)\tilde{\mu}_3(r,s) - W_A(r)\tilde{v}_3(r,s)]d^3r \qquad (3.116a)$$

$$\langle \tilde{R}(s)\rangle = \left(s + \frac{1}{\tau_A}\right) \int [W_A(r)\tilde{v}_2(r,s) - W_B(r)\tilde{\mu}_2(r,s)]d^3r \qquad (3.116b)$$

The auxiliary equations for the dyads of pair distributions v_3, μ_3 and v_2, μ_2 are the same

$$\dot{v} = -\left(\frac{1}{\tau_A} + W_A\right)v + W_B\mu + D\Delta v \qquad (3.117a)$$

$$\dot{\mu} = -\left(\frac{1}{\tau_B} + W_B\right)\mu + W_A v + D\Delta\mu \qquad (3.117b)$$

but the initial conditions for them are different:

$$v_2(r,0) = 1, \quad \mu_2(r,0) = 0; \qquad v_3(r,0) = 0, \quad \mu_3(r,0) = 1 \qquad (3.118)$$

The global structure of IET formalism is similar to that of DET. The integral kinetic equations substitute for their differential analogs. The kernels of these equations are defined by the transfer rates and pair distribution functions. The auxiliary equations for these functions are also similar but not identical to those in DET. In the next section we will see that the integral theory may sometimes be reduced to the differential one, albeit under rigid conditions and with some losses.

C.   Contact Approximation

If excitation is transferred from one reactant to another by dipole–dipole interaction (3.44), then one should use it as input data for solving equations (3.117) and taking the integrals (3.116). However, if this is triplet excitation, the dipole transitions are forbidden and transfer is carried out by the so called "exchange interaction." In this case a good approximation for the transfer rate is the exponential model (3.53). Energy quenching by exchange interaction, as well as electron tunneling in the normal Marcus region, takes place in the narrow reaction layer $l \ll \sigma$ and may be considered as contact. The forward and backward transfer reactions are represented by the kinetic rate constants

$$k_a = \int W_A(r)d^3r, \qquad k_b = \int W_B(r)d^3r \qquad (3.119)$$

that substitute for k_0 in Eq. (3.40) when $W_A(r)$ and $W_B(r)$ are considered contact.

If there are no reactions in the whole space except those at contact, the transfer terms can be excluded from the auxiliary equations (3.117):

$$\dot{v} = -\frac{v}{\tau_A} + D\Delta v \qquad (3.120a)$$

$$\dot{\mu} = -\frac{\mu}{\tau_B} + D\Delta\mu \qquad (3.120b)$$

but included in the boundary conditions for them:

$$4\pi\sigma^2 D \frac{\partial v}{\partial r}\bigg|_{r=\sigma} = k_a v - k_b \mu \tag{3.121a}$$

$$4\pi\sigma^2 D \frac{\partial \mu}{\partial r}\bigg|_{r=\sigma} = k_b \mu - k_a v \tag{3.121b}$$

The Laplace transformation of Eqs. (3.120) with the initial conditions (3.118), results in two sets of auxiliary equations, one for $\langle \tilde{R}(s) \rangle$

$$D\Delta \tilde{v}_2 - \left(s + \frac{1}{\tau_A}\right)\tilde{v}_2 = -1 \qquad D\Delta \tilde{\mu}_2 - \left(s + \frac{1}{\tau_B}\right)\tilde{\mu}_2 = 0 \tag{3.122}$$

and another for $\langle \tilde{S}(s) \rangle$:

$$D\Delta \tilde{v}_3 - \left(s + \frac{1}{\tau_A}\right)\tilde{v}_3 = 0 \qquad D\Delta \tilde{\mu}_3 - \left(s + \frac{1}{\tau_B}\right)\tilde{\mu}_3 = -1 \tag{3.123}$$

The boundary conditions for both are the same: the Laplace transformed Eqs. (3.121). To obtain the kernels from the solutions of Eqs. (3.122) and (3.123), one should insert them into the memory function definitions (3.116) adopted to the contact reactions:

$$\langle \tilde{S}(s) \rangle = \left(s + \frac{1}{\tau_B}\right)[k_b \tilde{\mu}_3(\sigma, s) - k_f \tilde{v}_3(\sigma, s)] \tag{3.124a}$$

$$\langle \tilde{R}(s) \rangle = \left(s + \frac{1}{\tau_A}\right)[k_f \tilde{v}_2(\sigma, s) - k_b \tilde{\mu}_2(\sigma, s)] \tag{3.124b}$$

Details of the calculations were presented in Ref. 50. The final results are:

$$\langle \tilde{R}(s) \rangle = k_a \tilde{\mathscr{F}}(s) \qquad \langle \tilde{S}(s) \rangle = k_b \tilde{\mathscr{F}}(s) \tag{3.125}$$

where

$$\tilde{\mathscr{F}}(s) = \left[1 + \frac{k_a/k_D}{1 + \sqrt{(s + 1/\tau_A)\tau_d}} + \frac{k_b/k_D}{1 + \sqrt{(s + 1/\tau_B)\tau_d}}\right]^{-1} \tag{3.126}$$

With these kernels, the general integral equations (3.114) take the particular form dedicated to the contact reactions:

$$\dot{N}_A^* = k_b N_A \int_0^t \mathscr{F}(\tau) N_B^*(t-\tau) d\tau - k_a N_B \int_0^t \mathscr{F}(\tau) N_A^*(t-\tau) d\tau - \frac{N_A^*}{\tau_A} \qquad (3.127\text{a})$$

$$\dot{N}_B^* = -k_b N_A \int_0^t \mathscr{F}(\tau) N_B^*(t-\tau) d\tau + k_a N_B \int_0^t \mathscr{F}(\tau) N_A^*(t-\tau) d\tau - \frac{N_B^*}{\tau_B} \qquad (3.127\text{b})$$

After instantaneous excitation of A, the normalized decay of A^* is given by $R(t) = N_A^*(t)/N_A 0)$. According to (3.10), its Laplace transformation, related to the lifetime τ_A, defines the relative quantum yield of luminescence:

$$\eta = \frac{\tilde{N}_A^*(0)}{N_A(0)\tau_A} = \frac{1}{1 + N_B k_q \tau_A} \qquad (3.128)$$

which resembles (3.11) provided $c\tau$ is substituted by $N_B \tau_A$. But there is an essential difference between the Stern–Volmer constant κ and k_q. To specify k_q, one should make the Laplace transformation of Eqs. (3.127), resolve them regarding $\tilde{N}_A^*(0)$, and substitute the latter into Eq. (3.128). Then it emerges that, unlike κ, the quenching constant k_q depends on the luminophor concentration N_A

$$k_q = \frac{K_{AB}}{1 + K_{BA} N_A \tau_B} \qquad (3.129)$$

where $K_{AB} = k_a \tilde{\mathscr{F}}(0)$ and $K_{BA} = k_b \tilde{\mathscr{F}}(0)$. The numerator in (3.129) accounts for the reversible transfer from particle A^* to particle B at their first encounter, while the denominator includes the contribution of backward transfer from separated B^* to other As met in the bulk. Therefore this term is proportional to N_A.

The generalized reaction constants of the forward and backward energy transfer relate to each other as follows:

$$\frac{K_{AB}}{k_a} = \frac{1}{1 + \dfrac{k_a/k_D}{1+\sqrt{\tau_d/\tau_A}} + \dfrac{k_b/k_D}{1+\sqrt{\tau_d/\tau_B}}} = \frac{K_{BA}}{k_b} \qquad (3.130)$$

Both of them are expressed via the function (3.126) taken at $s = 0$. This is exactly the result first obtained in Ref. 48. It is obvious that $K_{AB}/K_{BA} = k_a/k_b = W_A(\sigma)/W_B(\sigma) = K$ in accordance with the detailed balance principle (3.94). Taking this into account, Eq. (3.129) may be represented as follows:

$$\frac{1}{k_q} = \frac{1}{K_{AB}} + \frac{N_A \tau_B}{K} = \frac{1}{K_{AB}} + N_A \tau_B \cdot e^{\Delta G/T} \qquad (3.131)$$

The transfer from A to B becomes irreversible if it is highly exergonic ($-\Delta G \gg T$) or the product is very unstable ($\tau_B \to 0$). In both cases not only is the last term in Eq. (3.131) negligible but the term in the denominator of Eq. (3.130) containing k_b and τ_B is also zero. Only in this case $k_q = K_{AB}$ and

$$\frac{1}{K_{AB}} = \frac{1}{k_a} + \frac{1}{k_D\left(1 + \sqrt{\tau_d/\tau_A}\right)} \tag{3.132}$$

takes the form of DET expression (3.27) obtained for the Stern–Volmer constant of the irreversible quenching.

A quite different result follows from (3.131), if the same terms in these equalities are negligible because of a small concentration of active particles ($N_A \to 0$). As a limiting case, one may consider a single A particle surrounded by numerous Bs uniformly distributed in the whole space. Then the subsequent encounters are excluded and $k_q = K_{AB}$ as well, but the transfer in the geminate pair remains reversible and therefore

$$\frac{1}{K_{AB}} = \frac{1}{k_a} + \frac{1}{k_D\left(1 + \sqrt{\tau_d/\tau_A}\right)} + \frac{\exp(\Delta G/T)}{k_D\left(1 + \sqrt{\tau_d/\tau_B}\right)} \tag{3.133}$$

It should be stressed that the Markovian approach, taking into account the reversibility of transfer, leads to the simplest result

$$\frac{K_{AB}}{k_a} = \frac{k_D}{k_D + k_a + k_b} = \frac{K_{BA}}{k_b} \tag{3.134}$$

which follows from the non-Markovian IET equation (3.130) at $\tau_d \ll \tau_A, \tau_B$. The difference between Eq. (3.130) and its Markovian analog (3.134) is most pronounced at slow diffusion [48].

The problem under study has been widely recognized as an important test for the validity of different theoretical approaches competing with IET. Those who apply the popular superposition approximation (SA) to chemical kinetics were forced to essentially revise their theories to reproduce the results presented above. One of the revisions [52] enables us to obtain the right Stern–Volmer constant (3.133) for a particular case of a single A particle, when the general Eqs. (3.127) are reduced to the following set:

$$\dot{N}_A^* = -k_a N_B \int_0^t \mathscr{F}(\tau) N_A^*(t-\tau)d\tau - \frac{N_A^*}{\tau_A} \tag{3.135a}$$

$$\dot{N}_B^* = k_a N_B \int_0^t \mathscr{F}(\tau) N_A^*(t-\tau)d\tau - \frac{N_B^*}{\tau_B} \tag{3.135b}$$

These equations describe only the geminate reaction, which is not followed by bimolecular reaction of products in the bulk. On the other hand, the Naumann theory was also developed for noncontact (dipole–dipole) energy transfer considered in Section III.B.

Another revision undertaken with *many-particle kernel* theory, later named MPK1 [51], was even more successful in reproducing the general set of equations (3.127). Moreover, an attempt was made to account for the higher-order corrections in the particle concentrations. In the later version of their theory, named MPK 3 [126], the authors reached a one-to-one correspondence between their results and those obtained with MET. The relationship between IET, MET and other approaches will be discussed further in Section XII.

IV. IET VERSUS DET

The reaction rate in differential theories of bimolecular reactions is always the product of the reactant concentrations and the "rate constant," does not matter whether the latter is truly the constant or the time-dependent quantity. In integral theories there are no such constants at all; they give way to kernels (memory functions) of integral equations. However, there is a regular procedure that allows reduction the integral equations to differential equations under specificconditions [34,127]. This reduction can be carried out in full measure or partially, but the price for it should be well recognized.

A.   Reversible Energy Transfer

For the present goal it is enough to consider a very dilute solution or even a single excited A^* surrounded by Bs. This is the case when the separated products of the geminate reaction never meet again and the whole process is described by the reduced set of Eqs. (3.114):

$$\dot{N}_A^* = -N_B \int_0^t \langle R(t-\tau)\rangle N_A^*(\tau)d\tau - \frac{N_A^*}{\tau_A} \qquad (3.136a)$$

$$\dot{N}_B^* = N_B \int_0^t \langle R(t-\tau)\rangle N_A^*(\tau)d\tau - \frac{N_B^*}{\tau_B} \qquad (3.136b)$$

The first equation, (3.136a), describes the impurity quenching of A^* and may be considered separately from another one intended for the description of energy accumulation and dissipation at reaction partners, Bs.

To transform the integral equation (3.136a) to its differential analog, one should first make the substitution

$$N_A^*(t) = e^{-t/\tau_A} P_A^*(t) \qquad (3.137)$$

This eliminates fast exponential decay from the equation for P_A^* but changes its kernel:

$$\dot{P}_A^* = -N_B \int_0^t \langle R(\tau) \rangle e^{\tau/\tau_A} P_A^*(t - \tau) d\tau \qquad (3.138)$$

The right-hand side (RHS) of this equation may be approximated by a differential one if P_A^* varies slowly in comparison with all the other functions under the integral. If this is the case, the well-known recipe [34,127] recommends the following replacement:

$$P_A^*(t - \tau) \longrightarrow P_A^*(t) \qquad (3.139)$$

This turns the integral kinetic equation (3.138) into a differential one. In the original variables the latter takes the following form:

$$\dot{N}_A^* = -k(t)N_B N_A^* - \frac{N_A^*}{\tau_A} \qquad (3.140)$$

Here the rate constant

$$k(t) = \int_0^t \langle R(\tau) \rangle e^{\tau/\tau_A} d\tau \qquad (3.141)$$

is expressed through the kernel $\langle R \rangle$, which is the original of the Laplace transform $\tilde{R}(s)$ presented in Eq. (3.116). Since the original of $[s + (1/\tau_A)]\tilde{v}_2(r,s)$ is $\dot{v}_2 + (v_2/\tau_A) + \delta(t)$ and that of $[s + (1/\tau_A)]\tilde{\mu}_2(r,s)$ is $\dot{\mu}_2 + (\mu_2/\tau_A)$, this kernel can be easily recovered:

$$\langle R(\tau) \rangle = \left\langle W_A \left(\dot{v}_2 + \frac{v_2}{\tau_A} + \delta(\tau) \right) - W_B \left(\dot{\mu}_2 + \frac{\mu_2}{\tau_A} \right) \right\rangle$$
$$= [\langle W_A[\dot{n} + \delta(\tau)] \rangle - \langle W_B \dot{m} \rangle] e^{-\tau/\tau_A} \qquad (3.142)$$

The auxiliary pair distributions, v_2 and μ_2, which obey equations (3.117), simply relate to two others, $n(t) = v(t)e^{t/\tau_A}$ and $m(t) = \mu(t)e^{t/\tau_A}$. For them another set of equations is valid:

$$\dot{n} = -W_A n + W_B m + D\Delta n \qquad (3.143a)$$
$$\dot{m} = -W_B m + W_A n + D\Delta m + \left(\frac{1}{\tau_A} - \frac{1}{\tau_B} \right) m \qquad (3.143b)$$

The relaxation rate $1/\tau_A$ appears here to be less natural than in the original set (3.117) for v and μ. We will see here that it is precisely this rate that makes the transition from the integral to differential kinetic equations impossible when it is faster than $1/\tau_B$. Substituting Eq. (3.142) into Eq. (3.141), we obtain after integration the final expression for the time-dependent reaction constant:

$$k(t) = \langle W_A n \rangle - \langle W_B m \rangle \tag{3.144}$$

Here we took into account that $n(r,0) = 1, m(r,0) = 0$. According to such initial conditions

$$k(0) = \langle W_A(r) \rangle = k_a \tag{3.145}$$

which is initially the rate constant is equal to the kinetic one. The further evolution of $k(t)$ depends on the relationship between τ_A and τ_B.

If the energy transfer is irreversible ($W_B = 0$), the definition of the rate constant in (3.144) is exactly the same as that given by DET in Eqs. (3.34) and (3.35):

$$k(t) = \int W_A(r) n(r,t) d^3 r \qquad \text{where} \qquad \dot{n} = -W_A(r) n + D \Delta n \tag{3.146}$$

At $W_B \neq 0$ there is the nonzero term in Eq. (3.144), which contains $m(r,t)$ and accounts for the backward energy transfer during the encounter. This term is negligible only at large negative ΔG. Otherwise, it makes the essential difference between the general definition of the rate constant (3.144) and its DET analog (3.146). The difference becomes crucial at $\tau_A < \tau_B$ when $m(r,t)$ diverges with time, as well as

$$k(t) = \int W_A(r) n(r,t) d^3 r - \int W_B(r) m(r,t) d^3 r \to -\infty \qquad \text{at} \qquad \tau_A < \tau_B \tag{3.147}$$

This is a clear demonstration that a differential encounter theory is not a reasonable alternative to an integral one, when the energy transfer is reversible and occurs between the unstable reactants. Since the lifetimes are not equal, the rate constant of the energy transfer from the short-lived to the long-lived particle does not exist (diverges) at long times. In contrast, in integral formalism we do not encounter any difficulties. This is illustrated by the simplest example of contact transfer considered below.

B. Contact Intermolecular Transfer

If Eq. (3.135) is subjected to the same transformation as Eq. (3.136a), including replacement (3.139), then we again arrive at the DET equation (3.140) but with a contact rate constant

$$k(t) = k_a \int_0^t \mathscr{F}(\tau) e^{\tau/\tau_A} d\tau \qquad (3.148)$$

In differential contact theory $k(t)$ monotonously decreases with time, starting from the maximum kinetic value (3.145) and approaching the minimal stationary value:

$$k = k(\infty) = k_a \tilde{\mathscr{F}} \left(-\frac{1}{\tau_A} \right) \qquad (3.149)$$

By inserting Eq. (3.126) into this expression, we obtain

$$k = k_a k_D \left(k_D + k_a + \frac{k_b}{1 + [(1/\tau_B - 1/\tau_A)\tau_d]^{1/2}} \right)^{-1} \qquad (3.150)$$

In the case of irreversible energy transfer $(k_b = 0)$, this result reduces to the conventional one, given by Eq. (3.22) with $k_0 \equiv k_a$. However, when energy transfer is reversible $(k_b > 0)$, it comes to the stationary (Markovian) limit only at

$$\tau_B \le \tau_A \qquad (3.151)$$

In the opposite case, $\tau_B > \tau_A$, the rate constant (3.150) becomes complex, which is evidently senseless. This is the manifestation of the failure of the rate constant approach.

By criterion (3.151), the region is established where the energy transfer results in excitation quenching. At the border of this region, at $\tau_B = \tau_A$, the stationary constant (3.150) reaches the minimal value at given k_a and k_b

$$k = \frac{k_a k_D}{k_D + k_a + k_b} = \frac{k_a}{1 + \aleph} \qquad (3.152)$$

where $\aleph = (k_a + k_b)/k_D$. In other words, this is the maximal reduction of the stationary rate constant due to reversibility of the transfer. The stationary value k is monotonously approached by $k(t)$ from above [49]

$$k(\tau) = \frac{k_a}{1 + \aleph} \left\{ 1 + \aleph e^{(1+\aleph)^2 \tau} \operatorname{erfc}\left[(1 + \aleph)\sqrt{\tau} \right] \right\} \qquad (3.153)$$

where $\tau = t/\tau_d$. This is the straightforward extension of Eq. (3.21) for the reversible contact quenching that is for any $\aleph \ge k_a/k_D$.

As soon as τ_B exceeds τ_A, the energy quenching gives way to energy conservation. It is carried out by the reaction partner that lives longer than the initially excited one. After fast decay of the latter, the excitation returns back to the point from where it was conserved during the encounter and contributes to the delayed fluorescence. Hence, the presence of a long-lived partner first facilitates the excitation decay but finally makes it longer than the natural one. This phenomenon is reflected in the sign alteration of $k(t)$ if Eq. (3.141) is still used to cover its time dependence. The rate constant becomes negative when the direction of the energy flux is reversed.

This is a common feature for DET and IET, provided the solution to the integral equation is used to introduce the effective rate constant for the latter:

$$k_{\text{eff}}(t) = -\frac{d\ln(N_A^* e^{t/\tau_A})}{N_B\,dt}.\qquad(3.154)$$

This constant also becomes negative at long times, but remains finite while the $k(t)$ of DET diverges (Fig. 3.18). The astonishing singularity is inherent in $k(t)$ from (3.148), due to the long time asymptote of the integrand. In Ref. 48 the

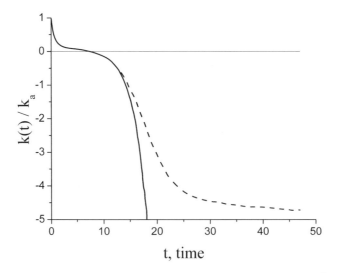

Figure 3.18. The sign alteration of the reaction constants k_{eff} (dashed line) and $k(t)$ (solid line) at $\tau_B = \infty$ ($k_a N_B \tau_d = 0.1, \tau_A = 2\tau_d$). Within the differential theory the rate constant, which is negative, diverges while the effective rate constant of the integral theory approaches its limit $k_{\text{eff}}/k_a = -1/k_a N_B \tau_A = -5$ (for $\tau_d/\tau_A = 2$ and $k_a N_B \tau_d = 0.1$). (From Ref. 49.)

latter was found to be different for $\tau_A > \tau_B$ and $\tau_A < \tau_B$ and was represented in Ref. 50 by a reasonable interpolation

$$\mathscr{F}(t)e^{t/\tau_A} = -\frac{1}{2\pi^{1/2}k_D\tau_d}\left(\frac{\tau_d}{t}\right)^{3/2}\left[\frac{K_{AB}^2}{k_a} + \frac{K_{BA}^2}{k_b}\exp\left(\frac{t}{\tau_A} - \frac{t}{\tau_B}\right)\right] \qquad \text{at} \qquad t \gg \tau_d$$

$$(3.155)$$

where K_{AB} and K_{BA} were as defined in Eq. (3.130). This integrand experiences the qualitative change of the long-time behavior when τ_A becomes shorter than τ_B. As a result, the integral (3.148) converging at $\tau_A \geq \tau_B$ diverges at $\tau_A < \tau_B$. In the latter case the concentration of active molecule $N_A^*(t)$ also diverges increasing to infinity, thus breaking the particle conservation law.

Nothing of the kind happens to $N_A^*(t)$ in IET. It approaches zero as $t \to \infty$, according to the asymptotic behavior established in Ref. 49:

$$\frac{N_A^*(t)}{N_A^*(0)} \propto \left(\frac{\tau_d}{t}\right)^{3/2} e^{-t/\tau_B} \qquad (3.156)$$

As this is used in Eq. (3.154), it gives a rate that is negative at $\tau_A < \tau_B$ but remains finite:

$$k_{\text{eff}}(t)N_B = \frac{3}{2t} - \left(\frac{1}{\tau_A} - \frac{1}{\tau_B}\right) \qquad (3.157)$$

The results of the numerical calculations shown in Figure 3.18 confirm that the effective IET rate constant levels off with time at $k_{\text{eff}}(\infty) = -1/\tau_A + 1/\tau_B \geq -1/\tau_A$, while its differential analog diverges: $k(\infty) = -\infty$.

Up to now we have discussed only the first equation of the set (3.135), which describes the quenching of reactant A^*. However, the physical picture outlined above would be incomplete if the time evolution of the product, B^*, was not taken into account. The integral equations (3.135) solved by the Laplace transformation result in the following:

$$\tilde{N}_A^*(s) = \frac{N_A^*(0)}{s + 1/\tau_A + k_a N_B \tilde{\mathscr{F}}(s)} \qquad \tilde{N}_B^*(s) = \frac{k_a N_B \tilde{\mathscr{F}}(s)N_A^*(0)}{(s + 1/\tau_B)[s + 1/\tau_A + k_a N_B \tilde{\mathscr{F}}(s)]}$$

$$(3.158)$$

The numerically calculated originals of $\tilde{N}_A^*(s)$ and $\tilde{N}_B^*(s)$ presented in Figure 3.19 are calculated with the initial conditions $N_A^*(0) = 1$ and $N_B^*(0) = 0$. While the

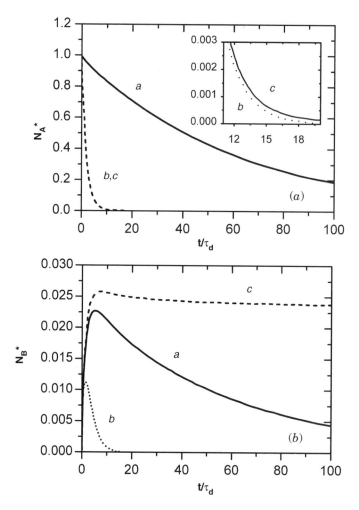

Figure 3.19. Kinetics of dissipation and accumulation of energy $N_A^*(t)$ and $N_B^*(t)$ at $N_A^*(0) = 1, N_B^*(0) = 0$, and $k_a = k_b = 5k_D$: (a) $\tau_A = \infty, \tau_B = 2\tau_d$; (b) $\tau_A = \tau_B = 2\tau_d$; (c) $\tau_A = 2\tau_d, \tau_B = \infty$. (From Ref. 50.)

monotonous quenching kinetics $N_A^*(t)$ [Fig. 3.19] describes the excitation decay due to energy quenching (at $\tau_A > \tau_B$) or trapping (at $\tau_A < \tau_B$), the partner dynamics $N_B^*(t)$ are composed from the ascending and descending branches [Fig. 3.19]. Initially the energy is accumulated in B^* due to the forward transfer, but in the final stage it dissipates naturally and via reverse transfer. The latter results in a delayed luminescence via the donor when $\tau_A < \tau_B$.

Three typical situations are illustrated by Figure 3.19:

1. When A^* is stable ($\tau_A = \infty$), its decay brought only by bimolecular quenching is slower than that of B^*.

2. When $\tau_A = \tau_B$, the reversible transfer results in equilibration of the excited-state populations, but after that both of them decay with the same time.

3. When B^* is stable ($\tau_B = \infty$), part of the accepted energy returns from it to A^*, making its final decay longer than natural. It is seen in the insert as delayed fluorescence. Since this part of the energy accumulated by B^* is lost during the geminate stage of the reaction, the final amount of B^*s separated from their partners is less than maximal. Therefore a small maximum is seen on the dashed line $N_B^*(t)$ in Figure 19 in case (c).

The last situation is extreme and rare for the excitations, but entirely natural for the reactions (3.52) if they are reversible and their charged products are stable. The quantum yield of these products following from Eq. (3.158) is

$$\psi = \frac{N_B^*(\infty)}{N_A^*(0)} = \frac{\kappa_0 N_B \tau_A}{1 + \kappa_0 N_B \tau_A} = 1 - \eta \qquad (3.159)$$

where η was defined in Eqs. (3.128) and (3.129) and $\kappa_0 = k_a \tilde{\mathscr{F}}(0)$:

$$\kappa_0 = \frac{k_a k_D}{k_D + k_b + \dfrac{k_a}{1 + \sqrt{\tau_d/\tau_A}}} = K_{AB}|_{\tau_B = \infty} \qquad (3.160)$$

The last equality follows from Eq. (3.133) at $\tau_B = \infty$.

The stable products that escaped the geminate backward reaction are not survived forever. They meet in the bulk with partners separated from other pairs, and in these encounters the reverse transfer of energy or charge continues to the point of their disappearance. This bimolecular reaction is represented by terms that are proportional to N_A in Eqs. (3.114) and (3.127). These terms omitted in Eqs. (3.135) as insignificant during the geminate stage of transfer, will be restored and discussed in Section VIII.

C. Induced Intramolecular Transfer

The situation is different from that with the intramolecular energy transfer catalyzed by solute encounters. If an active molecule with two excited states, A^* and B^*, say, singlet and triplet, meets quencher C considered as structureless but carrying heavy atom (halogen substituent), then the latter catalyzes the

intersystem crossing [101,128–131]. The latter may be represented as
$A^* + C \longleftrightarrow B^* + C$ or in more detail in a scheme such as (3.92):

$$[A^* \cdots C] \underset{W_B}{\overset{W_A}{\rightleftharpoons}} [B^* \cdots C] \qquad (3.161)$$

$$\downarrow \tau_A \qquad\qquad \downarrow \tau_B$$

Since it is essentially the same as its intermolecular analog, the definitions of the
kernels (3.116) as well as the auxiliary equations (3.117) do not change. Only the
integral equations become different:

$$\dot{N}_A^* = c\int_0^t \langle S(t-\tau)\rangle N_B^*(\tau)d\tau - c\int_0^t \langle R(t-\tau)\rangle N_A^*(\tau)d\tau - \frac{N_A^*}{\tau_A} \qquad (3.162a)$$

$$\dot{N}_B^* = -c\int_0^t \langle S(t-\tau)\rangle N_B^*(\tau)d\tau + c\int_0^t \langle R(t-\tau)\rangle N_A^*(\tau)d\tau - \frac{N_B^*}{\tau_B} \qquad (3.162b)$$

Here $c = [C]$ is the concentration of solutes that induce the interconversion but
do not participate in it. Their number does not change ($c = $ const) and cannot
serve as a small parameter because it enters equally into all the terms. Since
neither of them may be neglected, the route to the reduced set of equations
similar to (3.135) is closed and either equation from the set cannot be considered
separately.

Hence, the complete DET description of the intramolecular transfer is not
possible in principle at $\tau_A \neq \tau_B$ because the forward and backward transfers are
not separable and the rate of one of them diverges. To be sure of this, let us
make the substitution (3.137) but for both quantities:

$$N_A^*(t) = e^{-t/\tau_A}P_A^*(t) \qquad \text{and} \qquad N_B^*(t) = e^{-t/\tau_B}P_B^*(t) \qquad (3.163)$$

Then using for both of them, the same replacement as in Eq. (3.139)

$$P_A^*(t-\tau)\longrightarrow P_A^*(t), \qquad P_B^*(t-\tau)\longrightarrow P_B^*(t) \qquad (3.164)$$

we immediately reduce the integral equations to differential ones:

$$\dot{P}_A^* = k_B(t)cP_B^* - k_A(t)cP_A^* \qquad (3.165a)$$

$$\dot{P}_B^* = -k_B(t)cP_B^* + k_A(t)cP_A^* \qquad (3.165b)$$

The final definition of the time dependent constants is similar to that given in
Eq. (3.144):

$$k_A(t) = \int [W_A(r)n(r,t) - W_B(r)m(r,t)]d^3r \qquad (3.166a)$$

$$k_B(t) = \int [W_A(r)n'(r,t) - W_B(r)m'(r,t)]d^3r \qquad (3.166b)$$

Here we have two couples of pair distributions: (n, m) and (n', m'). One of them obeys the auxiliary equations (3.143) with initial conditions $n(r, 0) = 1$, $m(r, 0) = 0$, while for another one, we have a different set:

$$\dot{n}' = -W_A n' + W_B m' + D\Delta n' + \left(\frac{1}{\tau_B} - \frac{1}{\tau_A}\right) n' \qquad (3.167a)$$

$$\dot{m}' = -W_B m' + W_A n' + D\Delta m' \qquad (3.167b)$$

with initial conditions $n'(r, 0) = 0$, $m'(r, 0) = 1$, identical to those in Eq. (3.118).

The qualitative delimitation of the situations remains the same as before: energy quenching (1), redistribution (2), and conservation (3). If $\tau_B < \tau_A$, the induced transition $A \rightarrow B$ results in quenching of A^*, but the reverse transition $B \rightarrow A$, conserves the energy transferred from B^* to the more stable A^*. According to Eqs. (3.143) the rate constant $k_A(t)$ is finite and positive, while the rate constant $k_B(t)$ changes the sign with time and diverges as well as n', since the relaxation term in Eq. (3.167a) is positive. If $\tau_B > \tau_A$, the situation is quite the opposite: $k_B(t)$ is finite while $k_A(t)$ diverges. In fact, the transition to DET is available now for only the border case (1): when precisely $\tau_A = \tau_B$. Since there are no relaxation terms in this case, in neither (3.143) nor (3.167), it allows for an exact solution of the problem that will be considered in Section IV.E.

D. Contact Induction of Interconversion

In the contact approximation Eqs. (3.167) take the form [45]

$$\dot{N}_A^* = ck_b \int_0^t \mathscr{F}(\tau) N_B^*(t - \tau) d\tau - ck_a \int_0^t \mathscr{F}(\tau) N_A^*(t - \tau) d\tau - \frac{N_A^*}{\tau_A} \qquad (3.168a)$$

$$\dot{N}_B^* = -ck_b \int_0^t \mathscr{F}(\tau) N_B^*(t - \tau) d\tau + ck_a \int_0^t \mathscr{F}(\tau) N_A^*(t - \tau) d\tau - \frac{N_B^*}{\tau_B} \qquad (3.168b)$$

where $\mathscr{F}(\tau)$ was defined in Eq. (3.126) by its Laplace transformation.

The numerical solution to these equations with the initial conditions $N_A^*(0) = 1, N_B^*(0) = 0$ is shown in Figure 3.20. If the transfer to B^* is irreversible $(k_b = 0)$, the quenching proceeds exponentially up to the end [dashed line in Fig. 3.20(a)]. Otherwise, the energy coming back from B^* retards this decay and supports the delayed fluorescence from state A^*. Initially the density of this state decreases faster than in the absence of C, but later it decreases much more slowly. The smaller k_b, the later and slower should be the delayed fluorescence. This qualitative picture is supplemented with the corresponding kinetics of B^* accumulation and subsequent decay. The latter takes place even at $\tau_B = \infty$, due to the bimolecular backward transfer to unstable A^* [Fig. 3.20(b)]. Such a decay is absent only at $k_b = 0$ (upper curve) and is faster at higher k_b.

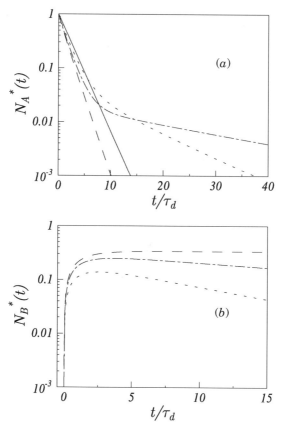

Figure 3.20. Population relaxation at $N_A^*(0) = 1, N_B^*(0) = 0$, $\tau_B = \infty$, $k_a = 5k_D$, and $ck_a\tau_d = 1.0$: (a) decay of A^* with its own lifetime $\tau_A = 2\tau_d$ (thick straight line) accompanied by the intersystem crossing at $k_b = 5k_D$ (dotted line), $k_b = k_D$ (dashed–dotted line), and $k_b = 0$ (dashed line); (b) accumulation and dissipation of the initially empty state B^*, which decays through A^* faster, when k_b is higher. (From Ref. 45.)

It should be noted that the integrodifferential equations equivalent to Eqs. (3.168) were first obtained with the "fully renormalized" kinetic theory of Yang, Lee, and Shin (YLS) [132]. However, they had not confined themselves to the original integral equations (3.168) but reduced them to the differential rate equations by the incorrect substitution used instead of (3.164):

$$N_A^*(t - \tau) \longrightarrow N_A^*(t), \qquad N_B^*(t - \tau) \longrightarrow N_B^*(t)$$

In this way the population densities N_A^* and N_B^* were factored outside the integral signs without preliminary exclusion of the excited-state decay. The resulting equations become differential

$$\dot{N}_A^* = ck_B^{\text{ph}}(t)N_B^* - ck_A^{\text{ph}}(t)N_A^* - \frac{N_A^*}{\tau_A} \tag{3.169a}$$

$$\dot{N}_B^* = -ck_B^{\text{ph}}N_B^* + ck_A^{\text{ph}}(t)N_A^* - \frac{N_B^*}{\tau_B} \tag{3.169b}$$

with the "phenomenological" rate constants

$$k_A^{\text{ph}}(t) = k_a \int_0^t \mathscr{F}(\tau)d\tau \qquad \text{and} \qquad k_B^{\text{ph}}(t) = k_b \int_0^t \mathscr{F}(\tau)d\tau \tag{3.170}$$

which are positive and finite. The divergency problem is thus eliminated in the "phenomenological" YLS theory, but the essential effects of energy conservation (trapping) and delayed fluorescence are lost. Even in the Markovian limit the phenomenological constants $k_A^{\text{ph}}(\infty) = K_{AB}$ and $k_B^{\text{ph}}(\infty) = K_{BA}$ differ from their DET analogs. For instance, in the simplest case of the irreversible transfer, $k_A^{\text{ph}}(\infty)|_{k_b=0} = K_{AB}|_{k_b=0} = \kappa_0$ is actually the Stern–Volmer constant (3.27) but not the true Markovian constant k from Eq. (3.22).

The proper reduction of IET to DET leads to the same differential equations (3.169) but with the right rate constants

$$k_A(t) = k_a \int_0^t \mathscr{F}(\tau)e^{\tau/\tau_A}d\tau \qquad \text{and} \qquad k_B(t) = k_b \int_0^t \mathscr{F}(\tau)e^{\tau/\tau_B}d\tau \tag{3.171}$$

substituted for the phenomenological ones. At $\tau_A \neq \tau_B$, either $k_A(t)$ or $k_B(t)$ diverges after changing the sign. Applying it to a problem in hands, A^* may be a short-lived (say singlet) excitation, while B^* (say, triplet) lives much longer. Initially A^* is quenched by the interconversion, but after its decay is accomplished the energy goes back from triplet to singlet, resulting in a delayed fluorescence from the latter. At this last stage $k_A(t)$ changes its sign (becomes negative) as well as the energy flux, that changed the direction. This effect, illustrated by Figure 3.18 and confirmed in Ref. 45, is lost in the "phenomenological" theory whose constants are always positive and approach their stationary values as in the classical Smoluchowski theory. In fact, the normal Smoluchowski behavior is peculiar only to the backward DET constant $k_B(t)$ while its forward analog, $k_A(t)$, is sign-alternating and diverges as $t \to \infty$.

In contrast, there are no divergencies in IET. The effective rate constants introduced by the relationships

$$k_A^{\text{eff}}(t)N_A(t) = k_a \int_0^t \mathscr{F}(\tau)N_A(t-\tau)d\tau, \qquad k_B^{\text{eff}}(t)N_B(t) = k_b \int_0^t \mathscr{F}(\tau)N_B(t-\tau)d\tau,$$

$$\tag{3.172}$$

and calculated using the solutions to the IET equations (3.168), are always finite
[45]. When $\tau_B > \tau_A$, $k_A^{\text{eff}}(t)$ changes its sign with time, but then levels off, while
the corresponding DET constant diverges: $\lim_{t\to\infty} k_A(t) = -\infty$. On the other
hand, IET is also not ideal. The kernel of IET has the power dependence on the
time resulting from the diffusional separation of the reacting pair. This property
is passed on the long-time asymptote of $N_A^*(t)$ or $N_B^*(t)$ when the corresponding
lifetime is infinite [49]. In this case the rate constants of DET and YLS approach
the asymptotic Markovian value as $t \to \infty$, while the effective constant of IET
declines from this limit downward. As a result the quenching in IET proceeds
slower than it should be at long times (see Section IX) [51,126]. This is the
principal fault of IET, which does not include all necessary terms from the
concentration expansion [43]. This drawback was overcome in MET
[40,41,133,134] and some other theories [126,132,135] that will be compared
in Section XII.

One more feature of the kinetic behavior that is lost in the phenomenological
approach is represented in Figure 3.21. The final energy quenching in a bulk
running by the backward energy transfer from B^* to A^* proceeds exponentially
in both IET and the phenomenological theory. In Figure 3.21 these asymptotes
are shown by the parallel thin lines having the same slope. However, contrary to
the phenomenological theory, the IET curve approaches its asymptote from
above, after a hump peculiar to the geminate backward transfer. The similar

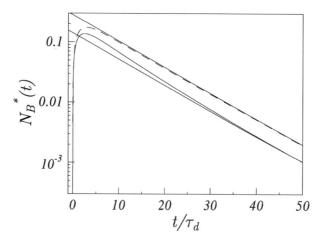

Figure 3.21. The same as in Figure 3.20(b) but in a wider dynamic region and for
$k_a = k_b = 5k_D$. The solid line represents the IET, which accounts for the initial geminate and
subsequent bimolecular process. The quasiexponential asymptote of the latter (thin straight line) is
approached from above. The dashed line is the same kinetics but obtained from the
phenomenological YLS theory, which also approaches the bimolecular asymptote (thin line), but
from below. (From Ref. 45.)

hump at the geminate stage is also seen in Figure 3.19, which demonstrates the intermolecular energy quenching after accumulation of some B^*. Such a phenomenon is completely lost in the phenomenological theory whose curve reaches the exponential asymptote from below, ignoring the backward geminate transfer. This evident drawback confirms that the reduction of the integral theory to any rate description is inappropriate when the lifetimes of the excited states are different. The sole exception is considered below.

E. Exact Solution at $\tau_A = \tau_B = \tau$

This is the unique case when the differential equations

$$\dot{N}_A^* = ck_B(t)N_B^* - ck_A(t)N_A^* - \frac{N_A^*}{\tau} \qquad (3.173a)$$

$$\dot{N}_B^* = -ck_B(t)N_B^* + ck_A(t)N_A^* - \frac{N_B^*}{\tau} \qquad (3.173b)$$

are not only correct but even better than their integral analogs. The auxiliary equations (3.143) and (3.167) in this case become identical, although their initial conditions remain different. Using their solutions in Eq. (3.166), one can express the forward and backward rate constants through the rate constant for the irreversible energy transfer $k(t)$ and the equilibrium constant (3.94) [136]:

$$k_A(t) = \frac{K}{1+K}k(t) \qquad \text{and} \qquad k_B(t) = \frac{1}{1+K}k(t) \qquad (3.174)$$

Thus, the problem of reversible reaction is reduced to the irreversible one, but with the effective rate constant replacing (3.34):

$$k(t) = \int [W_A(r) + W_B(r)]n_{eff}(r,t)d^3r \qquad (3.175)$$

Here $n_{eff}(r,t)$ continues to obey Eq. (3.35) for the irreversible remote quenching, but with the effective rate $[W_A(r) + W_B(r)]$ substituted for $W(r)$.

  Substituting the rate constants (3.174) into Eqs. (3.165), we get the well-known result that describes the redistribution of the energy between states A and B during their decay [137]

$$\frac{P_A^*(t)}{P_A^*(0)} = \frac{1}{1+K} + \frac{K}{1+K}P(t) = 1 - \frac{P_B^*(t)}{/P_A^*(0)} \qquad (3.176)$$

where

$$P(t) = \exp\left(-c\int_0^t k(t')dt'\right) \qquad (3.177)$$

The surprising thing is that this very result was shown to be exact for the "target problem" when energy redistribution in the immobile active molecule is induced by independently moving point quenchers [138]. Unlike IET, it is valid for their arbitrary concentrations. At $\tau_A = \tau_B$, there is no price for the reduction of the integral equations (3.162) to their differential analog (3.173) or (3.165). Quite the opposite—we got as a prize the most accurate description of the phenomenon, although only for a specific target problem.

Although the lifetimes are equal, the induced redistribution of the energy removes part of it from the initially excited state, producing the fluorescence quenching. Its relative quantum yield is

$$\eta = \int_0^\infty \frac{P_A^*(t)e^{-t/\tau}}{P_A^*(0)\,\tau}\,dt = \frac{1}{1+K} + \frac{K}{1+K}\,\eta_{eff} \qquad \eta_{eff} = \frac{1}{1+ck_q\tau} \qquad (3.178)$$

where k_q is the quenching constant, while

$$\eta_{\text{eff}} = \int_0^\infty P(t)e^{-t/\tau}dt/\tau = \frac{1}{1+c\kappa\tau} \qquad (3.179)$$

From Eqs. (3.178) and (3.179) it is inferred that the Stern–Volmer constants for the reversible and effective irreversible reactions are related to each other as follows:

$$k_q(c) = \frac{\kappa(c)K/(1+K)}{1+c\tau\kappa(c)/(1+K)} \qquad (3.180)$$

In the low-concentration limit $(c \to 0)$, one can obtain in contact approximation

$$\kappa(0) = \frac{1}{\tau}\int_0^\infty e^{-t/\tau}k(t)\,dt = \frac{k_a + k_b}{1 + (k_a + k_b)/k_D\left(1 + \sqrt{\tau_d/\tau}\right)} \qquad (3.181)$$

in full accordance with Eq. (3.13). Substituting this quantity into Eq. (3.180), we obtain

$$k_q(0) = \frac{K_{AB}}{1 + cK_{BA}\tau} \qquad (3.182)$$

where $K_{AB} = k_a\tilde{\mathscr{F}}(0)$ and $K_{BA} = k_b\tilde{\mathscr{F}}(0)$ are expressed through $\tilde{\mathscr{F}}(s)$ defined in Eq. (3.126). This is exactly the result that follows from Eqs. (3.168) at $\tau_A = \tau_B = \tau$; that is, the IET provides the right Stern–Volmer constant, but at the lowest concentrations, as usual.

A different situation arises if one calculates the same constant in the superposition approximation (SA) competing with encounter theory. This approximation was first employed in Refs. 139 and 140 and then applied to a number of particular problems [141–145]. However, the present problem is an exceptional one because it has an exact solution that allows the examination of all others. To make this possible, the original SA was employed for this very problem in Ref. 136, and the following Stern–Volmer constant was obtained:

$$k_q^{SA}(c) = \frac{K_A^s}{1 + cK_B^s\tau} \qquad (3.183)$$

It is very similar to IET result (3.182), but K_A^s and K_B^s do not coincide with neither K_{AB} and K_{BA} nor with stationary rate constants $k_A(\infty)$ and $k_B(\infty)$ from Eqs. (3.166). In superposition approximation $K_A^s = \lim_{t\to\infty} K_A(t)$ and $K_B^s = \lim_{t\to\infty} K_B(t)$ are the stationary values of the "reaction coefficients"

$$K_A(t) = \int W_A(r)\rho_A(r,t)d^3r \qquad \text{and} \qquad K_B(t) = \int W_B(r)\rho_B(r,t)d^3r \qquad (3.184)$$

which replace $k_A(t)$ and $k_B(t)$ in the differential SA equations analogous to Eqs. (3.182). Not only the definitions (3.184) differ from their DET analogs (3.175), but the auxiliary equations for ρ_A and ρ_B contain additional terms with quotients dependent on the macroscopic concentrations N_A^* and N_B^*. Thus the pair problem is not isolated from the multiparticle one, and the reaction coefficients acquire the concentration dependence.

In this respect the SA equations are rather unnatural and their distribution functions ρ_A and ρ_B describe pairs surrounded by other particles rather than isolated pairs as n and m in encounter theory. On the other hand, the macroscopic SA equations include the pumping term $I(t)$ that allows for their stationary solution, which is incorrect in differential encounter theory. The accuracy of this solution inspected in Ref. 133 was found to be satisfactory for the irreversible quenching. Unfortunately, for the reversible transfer, this is not the case.

Unlike stationary rate constants, $k_A(\infty)$ and $k_B(\infty)$, the reaction coefficients K_A^s and K_B^s are concentration-dependent. Even in the limit $c \to 0$, they tend to the following:

$$K_A^s \to \text{const}, \qquad cK_B^s \to \text{const} \qquad (3.185)$$

The divergency of K_B^s with $c \to 0$ is astonishing, but this does not complicate calculation of the Stern–Volmer constant (3.183) since it contains only the product cK_B^s, which remains finite. In the contact approximation and low-concentration limit these calculations furnished the following result [136]

$$k_q^{SA}(0) = \frac{k_a}{1 + (k_a + k_b)/k_D Q} \tag{3.186}$$

where Q obeys the following functional equation:

$$Q = 1 + \sqrt{\frac{\tau_d}{\tau}\left(1 + \frac{k_b}{k_D Q}\right)} \tag{3.187}$$

The SA expression (3.186) coincides with its IET analog K_{AB} only in the trivial case of irreversible energy transfer ($k_b = 0$) when $Q = \left(1 + \sqrt{\tau_d/\tau}\right)$ and the $K_{AB} = \kappa_0$.

It is relevant to note that the SA approach is based on a rather complex set of nonlinear equations. This is why the various additional approximations based on the timescale separation were made [145]. They eventually led to new equations differing from the original ones. Moving on to the low-density limit, Naumann and Molski have "simplified" the general SA equations [145], getting rid of terms in the pair equations that are directly proportional to c. Such a procedure is questionable since some of the reaction coefficients in the superposition approximation diverge as $c \to 0$, as in Eq. (3.185). Moreover, another simplification suggested by Naumann and Molski presumed that the quotient $N_A(t)/N_B(t)$ does not depend on time [145]. Only in this case the original pair distribution functions of SA can be represented through the conventional DET distributions, n and m, which do not depend on the macroscopic concentrations. As a result, the simplified version of SA reduces after certain additional manipulations to the differential encounter theory for reversible transfer, between the equally unstable reactants [136]. The corresponding ("phenomenological") theory for stable reactants does not differ from DET as well [146], because there is no difference between rate constants (3.170) and (3.171) when $\tau_A = \tau_B = \infty$. However, the successful reduction of these theories to DET reveals that the original superposition approximation, when not subjected to any "simplifications", is imperfect.

V. GEMINATE RECOMBINATION

Most of the experimental studies of irreversible bimolecular ionization (3.52) break down into two parts accordingly to the methods and goals of the investigation. One part concentrates on the stationary or time-resolved fluorescence of reactants that can be analyzed with a single kinetic equation for the excitation concentration like Eq. (3.2). The other part is confined to the reaction products only. Their geminate recombination and separation may be represented by the extended scheme substituting the RHS of Eq. (3.52):

$$[D^+ \cdots A^-] \longrightarrow D^+ + A^- \qquad (3.188)$$
$$\downarrow W_R$$
$$[D \cdots A]$$

This reaction is usually considered in its own right, separately from the precursor photoionization. The latter creates the initial conditions for the pair distribution $m(r,t)$ which describes the time and space evolution of the geminate pair $[D^+ \cdots A^-]$. All theories of this reaction differ from one another by choice of initial conditions and the model of the reaction layer where recombination takes place.

A. Exponential Model

The most primitive but popular "exponential model" (EM) implies that the recombination occurs within the transparent reaction sphere where the ions are born [Fig. 3.22(a)]. The backward electron transfer to the ground state proceeds there with the uniform rate k_{-et}, but some ions escape recombination leaving the sphere, due to encounter diffusion that finally separate them. EM ascribes to this process the rate

$$k_{\text{sep}} = \frac{3r_c\tilde{D}}{\sigma^3[e^{r_c/\sigma} - 1]} \qquad (3.189)$$

Here $\tilde{D}$ is the coefficient of the encounter diffusion of the ions and $r_c = e^2/\epsilon T$ is the Onsager radius of the Coulomb well in solvent with dielectric constant ϵ. By

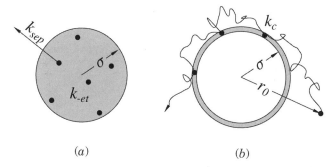

(a)   (b)

Figure 3.22. The schematic representation of (a) the exponential model (EM) and (b) contact approximation. In EM the reaction sphere of radius σ is transparent for particles, which leave it by a single jump with the rate k_{sep}. In contact calculations, the same sphere surrounds an excluded volume and recombination takes place only at its surface, or more precisely in a narrow spherical layer around it.

definition, k_{sep} is the inverse time of diffusional escape from whichever is larger, the Coulomb well (at $r_c > \sigma$) or the sphere of birth (at $r_c < \sigma$). Using such a "rate" in the EM, one actually implies that the ion separation occurs by a single jump from the reaction sphere to infinite distance. Thus EM ignores the fact that the diffusional charge separation proceeds nonexponentially and includes a number of re-contacts before definitive separation.

Against all odds, this concept constitutes the formal basis of the exponential model. The EM kinetic equations are written for the survival probabilities of ions in the reaction sphere Ω_{in} and out of it Ω_{out}:

$$\frac{d\Omega_{in}}{dt} = -k_{-et}\Omega_{in} - k_{sep}\Omega_{in} \tag{3.190}$$

$$\frac{d\Omega_{out}}{dt} = k_{sep}\Omega_{in} \tag{3.191}$$

The initial conditions are $\Omega_{in}(0) = 1$ and $\Omega_{out} = 0$ provided the photoionization is instantaneous. The total survival probability is given by

$$\Omega = \Omega_{in} + \Omega_{out} \tag{3.192}$$

This probability decreases exponentially with time, approaching the charge separation quantum yield $\varphi_0 = \Omega(\infty) = \Omega_{out}(\infty)$. The relaxation to this limit normalized to a fraction of recombined ions, $1 - \varphi_0$, can be represented as follows:

$$R_0(t) = \frac{\Omega - \varphi_0}{1 - \varphi_0} = e^{-t/t_e} \tag{3.193}$$

where $1/t_e = k_{-et} + k_{sep}$. Although the exponent $R_0(t)$ owes its name to the model, it can not be justified by a rigorous treatment of the geminate recombination, even in the contact approximation (see Section V.B).

A more complex situation arises with the charge separation quantum yield φ_0. In the exponential model, this is

$$\varphi_0 = \frac{1}{1 + k_{-et}/k_{sep}} = \frac{1}{1 + z/\tilde{D}} \tag{3.194}$$

where

$$z = \frac{k_{-et}\sigma^3}{3r_c}\left(e^{r_c/\sigma} - 1\right) \tag{3.195}$$

In most cases the only measured quantity is φ_0, which allows specifying z, provided the encounter diffusion coefficient of ions, $\tilde{D}$, is known. It is precisely z that is usually studied as a function of the recombination free energy at contact

$\Delta G_r = \Delta G_R(\sigma)$. Another question is, whether $z/\tilde{D}$ can be considered as a ratio k_{-et}/k_{sep}, because neither of this rates exists in reality. Despite this fact, the free energy dependence inherent in z is attributed to $k_{-et}(\Delta G_r)$ in the vast majority of experimental works. This is why the later theories also present the charge separation quantum yield in the same way:

$$\varphi = \frac{1}{1 + Z/\tilde{D}} \qquad (3.196)$$

although Z is not a constant like z but depends on diffusion and the initial charge separation.

B. Contact Approximation

In the contact approximation (CA) the electron transfer proceeds in the thin reaction layer adjacent to the nontransparent sphere of radius σ [Fig. 3.22(a)]. However, the ions do not necessarily start from there as in EM. When their initial separation exceeds σ, the ions do not recombine until they are delivered to the contact by encounter diffusion. Their distribution obeys the diffusional equation similar to (3.19) but for ions in the Coulomb well:

$$\dot{m} = \frac{\tilde{D}}{r^2}\frac{\partial}{\partial r}r^2\left(\frac{\partial m}{\partial r} + \frac{r_c}{r^2}m\right) = \hat{L}^+ m \qquad (3.197)$$

The reaction is switched on by the boundary condition equivalent to (3.20):

$$4\pi\tilde{D}r^2\left(\frac{\partial m}{\partial r} + \frac{r_c}{r^2}m\right)\Bigg|_{r=\sigma} = k_c m(\sigma, t) \qquad (3.198)$$

where the contact reaction constant

$$k_c = \int_\sigma^\infty W_R(r)4\pi r^2 dr \qquad (3.199)$$

Note that generally speaking k_c differs from the kinetic rate constant (3.18)

$$k_0 = \int_\sigma^\infty W_R(r)\exp\frac{r_c}{r}d^3r \approx k_c \exp\frac{r_c}{\sigma} \qquad (3.200)$$

1. Kinetics of Geminate Recombination

The set of EM equations (3.90) and (3.91) is replaced in CA by a single equation (3.197) which accounts for all the interior distances r. Instead of the sum

$\Omega_{in} + \Omega_{out}$, the total fraction of ions survived by time t is

$$\Omega(t) = \int m(r,t)d^3r \qquad (3.201)$$

provided their initial distribution $m(r,0) = f_0(r)$ is normalized: $\int f_0(r)d^3r = 1$. It is remarkable that Eqs. (3.197) and (3.198) for $m(r,t)$ are in fact analogs to DET equations (3.19) and (3.20) for $n(r,t)$. However, n is dimensionless and $n(r,0) = 1$ while m is the space distribution of the counterions normalized at the beginning.

If, for instance, the ions in all pairs are initially separated by the same starting distance r_0, their distribution at $t = 0$ is

$$m(r,0) = \frac{\delta(r - r_0)}{4\pi r^2} \qquad (3.202)$$

and $m(r,t)$ is actually the Green function of Eq. (3.197). The latter was obtained in the classical work of Hong and Noolandi [19] and used for the exhaustive kinetic study of geminate recombination in Ref. 20. Both works are well reviewed in Ref. 32, so that only the main features of the phenomenon will be quoted here.

First, the kinetics of charge recombination/separation $\Omega(t)$ is never exponential as in EM. In solvents of low polarity it is quasiexponential at the very beginning until the Coulomb well is devastated. The end of the process is universal and nonexponential at any polarity [19]:

$$\Omega(t) = \varphi(r_0)\left[1 + \frac{\exp(r_c/\sigma)}{\exp(r_c/\sigma) + 4\pi\tilde{D}r_c/k_c - 1}\sqrt{\frac{r_c^2}{\pi\tilde{D}t}}\right] \qquad \text{at } t \gg \rho^2/\tilde{D}$$

Here ρ is the greatest from σ, r_0, and r_c and

$$\varphi(r_0) = \frac{1 + (k_c/4\pi r_c\tilde{D})e^{r_c/\sigma}\left[e^{-r_c/r_0} - e^{-r_c/\sigma}\right]}{1 + (k_c/4\pi r_c\tilde{D})[e^{r_c/\sigma} - 1]} \qquad (3.203)$$

In highly polar solvents ($r_c = 0$) this expression is greatly simplified:

$$\varphi(r_0) = 1 - \frac{k_c}{k_D + k_c}\frac{\sigma}{r_0} = 1 - \frac{R_Q}{r_0} \qquad (3.204)$$

It should be emphasized that R_Q is the very same radius that enters the stationary rate constant $k = 4\pi R_Q D$ specified by the contact formula (3.22) with $k_0 = k_c$.

a. Contact Start. If not only recombination, but also the preceding ionization is contact, then $r_0 = \sigma$, and instead of the general expression (3.203) we have

$$\varphi(\sigma) = \frac{1}{1 + (k_c/4\pi r_c \tilde{D})[e^{r_c/\sigma} - 1]} = \varphi_c \qquad (3.205)$$

It is not surprising that this result can be identified with the EM expression (3.194), provided

$$z = \frac{k_c}{4\pi r_c}\left(e^{r_c/\sigma} - 1\right) \qquad (3.206)$$

can be set equal to the EM parameter (3.195). From such an equality the following relationship between EM and CA parameters emerges:

$$k_c = k_{-et}\frac{4\pi}{3}\sigma^3 \qquad (3.207)$$

From this relationship we can elucidate the physical meaning of the phenomenological parameter k_{-et}. If $W(r)$ is an exponential function (3.53) with $l \ll \sigma$, then according to Eq. (3.199) $k_c \approx W_c 2\pi\sigma^2 l$ and

$$k_{-et} = \frac{3l}{2\sigma}W_c \qquad (3.208)$$

The geometrical factor in this relationship is due to the different shape of the reaction zone in the contact and exponential models. This is the spherical layer of width $l/2$ in the former and the entire sphere of radius σ in the latter. The geometric factor, which is the ratio of these two values, accounts for an objective difference between the models, but does not affect the free-energy dependence, which is the same for k_{-et} and k_c as well as for W_c.

b. Subdivision of Geminate Reactions. Although the shape of the reaction zones in EM and CA varies, the recombination efficiency z is nevertheless diffusion independent constant in both theories provided the ions start from inside this zone. This is not the case if they start from the outside. The general CA expression (3.203) can be represented as in Eq. (3.196), but the recombination efficiency

$$Z = \frac{qz}{1 + (1 - q)z/\tilde{D}} \qquad (3.209)$$

depends on the starting distance through $q(r_0)$:

$$q = \frac{1 - \exp(-r_c/r_0)}{1 - \exp(-r_c/\sigma)} \qquad (3.210)$$

Moreover, it is not a constant at any $r_0 > \sigma$, but monotonously increases with diffusion and levels off approaching the plateau qz, which is lower the larger the initial separation of ions r_0.

This was the rationale for subdividing the geminate reactions into the same classes as bimolecular reactions: diffusional and kinetic [147]. The former are accelerated by diffusion, while the latter do not depend on it:

$$Z = \begin{cases} \frac{q}{1-q}\tilde{D} & \text{diffusional geminate recombination} & \tilde{D} \ll (1-q)z \\ qz & \text{kinetic geminate recombination} & \tilde{D} \gg (1-q)z \end{cases} \quad (3.211)$$

The geminate recombination is actually controlled by diffusion, if the initial separation of ions is so large that their transport from there to the contact takes more time than the reaction itself. The exponential model excludes such a situation from the very beginning, assuming that ions are born in the same place where they recombine. Thus, EM confines itself to the kinetic limit only and fixes $Z = z = $ const. The kinetic recombination in the contact approximation does not imply that the starts are taken from the very contact. If they are removed a bit and diffusion is fast, the recombination is also controlled by the reaction and its efficiency $Z = qz$ is constant although smaller than in EM.

The viscosity dependence of Z studied in Ref. 148 was for a long time the sole source of such information. As seen from Figure 3.23, Z is not a constant for sure. On this basis this dependence was attributed in Ref. 147 to the

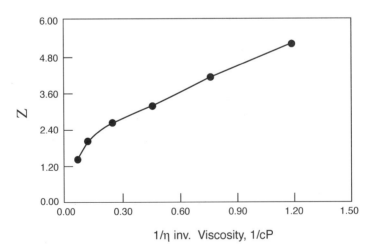

Figure 3.23. The viscosity dependence of Z (arbitrary units) that represents the diffusional acceleration of geminate recombination. The points are the experimental data for recombination of $Ru(bpy)^{2+}$ with MV^{2+}. (From Refs. 147 and 148.)

diffusional geminate recombination, and its nonlinearity was explained later as a result of the partial suppression of recombination by the spin conversion [28,149,150]. Then it appeared that the situation is much more complex and does not reduce to the simple alternative (3.211) provided by CA. At high exergonicity of electron transfer it is not contact (Section II.D), and the true viscosity dependence of Z is not monotonous [151]. Moreover, it can be subjected to strong distortion by a parallel change in the refraction index and dielectric constant accompanying the viscosity variation [152]. As we shall see in Section VII.D, these contingencies qualitatively change the interpretation of data shown in Figure 3.23.

2. Polar Solvents

The particular case of highly polar solvents is of special interest since it provides wider opportunities for analytical study of the phenomenon. If the Onsager radius is smaller than σ, it can be set to zero, thus switching off the electrostatic interaction. Then for polar solvents we obtain from (3.203)

$$\varphi = \frac{\varphi_c + \delta}{1 + \delta}, \qquad \varphi_c = \frac{1}{1 + k_c/k_D} \tag{3.212}$$

where $\delta = r_0/\sigma - 1$ and φ_c. This result is identical to the EM estimate of charge separation quantum yield, provided the relationship (3.207) holds. Instead of the exponential relaxation peculiar to EM, we now have a much more complex but exact result obtained with CA [20]

$$R(t) = \frac{\Omega(t) - \varphi}{1 - \varphi} = 1 + e^{-\delta^2/4\tau}\left[\mathrm{er}\left(\frac{\delta}{2\sqrt{\tau}} + \frac{\sqrt{\tau}}{\varphi_c}\right) - \mathrm{er}\left(\frac{\delta}{2\sqrt{\tau}}\right)\right] \tag{3.213}$$

where $\tau = \tilde{D}t/\sigma^2$, $\mathrm{er}(x) = \exp(x^2)\mathrm{erfc}(x)$ and $\mathrm{erfc}(x)$ is a complementary error function. Since there is no more Coulomb well, the relaxation kinetics is nonexponential from the very beginning.

Even for contact start, $r_0 = \sigma$, it differs qualitatively from the exponential law (3.193):

$$R(t) = e^{t/t_0}\mathrm{erfc}\sqrt{t/t_0} = \begin{cases} 1 - \sqrt{4t/\pi t_0} & \text{at } t \ll t_0 \\ \sqrt{t_0/\pi t} & \text{at } t \gg t_0 \end{cases} \tag{3.214}$$

The relaxation time of this process

$$t_0 = \tau_d \varphi_c^2 \tag{3.215}$$

is proportional to the square of the EM separation yield and to the encounter time (3.28). Not only is the time development of $R(t)$ nonexponential but the relaxation time t_0 also differs qualitatively from its EM analog t_e. From their comparison made in Refs. 20 and 32, it is clear that the EM is unsuitable for polar solvents. The situation is a bit different in nonpolar solvents because the kinetics of the particle escape from the deep Coulomb well is approximately exponential at the beginning [20]. However, the long-time kinetics, which is the same for any polarity, is essentially nonexponential. The best analytical treatment of contact charge recombination in solutions of an arbitrary polarity was given in Ref. 153.

Although the contact approximation is a good alternative to EM for low-viscosity solvents, at higher viscosities and especially in the static limit ($\tilde{D} = 0$), CA becomes inappropriate. In this limit both φ_0 and φ_c turn to zero, but $\varphi(r_0)$ does not, if $r_0 \neq \sigma$. Particles that are separated by the distance $r_0 > \sigma$ survive because there is no diffusion to bring them in contact where the reaction takes place. Of course, this is nonsense if electron transfer is remote. Sooner or later it involves in the reaction all immobile partners wherever they are, turning $\varphi(r_0)$ to zero.

C. Remote Geminate Recombination

Nonmodel formulation of the problem is given by the kinetic equation for remote electron transfer, which substitutes the CA equation (3.197):

$$\dot{m} = -W_R(r)m + \frac{\tilde{D}}{r^2}\frac{\partial}{\partial r}r^2 e^{r_c/r}\frac{\partial}{\partial r}e^{-r_c/r}m \qquad (3.216)$$

and reflecting boundary condition to it instead of (3.198):

$$\left(\frac{\partial m}{\partial r} + \frac{r_c}{r^2}m\right)\Bigg|_{r=\sigma} = 0 \qquad (3.217)$$

The share of radical ion pairs (RIPs) that survived by instant t is $\Omega(t|r_0) = \int m(r, r_0, t)d^3r$. It obeys the following equation conjugated to Eq. (3.216) [154,155]:

$$\frac{\partial}{\partial t}\Omega(t|r) = -W_R(r)\Omega(t|r) + \frac{\tilde{D}e^{-r_c/r}}{r^2}\frac{\partial}{\partial r}r^2 e^{r_c/r}\frac{\partial}{\partial r}\Omega(t|r) \qquad (3.218)$$

with the following initial and boundary conditions:

$$\Omega(0|r) = 1, \qquad \frac{\partial\Omega(t|r)}{\partial r}\Bigg|_{r=\sigma} = 0, \qquad \Omega(t|\infty) = 1 \qquad (3.219)$$

The quantum yield of the charge separation

$$\varphi(r_0) = \lim_{t\to\infty}\Omega(t|r_0) = \Omega(\infty|r_0) \qquad (3.220)$$

may be found as a stationary solution to Eq. (3.218):

$$\frac{\tilde{D}e^{-r_c/r_0}}{r_0^2}\frac{\partial}{\partial r_0}r_0^2 e^{r_c/r_0}\frac{\partial}{\partial r_0}\varphi(r_0) = W_R(r_0)\varphi(r_0) \tag{3.221}$$

with the following boundary conditions:

$$\left.\frac{\partial\varphi(r_0)}{\partial r_0}\right|_{r_0=\sigma} = 0 \qquad \varphi(\infty) = 1 \tag{3.222}$$

The general view of $\varphi(r_0)$ obtained numerically for the exponential $W_R(r_0)$ is shown in Figure 3.24 (top). It is only slightly less than 1 in the case of kinetic recombination but is much smaller in the case of diffusional recombination especially in the dip near the contact.

In the nearest vicinity of contact the recombination is static and most pronounced but in the rest of the space it can be described by the following universal formula:

$$\varphi(r_0) = 1 - \frac{R_Q}{r_0} \tag{3.223}$$

This is the straightforward generalization of Eq. (3.204) for noncontact transfer. Equation (3.221) for $\varphi(r_0)$ is identical to Eq. (3.39) for $n_s(r)$, and the boundary conditions for both of them are also the same. Therefore the asymptotic formula (3.42) holds also for the charge separation yield, rewritten as Eq. (3.223). This approximation, linear in the coordinates of Figure 3.24 (bottom), works well everywhere except in the static plateau. The effective radius R_Q found as a slope of the linear dependence is larger than $\sigma = 7\,\overset{\circ}{A}$ in the case of diffusion control ($R_Q = 10\,\overset{\circ}{A}$) and smaller than σ under kinetic control ($R_Q = 2\,\overset{\circ}{A}$).

1.   Normal Region

If the recombination is weakly exergonic ($-\Delta G_r < \lambda_c$), then the appropriate model for $W_R(r)$ is the exponential function (3.53). With this transfer rate the problem was rigorously solved at $r_c = 0$ [21,32,156]:

$$\varphi(r_0) = \frac{2\lambda}{1+\delta}\left\{K_0(\alpha) - I_0(\alpha)\frac{K_0(\alpha_0) - \sqrt{x/\lambda}K_1(\alpha_0)}{I_0(\alpha_0) + \sqrt{x/\lambda}I_1(\alpha_0)}\right\} \tag{3.224}$$

Here K_0, I_0, K_1, and I_1 are the modified Bessel functions [157] $\lambda = l/2\sigma$:

$$\alpha = 2\sqrt{\lambda x}e^{-\delta/2\lambda}, \qquad \alpha_0 = 2\sqrt{\lambda x}, \qquad \text{and} \qquad x = \frac{k_c}{k_D}$$

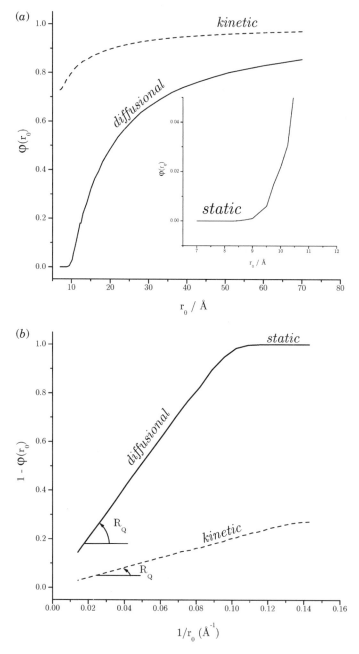

Figure 3.24. The charge separation quantum yield in polar solvents ($r_c = 0$) as a function of the starting distance r_0 for the exponential recombination rate ($W_c = 100\,\text{ns}^{-1}$, $l = 1\,\text{Å}$, $k_0 = 2.14 \times 10^{10}\,\text{m}^{-1}\text{s}^{-1}$) and the two diffusion coefficients: $\tilde{D} = 10^{-4}\,\text{cm}^2/\text{s}$ (- - -) and $\tilde{D} = 10^{-8}\,\text{cm}^2/\text{s}$ (——).

The latter parameter discriminates between the fast diffusion (kinetic) limit ($x \ll 1$, $\varphi \approx 1$) and the slow reaction (diffusional) limit ($x \gg 1$, $\varphi \ll 1$). In addition to these two limits, there is also the third one arising when

$$\tilde{D} \ll l^2 W_R(r_0)$$

In this limit geminate recombination becomes quasistatic and can be accomplished before the particles leave their initial positions. This is the static limit missed in the contact approximation:

$$\varphi(r_0) \approx \frac{\lambda}{1+\delta}\sqrt{\frac{2\pi}{\alpha}}e^{-\alpha} \quad \text{at} \quad \frac{k_c}{k_D} \gg \frac{e^{\delta/\lambda}}{4\lambda} \gg 1$$

As $\tilde{D} \to 0$, this quantity turns to zero unlike the CA expression (3.212), which remains finite at any $\delta \neq 0$.

The essential difference in the diffusional dependence of the true separation yield and its contact estimate is illustrated in Figure 3.25. It is convenient to use φ_c as an argument (instead of $\tilde{D}$) because it varies in a limited region. The geminate reaction is kinetic at $\varphi_c \approx 1$, diffusional at $\varphi_c \ll 1$, and static at

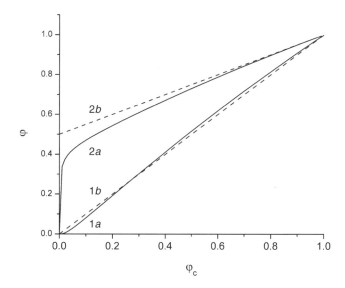

Figure 3.25. The dependence of the quantum yield $\varphi(r_0)$ on the contact quantum yield φ_c. The solid lines (*a*) indicate the remote electron transfer ($l = 0.2\sigma$); dashed lines (*b*) represent the same but in the contact approximation. Curves 1 account for the contact start ($r_0 = \sigma$); curves 2 relate to the start from the outside of the reaction layer ($r_0 = 2\sigma$). (From Ref. 21.)

$\varphi_c = 0$. As was shown, there is no difference between CA and EM if ions start from the contact: the identity $\varphi_0 = \varphi_c$ is represented by a dashed line $1b$ in Figure 3.25. At slow diffusion EM overestimates the charge separation yield whose true value calculated from Eq. (3.224) is shown in the same figure by the solid line $1a$. At fast diffusion everything is quite the reverse.

This is even more so for noncontact starts, especially for slow diffusion ($\varphi_c \to 0$). As was mentioned, CA leads to the finite value of the charge separation yield in the static limit when $\varphi_c = \tilde{D} = 0$. In highly polar solvents this quantity, $\lim_{\tilde{D}=0} \varphi(r_0) = \delta/(1 + \delta)$, is represented by the intersection of the dashed line $2b$ (drawn for $\delta = 1$) with ordinate (Fig. 3.25). Instead, the remote recombination theory provides the following formula, deduced in the quasistatic limit from the general Eq. (3.224):

$$\varphi(r_0) \approx \frac{\lambda}{1 + \delta} \sqrt{\frac{2\pi}{\alpha}} e^{-\alpha} \tag{3.225}$$

In full agreement with the well-grounded expectation, this quantity turns to zero at $\tilde{D} = \varphi_c = 0$ because $\alpha \to \infty$ (solid line $2a$ in Fig. 3.25). Although fundamental, this discrepancy between CA and the exact solution is confined to the very short strip in the highly viscous region. The rest of the curve is quasilinear and similar to the straight dashed line of CA.

It is also of importance that the curvature of curves $2a$ and $1a$ in Figure 3.25 is the opposite at slow diffusion. Calculating from them the recombination efficiency Z, one can see that there is a principal difference between the remote and contact starts [Fig. 3.26(a)]. When ions were born far from each other, their recombination is accelerated by fast diffusion, which brings them into the reaction zone. This is true for any D when ions are born outside the restricted (rectangular) reaction zone. In contrast, the contact-born ions are in reaction zone from the very beginning and diffusion can only hinder recombination bringing them out of there. Therefore the corresponding Z monotonically decreases with diffusion approaching the fast diffusion limit Z_0. At very slow diffusion this effect is inherent in a remote start as well, because the ions are in fact within the exponential reaction zone wherever they are. However, the descending branch of the curve gives way to the ascending one sooner, the larger the initial separation of ions. Only then does the contact approximation become reasonable.

2. Inverted Region

In the inverted region the recombination layer is first extended and then shifted from the contact as $|\Delta G_r|$ increases (Fig. 3.5). The bell-shaped $W_R(r)$ can be approximated by the rectangular model of the layer (3.55), whose inner border is $r_i = R - \Delta$ and external border is $r_e = R + \Delta$ (Fig. 3.6). In highly polar solvents

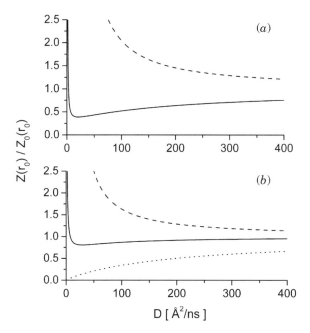

Figure 3.26. Efficiency of geminate recombination, related to its value at infinite diffusion Z_0 for the exponential (*a*) and rectangular (*b*) models of the reaction rate. Dashed lines—contact starts; solid lines—noncontact starts from reaction zones; dotted line—start from outside the rectangular zone.

$r_c = 0$ and the charge separation yield obeys the equation following from Eq. (3.221):

$$\frac{\tilde{D}}{r_0^2}\frac{\partial}{\partial r_0} r_0^2 \frac{\partial}{\partial r_0}\varphi(r_0) = W_R(r_0)\varphi(r_0) \qquad (3.226)$$

This provides an exact expression for charge separation yield [158]

$$\varphi(r_0) = \begin{cases} 2/\Phi_-(L) & r_0 < r_i \\ \Phi_+(r_0 - r_i)/\Phi_-(L)qr_0 & r_i < r_0 < r_e \\ 1 - r_e/r_0 + \Phi_+(L)/\Phi_-(L)qr_0 & r_e < r_0 \end{cases} \qquad (3.227)$$

where $r_e - r_i = 2\Delta = L$ and

$$\Phi_\pm(\xi) = (qr_i + 1)e^{q\xi} \pm (qr_i - 1)e^{-q\xi}, \qquad q = \sqrt{W_0/\tilde{D}} \qquad (3.228)$$

At $qL \gg 1$ these expressions reduce to those obtained in Ref. 159.

The remarkable feature of this phenomenon is that the quantum yield does not depend on the initial separation, as long as the latter is less than the inner radius of the remote recombination layer r_i. This layer screens the ions that start from its interior. The smaller is the separation quantum yield, the faster is the recombination inside the layer. However, the survival probability of ions sharply increases when the starting point is shifted outside (Fig. 3.27).

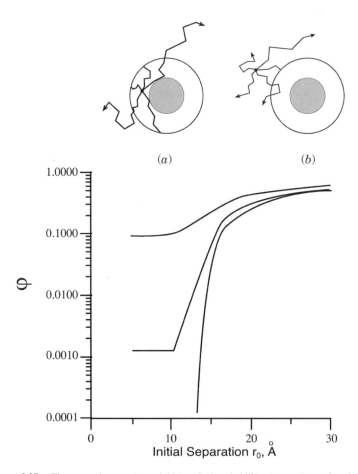

Figure 3.27. The separation quantum yield (survival probability at $t = \infty$) as a function of the initial distance between the ions for $\tilde{D} = 10^{-5}\,\text{cm}^2/\text{s}$ and three recombination rates (from top to bottom), $W_0 = 10, 100, 1000\,\text{ns}^{-1}$ ($\sigma = 5\,\text{Å}, r_i = 10\,\text{Å}, L = 5\,\text{Å}$). At the top: (a) the start from inside the recombination layer related to the left, horizontal branches of the curves; (b) the outside start related to the right branches approaching the maximum $\varphi = 1$ at $r_0 \to \infty$. (From Refs. 32 and 158.)

When diffusion becomes slower, the residence time in the recombination layer increases and the layer becomes nontransparent for particles that start from inside; that is, for inner starts Z universally decreases with diffusion [Fig. 3.26(b)]:

$$Z(r_0) = z \frac{\cosh(qL) + qr_i \sinh(qL) - 1}{q^2 LR}, \qquad r_0 < r_i \qquad (3.229)$$

Here $z = W_0 LR$ is identical to the EM parameter provided $W_0 = k_{-et}$. $Z(\tilde{D})$ behavior for inner starts is qualitatively similar for contact starts from the exponential reaction zone. The noncontact starts from the exponential reaction zone are also similar to the starts from inside the rectangular recombination zone. Only for starts outside this zone, $Z(\tilde{D})$ is a monotonically increasing function of the diffusion coefficient [Fig. 3.26(b)]. For thin layers ($qL \ll 1$) this dependence coincides with that predicted by the CA Eq. (3.209). This is a consequence of the artificially sharp borders of the rectangular recombination layer. Ions that are born outside do not react in principle unless diffusion delivers them into the reaction zone.

At any start, the recombination efficiency levels off with the diffusion approaching plateau $Z = Z_0$. This is actually the kinetic limit of recombination. It is reached when the fraction of ions that recombine in the reaction layer during the residence time there, $\tau_e = RL/\tilde{D}$, is small:

$$W_0 \tau_e = \frac{W_0 RL}{\tilde{D}} = \frac{z}{\tilde{D}} = \frac{k_c}{k_D} \ll 1 \qquad (3.230)$$

Under such a condition not only is the contribution of a single diffusional crossing of the layer small ($W_0 L^2/\tilde{D} \ll 1$) but the total composite effect of all R/L crossings is also small. This is the "gray" recombination layer, which is almost transparent for ions crossing it in both direction. However, with increasing viscosity the inequality (3.230) changes the sign and the layer becomes "black." For ions starting from inside, this leaves almost no chance to survive; when diffusion turns to zero, Z tends to infinity. However, for outsiders the limiting stage is the very delivery to the black layer. It is hindered in viscous solution reducing Z.

This is the reason for a qualitatively different diffusional dependence of Z for ions that are born in and out of the recombination layer. However, the initial separation of ions also depends on the diffusion when it controls the ionization. Therefore r_0 cannot be the same for all viscosities, as was implied above. Moreover, the initial distribution of ions is far from being similar to that in Eq. (3.202) and has to be specified separately. All this will be done in the next section.

D.   Unimolecular Reversible Dissociation

Exciplexes are the products of either bimolecular ionization (3.81), or the straightforward light excitation in the charge transfer band of already existing complexes [DA]. The alternative ways of exciplex creation were distinguished experimentally by the different FEG dependences of the resulting charge separation yield [113]. Here we focus on the monomolecular photogeneration of an exciplex, followed by its reversible dissociation to free ions. In the spinless theory the latter is represented by the following reaction scheme:

$$[D^+A^-] \underset{k_a}{\overset{k_d}{\rightleftharpoons}} [D^+ \cdots A^-] \longrightarrow D^+ + A^- \qquad (3.231)$$

$$h\nu \uparrow\downarrow \tau_{\text{exc}} \qquad \downarrow W_R$$

where k_d and k_a are the reaction constants of the contact dissociation/association and τ_{exc} is the time of exciplex decay due to fluorescence and radiationless transitions.

   The spinless theory that we are using here is approximate for electron transfer but absolutely correct for the proton reactions. The widely studied example is the reversible proton transfer from an excited photoacid to the solvent [160–165]

$$R^*OH \underset{k_a}{\overset{k_d}{\rightleftharpoons}} [R^*O^- \cdots H^+] \longrightarrow R^*O^- + H^+ \qquad (3.232)$$

$$h\nu \uparrow\downarrow u \qquad \downarrow u' \qquad\qquad \downarrow u'$$

where u and u' are the decay rates of the excited species. The similarity of the geminate reactions (3.231) and (3.232) allows us to consider them jointly in the framework of the abstract reaction scheme

$$C \Longleftrightarrow A + B \qquad (3.233)$$

where only C or both C and A are in excited electronic states. It is also possible that neither of the particles is excited ($u = u' = 0$) and the scheme (3.233) represents just a reversible dissociation of the ground-state complex. In the text that follows, we will first consider the exciplex dissociation, then the photoacid decay through proton transfer and after all this the dissociation of regular complexes.

1.   Exciplex Dissociation

The exciplex is formed with a particle that enters the first coordination sphere of the excited partner. This sphere is distinctive from others even in pure liquids where it is marked by the strongest peak in the pair distribution function. In this case $k_\pm$ are simply the rates of jumps in and out of the nearest shell. They are different from the exchange rates between other shells used in the theories that consider the encounter motion as a random-walk process. To account for this difference, the modified model of diffusion-controlled reactions (MMDCR) was once proposed [166]. Later on it was shown to be exact in the hopping limit of bimolecular reactions [167]. The main idea was that diffusion leads to the encounter of partners, but the last step is the hopping invasion into the first shell where the contact reaction takes place. If the latter creates the exciplex or other bounded pair the rate of its dissociation is greatly reduced, making $k_d \ll k_a$.

Although MMDCR was addressed to the bimolecular reactions in the bulk, the same ideas were successfully applied to the reversible geminate recombination of an ion pair into the exciplex [29–31]. The survival probability of the instantaneous created exciplex, $P_e(t)$, and the pair distribution at contact, $g(\sigma, t)$, were shown to obey the set of the following integral equations [31]

$$P_e = e^{-t/\tau_E} + k_a \int_0^t e^{-(t-t')/\tau_E} g(\sigma, t') dt' \qquad (3.234a)$$

$$g(\sigma, t) = k_d \int_0^t G(\sigma, t - \tau) P_e(\tau) d\tau \qquad (3.234b)$$

where $1/\tau_E = 1/\tau_{\mathrm{exc}} + k_d$. They are the integrals of the rate equations for reversible dissociation/association used in Refs. 29–32, which relate the whole process to the Green function of the geminate ion pair, $G(r, t)$. Since both creation and dissociation of the exciplex are contact the latter yields the conventional diffusional equation (3.216)

$$\dot{G} = -W_R(r)G + \frac{\tilde{D}}{r^2}\frac{\partial}{\partial r} r^2 \left(\frac{\partial G}{\partial r} + \frac{r_c}{r^2}G\right), \qquad G(r, 0) = \frac{\delta(r - \sigma)}{4\pi r^2} \qquad (3.235)$$

but with the boundary condition accounting for the exciplex formation from the ion pair:

$$4\pi\sigma^2 \tilde{D} \frac{\partial}{\partial r} e^{-r_c/r} G(r, t)\Big|_{r=\sigma} = k_a G(\sigma, t) \qquad (3.236)$$

From the set of equations (3.234), we obtain the single integrodifferential equation for exciplex concentration:

$$\dot{P}_e = -\frac{P_e}{\tau_E} + k_d \int_0^t \mathscr{F}(t - \tau) P_e(\tau) d\tau \qquad (3.237)$$

where the kernel

$$\mathscr{F}(t) = k_a G(\sigma, t) \qquad (3.238)$$

2. Contact Recombination of RIP

If recombination of ions not only into the exciplex but also to the ground state is contact, then the latter can be accounted for by omitting $W(r)G$ in Eq. (3.235) and including term $k_c G$ ($k_c = \int W_R(r)d^3r$) in the rhs of Eq. (3.236). This is the total contact approximation, which is better suited for the transfer of protons than electrons. It allows easy expression of the Green function $G(\sigma, t)$ through the Green function of the free diffusion $G_0(\sigma, t)$:

$$\tilde{G}(\sigma, s) = \frac{\tilde{G}_0(\sigma, s)}{1 + (k_a + k_c)\tilde{G}_0(\sigma, s)} \tag{3.239}$$

The latter obeys the "sinkless" diffusional equation

$$\dot{G}_0 = \frac{\tilde{D}}{r^2}\frac{\partial}{\partial r}r^2\left(\frac{\partial G_0}{\partial r} + \frac{r_c}{r^2}G\right), \qquad \frac{\partial}{\partial r}e^{-r_c/r}G_0(r, t)\bigg|_{r=\sigma} = 0 \tag{3.240}$$

In highly polar solvents ($r_c = 0$) this Green function takes the simplest form [20]:

$$\tilde{G}_0(\sigma, s) = \frac{1}{k_D\left(1 + \sqrt{s\sigma^2/\tilde{D}}\right)} \tag{3.241}$$

Using expressions (3.239) and (3.241) in the Laplace transformation of Eq. (3.238), we get the final expression for the kernel in the contact approximation:

$$\frac{1}{\tilde{\mathscr{F}}(s)} = 1 + \frac{k_D(1 + \sqrt{s\tau_d}) + k_c}{k_a} \tag{3.242}$$

3. Relationship to IET

The kinetic equation (3.237) may be transformed to a form identical to the IET equations [31]

$$\dot{P}_e = -\frac{P_e}{\tau_{exc}} - k_d\int_0^t F(t - \tau)P_e(\tau)d\tau \tag{3.243}$$

where the new kernel has the following Laplace transformation:

$$\tilde{F}(s) = 1 - \tilde{\mathscr{F}}(t) = 1 - \frac{k_a\tilde{G}_0(\sigma, s)}{1 + (k_a + k_c)\tilde{G}_0(\sigma, s)} \tag{3.244}$$

In the case of highly polar solvents ($r_c = 0$) this kernel has the following form:

$$\frac{1}{\tilde{F}(s)} = 1 + \frac{k_a}{k_D(1 + \sqrt{s\tau_d}) + k_c} \tag{3.245}$$

The Laplace transformation of Eq. (3.243) leads to the exact analytic expression identical to that obtained with a different method in Eq. (2.10b) of Ref. 135:

$$\frac{1}{\tilde{P}_e(s)} = s + \frac{1}{\tau_{exc}} + k_d \frac{1 + k_c \tilde{G}_0(\sigma, s)}{1 + (k_d + k_c)\tilde{G}_0(\sigma, s)} \tag{3.246}$$

In the same work Gopich and Agmon subjected to exhaustive analysis the original of $\tilde{P}_e(s)$ in the absence of ion interaction ($r_c = 0$).

4. Photoacid Dissociation

The theory extended to the proton transfer reaction (3.232) is essentially the same, although it uses different designations: $P_e(t) = p(*, t|*)$ and $g(\sigma, t) = p(\sigma, t|\sigma)$ [135,168,169]. It is distinctive in only a single respect—one of the dissociation products is excited and therefore unstable ($u' \neq 0$). This can be easily accounted for by the following modification of Eq. (3.240) valid at $r_c = 0$:

$$\dot{G}_0 = \frac{\tilde{D}}{r^2}\frac{\partial}{\partial r} r^2 \frac{\partial G_0}{\partial r} - u'G_0, \qquad \frac{\partial}{\partial r} G_0(r, t)\bigg|_{r=\sigma} = 0 \tag{3.247}$$

To make an extension of IET for this case, one should just do the following substitution:

$$\tilde{G}_0(\sigma, s) \Longrightarrow \tilde{G}_0(\sigma, s + u') \qquad \text{and} \qquad \frac{1}{\tau_{exc}} = u \tag{3.248}$$

With this rearrangement Eqs. (3.243) and (3.244) are also applicable to the photoacid dissociation kinetics $P_e(t) = [R^*OH] = [C]$ provided $[A] = [B] = 0$ at the beginning. An exact solution of Eq. (3.247) was obtained and exhaustively analyzed in Ref. 135. This is an extension of the solution for the same problem, for the particular case $u = u'$ and $k_c = 0$ [170,171].

5. Long-Time Asymptote of Dissociation

The asymptotic behavior of [C] was analyzed by Agmon and Gopich in the most general case of a different u and u' [135,172]. It was shown that at long times $P_e(t)$ experiences the transition from the power law for stable dissociation products ("the A regime") to an exponential decay for highly unstable products ("the AB regime"). More specifically, the A regime holds at $u' < u + k_{off}$ where $k_{off} = k_d k_D/(k_a + k_c + k_D)$ is the overall dissociation rate constant in highly polar solvents. In other words, A should live longer than C. Otherwise the situation changes to the AB regime. The lifetime interrelationship plays here a crucial role, as in the reversible bimolecular reaction studied in Section IV.

Since ions are stable ($u' = 0$), the exciplex theory deals with a pure "A regime" where the asymptotic behavior obeys the widely recognized power law deduced by Berg [120] and then confirmed by a number of different theoretical methods [135,168,172,173]:

$$P_e(t) \propto Z^2 K_{eq}\left(4\pi \tilde{D}t\right)^{-3/2} \qquad \text{at} \qquad t \to \infty \qquad (3.249)$$

where $K_{eq} = k_a/k_d$ and

$$Z = \frac{k_D k_d \tau_{exc}}{k_a + k_D[1 + k_d \tau_{exc}]}$$

At the beginning the dissociation enhances the exciplex decay, making its rate equal to $1/\tau_E = 1/\tau_{exc} + k_{off}$, but at the very end the survival probability $P_e(t)$ vanishes much more gradually following the power law (3.249). This is caused by the restoration of exciplexes from ion pairs in the course of their recombination/separation, which lasts much longer and backs the delayed fluorescence all this time. In the log–log coordinates of Figure 3.28(a) the power law is represented by the final linear dependence shown by the dotted line. The experimental verification of the power law for geminate dissociation was first made in Ref. 163.

If one uses instead the integral equation (3.243), the equivalent differential equation

$$\dot{P}_e = -\frac{P_e}{\tau_{exc}} - k_{eff}(t)P_e \qquad (3.250)$$

then the true solution of the former can be used in the latter to find an exact time dependence of the effective dissociation rate $k_{eff}(t)$. At any finite τ_{exc} the latter appears to be the sign-altering quantity [31] because the effective rate becomes negative when the ion association into the exciplex dominates over its dissociation [Fig. 3.28(b)]. Qualitatively, this is the same picture as established for the reversible reactions of energy transfer from unstable to stable reactants (Fig. 3.18). Over long times $P_e(t)$ from (3.149) is an integral of Eq. (3.150) provided that

$$k_{eff} = \frac{3}{2t} - \frac{1}{\tau_{exc}} \qquad (3.251)$$

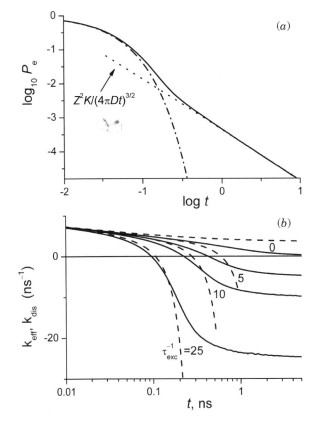

Figure 3.28. (a) The kinetics of exciplex dissociation at $\tau_{exc} = 0.2$ ns, $\sigma = 5$ Å, $k_d = 10^4$ Å^3/ns, $k_D = 5 \cdot 10^3$ Å^3/ns, $k_a = 10$ ns^{-1}. The power-law asymptote (dotted straight line) and the phenomenological kinetics of exciplex dissociation (dashed–dotted line) are shown for the comparison. (b) The time variation of the dissociation rate of DET, k_{dis} (dashed lines), in comparison with the effective rates of IET, k_{eff} (solid lines), at $\tau_{exc} = 0.04$ ns, 0.1, and 0.25 ns, ∞ (from bottom to top). All $k_{dis}(t)$ diverge with t except the upper one related to dissociation of the stable complex.

This rate approaches with time the constant negative value as well as the rate (3.157) at $\tau_B = \infty$.

6. Relationship to Differential Theories

The standard procedure used in Section IIIA can also be employed for the integral equation (3.243) to transform it into the DET equation

$$\dot{P}_e = -\frac{P_e}{\tau_{exc}} - k_{dis}(t)P_e \qquad (3.252)$$

where the time-dependent dissociation rate

$$k_{\text{dis}}(t) = k_d \int_0^t F(t - \tau) e^{t/\tau_{\text{exc}}} d\tau \qquad (3.253)$$

In view of Eq. (3.245), the Markovian limit of this rate, $k_{\text{dis}}(\infty) = k_d \tilde{F}(-1/\tau_{\text{exc}})$, is the complex quantity as well as the rate constant (3.150) at $\tau_B = \infty$. This means that the stationary dissociation of the exciplex does not exist. Unlike $k_{\text{eff}}(t)$, its DET analog, $k_{\text{dis}}(t)$ from Eq. (3.253), diverges after changing the sign, thus indicating that the rate description of the delayed fluorescence is not appropriate at long times.

This is even more true in regard to the "modified" or "phenomenological" rate equations obtained by Lee and Karplus [139] in the same way as their analogs (3.169), discussed in Section IV.D. The same equations were also deduced by Molski and Keizer assuming that the timescale for the chemical change is much shorter than the encounter time $\sigma^2/\tilde{D}$ [174]. Although the phenomenological equations are widely used, [145,175,176] they do not reproduce the long-time dissociation kinetics [see Fig. 3.28(a)]. The phenomenological dissociation rate is always positive and the delayed fluorescence is completely lost. This is the price for the incorrect reduction of the integral equations to differential ones. The critical comparison of the phenomenological and other simplified equations was made by Szabo [63].

7. Dissociation of Stable Complexes

As has been noted, the present theory includes as the simplest particular case the dissociation of the stable complex ($\tau_{\text{exc}} = \infty$) represented by the following reaction scheme:

$$[DA] \underset{k_a}{\overset{k_d}{\rightleftharpoons}} [D \cdots A] \Longrightarrow D + A \qquad (3.254)$$

For such an elementary reaction we have to set $1/\tau_E \equiv k_d$ and $k_c = 0$, reducing Eq. (3.237) to Eq. (3.257) of the classical Berg work [120]:

$$\dot{P}_e = -k_d P_e + k_d \int_0^t \mathscr{F}(t - \tau) P_e(\tau) d\tau \qquad (3.255)$$

Berg pointed out that the asymptotic dissociation obeys the power law identical to Eq. (3.249) with $Z = 1$. This law was confirmed experimentally [162,163] and reproduced a few times later [170,177]. In fact, not only the asymptotic behavior of $P_e(t)$ but also the original integral equation for it derived by Agmon and Szabo [170] was shown to be equivalent to the Berg equation (3.255) [31]. The same is true regarding the Molski–Keizer thermodynamic approach [174] that was developed further [145,176] and finally shown to be identical to IET [175]. When applied to a single complex decay, the corresponding equation is easier to obtain from the IET equation (3.243):

$$\dot{P}_e = -k_d \int_0^t F(t-\tau)P_e(\tau)d\tau \qquad (3.256)$$

This equation is also equivalent to that derived by Berg.

VI.   IRREVERSIBLE PHOTOIONIZATION

If one studies only the fluorescence quenching by irreversible bimolecular ionization (3.52), there is seldom any need to trace the fate of the charged products. On the contrary, those who are interested in photoinduced geminate recombination (3.188) rarely care about the kinetics of ionization, its quenching radius, and all the rest studied in Section III. All that they need to obtain the charge separation yield is the initial ion distribution $m_0(r)$, prepared by photoionization. However, the latter is scarcely so simple as in Eq. (3.201), which is usually favored. Even so, the initial separation r_0 is not a fitting parameter but the characteristic interion distance, which is dependent on the precursor reaction of photoionization.

To avoid any arbitrariness in the choice of this distance or distance distribution $m(r_0)$, one should consider the unified reaction scheme composed of sequential reactions of bimolecular ionization and geminate RIP recombination, followed by bimolecular free-ion recombination in the bulk:

$$D^* + A \Longrightarrow [D^+ \cdots A^-] \Big\langle \begin{array}{l} D^+ + A^- \Longrightarrow D + A \\ [D \cdots A] \end{array} \qquad (3.257)$$

The bimolecular ionization creates the initial conditions for the subsequent geminate reaction in the ion pair. This is the simplest example of the sequential reactions of electron transfer considered by means of IET.

A.   Integral Theory of the Phenomenon

In a two-state basis of donor $\langle D^*|$, $\langle D^+|$ and acceptor $\langle A|$, $\langle A^-|$, we have to define two vectors similar to (3.100):

$$\mathbf{N}_D = \begin{pmatrix} N^* \\ N^+ \end{pmatrix} \quad \text{and} \quad \mathbf{N}_A = \begin{pmatrix} A \\ N^- \end{pmatrix} \qquad (3.258)$$

Here

$$N^*(t) = [D^*], \; A(t) = [A] \qquad \text{and} \qquad N^+(t) = N^-(t) = N^{\pm}$$

are the concentrations of the counterions. The initial conditions for them correspond to an instantaneous excitation with a light pulse, whose integral intensity determines the total number of excited donors N_0:

$$N^*(0) = N_0, \; A(0) = c \qquad \text{and} \qquad N^{\pm}(0) = 0$$

The theory is in principle the same for either of the reactions (3.52). In any of them there is only one excited particle and only two transfer reactions: ionization with the rate $W_I(r)$ and subsequent recombination of ions with the rate $W_R(r)$.

For the particular reaction (3.257), we have in the individual basis

$$\hat{\mathbf{Q}}_D = \begin{pmatrix} -1/\tau_D & 0 \\ 0 & 0 \end{pmatrix} \qquad \text{and} \qquad \hat{\mathbf{Q}}_A = 0$$

while in collective basis

$$\mathbf{N}_D \times \mathbf{N}_A = \begin{pmatrix} AN^* \\ N^-N^+ \\ N^-N^* \\ AN^+ \end{pmatrix} \qquad \text{and} \qquad \hat{\mathbf{W}}(r) = \begin{pmatrix} -W_I(r) & 0 & 0 & 0 \\ W_I(r) & -W_R(r) & 0 & 0 \\ 0 & 0 & 0 & 0 \\ 0 & 0 & 0 & 0 \end{pmatrix}$$

From here we can significantly simplify the problem leaving only a two-state collective basis wherein

$$\mathbf{N}_D \times \mathbf{N}_A = \begin{pmatrix} AN^* \\ N^-N^+ \end{pmatrix}, \; \hat{\mathbf{W}}(r) = \begin{pmatrix} -W_I(r) & 0 \\ W_I(r) & -W_R(r) \end{pmatrix}, \; \hat{\mathbf{Q}} = -\frac{1}{\tau_D} \begin{pmatrix} 1 & 0 \\ 0 & 0 \end{pmatrix} \tag{3.259}$$

As a result, the kernel $\hat{\mathbf{M}}$ and the matrix $\hat{\mathbf{F}}$ from Eq. (3.104) take the following forms:

$$\hat{\mathbf{M}} = \begin{pmatrix} -R^* & 0 \\ R^{\dagger} & -R^{\ddagger} \end{pmatrix} \qquad \text{and} \qquad \hat{\mathbf{F}} = \begin{pmatrix} \nu & 0 \\ \mu & f \end{pmatrix} \tag{3.260}$$

The secular elements of these matrices are for bimolecular ionization and recombination, while the nonsecular elements account for the generation of the geminate ion pairs and their subsequent recombination and separation.

Since ionization was assumed irreversible, the elements containing the backward electron transfer rate $W_B(r)$ are zero. The assumption is actually acceptable for exclusively exergonic ionization

$$-\Delta G_I \gg T \tag{3.261}$$

where the equilibrium constant similar to Eq. (3.94) is

$$K = \frac{W_I}{W_B} = \exp\frac{-\Delta G_I}{T} \gg 1 \tag{3.262}$$

1. Acceptors in Large Excess

For further simplification, we address systems with a great excess of electron acceptors. Assuming that

$$N^-(t) = N^+(t) < N_0 \ll c \approx \text{const} \tag{3.263}$$

we leave the more general case for detailed consideration in Section IX. Under the simplifying condition (3.263), the equation for $\mathbf{N}_A$ is superfluous because $A(t) \approx c$ does not change while another of this vector component, N^-, is equal to N^+ already included in $\mathbf{N}_D$. The equation for the latter is similar to (3.10a):

$$\dot{\mathbf{N}}_D = Sp_A \int_0^t \mathbf{M}(t-\tau)\mathbf{N}_D(\tau) \otimes \mathbf{N}_A(\tau)d\tau + \hat{\mathbf{Q}}_D\mathbf{N}_D \tag{3.264}$$

Written in components taking into account (3.260), this transforms into the following set of integral equations:

$$\dot{N}^* = -c\int_0^t R^*(\tau)N^*(t-\tau)d\tau - \frac{N^*}{\tau_D} \tag{3.265a}$$

$$\dot{N}^\pm = c\int_0^t R^\dagger(\tau)N^*(t-\tau)d\tau - \int_0^t R^\ddagger(\tau)[N^\pm(t-\tau)]^2d\tau \tag{3.265b}$$

The Laplace transformations of the kernels are

$$\tilde{R}^*(s) = \left(s + \frac{1}{\tau_D}\right)\int W_I(r)\tilde{v}(r,s)d^3r \tag{3.266a}$$

$$\tilde{R}^\dagger(s) = \left(s + \frac{1}{\tau_D}\right)\int [W_I(r)\tilde{v}(r,s) - W_R(r)\tilde{\mu}(r,s)]d^3r \tag{3.266b}$$

$$\tilde{R}^\ddagger(s) = s\int W_R(r)\tilde{f}(r,s)d^3r \tag{3.266c}$$

where the originals of v, μ, and f yield the auxiliary equations

$$\dot{v} = -W_I(r)v - \frac{1}{\tau_D}v + \hat{L}v \tag{3.267a}$$

$$\dot{\mu} = W_I(r)v - W_R(r)\mu + \hat{L}^+\mu \tag{3.267b}$$

$$\dot{f} = -W_R(r)f + \hat{L}^+f \tag{3.267c}$$

with initial conditions

$$v(r,0) = 1, \qquad \mu(r,0) = 0, \qquad f(r,0) = 1$$

and the reflecting boundary conditions at the contact distance $r = \sigma$. The diffusional operators for the neutral reactants and charged products are

$$\hat{L} = D\Delta = \frac{D}{r^2}\frac{\partial}{\partial r}r^2\frac{\partial}{\partial r} \quad \text{and} \quad \hat{L}^+ = \frac{\tilde{D}}{r^2}\frac{\partial}{\partial r}r^2\left(\frac{\partial}{\partial r} + \frac{r_c}{r^2}\right) \tag{3.268}$$

but for highly polar solvents they are essentially the same: $\hat{L} = \hat{L}^+ = \tilde{D}\Delta$.

2. Only Geminate Recombination

Let us now turn to a particular case of the pure geminate reaction considered in the previous section. When the acceptor concentration c is rather small, the bulk recombination is much slower then the fast geminate recombination and the precursor charge accumulation. Within a restricted time interval from the very beginning, it can be completely ignored and the geminate stage of the reaction (3.257) is adequately described by the reduced set of homogeneous equations obtained from (3.265):

$$\dot{N} = -c\int_0^t R^*(\tau)N(t-\tau)d\tau - \frac{N}{\tau_D} \tag{3.269a}$$

$$\dot{P} = c\int_0^t R^\dagger(\tau)N(t-\tau)d\tau \tag{3.269b}$$

where $N = N^*/N_0$ is the survival probability of excitation and $P = N^\pm/N_0$ is the same for the ions. Since these equations are linear, they can be subjected to the Laplace transformation and resolved regarding $\tilde{N}(s)$ and $\tilde{P}(s)$:

$$\tilde{N}(s) = \frac{1}{s + c\tilde{R}^* + 1/\tau_D}, \qquad s\tilde{P}(s) = \frac{c\tilde{R}^\dagger}{s + c\tilde{R}^* + 1/\tau_D} \tag{3.270}$$

The fluorescence quantum yield η was defined in Eq. (3.10), where $R(t)$ is exactly the same quantity as $N(t)$ here. Being proportional to the Laplace transformation of the latter, the quantum yield of fluorescence obeys the truly linear Stern–Volmer law:

$$\frac{1}{\eta} = \frac{\tau_D}{\tilde{N}(0)} = 1 + c\kappa_0\tau_D \qquad (3.271)$$

In view of Eq. (3.270), the ideal Stern–Volmer constant from this equation

$$\kappa_0 = \tilde{R}^*(0) = \frac{1}{\tau_D}\int W_I(r)\tilde{v}(r,0)d^3r \qquad (3.272)$$

is an IET alternative to its DET definition (3.13). An attractive advantage of IET is that it relates the stationary characteristics of the reaction, such as κ_0, to the kernels of integral equations, thus eliminating their solution.

Using the contact approximation and polar solvents ($r_c = 0$), one can find for the irreversible ionization [38]:

$$\kappa_0 = \frac{k_0\tilde{v}(\sigma,0)}{\tau_D} = \frac{k_0k_D}{k_D + k_0\big/\left(1 + \sqrt{\tau_d/\tau_D}\right)}, \qquad (3.273)$$

which is exactly the same κ_0 as in Eq. (3.27), which is obtained by means of DET in the low-concentration limit.

The inverse Laplace transformation of expressions (3.270) provides the kinetics of the energy-quenching $N(t)$ and charge accumulation/separation $P(t)$. The latter ends approaching the free-ion quantum yield

$$\phi = P(\infty) = \lim_{s\to 0} s\tilde{P}(s) = \psi\bar{\varphi} \qquad (3.274)$$

which is a product of the ionization yield

$$\psi = \frac{c\kappa_0\tau_D}{1 + c\kappa_0\tau_D} = 1 - \eta \qquad (3.275)$$

and charge separation yield:

$$\bar{\varphi} = \frac{\tilde{R}^\dagger(0)}{\tilde{R}^*(0)} \qquad (3.276)$$

The bar over φ means that this quantity is averaged over the initial distribution of the interion distances prepared by the forward electron transfer. In IET this averaging is done implicitly and automatically, justifying the straightforward calculation of this quantity.

B. Unified Theory

Since the bimolecular ionization and the free-ion recombination are irreversible, there is no need to preserve the integral description for these reactions. It can be easily changed for the differential one without any losses and even with some gain. For this goal the conventional procedure applied to Eqs. (3.136a) and (3.137)–(3.139) should be employed for integral equation (3.265a), turning it into a differential equation (3.2) with N^* substituted for N and $k_I(t)$ substituted for $k(t)$. To get the $k_I(t)$ definition from the DET equations (3.34) and (3.35), one should just substitute $W(r)$ by $W_I(r)$. The last term in Eq. (3.265b) should be subjected to a similar transformation.

As a result, the set of semiintegral UT equations appears as [178]

$$\frac{dN^*}{dt} = -ck_I(t)N^* - \frac{N^*}{\tau_D} \tag{3.277a}$$

$$\frac{dN^\pm}{dt} = c \int \dot{m}(r,t)d^3r - k_R(t)(N^\pm)^2 \tag{3.277b}$$

where

$$k_I = \int W_I(r)n(r,t)d^3r, \qquad k_R = \int W_R(r)f(r,t)d^3r \tag{3.278}$$

and

$$\dot{n} = -W_I(r)n + \hat{L}n, \qquad \dot{f} = -W_R(r)f + \hat{L}^+ f \tag{3.279}$$

with initial conditions $n(r,0) = f(r,0) = 1$ and reflecting boundary conditions.

In Eqs. (3.277) only the bimolecular rates are transformed to the form that they have in DET. The term that remains integral is essentially the same as in Eq. (3.265b), although it is represented in a different way. By equalizing the Laplace transformations of these terms, we obtain

$$s \int \tilde{m}(r,s)d^3r = \tilde{R}^\dagger(s)\tilde{N}^*(s) \tag{3.280}$$

$\tilde{R}^\dagger(s)$ is defined through $\tilde{\mu}$ and $\tilde{v}$ whose originals are related to each other by Eq. (3.267). The Laplace transformation of the latter is the following:

$$s\tilde{\mu} + W_R(r)\tilde{\mu} - \hat{L}^+\tilde{\mu} = W_I(r)\tilde{v} \tag{3.281}$$

Using this in Eq. (3.266) and taking into account that $\int \hat{L}^+\mu\, d^3r = 0$, due to the reflecting boundary condition, we obtain

$$\tilde{R}^\dagger(s) = s\left(s + \frac{1}{\tau_D}\right)\int \tilde{\mu}(r,s)d^3r \tag{3.282}$$

The solution to Eq. (3.281) may be expressed via the Green function of its left-hand side (LHS), $\tilde{G}(r, r', s)$:

$$\tilde{\mu}(r, s) = \int d^3 r' \tilde{G}(r, r', s) W_I(r') \tilde{v}(r', s) \qquad (3.283)$$

By substituting (3.283) into (3.282) and using the result in (3.280), we obtain

$$\tilde{m}(r, s) = \int d^3 r' \tilde{G}(r, r', s) \tilde{j}(r', s) \qquad (3.284)$$

where

$$\tilde{j}(r, s) = \left(s + \frac{1}{\tau_D} \right) W_I(r) \tilde{v}(r, s) \tilde{N}^*(s) \qquad (3.285)$$

The similarity of integrals (3.283) and (3.284) indicate that $\tilde{m}$ obeys the same Eq. (3.281) as $\tilde{\mu}$ but with a different pumping term on the right-hand side:

$$s\tilde{m} + W_R(r)\tilde{m} - \hat{L}^+ \tilde{m} = \tilde{j}(r, s) \qquad (3.286)$$

The original of this equation

$$\frac{\partial}{\partial t} m = -W_R(r)m + \hat{L}^+ m + j(r, t) \qquad (3.287)$$

contains the original of the pumping term (3.285):

$$j(r, t) = W_I(r)n(r, t)N^*(t) \qquad (3.288)$$

Now we see that $m(r, t)$ is exactly the same distribution of ions that was studied in the previous section; in the absence of the pumping, Eq. (3.287) reduces to Eq. (3.216). However, the presence of a pumping term makes an essential extension of the theory, which now allows us to account not only for the recombination of ions but also for their accumulation. Using the initial condition $m(r, 0) = 0$, implies that there were no ion pairs at the beginning. All of them appear as ionization products in the encounters of excited donors with electron acceptors. Their accumulation and geminate recombination are described by the integral term in Eq. (3.277b), where the solution of Eq. (3.287) should be used.

The last term in the same equation describes the subsequent bimolecular recombination. The general solution to this equation shows the complex interplay between geminate and bimolecular recombination (Fig. 3.29). When

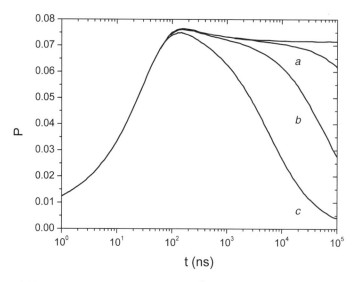

Figure 3.29. The relative charge density $P = N^{\pm}/N_0$ at different levels of light excitation: $N_0 = 5 \times 10^{-5}$ M (a), 5×10^{-4} M (b), 5×10^{-3} M (c). The line approaching the plateau is merely a geminate accumulation–recombination curve ($N_0 \to 0$). The concentration of the acceptors $c = 0.05$ M and $D = 10^{-6}$ cm^2/s. (From Ref. 178.)

ionization is fast, $N^{\pm}(t)$ goes through the maximum before reaching the plateau $N_0\phi = $ const. However, the plateau is not infinitely long, as soon as the bimolecular recombination is taken into account. As can be seen in Figure 3.29, the final bimolecular stage starts earlier and proceeds faster when more free ions are generated. This partially depends on their quantum yield ϕ, but mainly on the initial concentration of excited states N_0, which is higher the stronger the pumping light.

Let us again turn to the simplest case of a purely geminate reaction. Setting $k_R = 0$ in Eqs. (3.277), we obtain from these equations the reduced set of homogeneous equations for survival probabilities $N = N^*/N_0$ and $P = N^{\pm}/N_0$

$$\dot{N} = -ck_I(t)N - \frac{N}{\tau_D} \tag{3.289a}$$

$$P = c \int m(r,t)d^3r \tag{3.289b}$$

where $m(r,t)$ obeys the equation following from Eqs. (3.287) and (3.288):

$$\frac{\partial}{\partial t}m = -W_R(r)m + \hat{L}^+m + W_I(r)n(r,t)N(t) \qquad m(r,0) = 0 \tag{3.290}$$

The latter equation should be solved borrowing $n(r,t)$ and $N(t)$ from the solutions of the corresponding equations. One of them is

$$\dot{n} = -W_I(r)n + \hat{L}n \qquad n(r,0) = 1 \qquad \left.\frac{\partial n(r,t)}{\partial t}\right|_{r=\sigma} = 0 \qquad (3.291)$$

while another is Eq. (3.289), whose general solution is

$$N(t) = \exp\left(-\frac{t}{\tau_D} - c\int_0^t k_I(t')dt'\right) \qquad (3.292)$$

where $k_I(t) = \int W_I(r)n(r,t)d^3r$ is defined through $n(r,t)$ obeying Eq. (3.291).

The algorithm for solving the geminate problem with Eqs. (3.289)–(3.292) was first invented intuitively and independently in two simultaneously published works [22,23]. Later it was recognized as the separate unified theory (UT), which differs from DET because of the integral nature of the geminate term. Moreover, the charge distribution $m(r,t)$ that appears in this term had not appeared in bimolecular theories previously, but was used only for studying the geminate processes. Finally, the UT was derived from IET as we did here, following two original publications [38,178]. The main advantages of UT demonstrated in a number of specific applications were emphasized in our recently published review [32].

C. Initial Distribution of RIPs

The total process of photoionization can be divided into two sequential stages: accumulation of charges and their recombination/separation. The latter stage is represented by two first terms on the RHS of Eq. (3.290). Setting $W_R = \tilde{D} = 0$, one stops the recombination and conserves the ions at the very same place where they were produced, by forward electron transfer. In such a special case only the last term remains on the RHS of Eq. (3.290) and its integral represents merely the accumulation of ions over time at any given distance:

$$m(r,t) = W_I(r)\int_0^t n(r,t)N(t)dt \qquad (3.293)$$

The final distribution

$$m_0(r) = m(r,\infty) = W_I(r)\int_0^\infty n(r,t)N(t)dt = \int_0^\infty j(r,t)dt \qquad (3.294)$$

is actually the initial condition for the next stage of recombination/dissociation.

The total number of ions produced in the primary act of forward electron transfer ψ constitutes the quantum yield of ionization

$$\psi = c \int m_0(r)d^3r = c \int_0^\infty k_I(t)N(t)dt \qquad (3.295)$$

where $N(t)$ was given by Eq. (3.292). Using it here, we obtain, after integration by parts:

$$\psi = 1 - \int_0^\infty \frac{N(t)dt}{\tau_D} = 1 - \eta \qquad (3.296)$$

where η is the same as in Eq. (3.10) and its interrelation with ψ is the same as in the IET equation (3.275).

1. UT/DET Distribution

All that we need to study the shape of the UT distribution (3.294) is $n(r,t)$, the solution to the DET equation [Eq. (3.291)], which is also used in the DET equation [Eq. (3.292)] for specification $k_I(t)$. Hence, the UT initial distribution is also that of DET.

It is clear from the beginning that $n(r,t) \approx n(r,0) = 1$, where $r \gg R_Q$ and therefore

$$m_0(r) \approx W_I(r)\tilde{N}(0) \qquad \text{at} \qquad r \gg R_Q \qquad (3.297)$$

This is an asymptotic behavior for large r peculiar to any distribution, but in the kinetic control limit ($\tilde{D} \to \infty$), when $R_Q \ll \sigma$, Eq. (3.297) holds everywhere. In the opposite static limit ($\tilde{D} \to 0$), when $R_Q \to \infty$, the far asymptote (3.297) is not achievable at all and we obtain instead the product distribution in solids:

$$m_0(r) = W_I(r) \int_0^\infty e^{-W_I(r)t} R_0(t)dt \qquad \text{at} \qquad \tilde{D} = 0 \qquad (3.298)$$

The static quenching kinetics $N(t)|_{D=0} = R_0(t)$ is given by Eq. (3.45) with $W_I(r)$ substituted for $W(r)$.

In Figure 3.30 we see the family of initial distributions for the exponential ionization rate $W_I(r) = W_c \exp(-2[r - \sigma]/l)$ first obtained in Ref. 24. All of them are between the kinetic, purely exponential distribution (K), and that produced by the static ionization (S). The intermediate curves that relate to diffusional ionization (D) pass through the maximum located near R_Q, which is farther from contact when the diffusion is slower. Although all of them have the exponential tail (3.297), in all the rest their shape differ significantly from both

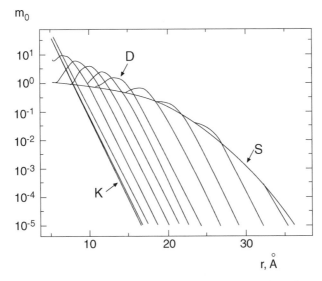

Figure 3.30. The initial charge distributions for $D = \infty, 10^{-3}, 10^{-4}, \ldots, 10^{-9}, 10^{-11},$ $10^{-13}, 10^{-16}, 10^{-20}, 0\,\mathrm{cm}^2/\mathrm{s}$. ($K$, D, and S denote kinetic, diffusional, and static ionization). The remaining parameters are $W_c = 10^3\,\mathrm{ns}^{-1}, L = 0.75\,\text{Å}, \sigma = 5\,\text{Å}, c = 0.1\,\mathrm{M}, \tau_D = \infty$. (From Ref. 24.)

the kinetic and static distributions, which are monotonously decreasing functions of r.

To emphasize this difference, it is better to compare the normalized distributions:

$$f_0(r, c) = \frac{m_0(r)}{\int m_0(r) d^3 r} \qquad (3.299)$$

Since $m_0(r)$ is expressed through $R_0(t)$, which is concentration-dependent, $f_0(r, c)$ also changes with concentration. This is not the case only in the kinetic limit. Under the kinetic control of ionization the distribution (3.297) holds in the whole space and its inclusion in Eq. (3.299) produces the following concentration independent distribution:

$$f_0^{\mathrm{kin}}(r) = \frac{W_I(r)}{\int W_I(r) d^3 r} \qquad (3.300)$$

This distribution duplicates the shape of the ionization rate; in the normal Marcus region it is quasiexponential, while in the inverted region it is bell-shaped and shifted from the contact as in Figure 3.5 at large λ.

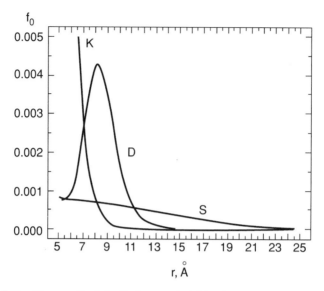

Figure 3.31. The normalized distribution over the initial charge separations under (K) kinetic control $(D = \infty)$, (D) diffusional control $(D = 10^{-6}\,\mathrm{cm^2/s})$ and for (S) static ionization $(D = 0)$. The remaining parameters are the same as in Figure 3.30. (From Ref. 24.)

When diffusion slows down, starting to control the ionization, the shape of the distribution gradually transforms from kinetic to static. This transformation is especially dramatic in the normal Marcus region. This is illustrated by the example of the exponential transfer rate (Fig. 3.31). Being exponential at the fastest diffusion, the distribution $f_0(r, c)$ shifts to higher distances, acquiring a bell shape at slower one. Finally it acquires a static shape monotonously decreasing with r.

In general such distributions are neither contact nor infinitely thin, as in Eq. (3.202), and their subsequent evolution in the course of geminate recombination can be essentially different. In particular, when the backward transfer occurs in the remote recombination layer (see Fig. 3.27), the starting positions of ions can be either inside or outside it and their separation crucially depends on this initial distance. The initial distribution (3.299) allows one to specify the fractions of ions that are in and out and estimate their true contribution to the total charge separation yield.

The initial charge distributions, which were recognized as very important, were analyzed in more detail in a number of works [32]. The analysis was extended to multichannel ionization [179], and the role of the excitation lifetime was specified [25]. A few articles devoted to them were published by Murata and Tachiya [26,27]. Earlier these authors studied experimentally the non-stationary kinetics of fluorescence quenching, which was fitted numerically

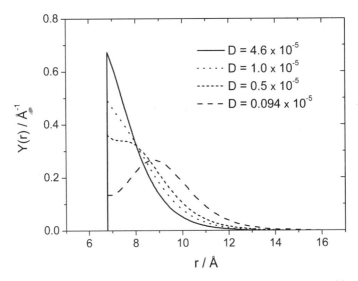

Figure 3.32. A change in the initial distribution of electron transfer distances $Y(r) = cm_0(r)$ with the encounter diffusion coefficient D. Ions are produced in 9,10-dicyanoanthracene (DCA, fluorescer) + p-anisidine (ANS, quencher) encounters at $c = $ [ANS] $= 0.3$ M. The value of D is 4.6×10^{-5} cm²/s in acetonitrile and 0.094×10^{-5} cm²/s in ethylene glycol. The parameters of $W_I(r)$ were determined experimentally [90] and used in Ref. 27 to obtain these distributions.

using the Marcus rate $W_I(r)$ from (3.47) and $N(t)$ calculated by means of DET [66,90]. The best fit was reached at $L = 1.8$ Å and $V_0 = 40 \, \text{cm}^{-1}$, and these data were used for calculating the corresponding initial charge distributions [90]. The results are essentially the same as for the exponential $W_I(r)$ (Fig. 3.32). At slower diffusion (or at smaller V_0) the distributions have the maxima shifted $1 \div 3$ Å out of the contact, as it should be under diffusional control.

2. IET Distribution

So far we discussed only the initial distributions resulting from Eq. (3.294) obtained with the unified theory. But this definition, which is peculiar only for UT and DET, is not unique. The distribution $m(r, t)$ can be introduced advantageously in IET if we identify Eqs. (3.289) and (3.269):

$$\dot{P} = \int \dot{m} d^3 r = c \int_0^t R^\dagger(\tau) N(t - \tau) d\tau \qquad (3.301)$$

According to Eq. (3.266b), the original of $\tilde{R}^\dagger(s)$ in the absence of recombination is

$$R^\dagger(\tau) = \int W_I(r) \left[\dot{v} + \delta(\tau) + \frac{v}{\tau_D} \right] d^3 r \qquad (3.302)$$

where $v(r, \tau)$ obeys Eq. (3.267a):

$$\dot{v} = -\left(\frac{1}{\tau_D} + W_I\right)v + D\Delta v \qquad (3.303)$$

Taking off the integrals over space in Eq. (3.301) and equalizing the integrands, we obtain the IET definition for the pumping term alternative to its UT/DET analog (3.288):

$$\dot{m} = W_I(r)\left\{N(t) + \int_0^t \left[\dot{v}(r, \tau) + \frac{v(r, \tau)}{\tau_D}\right]N(t - \tau)d\tau\right\} = j_{iet}(r, t) \qquad (3.304)$$

Obtained by the straightforward integration of Eq. (3.304), the initial charge distribution $m_0(r) = m(r, \infty)$ also differs from the previous result (3.294) common for UT and DET

$$m_0^{iet}(r) = \tilde{j}_{iet}(r, 0) = W_I \tilde{v}(r, 0)\eta \qquad (3.305)$$

where η obeys the Stern–Volmer law (3.271) with κ_0 from Eq. (3.272). From Eqs. (3.271) and (3.305), we obtain the final IET definition of the initial charge distribution [133]:

$$m_0^{iet}(r) = \frac{W_I \tilde{v}(r, 0)}{1 + c\kappa_0 \tau_D} \qquad (3.306)$$

Unlike that of UT/DET, it is concentration-independent.

It is instructive to separately consider two alternative limits of the very fast and very slow excitation decay. In the former case $v \approx \exp(-t/\tau_D)$ and $\tilde{v}(r, 0) \approx \tau_D$, so that

$$m_0^{iet}(r) = W_I(r)\tau_D \qquad \text{at} \qquad \tau_D \to 0 \qquad (3.307)$$

In the alternative case of slow excitation decay, one can pass to the limit of infinitely large τ_D, after substitution an identity $\tilde{v}(0) = \tilde{n}(1/\tau_D)$ into Eq. (3.306):

$$m_0^{iet}(r) = \frac{W_I(r)}{c}\lim_{\tau_D \to \infty}\frac{1}{\kappa_0 \tau_D}\tilde{n}\left(\frac{1}{\tau_D}\right) = \frac{W_I n_s(r)}{ck_i} \qquad \text{at} \qquad \tau_D \to \infty \qquad (3.308)$$

Here $n_s = n(r, \infty)$ is the stationary pair distribution that obeys Eq. (3.39) and $k_i = k_I(\infty)$ is the stationary (Markovian) rate constant defined through κ_0 in Eq. (3.14). Since at short τ_D the distribution (3.307) simply reproduces the shape of the ionization rate $W_I(r)$, it is the same for all theories, but this identity is violated in the opposite limit of large τ_D.

3. UT/DET versus IET and MET

The most general IET distribution (3.306) takes the following form after normalization:

$$f_0^{iet}(r) = \frac{W_I(r)\tilde{v}(r,0)}{\int W_I(r)\tilde{v}(r,0)d^3r} \qquad (3.309)$$

This differs essentially from the normalized UT/DET distribution (3.299) with $m_0(r)$ from (3.294)

$$f_0(r,c) = \frac{W_I(r)\int_0^\infty n(r,t)N(t)dt}{\int d^3r W_I(r)\int_0^\infty n(r,t)N(t)dt} \qquad (3.310)$$

where n is the solution of Eq. (3.291) and N is as defined in Eq. (3.292). At $c = 0$ one should use $N = \exp(-t/\tau_D)$ in Eq. (3.310), getting a much simpler distribution that is exactly the same as in IET:

$$f_0(r,0) = \frac{W_I(r)\tilde{n}(r,1/\tau_D)}{\int W_I(r)\tilde{n}(r,1/\tau_D)d^3r} = f_0^{iet}(r) \qquad (3.311)$$

This identity follows from the relationship $\tilde{v}(r,0) = \tilde{n}(r,1/\tau_D)$. Once again we see that the IET is the lowest order approximation with respect to c.

In the fast decay limit $\tilde{n}(r,1/\tau_D) = \tau_D$, while at slow decay $\lim_{s\to 0} s\tilde{n}(r,s) = n_s(r)$. Correspondingly

$$f_0^{iet}(r) = \begin{cases} W_I(r)/\int W_I(r)d^3r = f_0^{kin}(r) & \text{at} \quad \tau_D \to 0 \\ W_I(r)n_s(r)/\int W_I(r)n_s(r)d^3r & \text{at} \quad \tau_D \to \infty \end{cases} \qquad (3.312)$$

At fast decay the ionization is always kinetic, as well as the resulting distribution $f_0^{kin}(r)$, which is the same in all theories. Only at slow decay can one trace the transition from the kinetic to diffusional limit with increasing viscosity.

The difference between the concentration-dependent UT/DET distribution $f_0(r,c)$ and concentration-independent $f_0^{iet}(r)$ is expected to be pronounced only at slow excitation decay, provided ionization is either diffusional or static. This conclusion is easier to verify in the limit $\tau_D \to \infty$ when it follows from Eqs. (3.310) and (3.311):

$$f_0(r,c)|_{\tau_D=\infty} = \frac{W_I(r)\int_0^\infty n(r,t)\exp\left(-c\int_0^t k(t')dt'\right)dt}{\int W_I(r)\int_0^\infty n(r,t)\exp\left(-c\int_0^t k(t')dt'\right)dt\,d^3r} \qquad (3.313a)$$

$$f_0^{iet}(r)|_{\tau_D\to\infty} = \lim_{\tau_D\to\infty} \frac{W_I(r)\tilde{n}(r,1/\tau_D)}{\int W_I(r)\tilde{n}(r,1/\tau_D)d^3r} = \frac{W_I n_s(r)}{\int W_I(r)n_s(r)d^3r} \qquad (3.313b)$$

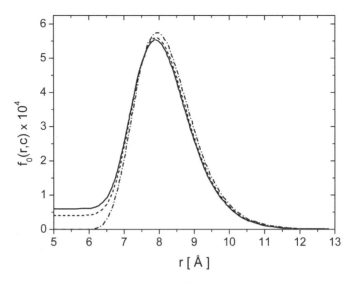

Figure 3.33. The normalized distributions of ionization products over their separation. The dashed–dotted line relates to the simplest integral theory (IET) and the dashed line, to its modified version (MET). The thick line represents the same distribution calculated with DET/UT. The ionization rate was assumed to be exponential, $W_I(r) = W_c \exp\{-[2(r-\sigma)]/l\}$ ($W_c = 10^3\,\mathrm{ns}^{-1}$, $l = 1\,\text{Å}, D = 10^{-7}\,\mathrm{cm}^2/\mathrm{s}$, and $c = 10^{-3}\,\mathrm{M}$. From Ref. [133].

As shown in Figure 3.33, two distributions are almost the same everywhere but markedly different in the narrow strip near the contact. IET accounts for the more dense ion population near the contact (resulting from nonstationary quenching) less well than does DET. The zero value of $f_0^{iet}(\sigma)$ in this region is really nonsense. This is because the pumping rate $j_{iet}(r,t)$ changes the sign with time and what was collected during nonstationary quenching is subsequently pumped out.

Although small, this is a principal disadvantage of the simplest integral theory. The near-contact density of the products nonlinear in c is lost in the lowest-order approximation to this parameter. However, the nonzero contribution to this region is provided by a modified encounter theory outlined in Section XII. The chief merit of MET is that the argument of the Laplace transformation of $n(r,t)$ in (3.311) is shifted from $1/\tau_D$ to $1/\tau_D + ck$. As a result, in the limit $\tau_D = \infty$ we have instead of (3.313) [133]:

$$f_0^{met}(r,c)\big|_{\tau_D=\infty} = \frac{W_I \tilde{n}(r,ck)}{\int W_I(r)\tilde{n}(r,ck)d^3r} \tag{3.314}$$

Due to such a shift in the argument, the gap near the contact in the IET distribution is almost removed (Fig. 3.33). This effect is a bit stronger in the recent theory of Szabo and Gopich (SCRTA) [180], which confirms the distribution (3.314) but with slightly higher k defined by equation $c \int W_I(r)\tilde{n}(r,ck)d^3r = 1$ [181]. However, the total contribution of the nonstationary and static quenching is correctly estimated in only DET and UT. This is their advantage over integral theories.

D. Accumulation and Geminate Recombination of RIPs

The main achievement of UT was the incorporation of a distribution of ionization products $m(r,t)$. The latter was not inherent to the original DET and was introduced only a posteriori in IET. The UT kinetic equation for this quantity, (3.290), is actually a symbiosis of Eq. (3.216) for the remote RIP recombination and the pumping term (3.288) responsible for their accumulation. This combination allows tracing the photogeneration of ions after δ-pulse excitation and their subsequent recombination and separation.

In the absence of recombination the simple integration over space transforms Eq. (3.290) to the conventional rate equation:

$$\dot{P} = ck_I(t)N \qquad \text{at} \qquad W_R = 0 \qquad (3.315)$$

Together with Eq. (3.289), these two constitute the formal basis for DET. This provides us with the rate description of the quenching kinetics $N(t)$ and charge accumulation $P(t)$ in the absence of geminate recombination. In contrast, the solution to Eq. (3.290) with $W_R \neq 0$ describes both the RIP accumulation and their recombination, but inclusion in Eq. (3.289) does not transform it to the rate equation. As long as Eq. (3.289) in the basic set (3.289) remains integral, UT is actually a "half-integral" theory: differential for forward electron transfer and integral for backward electron transfer.

The general solution to Eq. (3.290) can be expressed via the Green function of Eq. (3.216), $G(r,r',t-t')$

$$m(r,t) = \int d^3r' \int_0^t G(r,r',t-t')j(r',t')dt' \qquad (3.316)$$

where $j(r,t)$ was defined in Eq. (3.288). Substituting Eq. (3.316) into Eq. (3.289b) and taking into account that the survival probability $\Omega(t-t'|r') = \int G(r,r', t-t')d^3r$, we have

$$P(t) = c \int d^3r' \int_0^t \Omega(t-t'|r')j(r',t')dt' \qquad (3.317)$$

where $\Omega(t|r')$ obeys Eqs. (3.218) and (3.219). As $t \to \infty$, the survival probability $\Omega(t|r')$ turns to the charge separation quantum yield $\varphi(r') = \Omega(\infty|r')$, and the remainder of the time integral transforms into the initial charge distribution (3.294). The resulting free-ion quantum yield is

$$\phi = P(\infty) = c \int d^3 r' \varphi(r') m_0(r') = \psi \bar{\varphi} \qquad (3.318)$$

According to definition (3.295), the quantum yield of ionization is the same as in the IET equation (3.275) and UT Eq. (3.296):

$$\psi = c \int m_0(r) d^3 r = 1 - \eta \qquad (3.319)$$

and $\bar{\varphi}$ is the charge separation yield, averaged over the normalized initial distribution of ions (3.299):

$$\bar{\varphi} = \int \varphi(r) f_0(r, c) d^3 r \qquad (3.320)$$

This is the UT analog of Eq. (3.276), where the same averaging was implicitly done over the IET distribution f_0^{IET}.

1. Photoionization Kinetics

The shape of the kinetic curve $P(t)$ strongly depends on how fast the primary ionization is compared to the geminate recombination. At slow ionization, $P(t)$ increases monotonously, approaching the plateau $P(\infty) = \phi$ from below (Fig. 3.34). Qualitatively different behavior is peculiar to the opposite case of fast ionization. This is the most frequent case that is worthy of special attention.

Since the pumping is much faster than the recombination given by $\Omega(t - t'|r')$, the latter can be factored outside the integral (3.317) at time t. Then, accounting for Eq. (3.288) we obtain

$$P(t) = c \int W_I(r') \Omega(t|r') d^3 r' \int_0^t n(r', t') N(t') dt' \qquad (3.321)$$

The pumping can be fast because of the high ionization rate at large acceptor concentration c, but it is more common because of the abrupt decay of the excited state. The latter interrupts the ionization at its earliest stage when it proceeds exponentially with the kinetic rate constant $k_0 = k_I(0) = \int W_I(r) d^3 r$. Under these conditions at times $t \sim \tau_D$, we have

$$n(r, t) \approx n(r, 0) = 1 \qquad \text{and} \qquad N(t) \approx e^{-t/\tau_D - ck_0 t} \qquad (3.322)$$

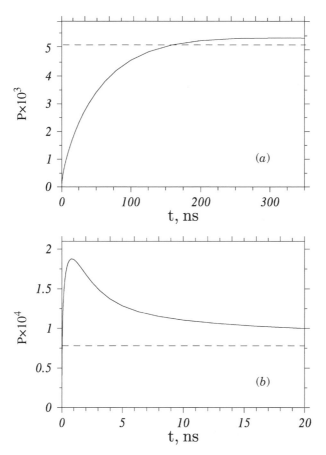

Figure 3.34. The kinetics of ion accumulation at slow (*a*) and fast (*b*) ionization, assuming the rates of electron transfer are exponential with $W_c = 1$ ns^{-1} for the forward and $W_c = 10^3$ ns^{-1} for the backward transfer ($c = 0.15$ M, $D = \tilde{D} = 10^{-6}$ cm^2/s, $\sigma = 5$ Å). Dashed horizontal lines indicate the free ion quantum yields ϕ at $\tau_D = \infty$ (*a*) and $\tau_D = 1$ ns (*b*).

Using these approximations in Eq. (3.321), we obtain

$$P(t) \approx \psi_{\text{kin}} \int \Omega(t|r')f_0^{\text{kin}}(r')d^3r'\left[1 - e^{-t/\tau_D - ck_0 t}\right] \qquad (3.323)$$

where the kinetic distribution of ions, $f_0^{\text{kin}}(r)$, is the same as in Eq. (3.312) and the ionization quantum yield in the kinetic limit, $\psi_{\text{kin}} = ck_0\tau_D/(1 + ck_0\tau_D)$, follows from Eq. (3.296) when $N(t)$ from (3.322) is used there. According to Eq. (3.323), the time dependence of $P(t)$ is not monotonous. It consists of a short ascending branch, determined by the expression in brackets, followed by a much longer descending tail, given by the averaged quantity $\Omega(t|r)$. The ascending branch describes the fast accumulation of ions, while the descending one represents their subsequent geminate recombination (Fig. 3.34). Only in the limit of instantaneous ionization (at $\tau_D \to 0$) does this complex kinetics degenerate into a single descending branch, which we studied in the previous section.

If ionization is neither fast nor slow, one has to solve numerically Eq. (3.290) together with the supplementary Eqs. (3.291) and (3.292) and use the result in Eq. (3.289). In Ref. 24 the numerical procedure based on the expanded DCR (diffusion-controlled reaction) program [182] was used to evaluate $P(t)$. From the simultaneously calculated initial distribution the root-mean-square interion distance was found:

$$\bar{R}^2 = \int_\sigma^\infty r^2 f_0(r,c) d^3 r \tag{3.324}$$

It is useful as an estimate of the initial ion separation that may be identified with r_0, if the initial condition (3.201) is used.

There is a complex interrelationship between the two-stage photoionization kinetics and the kinetics of geminate recombination/separation following instantaneous ionization. The descending branch of the former can be successfully approached by the latter if the initial ion separation r_0 is designed so that the charge separation yields in both theories are the same. At $\tau_D = \infty$ and $r_c = 0$, this was done in Ref. 24 using the exponential rates of transfer in UT and contact approximation for geminate recombination (Fig. 3.35). The best coincidence of the long-time tails was reached at $r_0 = \bar{R} = 7.3\,\text{Å}$, which is essentially larger than the contact distance $\sigma = 5\,\text{Å}$. Hence, the exponential model is completely inadequate while diffusional models of geminate recombination may be consistent with the UT asymptote provided the choice of r_0 is made a posteriori.

2. Recombination through Proton Transfer

It does not always happen that the ion recombination proceeds through the backward electron transfer. Sometimes the ions are discharged because of proton transfer between them, resulting in their transformation to free radicals $\dot{D}$ and $\dot{A}$:

$$D^* + A \longrightarrow [D^+ \cdots A^-] \to D^+ + A^-$$
$$W_R \diagup\diagdown k_p \tag{3.325}$$
$$[D \cdots A] \qquad [\dot{D} \cdots \dot{A}] \to \dot{D} + \dot{A}$$

The generation of the free radicals in the photoreduction of anthraquinone by triethylamine was studied in alcoholic solutions by means of FT-EPR (Fourier

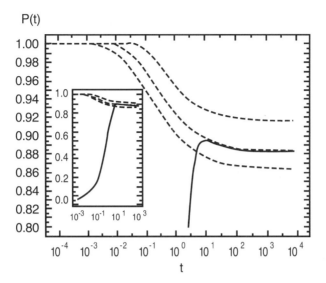

Figure 3.35. The kinetics of photoionization (solid lines) at $\tau_D = \infty (\phi \equiv \bar{\phi})$ and $r_c = 0$. The dashed lines indicate the kinetics of the geminate recombination of ions started from different initial distances $r_0 = 6; 7.3; 10$ Å (from bottom to top) while the average one is $\bar{R} = 7.3$ Å. The remaining parameters are $k_c/k_D = 0.2$, $W_R(\sigma) = 3.96 \, \text{ns}^{-1}$, $\tilde{D} = D = 10^{-5} \, \text{cm}^2/\text{s}$, $c = 0.1 \, \text{M}$, $W_I(\sigma) = 10^3 \, \text{ns}^{-1}, l = 0.75, \sigma = 5$ Å. (From Ref. 24.)

transform–electron paramagnetic resonance) [183,184]. It was expected that the Coulomb attraction that hinders ion separation works in favor of the neutral products of proton transfer. The quantitative theory of the phenomenon developed in Ref. 185 incorporates the radical pair distribution $g(r)$ in the set of UT equations

$$\frac{\partial}{\partial t}m = -W_R(r)m + \frac{\tilde{D}}{r^2}\frac{\partial}{\partial r}r^2\left(\frac{\partial m}{\partial r} + \frac{r_c}{r^2}m\right) + W_I(r)nN(t) \tag{3.326a}$$

$$\frac{\partial}{\partial t}g = \bar{D}\frac{1}{r^2}\frac{\partial}{\partial r}r^2\frac{\partial g}{\partial r} \tag{3.326b}$$

where $\bar{D}$ is an encounter diffusion coefficient of neutral radicals. Since proton transfer is essentially contact, it can be accounted for in the boundary conditions

$$4\pi\tilde{D}r^2\left(\frac{\partial m}{\partial r} + \frac{r_c}{r^2}m\right)\Bigg|_{r=\sigma} = k_p m(\sigma, t) = -4\pi\bar{D}r^2\frac{\partial g}{\partial r}\Bigg|_{r=\sigma} \tag{3.327}$$

where k_p is analogous to k_c from Eq. (3.197).

The set of equations (3.326) was numerically solved neglecting the ion recombination to the ground state ($W_R = 0$) [185]. The solutions were used to calculate the total amount of charged and neutral products of photoionization that can be detected experimentally:

$$P(t) = c \int_\sigma^\infty m(r,t) 4\pi r^2 dr, \qquad R(t) = c \int_\varsigma^\infty g(r,t) 4\pi r^2 dr \qquad (3.328)$$

Here we took into account that only distantly separated free radicals are observable with the FT-EPR technique. Those that are too close cannot be detected because of a dominant exchange coupling, resulting in splitting or broadening of the EPR spectra. Therefore, the EPR registration was "successful only after complete separation of the radicals," out of the "invisible cage" of radius ς, which is essentially larger than σ [183,184].

The calculated kinetics of charge and radical accumulation was in qualitative agreement with what was expected. However, the theory does not fit the experimental data obtained by Beckert et al. [183,184]. The reaction time reported there is 0.8 ns, while the calculated buildup time of radical accumulation varies from 60 to 560 ns. The alcohols used by Dinse et al. can be scarcely considered as nonpolar solvents. Their polarity is so high as to render the Coulomb well insignificant and facilitates evacuation of products. The accumulation of radicals must end in a few nanoseconds, but in fact it lasts about 100 ns. This is actually an indication that there is some binding potential for ions or radicals that is deeper than that originating from the Coulomb attraction. There is room for the exciplex formation discussed in a recent review in 2000 [32].

VII. AVERAGING CHARGE SEPARATION YIELD

If excitation is weak or partner concentration is small, then the free ions are produced in low concentration and their bimolecular recombination is too slow to be seen in the timescale of the geminate reaction. Therefore the kinetics of the latter is often studied separately with a fast time-resolved technique. Alternatively, the free-ion quantum yield found from the initial concentration of ions participating in the slow bimolecular recombination can be used to calculate $\bar{\varphi}$ from Eq. (3.274) provided $\psi = 1 - \gamma$ is known. Anyhow, the charge separation quantum yield $\bar{\varphi}$ is the usual subject of numerous investigations. Here we will concentrate only on two of them, where this quantity was studied as a function of not only the recombination free energy but of the solvent viscosity as well. These investigations were carried out on the following systems:

1. A ruthenium complex quenched by methylviologen (MV^{2+}) [148]:

$$\left(^3Ru^{2+}\right)^* + MV^{2+} \overset{W_I}{\Longrightarrow} \left[\dot{R}u^{3+} \cdots \dot{M}V^+\right] \longrightarrow \dot{R}u^{3+} + \dot{M}V^+ \qquad (3.329)$$
$$\Downarrow W_R$$
$$\left[Ru^{2+} \cdots MV^{2+}\right]$$

2. Perylene (Per) quenched by some aromatic amines (A) [151,186]:

$$^1Per^* + D \overset{W_I}{\Longrightarrow} \left[\dot{P}er^- \cdots \dot{D}^+\right] \longrightarrow \dot{P}er^- + \dot{D}^+ \qquad (3.330)$$
$$\Downarrow W_R$$
$$\left[Per \cdots D\right]$$

Both of these formulas provide important data for the viscosity dependence of geminate recombination, which will be analyzed later.

A. FEG Law for Recombination Efficiency

According to Eqs. (3.320) and (3.195), the averaged charge separation quantum yield is expressed through the mean recombination efficiency $Z(\Delta G, D, \tilde{D})$:

$$\bar{\varphi} = \int_\sigma^\infty \varphi(r) f_0(r) 4\pi r^2 dr = \frac{1}{1 + Z/\tilde{D}} \qquad (3.331)$$

Let us first specify Z in the fast diffusion limit and only then clarify how this limit is approached.

At fast diffusion the recombination is kinetic and so weak that in the zero approximation we obtain the following equation from (3.170): $\varphi(r_0) = 1$. However, in the next (first-order) approximation it follows from the same equation that [147]

$$\varphi(r) = 1 - \frac{1}{\tilde{D}} \int_r^\infty \frac{dr'}{r'^2} e^{-r_c/r'} \int_\sigma^{r'} e^{r_c/r} W_R(r) r^2 dr \qquad (3.332)$$

Substituting this result into Eq. (3.331), we obtain

$$\bar{\varphi} = 1 - \frac{1}{\tilde{D}} \int Z_0(r) f_0(r) d^3 r = 1 - \frac{Z_0}{\tilde{D}} \qquad (3.333)$$

where $Z_0 = \lim_{\bar{D} \to \infty} Z$ is a kinetic limit of recombination efficiency whose value is

$$Z_0(r) = \frac{1 - e^{-r_c/r}}{r_c} \int_{\sigma}^{r} e^{r_c/R} \, W_R(R) R^2 dR$$
$$+ \int_{r}^{\infty} \frac{e^{r_c/R} - 1}{r_c} \, W_R(R) R^2 dR \qquad (3.334)$$

The rates $W_I(r)$ and $W_R(r)$, which control $f_0(r)$ and $Z_0(r)$, are given by the general expression identical to (3.50):

$$W_{I,R} = W_0 \exp\left(-\frac{(\Delta G_{I,R} + \lambda)^2}{4\lambda T}\right) \qquad (3.335)$$

The free energies and the preexponent are specific for ionization and recombination, while $\lambda(r)$ defined in Eqs. (3.48) and (3.49) is universal. The reaction layers determined by the rates (3.335) can be either contact or remote, exponential or bell-shaped as in Fig. 3.5. Their shapes and relative positions depend on the corresponding free energies, ΔG_i and ΔG_r, whose sum is fixed equal to $-\mathscr{E}$, where $\mathscr{E}$ is the energy of the excited reactant from Eq. (3.89). By changing the reaction partner, the energy of the ion pair can be varied over a wide range, changing simultaneously the free energies of ionization and recombination. In such a case the FEG laws for the forward and backward electron transfers are correlated, and their relationship depends on the ratio $\mathscr{E}/2\lambda$.

As shown in Figure 3.36, there are two alternative cases: (a) small excitation energy ($\mathscr{E} < 2\lambda$) and (c) large excitation energy ($\mathscr{E} > 2\lambda$). They are separated by the border case (b), where $\mathscr{E} = 2\lambda$. Both ionization and recombination can occur in either the normal (N) or inverted (I) regions, depending on where is the RIP energy level situated with respect to the energy levels of the excited and ground states. If it is very close to the excited level, then ionization is in the normal region while the recombination is in the inverted one (NI). If the energy of the ion pair is much closer to the ground state than vice versa, the ionization is inverted while the recombination is a normal transfer (IN). Case (b) is exhausted only by these two opposite situations. However, in case (a) both transfers can be simultaneously normal (NN) when the ionized state is approximately in the middle. In the same situation, but in case (c) they are simultaneous in the inverted region (II).

Now we are ready to establish the relationship between Z_0 defined in Eq. (3.333) and the approximate estimates of this quantity made earlier. In the NI situation the ionization may be considered as contact (unlike recombination) and the initial charge distribution is also contact when ionization is under

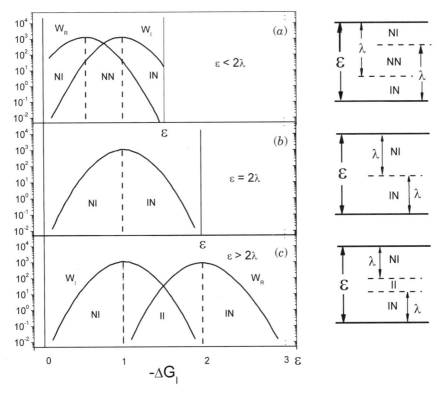

Figure 3.36. The dependence of $\ln W_I$ and $\ln W_R$ on ΔG_I in the range between 0 and $\mathscr{E}$ for small (a), intermediate (b), and large (c) $\mathscr{E}$ as compared to 2λ. The corresponding energy diagrams show the separation of the normal regions from the inverted regions for ionization and recombination. The normal region is within λ strips from the top (for ionization) or bottom (for recombination). Outside these strips the corresponding reactions proceed in the inverted regions.

kinetic control: $f_0(r) = \delta(r - \sigma)/4\pi\sigma^2$. In such a case, it follows from Eqs. (3.333) and (3.334) that

$$Z_0 = Z(\sigma) = \int_\sigma^\infty \frac{e^{r_c/r} - 1}{r_c} W_R(r) r^2 dr = z, \qquad \text{NI} \qquad (3.336)$$

In highly polar solvents ($r_c \ll \sigma$)

$$z = \int_\sigma^\infty W_R(r) r\, dr \qquad (3.337)$$

In the alternative IN situation the recombination may be considered contact (unlike ionization), so that by setting $W_R(r) = (k_c/4\pi\sigma^2)\delta(r - \sigma)$ one can obtain from Eqs. (3.333) and (3.334)

$$Z_0 = \bar{q}z, \qquad \text{IN} \tag{3.338}$$

where $\bar{q}$ is the averaged value of partial $q(r)$ defined in Eq. (3.209) and

$$\bar{q} = \int \frac{1 - \exp(-r_c/r)}{1 - \exp(-r_c/\sigma)} f_0(r)\, d^3r, \quad \text{and} \quad z = \frac{k_c}{4\pi r_c}\left[\exp\left(\frac{r_c}{R}\right) - 1\right] \tag{3.339}$$

is the same as given by Eqs. (3.194) and (3.206). In highly polar solvents ($r_c \ll \sigma$), we have

$$\bar{q} = \int \left(\frac{\sigma}{r}\right) f_0(r) d^3r \leq 1 \tag{3.340}$$

The mean recombination efficiencies (3.336) and (3.338) are just the opposite limits of the general expression valid at $D = \tilde{D} = \infty$:

$$Z_0 = \int Z_0(r) f_0^{\text{kin}}(r) d^3r = \begin{cases} z & \text{for inner starts} \\ \bar{q}z & \text{for outer starts} \end{cases} \tag{3.341}$$

In the NI case most of the starts are from inside the recombination zone, while in the IN region ions always start from outside. In the latter case Z_0 is a bit smaller than z, but the difference is insignificant and the free-energy dependence of Z_0 results mainly from the $z(\Delta G_r)$ dependence. According to the EM equations (3.194) and (3.207), z should reproduce the FEG law peculiar to $W_R(\sigma)$. This is the symmetric curve represented by the dotted line in Figure 3.37, while the true kinetic efficiency represented by the thick solid line in the same figure is asymmetric and wider. This is due to the $\lambda(r)$ dependence, which broadens the FEG law for recombination as it did for ionization [compare the dashed and solid lines (a) in Fig. 3.16]. This dependence is accounted for when z is calculated from the integral (3.337).

In practice all the free-energy dependencies shown in Fig. 3.37 were obtained in another way. The expanded DCR program [182] was employed to calculate $P(t)$ at any given viscosity and then $P(\infty) = \bar{\varphi}$ was used to find $Z = (1 - \bar{\varphi})\tilde{D}/\bar{\varphi}$. The results presented in Figure 3.37 relate to the border case b ($\mathscr{E} = 2\lambda_c$) when only two situations are possible: either NI (left half) or IN (right half). The FEG law for kinetic controlled recombination $Z_0(\Delta G_r)$ is given by the solid line. It is qualitatively the same as the FEG dependence of the EM recombination parameter z shown by the dotted line. In the IN region the latter is a bit higher than $Z_0 = \bar{q}z$ because $\bar{q} < 1$. However, in the NI region Z_0 essentially exceeds z as a result of the space dispersion of $W_0(r)$ and $\lambda(r)$.

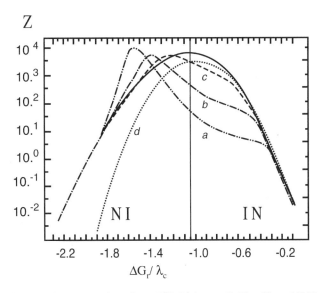

Figure 3.37. The free-energy dependence $Z(\Delta G_r)$ in case B ($\mathscr{E} = 2\lambda_c = 110\,T$) at $\tau_D = \infty$. Assuming that the transfer rates (3.335) have the same $W_0(\sigma) = 10^{12}\,\mathrm{s}^{-1}$, the four curves for the different diffusion coefficients are calculated: $D = \tilde{D} = 10^{-7}\,\mathrm{cm}^2/\mathrm{s}$ (a), $10^{-6}\,\mathrm{cm}^2/\mathrm{s}$ (b), $10^{-5}\,\mathrm{cm}^2/\mathrm{s}$ (c), and ∞ (solid line). The latter is the kinetic curve $Z_0(\Delta G_r)$. For comparison, the exponential model parameter $z(\Delta G_r)$ (dotted line d) is also depicted. (From Ref. 111.)

Setting for simplicity $\tau_D = \infty$, the FEG law was inspected in Ref. 111, varying diffusion coefficients $D = \tilde{D}$ over a wide range. At all $\tilde{D}$ the far wings of the FEG curve in the NI and IN regions are the same as the kinetic ones. Since the recombination there is weak, it is under kinetic control. In contrast, the top of the curve where the transfer is fastest is essentially distorted by diffusion. When the latter slows down, the initial distribution of ions is shifted out of contact (Figs. 3.31 and 3.32). In the IN region this lengthens the path to the contact reaction zone, thus hindering diffusion controlled recombination. The slower is diffusion, the smaller is Z comparable to the kinetic value Z_0. The kinetic limit is approached from below when diffusion accelerates. A different situation arises in the NI region, where the kinetic distribution of ions is near-contact, located inside the remote reaction layer. With decreasing diffusion, this distribution also spreads out of the contact, but the redistribution inside the reaction sphere does not significantly affect the charge separation yield (see Fig. 3.27). Much more important is the lengthening of the residence time in the reaction layer. The slower the crossing of this layer from inside, the stronger is the recombination there, which makes $Z > Z_0$ at lower $\tilde{D}$. The opposite behavior of $Z(\tilde{D})$ in the NI and IN regions reflects the opposite diffusional dependence of the same quantity at the inner and outer starts obtained earlier (Fig. 3.26).

Such behavior is a very general phenomenon inherent in $Z(\tilde{D})$ even when transfer is multichannel, as given by Eq. (3.51) for $W_{I,R}$. This fact was confirmed in Ref. 179, where the transfer was studied assuming $\tau_D = \infty$ as above. However, the role of the finite lifetime was then subjected to a separate investigation and not only in case b but in cases a and c as well [25]. As was expected in all cases, the FEG curves approach their kinetic limit with reduction of τ_D as they do with increasing D [25,32].

B. Extremal Viscosity Dependence of Z

A more intent examination of Figure 3.37 displays even more strikingly the nonmonotonous $Z(\tilde{D})$ behavior near $\Delta G_r/\lambda_c = -1.4$. There one can see that Z first increases with $\tilde{D}$ going from the a curve to the b curve and only then decreases from b to c, going lower up to the kinetic limit. The corresponding maximum in the viscosity dependence of Z is clearly seen in Figure 3.38,

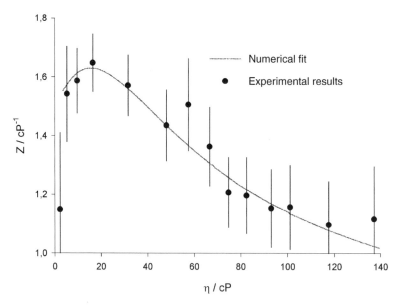

Figure 3.38. The experimental viscosity dependence of Z in mixtures of dimethylsulfoxide (DMSO) with glycerol. The excitation energy of perylene $\mathscr{E} = 2.83\,\text{eV}$, $\Delta G_i = -0.56\,\text{eV}$, $\Delta G_r = -2.27\,\text{eV}$. The contact reorganization energy of solvent $\lambda_c = 0.885\,\text{eV}$, $L = 1\,\text{Å}$, $\tau_D = 4.3\,\text{ns}$. The theoretical fit (solid line) was done with the rates (3.343), where $w_i = 230\,\text{ns}^{-1}$, $\lambda_i = 0.1298\,\text{eV}$ for ionization and $w_r = 1320\,\text{ns}^{-1}$, $\lambda_i = 0.2346\,\text{eV}$ for recombination (λ_i—inner sphere reorganization energy). (From Ref. 151.)

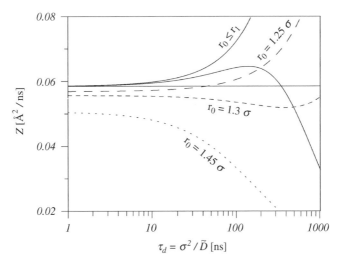

Figure 3.39. The viscosity dependence of the recombination efficiency Z at fixed starts from the interior of the remote rectangular recombination layer (thin solid line), from inside of this layer (long dashed and dashed lines) and outside of it (dotted line). The true $Z(\tilde{D})$ dependence, calculated from $\bar{\varphi}$ by means of IET, is indicated by the thick solid line. The horizontal line represent $z = 0.0585 \, \text{Å}^2/\text{ns}$. The parameters, defining the rectangular recombination layer are the following: $r_i = 1.1 \, \sigma$, $r_e = 1.4 \, \sigma$, $W_0 = 0.156 \, \text{ns}^{-1}$, $\sigma = 10 \, \text{Å}$. (From Ref. 151.)

exhibiting the recombination efficiency of the ion pair produced by electron transfer from the excited perylene to N,N-dimethylaniline (DMA). This extremum was discovered experimentally, and first explained by Neufeld et al. [151].

For the qualitative understanding of the phenomenon, let us reconsider it in the framework of the rectangular model of $W_R(r)$. Figure 3.39 shows the recombination efficiency as a function of viscosity, measured by the mean encounter time $\tau_d = \sigma^2/\tilde{D}$. At any fixed r_0, indicated in the figure, this dependence is different and even the opposite for the smallest and largest initial separations, as in Figure 3.26. The rectangular approximation of the real recombination layer, used in our calculations, is shown by the dotted square on Figure 3.40(a). For ions, starting from inside the rectangular layer, the recombination efficiency increases from z to infinity with increasing viscosity and does not depend on the initial separation r_0 until it is less than r_1 (two upper curves in Fig. 3.39). This effect has been attributed earlier in this review to an increase in the residence time in the reaction layer, which makes the recombination more efficient. The opposite dependence of Z on viscosity, down from $\bar{q}z$, is observed for outer starts. At high viscosities the reaction is controlled by diffusion and Z is reduced when diffusion slows down (the lowest curve in Fig. 3.39).

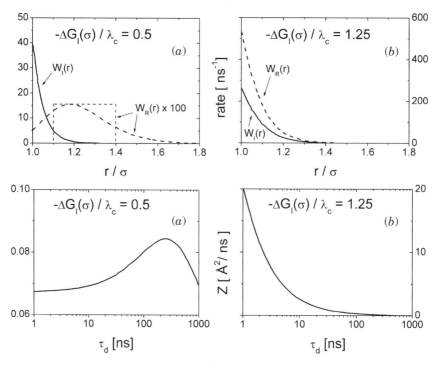

Figure 3.40. The transfer rates (3.343) (above) and the corresponding recombination efficiencies (below) in NI situation (a) and II situation (b) at $w_i = 500\,\mathrm{ns}^{-1}$, $w_r = 1000\,\mathrm{ns}^{-1}$; $L/\sigma = 0.142$ ($\sigma = 10\,\mathrm{\mathring{A}}$), $\lambda_c = 1\,\mathrm{eV}$, $r_c = 0$, $T = 293\,\mathrm{K}$. The viscosity is measured by the mean encounter time for the neutral reactants $\tau_d = \sigma^2/\tilde{D}$. The excitation energy of the donor $\mathscr{E} = 2.5\lambda_c$. (From Ref. 151.)

Thus, the recombination efficiency parameter $Z(r_0; \tilde{D})$ has the opposite viscosity dependence for ions, starting from inside and outside the remote reaction layer. However, the initial charge separation r_0 identified with $\bar{R}$ from Eq. (3.324) is not fixed forever but changes with diffusion as well as the initial distributions shown in Figures 3.31 and 3.32. As both diffusion coefficients $D \approx \tilde{D}$ decrease simultaneously with viscosity, the latter affects $Z(r_0(D); \tilde{D})$ explicitly and implicitly. This quantity calculated with IET is represented in Figure 3.39 by the thick solid line, which exhibits an extremum of the same origin as in Figure 3.38. The thick line crosses all the other lines so that the crossing point moves from one curve to another, reflecting the starting distance change from smaller to larger r_0.

The very existence of the maximum in the viscosity dependence of Z seen in Figure 3.40(a) (bottom) is conditioned by a clearcut separation of the near contact ionization and remote recombination zones in the NI situation

[Fig. 3.40(a), top]. In situation II ($|\Delta G_I| = |\Delta G_R| > \lambda_c$) the effect is completely lost since both reaction zones have exactly the same shape [Fig. 3.40(b), top]. Thus the initial ion distribution, even when it coincides with one of them, cannot be inside the other. As a consequence, only the descending (diffusion-controlled) branch of this dependence is seen in Figure 3.40(a) (bottom). Such a high sensitivity of the results to the shape and relative location of the ionization and recombination zones makes any model simplifications of these zones undesirable.

Therefore, from this point on we prefer to use only the nonmodel IET approach to the problem. This is much easier than performing a complete UT calculation of the whole kinetics of $P(t)$ just for the retrieval of the final value $P(\infty) = \psi\bar{\varphi}$. In IET $\bar{\varphi}$ could be obtained directly from Eq. (3.276) after inserting Eqs. (3.266a) and (3.266b) there

$$\bar{\varphi} = 1 - \frac{\int W_R(r)\tilde{\mu}(r,0)d^3r}{\int W_I(r)\tilde{\nu}(r,0)d^3r} \tag{3.342}$$

if the solution to the set (3.267) is found using the Marcus rates for ionization and recombination:

$$W_{I,R}(r) = w_{i,r}\sqrt{\frac{\lambda_c}{\lambda(r)}}e^{-2(r-\sigma)/L} \times \exp\left[-\frac{[\Delta G_{I,R}(r) + \lambda(r)]^2}{4\lambda(r)T}\right] \tag{3.343}$$

Here w_i and w_r are quadratic in the corresponding V_0 values, V_i and V_r:

$$w_{i,r} = \frac{\sqrt{\pi}}{\sqrt{\lambda_c T}}\frac{V_{i,r}^2}{\hbar} \tag{3.344}$$

This method was used to fit the experimental results shown in Figure 3.38, except that the multichannel $W_R(r)$ was substituted for the single-channel one because the highly exergonic recombination excites the vibronic mode (whose frequency $\omega = 43.5 \times 10^{-3}\,\text{eV}$ and corresponding reorganization energy $0.61\,\text{eV}$ were taken from Ref. 187). IET fits well the experimental data at a reasonable $L = 1.38\,\text{Å}$ and other parameters [98] and reinforces the fact of principal importance: the very existence of the maximum in $Z(\tilde{D})$ dependence.

This maximum separates the branch where Z increases with diffusion from that where Z decreases with it. As we have seen in Figure 3.40(b), the latter can never be reached if the forward and backward transfer rates are the same. In such a case the initial charge distribution is always more remote than the reaction zone, so that ions enter it mainly from outside, only in this case the popular contact approximation for recombination is a reasonable alternative provided

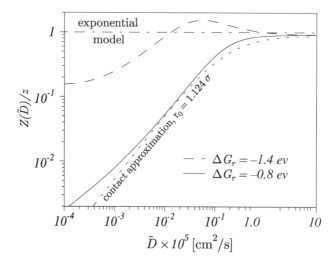

Figure 3.41. The diffusional dependence of the recombination efficiency Z in the contact approximation (dotted line) at starting distance $r_0 = 1.124\,\sigma$ and the same for the remote recombination in a normal (solid line) and inverted (dashed line) Marcus region, in highly polar solvents. The horizontal dashed–dotted line represents the exponential model result, $Z = z = \text{const}$. (From Ref. 152.)

$r_0 - \sigma$ is large enough. The proper choice of r_0 brought closer together the results obtained with this approximation and IET. However, the similarity takes place only in the intermediate, diffusion-controlled, and far kinetic regions (delineated by solid and dotted lines in Fig. 3.41) but not in the static limit (at $\tilde{D} \to 0$). There the true Z is finite or even infinite while the contact estimate tends to zero. Such an evident deficiency of the contact approximation was discussed in text and illustrated in Figures 3.25 and 3.26.

At larger $|\Delta G_r|$ the difference between the IET and contact approximation is more pronounced and not only at slow diffusion, but also in the opposite limit where the IET curve passes through the maximum. This maximum cannot be reproduced either with contact or with the exponential model of the rates. The variation of free energies does not change the exponential shape of the rate, affecting only the preexponential factor chosen from the relationship

$$W_0 \int \exp\left(-\frac{2(r - \sigma)}{L}\right) d^3 r = \int W_{I,R}(r) d^3 r$$

As they are normalized, the exponential rates have the same shape for forward and backward transfers at any ΔG_I and ΔG_R. On the contrary, their Marcus analogies can be of the same shape only if ΔG_r and ΔG_i are equal

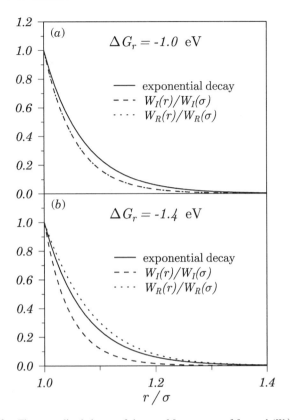

Figure 3.42. The normalized shapes of the true Marcus rates of forward (W_I) and backward (W_R) transfer in a highly polar solvent $(r_c = 0)$, compared to their exponential simplification (solid line). The recombination free energy is either equal to that of ionization (a) or larger than it (b), provided that the excitation energy of donor $\mathscr{E} = 2\,\text{eV}$ is the same in both cases. The other parameters: $\lambda_i = 0.2\,\text{eV}$, $\lambda_0 = 1.0\,\text{eV}$, preexponential factors $w_i = w_r = 100\,\text{ns}^{-1}$, $L = 1\,\text{Å}$, $\sigma = 7\,\text{Å}$. (From Ref. 152.)

[Fig. 3.42(a)], but at $|\Delta G_r| > |\Delta G_i|$ the recombination zone is stretched while the ionization zone is squeezed as compared to the exponential one [Fig. 3.42(b)]. If the exponential rates are used in IET, the recombination can be accelerated only by speeding up the encounter diffusion. For the Marcus rates this is true only for $\Delta G_r \lesssim \Delta G_i$, while for $|\Delta G_r| \gg |\Delta G_i|$ there should be an extremal dependence with a well-defined maximum. These qualitative expectation were proved in Ref. 152, where the straightforward calculation of $Z(\tilde{D})$ with the exact Marcus rates (3.343) and their exponential simplification (3.53) were undertaken. Figure 7 in Ref. 152 demonstrates that all curves representing the exponential transfer rates monotonously increase with diffusion approaching the kinetic limit from below. The exact results appear similar only for the lowest recombination free energy. At larger exergonicity of recombination, the curves calculated with the Marcus rates pass through the maxima and approach the kinetic limit from above.

The Marcus rates are essentially better than their exponential models because they include the space-dependent Arrhenius factor, which causes them to differ for ionization and recombination. However, the position of the maximum on the curve $Z(\tilde{D})$ and its inherent existence depends on the relationship between the free energies of forward and backward electron transfers. As a rule, the situation is favorable for the emergence of a maximum because the backward transfer is usually more exergonic than the forward one. Nonetheless, so far the extremum has been obtained in only one system, and we have to ascertain why it is so exceptional.

C. Initial Correlations

To our knowledge, there are only two reactions of electron transfer, (3.329) and (3.330), studied in solvents of different viscosity. They were investigated separately and presented differently in Figures 3.23 and 3.38. These data, brought together in Figure 3.43, indicate an essential difference between the results related to reactions I and II. The recombination efficiency in reaction II

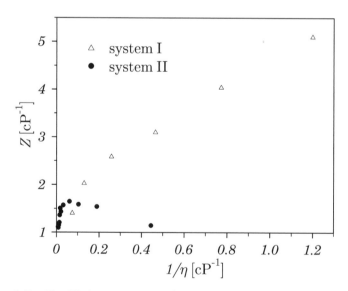

Figure 3.43.   The diffusional dependence of the recombination efficiency Z in reactions (I) and (II) found experimentally in Refs. 148 and 151, respectively.

increases with diffusion in highly viscous solvents, but decreases in solvents of low viscosity. However, in reaction I, studied only in low-viscosity solvents, the behavior of Z is quite the opposite: it monotonously increases with diffusion. Earlier, this phenomenon was attributed to the diffusional delivery of remote RIPs to the contact where they recombine with efficiency (3.210) [147,149,150]. Then this concept was revised and rejected, because neither the RIPs photogenerated in nonviscous solutions are distant nor the recombination zone is contact at electron transfer exergonicities peculiar to reaction I [152]. To explain the difference between reactions I and II, a comprehensive comparison is necessary.

The reaction schemes testify that reactions I and II differ in several aspects. The electron transfer in reaction I occurs between positively charged reactants, while in reaction II they are neutral for forward transfer and oppositely charged for backward transfer. In I the excitation participating in the forward transfer is triplet while in II it is the singlet excited state. Also, there is a pronounced difference in free energies (Fig. 3.44); in reaction I both forward and backward electron transfers are less exergonic than in reaction II. Finally, only in reaction I is there the systematic change of reorganization energy in solvent mixtures along with the viscosity variation.

At first glance the most serious difference is the electrostatic repulsion between similarly charged reactants in reaction I, which is completely absent in reaction II. This is a new feature that was not considered so far in either DET or IET, but can be easily accounted for in both theories if the potential of

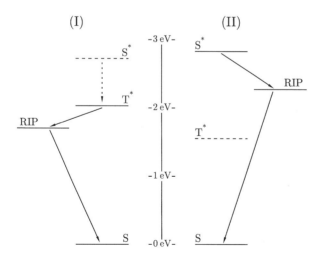

Figure 3.44. The energetic diagram for reaction (I) is to the left and for reaction (II) to the right. States not participating in the transfer reactions are shown by dashed lines.

electrostatic interaction $V(r)$ is known. In the present case this is the standard Debye–Hückel potential (which is zero for reaction II):

$$V(r) = \frac{q_1 q_2 e^2}{\epsilon r(1 + \sigma/\Lambda)} \exp\left(-\frac{r - \sigma}{\Lambda}\right) \qquad (3.345)$$

Here q_1 and q_2 are the ion charges in units of electron charge e, and the so-called Debye screening length

$$\Lambda = \sqrt{\frac{\epsilon T}{4\pi e^2 n_e}} \qquad (3.346)$$

is determined by the number density of ions $n_e = 2 N_A \mu / 1000$, where N_A is Avogadro's number and μ (mol/L) is the ionic strength of the solution.

The Coulomb interactions affect both forward and backward electron transfers, changing their free energies

$$\Delta G_I(r) = I - A - \Delta V(r) = -\mathcal{E} - \Delta G_R(r) \qquad (3.347)$$

where I and A respectively are the ionization potential and electron affinity of the reaction partners, while $\Delta V(r)$ is the difference in their electrostatic interaction before and after forward electron transfer. In both reactions I and II, ΔG_I increases with r as a result of Coulombic interactions, but ΔG_R decreases correspondingly.

The effect of the reactant motion in the electrostatic field is also quite transparent. The definition given in Eq. (3.268) to diffusional operators for the reactants and their products should be generalized

$$\hat{L} = \frac{1}{r^2}\frac{\partial}{\partial r} r^2 D(r) \left[\frac{\partial}{\partial r} + \frac{1}{T}\frac{dV(r)}{dr}\right] \quad \text{and} \quad \hat{L}^+ = \frac{1}{r^2}\frac{\partial}{\partial r} r^2 \tilde{D}(r) \left[\frac{\partial}{\partial r} + \frac{1}{T}\frac{dV^+(r)}{dr}\right]$$

$$(3.348)$$

where V and V^+ are proportional to the product $q_1 q_2$, which differs before and after forward electron transfer. With this generalization the set of UT equations (3.290) and (3.291) remain the same, but the initial condition to the latter must give way to that from (3.17):

$$n(r,0) = e^{-V(r)/T} \qquad (3.349)$$

This is the equilibrium distribution of ions before excitation, which is evidently the eigenfunction of $\hat{L}$.

Due to electrostatic repulsion between reactants, the equilibrium distribution in case I is not homogeneous as in case II. The density of reacting ions is smaller at contact than farther away. This property is transferred to the initial distribution of reactants (3.349), which arises after instantaneous light excitation at $t = 0$. It is qualified as the "initial correlation" between reactants.

In principle, the Coulomb repulsion can affect the distribution of transfer products preventing reactants from coming closer to each other and making the shape of this distribution, $f_0(r)$, more remote than the recombination layer. This could explain why in the fast diffusion limit similarly charged reactants give a viscosity dependence of Z opposite that of the neutral ones, as follows from Figure 3.43. Unfortunately, this hypothesis collapsed when tested numerically: $f_0(r)$ is rather weakly affected by the electrostatic interaction [152]. It is true that the layer for the low exergonic forward electron transfer is narrow and that repulsive Coulomb forces reduce the probability of approaching it. However, the spatial distribution of products is left smooth and after normalization it acquires almost the same shape as in the absence of any electrostatics. Hence, for the opposite low-viscosity dependence of $Z(\tilde{D})$ in reactions I and II, other explanation must be invoked.

D.   Fitting $Z(\tilde{D})$ with IET

Thus, the main question is why the maximum of $Z(\tilde{D})$ and its descending branch are inherent only to reaction II and not reaction I. The efficiency of the ion recombination in reaction I was found to increase monotonously with diffusion (see the triangles in Fig. 3.43). However, this dependence cannot originate from "outer" starts. With the exergonicities of electron transfer specific to this system ($\Delta G_r = -1.7\,\mathrm{eV}$, $\Delta G_i = -0.33\,\mathrm{eV}$), the recombination layer should be more remote than the ionization layer, as in Figure 3.42(b). This leads to "inner" starts at fast diffusion, because the initial RIP distribution in this limit reproduces the ionization layer, which is much sharper. At such starts, Z is expected to decrease with diffusion, but the observed $Z(\tilde{D})$ dependence in reaction I is the opposite. Since the recombination of contact-born RIPs cannot be diffusional, there should be an alternative explanation for this dependence. A similar dependence may arise from the reinforcement of the recombination with an increase in the outer-sphere reorganization energy in solvent mixtures of lower viscosity.

It is a matter of general knowledge that the viscosity varies by changing either the pure solvents or the composition of mixtures being used. Apart from the viscosity, some macroscopic parameters of the solvents also vary in this way. These are the dielectric constant ϵ and refraction index n. Variation of ϵ changes a bit the Onsager radius and the residence time inside the Coulomb well. What is even more important, the variation of n and ϵ affects the outer-sphere reorganization energy, through its contact value λ_0. The latter is expressed in Eq. (3.49) via Pekar's factor

$$\gamma = \left(\frac{1}{n^2} - \frac{1}{\epsilon} \right) \tag{3.350}$$

As a result, the total reorganization energy (3.48) changes with the composition of the solvent, together with its viscosity. In other words, λ depends on $\tilde{D}$, but this important factor was rarely taken into account. Meanwhile, the inverted region can become normal with a sufficient increase in λ_0, making remote recombination contact. With a moderate increase in λ, less dramatic changes occur, but still the maximum of the $Z(\tilde{D})$ curve shifts to a higher $\tilde{D}$ and can be driven out of the available viscosity interval.

To prove this statement an alternative attempt to fit the experimental results of Ref. 148 was made, factoring in the change in the reorganization energy. The set of parameters used for numerical simulations was borrowed from Ref. 148: the free energies of electron transfer, $\Delta G_i = -0.33$ eV and $\Delta G_r = -1.7$ eV, the inner-sphere reorganization energy $\lambda_i = 0.3$ eV, the ionic strength of the solution $\mu = 0.2$ M, the radii of reactants $r_1 = 6$ Å (for the Ru complex), and $r_2 = 4$ Å (for MV^{2+}), which constitute the contact distance $\sigma = r_1 + r_2 = 10$ Å. It follows from this set of data that the Debye screening length $\Lambda = 2.38\sqrt{\epsilon}$ [Å], Onsager radius $r_c = q_1 q_2 570.3/\epsilon$ [Å], and contact reorganization energy

$$\lambda_0 = 1.56\gamma \, [\text{eV}] \tag{3.351}$$

The Stokes–Einstein relationship was used to express the diffusion coefficient through the viscosity of solvent: $\tilde{D}[\text{cm}^2/\text{s}] = 0.894 \times 10^{-5}/\eta[\text{cP}]$. All other parameters that change with viscosity, including Pekar's factor γ, were borrowed or calculated from the same data sources [148]. They are listed in Table I.

Thus, the medium dielectric constant ϵ is varied by 40%, while Pekar's factor γ is changed by 14% in the whole range of available viscosities. The change in ϵ affects the Coulomb interactions in the [Ru$^{3+}\cdots$MV$^+$] pair, while change in γ affects the outer-sphere reorganization energy. The viscosity dependence of both ϵ and γ is accurately interpolated by the two-exponential model function

$$y(\eta) = y_0 + ae^{-(\eta-\eta_0)/x_1} + be^{-(\eta-\eta_0)/x_2} \tag{3.352}$$

with the following parameters listed in Table II.

The accuracy of the interpolation inspected in Ref. 152 was acknowledged as quite satisfactory. Using this interpolation, the authors were able to get smooth curves that approximate the experimental data in the whole region of interest (Fig. 3.45).

Now we are ready to explain the qualitative difference between reaction I, where the recombination efficiency monotonously increases with diffusion, and reaction II, where it passes through the maximum. This difference could not be attributed to either the different charge or spin states of the excited reactants nor

TABLE I
Viscosity Dependence of Electrostatic Parameters

η (cP)	0.834	1.295	2.15	3.884	7.68	13.4
n	1.347	1.365	1.382	1.40	1.416	1.427
ϵ	57.8	53.5	49.1	44.8	40.5	38.3
γ	0.534	0.518	0.503	0.488	0.474	0.465

TABLE II
Two-Exponential Model Parameters

	y_0	a	b	η_0	x_1	x_2
ϵ	36.2	11.98	11.69	0.627	1.21	7.33
γ	0.457	0.04	0.046	0.606	0.98	7.38

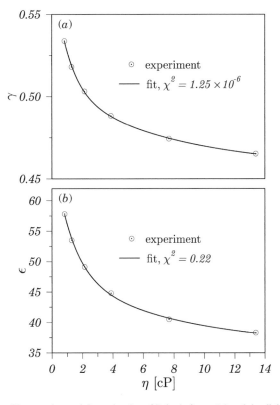

Figure 3.45. The experimental dependencies of Pekar's factor (a) and the dielectric constant ϵ (circles) on the viscosity and their interpolations by Eq. (3.352) (solid lines). (From Ref. 152.)

to the slightly different energetics. The origin of the contradiction lies in the different macroscopic properties of the solvents used in these two systems. In reaction II neither of these properties (like the refraction index and dielectric constant) are changed, while in reaction I they vary together with the solvent viscosity.

Using original programs for solving the integrodifferential equations of IET, all these properties were at first taken into account in Ref. 152. When they are invariant, the non-monotonous diffusional dependence of the recombination efficiency is confirmed. Alternatively, the reorganization energy varies in parallel with the viscosity of the solvent whose composition is varied. There the observable diffusional dependence of the recombination efficiency can be monotonous and well fitted by taking into account just this variation.

In fact, only three parameters have to be varied to fit the diffusional dependence of Z: the preexponents of the transfer rates for forward and backward transfers, w_i and w_r, and the decay length of the tunnelling L. A good fit (solid line on Fig. 3.46) was obtained at rather realistic values: $w_i = 10^3\ \mathrm{ns}^{-1}$, $w_r = 1.1 \times 10^3\ \mathrm{ns}^{-1}$, and $L = 1.35$ Å.

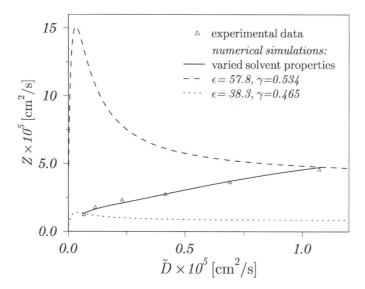

Figure 3.46. The dependence of the recombination efficiency parameter Z on the diffusion coefficient $\tilde{D}$. The experimental data are shown by symbols (triangles). The solid line corresponds to the calculation when the change in dielectric constant ϵ and refraction index n along with diffusion was taken into account. The dashed line corresponds to the calculation with fixed $\epsilon = 57.8$ and $n = 1.347$, related to the fastest diffusion $\tilde{D} = 1.072 \times 10^{-5}\ \mathrm{cm}^2/\mathrm{s}$, and the dotted line corresponds to the fixed $\epsilon = 38.3$ and $n = 1.427$ related to the slowest diffusion $\tilde{D} = 6.67 \times 10^{-7}\ \mathrm{cm}^2/\mathrm{s}$. (From Ref. 152.)

If γ and ϵ were not changed with the solvent viscosity (diffusion), the result would be qualitatively different. To illustrate this point, two other calculations were performed with invariable γ and ϵ. One time their values were fixed, equal to those that correspond to the slowest diffusion ($\tilde{D} = 6.67 \times 10^{-7}\,\text{cm}^2/\text{s}$, $\epsilon = 38.3$, $\gamma = 0.465$), while another time, conversely, they were the same as for the fastest diffusion ($\tilde{D} = 1.072 \times 10^{-5}\,\text{cm}^2/\text{s}$, $\epsilon = 57.8$, $\gamma = 0.5338$). In Figure 3.46, the former is shown by a dotted line and the latter by a dashed line. It can be seen readily that both curves calculated at fixed γ and ϵ pass through the maximum and their viscosity dependence at faster diffusion is opposite that obtained experimentally. The ascending branches of these curves can be attributed to the diffusion delivery of remote RIP into the recombination zone, as was stated in Refs. 149 and 150. However, this does not explain the experimental $Z(\tilde{D})$ dependence for reaction I, as was thought previously. All experimental data for this reaction lie in the range of relatively fast diffusion, when the forward electron transfer is under kinetic control. Such a transfer produces an initial RIPs distribution nearest to contact and almost invariable with further diffusion acceleration [152]. Moreover, their sensitivity to the electrostatic interaction of reactants is very weak, as was stated above. The RIPs do not draw significantly back from contact due to repulsion, and they are from the very beginning almost completely inside the recombination layer. Hence, the diffusional delivery of RIPs into this layer is not necessary and cannot be responsible for the pronounced increase in Z with diffusion in reaction I as is seen in Figure 3.43.

There is now an alternative explanation for this phenomenon: acceleration of RIP recombination with an increase in the reorganization energy λ_0 caused by growing γ. At larger λ, the recombination is not as far in the inverted region. Therefore, the reaction layer is closer to contact and the recombination there is stronger. Consequently, increasing Pekar's factor γ by changing the solvent composition accelerates the recombination as well as its efficiency Z. For a number of reasons the set of parameters used in our calculations cannot be considered unique. However, with any reasonable variation in them, the observed increase in Z with diffusion in reaction I is more likely caused by the collateral change in ϵ and n with solvent viscosity.

VIII. REVERSIBLE PHOTOIONIZATION

The vast majority of works that study the impurity ionization of excited molecules are confined to highly exergonic electron transfer specified by inequality (3.261). Under this condition the reverse electron transfer regenerating the excited state can be forgotten. All photogenerated ions recombine uniquely into the ground state of the neutral products. An important exception to this rule was demonstrated in the pioneering work of Rehm and Weller [53]. This experimental study, discussed in Section III.G (see Figs. 3.13 and 3.14), covers a wide range of free energies, including zero and the positive values of this quantity, wherein the rate of reverse transfer to the excited state, $W_B(r)$, is not negligible. Contrary to Eqs. (3.261) and (3.262) in this region

$$-\Delta G_I \ll T \qquad \text{and} \qquad \frac{W_B}{W_I} = \frac{1}{K} \gtrsim 1 \qquad (3.353)$$

that is, the backward (reverse) transfer to the excited state should be taken into account in both geminate and bimolecular stages of the reaction:

$$D^* + A \underset{W_B}{\overset{W_I}{\rightleftarrows}} [D^+ \cdots A^-] \overset{\bar{\phi}}{\longrightarrow} D^+ + A^- \underset{W_I}{\overset{W_B}{\rightleftarrows}} [D^* \cdots A] \longrightarrow D^* + A$$

$$
\begin{array}{ccccc}
I_0 \uparrow\downarrow \tau_D & \Downarrow W_R & \Downarrow W_R & \downarrow \tau_D & \downarrow \tau_D \\
D & [D \cdots A] & D + A & D \cdots A] & D
\end{array}
\qquad (3.354)
$$

We have seen in Section IV that the study of the reversible reaction of energy transfer was made possible only by means of integral encounter theory. The same is true for reversible electron transfer (3.354) that was first considered with IET in Ref. 188 and then in a much wider context in subsequent publications [107,189].

In dealing with the reversible photoionization, it is very important to differentiate between the steady-state excitation accompanied by stationary fluorescence and free-ion production and the δ-pulse excitation followed by geminate recombination/separation of ions whose yields and subsequent time evolution are studied. They should be considered separately, step by step.

A.   General IET of Photoionization

The main distinction between the reversible photoionization from the irreversible one is brought by the non-zero-element $W_B(r)$ arising in the rate matrix $\hat{\mathbf{W}}(r)$ from Eq. (3.259):

$$\hat{\mathbf{W}}(r) = \begin{pmatrix} -W_I(r) & W_B(r) \\ W_I(r) & -W_R(r) \end{pmatrix} \qquad (3.355)$$

As a result, the kernel $\hat{\mathbf{M}}$ and the matrix $\hat{\mathbf{F}}$ from Eq. (3.260) take the following form:

$$\hat{\mathbf{M}} = \begin{pmatrix} -R^* & R^{\#} \\ R^{\dagger} & -R^{\ddagger} \end{pmatrix} \qquad \text{and} \qquad \hat{\mathbf{F}} = \begin{pmatrix} \nu & g \\ \mu & f \end{pmatrix} \qquad (3.356)$$

In the case of great excess of electron acceptors, condition (3.263) remains valid but the generalized integral equations (3.265) accounts for the reversibility of transfer:

$$\dot{N}^*(t) = -c \int_0^t R^*(\tau) N^*(t-\tau) d\tau + \int_0^t R^{\#}(\tau) [N^{\pm}(t-\tau)]^2 d\tau - \frac{N^*(t)}{\tau_D} + I(t)N$$

$$(3.357a)$$

$$\dot{N}^{\pm}(t) = c \int_0^t R^{\dagger}(\tau) N^*(t-\tau) d\tau - \int_0^t R^{\ddagger}(\tau) [N^{\pm}(t-\tau)]^2 d\tau \qquad (3.357b)$$

Here N^* is the density of the excited molecules while $N^{\pm} = [D^+] = [A^-]$ is the total concentration of radical ions of each sign, in pairs and in the bulk. Equation (3.357a) also includes the excited-state decay with time τ_D and the light pumping with the rate I_0. It implies that the light pumping is too weak to affect the kernels of integral equations as it does at higher light power (see Section X). Even more so, the pumping is assumed to be so weak that it does not exhaust the ground-state population N, which remains the same throughout, approximately equal to the total concentration of donors, $\mathcal{N}$.

Here the kernels R^* and $R^{\dagger}$ are different from their analogs in (3.266a) and (3.266b) because they account not only for forward (W_I) but also for backward transfer to the excited state W_B [in line with recombination to the ground state with a rate $W_R(r)$]. Both kernels are given by their Laplace transformations:

$$\tilde{R}^*(s) = \left(s + \frac{1}{\tau_D}\right) \int [W_I(r)\tilde{v}(r,s) - W_B(r)\tilde{\mu}(r,s)] d^3r \qquad (3.358a)$$

$$\tilde{R}^{\dagger}(s) = \left(s + \frac{1}{\tau_D}\right) \int [W_I(r)\tilde{v}(r,s) - W_B(r)\tilde{\mu}(r,s) - W_R(r)\tilde{\mu}(r,s)] d^3r \quad (3.358b)$$

The auxiliary equations for pair correlation functions also differ from previous ones in the very same respect:

$$\dot{v} = -W_I(r)v + W_B(r)\mu - \frac{1}{\tau}v + \hat{L}v \qquad (3.359a)$$

$$\dot{\mu} = W_I(r)v - W_B(r)\mu - W_R(r)\mu + \hat{L}^+\mu \qquad (3.359b)$$

but the initial conditions for them, $v(r,0) = 1$ and $\mu(r,0) = 0$, as well as the reflecting boundary conditions, remain the same.

The Laplace transformations of the kernels, representing the bimolecular recombination to the ground and excited states, are

$$\tilde{R}^{\ddagger}(s) = s \int [W_R(r)\tilde{f}(r,s) + W_B(r)\tilde{f}(r,s) - W_I(r)\tilde{g}(r,s)]d^3r \qquad (3.360a)$$

$$\tilde{R}^{\#}(s) = s \int [W_B(r)\tilde{f}(r,s) - W_I(r)\tilde{g}(r,s)]d^3r \qquad (3.360b)$$

where the pair distributions obey the following set of auxiliary equations:

$$\dot{f} = W_I(r)g - W_B(r)f - W_R(r)f + \hat{L}^+f \qquad (3.361a)$$

$$\dot{g} = -W_I(r)g + W_B(r)f - \frac{1}{\tau}g + \hat{L}g \qquad (3.361b)$$

There are the initial conditions, $f(r,0) = 1, g(r,0) = 0$, and the reflecting boundary conditions.

B. Stationary Phenomena

The principal advantage of IET as compared to DET is the inclusion of the pumping term into the integral equations. This allows us to find their stationary solutions at steady-state pumping, N_s^* and $N_s^{\pm}$. Setting $I_0 = $ const and $\dot{N}^*(\infty) = \dot{N}^{\pm}(\infty) = 0$, we obtain the following from Eqs. (3.357):

$$N_s^* = \frac{I_0 N \tau_D}{1 + c\kappa_0 \tau_D} = \frac{\tilde{R}^{\ddagger}(0)}{c\tilde{R}^{\dagger}(0)}(N_s^{\pm})^2 \qquad (3.362)$$

Using this result in Eq. (3.8), we see that

$$\eta = \frac{N_s^*(c)}{N_s^*(0)} = \frac{1}{1 + c\kappa_0 \tau_D} \qquad (3.363)$$

obeys the Stern–Volmer law whose constant acquires the following definition [188]

$$\kappa_0 = \kappa_g[1 - \chi\bar{\varphi}] \qquad (3.364)$$

where

$$\kappa_g = \tilde{R}^*(0), \qquad \chi = \frac{\tilde{R}^{\#}(0)}{\tilde{R}^{\ddagger}(0)}, \qquad \bar{\varphi} = \frac{\tilde{R}^{\dagger}(0)}{\tilde{R}^*(0)} \qquad (3.365)$$

Here κ_g is the rate constant of the energy quenching during the primary geminate process, which is not completely irreversible. The ions that avoid geminate recombination and separate with a quantum yield $\bar{\varphi}$ can have the encounters later in the bulk and restore the excitation with a probability χ. The product $\chi\bar{\varphi}$ is a fraction of the excitations restored in subsequent encounters that contribute to the delayed fluorescence.

The stationary photocurrent is determined by the product of the free-ion mobility and their concentration $N_s^{\pm}$ specified in Eq. (3.362). With the notations made above, we can represent the latter in the following form

$$N_s^{\pm} = \sqrt{\frac{c\kappa_g}{\kappa_r}} \, \bar{\varphi} \, N_s^* = \sqrt{\frac{\psi\bar{\varphi}}{\kappa_r}} I_0 N \qquad (3.366)$$

where $\kappa_r = \tilde{R}^{\ddagger}(0)$ is the recombination constant of free ions in a bulk to either the ground or excited state of neutral products. The total ionization quantum yield is

$$\psi = \frac{c\kappa_g\tau_D}{1 + c\kappa_0\tau_D} = c\kappa_g\tau_D\eta \qquad (3.367)$$

but only a fraction of ions, $\bar{\varphi}$, become free. It should be emphasized that $\psi \neq 1 - \eta$ any more. The equality (3.275) holds only in the case of irreversible photoionization when $\chi \propto \tilde{R}^{\#} \propto k_b = 0$ and $\kappa_g = \kappa_0$.

1. Contact Transfer in Both Directions

To be more particular, let us consider the contact electron transfer when all the rates are defined as in Eq. (3.40):

$$W_I = \frac{k_0\delta(r - \sigma)}{4\pi\sigma^2}, \qquad W_B = \frac{k_b\delta(r - \sigma)}{4\pi\sigma^2}, \qquad W_R = \frac{k_c\delta(r - \sigma)}{4\pi\sigma^2} \qquad (3.368)$$

In fact, either the forward or backward transfer (or both) is not contact, but there is no other way to get the results that can be compared with the elementary rate theory of Rehm and Weller considered in Section II G.

In the contact approximation, one of the kernels, $\tilde{R}^*(0)$, was obtained in Ref. 38 and all the rest in Ref. 107

$$\tilde{R}^* = \frac{k_0(1 + k_c\tilde{g}_2)}{Z}, \qquad \tilde{R}^{\#} = \frac{k_b}{Z}, \qquad \tilde{R}^{\dagger} = \frac{k_0}{Z}, \qquad \tilde{R}^{\ddagger} = \frac{k_b + k_c + k_0k_c\tilde{g}_1}{Z} \qquad (3.369)$$

where $Z = (1 + k_0\tilde{g}_1)(1 + k_c\tilde{g}_2) + k_b\tilde{g}_2$

$$\tilde{g}_1(s) = \left[k_D\left(1 + \sqrt{s\tau_d + \tau_d/\tau_D}\right)\right]^{-1}, \qquad \tilde{g}_2(s) = [k_D(1 + \sqrt{s\tau_d})]^{-1} \qquad (3.370)$$

and $\tau_d = \sigma^2/D$ is the encounter time. Using these results in Eqs. (3.365) and the kernels obtained in Eq. (3.364), we obtain the Stern–Volmer constant in the contact approximation:

$$\kappa_0 = \frac{k_0 k_c}{k_b + k_c + k_0 k_c \big/ k_D \left(1 + \sqrt{\tau_d/\tau_D} \right)} \qquad (3.371)$$

According to Eq. (3.273), the same constant for the irreversible transfer is

$$\kappa_i = \frac{k_0}{1 + k_0 \big/ k_D \left(1 + \sqrt{\tau_d/\tau_D} \right)} \qquad (3.372)$$

Utilizing this parameter together with the contact estimates of $k_c = W_R v$ and $k_b = W_B v$, one can bring Eq. (3.371) into a form similar to the Rehm–Weller expression (3.86)

$$\kappa_0 = \frac{\kappa_i}{1 + \kappa_i k_b/k_c k_0} = \frac{\kappa_i}{1 + \kappa_i/k_c K} \qquad (3.373)$$

where $K = k_0/k_b = W_i/W_B$ is as introduced in Eq. (3.262).

Since in the contact approximation

$$k_c K = W_R v K = W_R K_{eq}$$

the last expression in Eq. (3.373) is in fact the non-Markovian generalization of Eq. (3.86) extended to reversible reaction of a different sort: $A + B \Longleftrightarrow C + D$. Substitution of the irreversible non-Markovian Stern–Volmer constant (3.372) for the stationary (Markovian) rate constant $k_i = k_0/(1 + k_0/k_D)$ shifts the diffusional plateau up as shown by the circles in Figure 3.47. At $\tau_D = \infty$, $\kappa_i \equiv k_i$ and our contact IET reduces to Rehm–Weller theory outlined in Section II.F.

C. Charge Separation after Photoionization

The relatively weak pumping produces a moderate number of initially excited donors N_0^* and even a lesser amount of charged products of their ionization. The fast geminate accumulation/recombination of the RIPs in a short run gives way to much longer free-ion recombination in the bulk (see Fig. 3.29). The free ions react during random encounters, with a rate $\tau_f^{-1} \propto [N^\pm]^2$. At $\tau_f \gg \tau_d = \sigma^2/\tilde{D}$ there is an essential time separation between the geminate and bulk recombination that permits us to treat them separately. At stronger pumping τ_f is shorter and such a separation is hardly possible. This situation is more favorable for the observation of delayed fluorescence caused by bimolecular recombination.

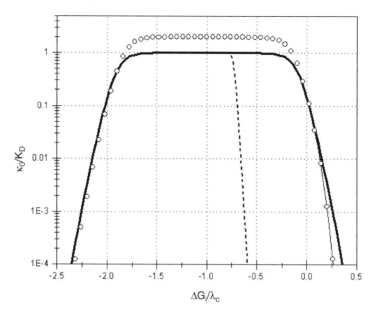

Figure 3.47. The free-energy dependence of the stationary rate constant for irreversible ionization $k \equiv k_i$ defined in Eq. (3.22) (thick line) and the Markovian Stern–Volmer constant for reversible ionization κ_0 from Eq. (3.85) (dashed line). The open circles represent the non-Markovian Stern–Volmer constant of irreversible ionization $\kappa_0 \equiv \kappa_i$ from Eq. (3.27) or (3.372) for $\tau = \tau_d$. The energy of the excited singlet state is $\mathscr{E} = 3.5\lambda_c$ and $\lambda_c = 35T$. (From Ref. 107.)

Turning back to geminate recombination after weak and instantaneous excitation, we simply have to omit the bimolecular terms in Eqs. (3.357):

$$\dot{N}^*(t) = -c \int_0^t R^*(\tau) N^*(t-\tau) d\tau - \frac{N^*}{\tau_D} \qquad (3.374a)$$

$$\dot{N}^{\pm}(t) = c \int_0^t R^{\dagger}(\tau) N^*(t-\tau) d\tau \qquad (3.374b)$$

In addition, the following initial conditions should be added in lieu of the pumping term:

$$N^*(0) = N_0, \qquad N^+(0) = 0$$

From Eqs. (3.374) one can find the short range kinetics, as well as the quantum yields of ionization and charge separation.

In particular, one can obtain the survival probability of excitation $R(t) = N^*(t)/N_0$ that can be used to calculate the quantum yield of nonstationary (geminate) fluorescence (3.10):

$$\eta_0 = \frac{1}{N_0 \tau_D} \int_0^\infty N^*(\tau)\,dt = \frac{1}{1 + c\kappa_g \tau_D} \tag{3.375}$$

which enables us to find the quenching constant $\kappa_g = \tilde{R}^*(0)$ if τ_D is known. However, η_0 should not be confused with the true quantum yield of stationary fluorescence η defined in Eq. (3.8) and disclosed in (3.363), which is expressed through another constant, κ_0 from Eq. (3.364). The latter accounts for the restoration of excitations in bimolecular encounters of ions, while the former is purely a geminate parameter.

This parameter determines not only η_0 but also the quantum yield of free ions

$$\phi = \frac{N^{\pm}(\infty)}{N_0} = c\tilde{R}^\dagger(0)\frac{\tilde{N}^*(0)}{N_0} = \frac{c\,\tau_D\tilde{R}^\dagger(0)}{1 + c\,\tau_D\tilde{R}^*(0)} = \psi_0\bar{\phi} \tag{3.376}$$

where

$$\psi_0 = \frac{c\kappa_g \tau_D}{1 + c\kappa_g \tau_D} = 1 - \eta_0 \tag{3.377}$$

Although this relationship looks similar to Eq. (3.257) for irreversible transfer, the Stern–Volmer constant of the latter ($\kappa_0 \equiv \kappa_i$) is different from κ_g, which accounts for the reversibility of ionization during the geminate stage. The difference between $\kappa_g = \tilde{R}^*(0)$ and its irreversible analog κ_i from (3.372) is worthy of special investigation based on the analysis of pair distribution functions obeying Eqs. (3.359).

1. Geminate Ionization by Reversible Transfer

The general solution of Eq. (3.359b) in the Laplace presentation is

$$\tilde{\mu}(r;s) = \int \tilde{G}(r, r_0; s)W_I(r_0)\tilde{v}(r_0; s)\,d^3r_0 \tag{3.378}$$

It is expressed through the Green function, obeying the following equation:

$$[s + W_B + W_R - \hat{L}^+]\tilde{G}(r, r_0; s) = \frac{\delta(r - r_0)}{4\pi r r_0} \tag{3.379}$$

Although this equation involves the rate W_B of backward electron transfer, it treats this transfer as the irreversible recombination of ions through an additional reaction channel. The general solution of Eq. (3.379) can be represented as follows:

$$\tilde{G}(r, r_0; s) = \tilde{G}_0(r, r_0; s) - \int \tilde{G}_0(r, r'; s)[W_B(r') + W_R(r')]\tilde{G}(r', r_0; s)d^3r'$$

(3.380)

where $\tilde{G}_0(r, r_0; s)$ is the Green function of non-reacting ions obeying the much simpler equation:

$$[s + \hat{L}^+]\tilde{G}_0(r, r_0; s) = \frac{\delta(r - r_0)}{4\pi r r_0}$$

(3.381)

Substituting Eq. (3.378) into Eq. (3.358a), we obtain the geminate quenching constant

$$\kappa_g = \tilde{R}^*(0) = Q\bar{\varphi}_\alpha$$

(3.382)

which is a product of

$$Q = \frac{1}{\tau_D}\int W_I(r)\tilde{v}(r; 0)d^3r \qquad \text{and} \qquad \bar{\varphi}_\alpha = \int \varphi_\alpha(r_0)f_0(r_0)d^3r_0$$

(3.383)

The partial charge separation yields

$$\varphi_\alpha(r_0) = 1 - \int W_B(r)\tilde{G}(r, r_0; 0)d^3r$$

(3.384)

which is the yield of ions that start from distance r_0 and are separated or irreversibly recombine to the ground state. In other words, this is the probability of avoiding reverse electron transfer into the excited state for ions initially separated by distance r_0. If ionization is irreversible ($W_B = 0$), then $\bar{\varphi}_\alpha \equiv 1$ and $\kappa_g \equiv Q = \kappa_0$ from (3.272) is the rate constant of RIP production when RIPs can only recombine to the ground state or separate. However, in the case of reversible ionization ($W_B \neq 0$), all RIPs produced with the rate constant Q and distributed with the normalized density $f_0(r)$ are subjected to backward electron transfer to the excited state. Therefore κ_g is reduced by a multiplier $\bar{\varphi}_\alpha < 1$, which is the averaged quantum yield of irrevocable products of ionization.

The averaging of φ_α in Eq. (3.383) is performed over the normalized density

$$f_0(r) = \frac{W_I(r)\tilde{v}(r, 0)}{\int W_I(r)\tilde{v}(r, 0)d^3r}$$

(3.385)

which is actually the IET distribution of starting interion distances (3.309) but for reversible ionization. In fact, this is an alternative to the UT density $f_0(r, c)$ defined in Section VI.C through the distribution of immobile and non-reacting charge products. When these products are subjected to reverse electron transfer, they totally disappear with time, rendering such a definition impossible. However, the mobile and reactive ions may be ranged over their initial separation, which is an argument of φ_α. Although the distribution over this argument, given in Eq. (3.385) and in its irreversible analog (3.309) have exactly the same form, they are essentially different. In Eq. (3.309) one should use the solution of a single equation (3.267), while to specify the distribution (3.385), one needs to get $\tilde{v}(r,0)$ from the set of equations (3.359) accounting for the reversibility of the reaction.

These two coupled equations involve the rates of ionization, recombination, and reverse transfer to the excited state. However, reverse transfer does not affect the shape of the distribution when ionization is under kinetic control. In such a case $v \approx \exp(-t/\tau_D), \mu \approx 0$, and there is nothing to transfer back from μ to v. Kinetic ionization creates a distribution identical in form to that of $W_I(r)$, regardless of the rate of reverse transfer. Hence, $f_0(r)$ may be affected only if ionization is under diffusional control. The maximal effect is expected to be at $\tau_D = \infty$ when there are stationary functions of distance $n_s(r) = \lim_{s \to 0} s\tilde{v}(s)$ and $m_s(r) = \lim_{s \to 0} s\tilde{\mu}(s)$, with a large dip in $n_s(r)$ near the contact and a hump in $m_s(r)$ at the same place $(n_s + m_s = 1)$.

The change in product distribution resulting from diffusional distortion of n_s and m_s is demonstrated in Figure 3.48 by the example of a system with $\tau_D = \infty$,

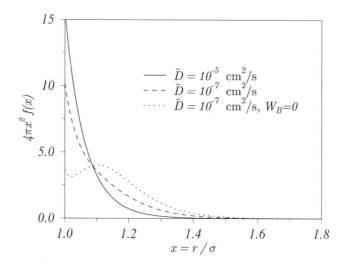

Figure 3.48. The initial distributions of ions in the presence of reverse transfer (dashed line) and in the absence of it (dotted line). Solid line—both of them at faster diffusion. The parameters are: $\Delta G_i = -0.1\,\text{eV}$, $\lambda_c = 0.4\,\text{eV}$, $w_I = 100\,\text{ns}^{-1}$, $L = 1.4\,\text{Å}$, $\sigma = 10\,\text{Å}$, $T = 293\,\text{K}$, $\tau = \infty$. The energy of recombination $|\Delta G_r|$ is assumed so large that $W_R = 0$. (From Ref. 189.)

small $\lambda_c = 0.4\,\text{eV}$, and almost resonant ionization ($\Delta G_i = -0.1$ eV). The excitation energy is assumed to be high enough to exclude the recombination of RIPs to the ground state ($W_R = 0$). To emphasize the effect of reverse transfer, we present calculations in its presence ($W_B = W_I \exp\{\Delta G_I/T\}$) and absence ($W_B = 0$). At fast diffusion ($\tilde{D} = 10^{-5}\,\text{cm}^2/\text{s}$), the ionization is kinetic and, therefore, there is no visible difference between distributions accounting for and not accounting for backward transfer. Both of them have the same shape identical to that of $W_I(r)$ (solid line). At much slower diffusion ($\tilde{D} = 10^{-7}\,\text{cm}^2/\text{s}$), the real initial distribution (long dashed line) even qualitatively differs from that calculated neglecting the reverse process (dotted line). The latter has a maximum shifted from the contact as in Figure 3.33, while the former does not have it. Due to reverse transfer and repeated ionization of the excited state, there is an accumulation of charged products near the contact. As a result, the maximum is smoothed off.

The reverse transfer affects each quantity in Eq. (3.382) but mainly in the region of resonance ($\Delta G_i \approx 0$). This is the favorable interval for reverse transfer which manifests itself by the high but narrow peak in the free energy dependence of Q [solid line in Fig. 3.49(a)]. There $Q > \kappa_g$ because it accounts not only for primary, but all subsequent ionization acts as well. These acts follow reverse electron transfer, which occurs much more often within than outside resonant interval. However, for the same reasons the quantum yield of irrevocable electron transfer $\bar{\varphi}_\alpha$ rapidly drops near the middle of the same interval [Fig. 3.49(b)]. Therefore, the product $Q\bar{\varphi}_\alpha = \kappa_g$ does not experience equally dramatic changes. However, it vanishes with ΔG_i much sharper than its irreversible analog $\kappa_i \equiv \kappa_0$ from (3.254) shown by the short dashed line in Figure 3.49(c).

As was expected, all these effects are almost eliminated when λ_c is large (long dashed lines in Fig. 3.49). In this case the whole curve $\kappa_g(\Delta G_i)$ is shifted to higher exergonicity (where the transfer is irreversible), leaving the resonant strip at far periphery (where the effect is small both with and without reverse transfer).

2. Separation of Reversibly Produced RIPs

The RIP separation quantum yield $\bar{\varphi}$ is expressed in Eqs. (3.365) and (3.276) as a ratio of two kernels. One of them has been already specified in (3.382). Another can be represented in the same way using Eqs. (3.358) and (3.378):

$$\tilde{R}^\dagger(0) = Q\bar{\varphi}_\beta \tag{3.386}$$

where

$$\bar{\varphi}_\beta = \int \varphi_\beta(r_0)f_0(r_0)d^3r_0 \tag{3.387}$$

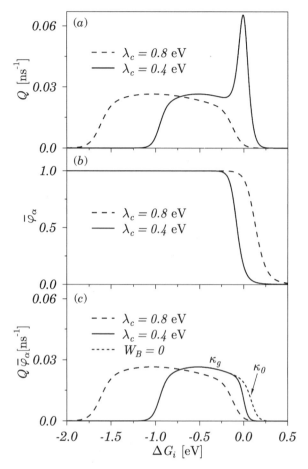

Figure 3.49. The effect of transfer reversibility on the free-energy dependence of the quenching rate components Q and $\bar{\varphi}_\alpha$ as well as on their product $\kappa_0 = \tilde{R}^*(0) = Q\bar{\varphi}_\alpha$. Solid lines indicate $\lambda_c = 0.4$ eV while dashed lines represent $\lambda_c = 0.8$ eV. Other parameters are the same as in Figure 3.48, except $D = \tilde{D} = 10^{-7}$ cm^2/s for all the curves. (From Ref. 189.)

and

$$\varphi_\beta(r_0) = 1 - \int [W_R(r) + W_B(r)]\tilde{G}(r, r_0; 0)d^3r \qquad (3.388)$$

Substituting Eqs. (3.387) and (3.382) into Eq. (3.365), we can express the RIP separation quantum yield as the ratio of $\bar{\varphi}_\beta$ and $\bar{\varphi}_\alpha$:

$$\bar{\varphi} = \frac{\tilde{R}^\dagger(0)}{\tilde{R}^*(0)} = \frac{\bar{\varphi}_\beta}{\bar{\varphi}_\alpha} \qquad (3.389)$$

When the ionization is irreversible, $\bar{\varphi}_\alpha = 1$, the charge separation quantum yield is equal to $\bar{\varphi}_\beta$, which becomes the share of RIPs that escaped geminate recombination into the ground state. In more general case, when the reverse electron transfer into the excited state can not be neglected, this is the fraction of ions that escaped any recombination, in either the ground or excited states. To clarify this point, let us illustrate it by an example of contact electron transfer.

D. Contact Approximation

As follows from the sets of Eqs. (3.383)/(3.384) and (3.387)/(3.388), the contact quantities

$$\bar{\varphi}_\alpha = \varphi_\alpha(\sigma) = 1 - k_b \tilde{G}(\sigma,\sigma;0) \quad \text{and} \quad \bar{\varphi}_\beta = \varphi_\beta(\sigma) = 1 - (k_b + k_c)\tilde{G}(\sigma,\sigma;0) \tag{3.390}$$

are expressed via the single Green function $\tilde{G}(\sigma,\sigma;0)$. Using the contact rates (3.368) in Eq. (3.380), we transform it into an algebraic equation, leading to the following relationship between the contact Green functions:

$$\tilde{G}(\sigma,\sigma;s) = \frac{\tilde{G}_0(\sigma,\sigma;s)}{1 + [k_b + k_c]\tilde{G}_0(\sigma,\sigma;s)} \tag{3.391}$$

In highly polar solvents $\tilde{G}_0(\sigma,\sigma;s) \equiv \tilde{G}_0(\sigma;s)$ obeys Eq. (3.240), from which it follows that $\tilde{G}_0(\sigma,\sigma;0) = 1/k_D$. Using this value in Eq. (3.391), it is easy to obtain the following equation from Eq. (3.390):

$$\varphi_\alpha(\sigma) = \frac{k_D + k_c}{k_D + k_c + k_b}, \quad \text{and} \quad \varphi_\beta(\sigma) = \frac{k_D}{k_D + k_c + k_b} \tag{3.392}$$

1. Charge Separation

Substituting these results into Eq. (3.389) one obtains the contact estimate of the charge separation yield, which is exactly the same as φ_c in Eq. (3.211) [38]:

$$\varphi(\sigma) = \frac{\varphi_\beta(\sigma)}{\varphi_\alpha(\sigma)} = \frac{1}{1 + k_c/k_D} \tag{3.393}$$

From Figure 3.50 we see how the reverse transfer to the excited state reduces $\varphi_\beta(\sigma)$ in relationship to $\varphi(\sigma)$, and how small $\varphi_\alpha(\sigma)$ is compared to the value 1, which $\varphi_\alpha(\sigma)$ takes in the case of irreversible transfer. However, the ratio of these quantities (3.393) remains unchanged at any rate of reverse transfer. This is a fraction of the free ions from the total amount of irrevocable products of reversible ionization. Each portion of photogenerated RIPs adds some free ions

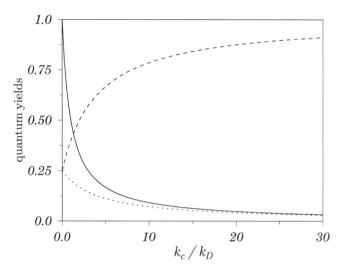

Figure 3.50. The partial quantum yields $\varphi_\alpha(\sigma)$ (dashed line) and $\varphi_\beta(\sigma)$ (dotted line) in the contact approximation and their ratio $\varphi(\sigma)$ (solid line) as functions of the recombination rate constant k_c (at $k_b/k_D = 3$). (From Ref. 189.)

and neutral particles to those produced earlier and partially restores the excitations from which the next portion of RIPs is generated, later on. This cycle is repeated many times until the excitations completely vanish. Each time, exactly the same fraction of irrevocable ionization products escape recombination. This fraction is permanent for each portion of photogenerated RIPs and thus remains the same for all of them. Therefore, in the contact approximation $\bar{\varphi}$ does not depend on reverse transfer at all. It was demonstrated in Ref. 189 that this conclusion holds for noncontact reversible ionization as well.

2. Geminate Quenching

However, the yields of nonstationary fluorescence and ionization, (3.375) and (3.377), are affected as much as

$$\kappa_g = \frac{\kappa_i}{1 + \kappa_i / K(k_D + k_c)} \tag{3.394}$$

where $K = k_0/k_b$ was defined in Eq. (5.6) and κ_i in Eq. (3.262) [38]. Let us compare κ_g and κ_i as functions of the ionization free energy (Fig. 3.51). As always, κ_i originates from the quasiparabola $k_0(\Delta G_i)$ cut from the top by a diffusional plateau. As far as $\kappa_g \approx \kappa_i$, it has the same shape but this is not the case where $\Delta G_i \approx \lambda_c$. There the forward transfer is under kinetic control [i.e.,

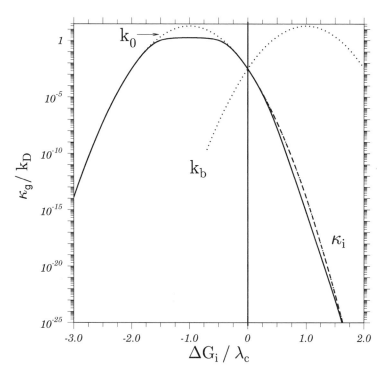

Figure 3.51. The quenching constant for geminate reversible ionization κ_g from Eq. (3.394) (solid line), compared to that for the irreversible one, κ_i (dashed–dotted thick line). Dotted lines represent the FEG laws for the forward and backward kinetic constants, k_0 and k_b.

$\kappa_i(\lambda_c) \approx k_0(\lambda_c) \ll k_D]$, but the reverse transfer to the excited state becomes diffusional near the maximum of $k_b(\Delta G_i)$: $k_b(\lambda_c) = k_0(-\lambda_c) \gg k_D$. Therefore at small k_c the denominator in Eq. (3.394) exceeds 1, making κ_g a bit smaller than κ_i [189]:

$$\kappa_g \approx \frac{k_0}{1 + k_b/(k_D + k_c)} \approx K(\lambda_c)\,k_D \qquad \text{at} \qquad k_b \gg k_D \gg k_0, k_c \qquad (3.395)$$

However, no dramatic changes occur in κ_g, even at $k_c = 0$. In this case the electron transfer produces only free ions with a quantum yield $\psi_0 = 1 - \eta_0$ and the quenching constant (3.394), which remains finite at any k_c.

3. Stationary Quenching

Unlike κ_g, the Stern–Volmer constant κ_0 from Eq. (3.373) turns to zero as $k_c \to 0$. If there is no recombination to the ground state, then all separated ions sooner or later recombine to the excited state in the bulk. Since all ions are subjected to recombination, either geminate or bimolecular, all of them contribute to the stationary fluorescence. Therefore the fluorescence quantum yield becomes 1 and κ_0 turns to zero.

At finite k_c there are two sharply separated regions. Where ionization is highly exergonic, the quenching can be considered as irreversible insofar as $\kappa_0 \approx \kappa_i \approx \kappa_g$. On the contrary, in the highly endothermic region κ_0 turns to zero because the transfer becomes fully reversible. The transition from one region to another is governed by the probability of the excited-state restoration χ defined in Eq. (3.365). Using the contact definitions (3.369), one can find that

$$\chi = \frac{k_b}{k_b + k_c \left[1 + \frac{k_0}{k_D\left(1+\sqrt{\tau_d/\tau_D}\right)}\right]} = \begin{cases} 1 & \text{at} \quad k_b \gg k_c \\ 0 & \text{at} \quad k_b \ll k_c \end{cases} \qquad (3.396)$$

The border between the irreversible and reversible ionization lies at a ΔG_i value where the rate of the backward transfer to the ground state equals the rate of reverse transfer to the excited state: $k_b(\Delta G_i^\circ) = k_c(\Delta G_i^\circ)$. As a rule ΔG_i° is negative and not small, so that the border (indicated by the dashed line in Fig. 3.47) is far below the resonance reached at $\Delta G_i = 0$. To shift it up another channel of charge, recombination should be opened and its rate must be faster than k_c. This ionic reaction may be parallel to that included in scheme (3.90) or recombination through the triplet channel proposed in Refs. 107 and 150. The latter is discussed in Section XI among other reactions affected by the spin conversion.

E. Delayed Fluorescence

When geminate recombination is completed and gives way to a bimolecular reaction, the latter restores either the excited- or ground-state neutral products. Since the density of restored states at a late stage of excitation decay is quadratic in the concentration of free ions, the latter should be large enough to make the delayed fluorescence detectable. To reach this goal, one should use strong pumping and fluorophores with long-lived excited states. Then their fluorescence at times $t > \tau_D$ will be stronger than the natural one [Fig. 3.52(a)].

1. Charge Recombination

The long-time asymptote of charge recombination can be described by a single Eq. (3.357), setting the geminate term to zero and changing the initial condition from $N^\pm = 0$ to another one following from Eq. (3.376):

$$N^\pm(0) = N_0 \phi = N_0 \psi_0 \bar\varphi \qquad (3.397)$$

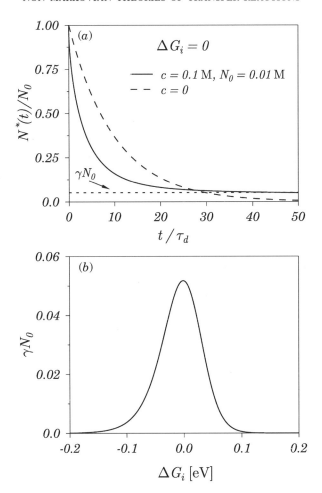

Figure 3.52. The kinetics of excitation quenching followed by delayed decay [solid line in (a)] in comparison to the free exponential decay with time τ_D [long dashed line in (a)]. The short dashed horizontal line in (a) indicates the amplitude of the delayed fluorescence γN_0, whose free energy dependence is outlined in (b). The initial concentration of the excitations $N_0 = 0.01$ M, $\tau_D = 1\,\mu$s. The other parameters are $D = \tilde{D} = 10^{-7}\,\text{cm}^2/\text{s}$, $r_c = 0$, $\lambda_c = 0.4\,\text{eV}$, $c = 0.1$ M, $N_0^* = 10^{-4}$ M, $w_i = 10^3\,\text{ns}^{-1}$, $L = 1.4\,\text{Å}$, $\sigma = 10\,\text{Å}$, $T = 293$ K. (From Ref. 189.)

The recombination term of the reduced integral equation can be transformed into its differential analog as in Eq. (3.278). In the Markovian limit this differential equation reduces to the conventional one with the rate constant $k_r = k_R(\infty)$ [48]:

$$\dot{N}^{\pm} = -k_r[N^{\pm}]^2 \qquad (3.398)$$

Since the ions are stable particles, $k_r \equiv \kappa_r = \tilde{R}^{\ddagger}(0)$ is the same constant as in Eq. (3.366). In the contact approximation we obtain from Eqs. (3.369) and (3.370):

$$k_r = \frac{(k_c + \kappa_b)k_D}{(k_c + \kappa_b) + k_D} \tag{3.399}$$

Here $k_c + \kappa_b$ is an analog of the kinetic rate constant for the two-channel recombination. The augend is for the ground state, while the addend is for the excited state:

$$\kappa_b = \frac{k_b}{1 + \frac{k_0/k_D}{1 + \sqrt{\tau_d/\tau_D}}} \tag{3.400}$$

Here exactly the same correction for the nonstationary ionization appears in the denominator as in Eq. (3.372). As a result, an equilibrium relationship between these constants is preserved:

$$\frac{\kappa_i}{\kappa_b} = \frac{k_0}{k_b} = K = e^{-\Delta G_i/T}$$

The solution to Eq. (3.398) reproduces the usual second-order kinetics

$$\frac{1}{N^{\pm}(t)} = \frac{1}{N^{\pm}(0)} + k_r t \tag{3.401}$$

where $N^{\pm}(0)$ is specified in Eq. (3.397).

2. Fluorescence

As $t \to \infty$, there should be the quasistationary solution of Eq. (3.357) transformed in this limit to the following one:

$$0 = -c\tilde{R}^*(0)N^*(t) + \tilde{R}^{\#}(0)[N^{\pm}(t)]^2 - \frac{N^*(t)}{\tau_D}$$

The delayed fluorescence is proportional to the quasistationary concentration of excitations

$$N^*(t) = \kappa^* \tau^* [N^{\pm}]^2(t) \tag{3.402}$$

where $\kappa^* = \tilde{R}^{\#}(0)$, $1/\tau^* = 1/\tau_D + c\kappa_g$, and $\kappa_g = \tilde{R}^*(0)$ was specified in Eq. (3.394). By substituting Eq. (3.401) into Eq. (3.402), we obtain the long-time asymptote

$$N^*(t) = \frac{A}{t^2} \tag{3.403}$$

where $A = \kappa^* \tau^* / k_r^2$.

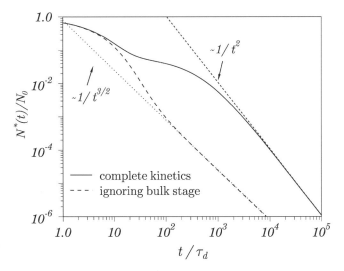

Figure 3.53. The quenching kinetics at long times with and without bulk recombination of ions (solid and long dashed lines, respectively). The false IET asymptote ($t^{-3/2}$) is indicated by a dotted line, while the true asymptotic behavior of delayed fluorescence (t^{-2}) is shown by a short dashed line. All the parameters are the same as for Figure 3.52. (From Ref. 189.)

The experimental registration of delayed fluorescence is problematic because of the very rigid conditions for its observation. According to Eqs. (3.402) and (3.397), the number of second-time excited molecules does not exceed $\kappa^*\tau^*[N^\pm(0)]^2 = \gamma N_0^2$, where $\gamma = \kappa^*\tau^*(\psi_0\bar{\varphi})^2$. Hence, the amplitude of the delayed fluorescence is smaller than the initial one by a factor γN_0. This factor, studied in Ref. 189, was shown to be nonzero only near $\Delta G_i \approx 0$ [Fig. 3.52(b)].

In Figure 3.53 the long-time decay of the excited state is shown with and without factoring in the charge recombination in the bulk, which is negligible at a low concentration of ions (weak excitation). In the latter case there is a false asymptotic behavior at long times, noticed in Refs. 48 and 51 but known even earlier [190]:

$$N^*(t) \propto t^{-3/2} \qquad (3.404)$$

This behavior, inherent to the IET description of either reversible or irreversible transfer, can be eliminated using modified integral encounter theory (MET) [41,44], or an improved superposition approximation [51,126].

The substitution of MET for IET is inevitable for studying the long-time asymptote of highly exergonic ionization, which is practically irreversible. However, in the case of reversible ionization, the restoration of excitations through ion recombination in the bulk hinders their decay, making it a much slower quasi-stationary evolution (3.402). At strong pumping this may occur before the false asymptote is reached (Fig. 3.53). As a result, the whole excitation decay from the very beginning until the end is perfectly described by IET. As shown in Figure 3.53, the delayed fluorescence at long times follows the power law (3.403), which is easy to distinguish from anything else.

F. Electrochemiluminescence

The right half of reaction (3.354) can be realized apart from the left half. This can be done by the electrochemical injection of ions, subjected to the subsequent recombination to either the ground or excited states of neutral products:

$$
\begin{array}{ccccc}
& & W_B & & \\
D^+ + A^- & \xrightleftharpoons[W_I]{} & [D^* \cdots A] & \longleftrightarrow & D^* + A \\
\Big\downarrow W_R & & \Big\downarrow \tau_D & & \Big\downarrow \tau_D \\
D + A & & [D \cdots A] & & D
\end{array}
\qquad (3.405)
$$

This is the process studied experimentally in Refs. 191 and 192 through the luminescence coming from the triplet excited state D^*. The quantum yield of excitations ϕ_{es} can be extracted from the electro-chemiluminescence quantum yield ϕ_{ecl} if the emission quantum yield from the excited state ϕ_e is known [191]:

$$
\phi_{ecl} = \phi_e \phi_{es} \qquad (3.406)
$$

In Refs. 191 and 192 the ϕ_{es} dependence on the free energy of ionization, $\Delta G_i = -\Delta G_b$, was measured for a number of systems.

To specify this dependence, we have to calculate

$$
\phi_{es} = \frac{\int_0^\infty N^*(t)dt}{N^\pm(0)\tau_D} \qquad (3.407)
$$

borrowing $N^*(t)$ from the solution of Eqs. (3.357), where we set $I_0 = 0$ and use the new initial conditions created by the external injection of ions into the solution:

$$
N^*(0) = 0, \quad N^\pm(0) = N_0^\pm \qquad (3.408)
$$

Making the Laplace transformation of Eqs. (3.357) under these conditions and excluding $\int_0^\infty [N^\pm]^2 \, dt$, the following result was obtained [189]:

$$
\phi_{es} = \frac{\tilde{N}^*(0)}{N_0^\pm \tau_D} = \left\{ \frac{\tilde{R}^\ddagger(0)}{\tilde{R}^\#(0)} + c\tau_D \left[\frac{\tilde{R}^*(0)\tilde{R}^\ddagger(0)}{\tilde{R}^\#(0)} - \tilde{R}^\dagger(0) \right] \right\}^{-1} \qquad (3.409)
$$

Using expressions for the kernels listed in Eq. (3.369), we obtain from Eq. (3.409) the contact estimate of the excited-state quantum yield [189]

$$\phi_{es} = \frac{1}{1 + k_c(1 + c\tau_D\kappa_i)/\kappa_b} \qquad (3.410)$$

where κ_i and κ_b are as given by Eqs. (3.372) and (3.400).

In Refs. 191 and 192 the recombination into the excited state was considered as an irreversible reaction. This can be true if $\Delta G_i \gg T$. Then $k_0 \approx 0$ ($\kappa_i \approx 0$ and $\kappa_b \approx k_b$), and the general Eq. (3.410) reduces to the simplest one used in these works:

$$\phi_{es} = \frac{1}{1 + k_c/k_b} \qquad (3.411)$$

Using the contact estimates of the kinetic rate constants following from Eqs. (3.343)

$$k_c \approx W_R(\sigma)v = w_r v \exp\left[-\frac{(\Delta G_r + \lambda_c)^2}{4\lambda_c T}\right] \qquad (3.412a)$$

$$k_b \approx W_B(\sigma)v = W_I(\sigma)v e^{\Delta G_i/T} = w_i v \exp\left[-\frac{(\Delta G_i - \lambda_c)^2}{4\lambda_c T}\right] \qquad (3.412b)$$

we obtain the following from Eq. (3.411):

$$\phi_{es} = \left[1 + \frac{w_r}{w_i}\exp\left(-\frac{\mathscr{E}(\Delta G_i - \lambda_c + \mathscr{E}/2)}{2\lambda_c T}\right)\right]^{-1} \qquad (3.413)$$

This is a stepwise function that approaches unity with increasing ΔG_i, because the excited state becomes the unique reaction product when the recombination to the ground state is switched off. This function, shown in Figure 3.54, resembles the experimental results obtained in Refs. 191–193. A typical example of such data is shown in Figure 3.55. However, the plateau approached by most of the curves obtained experimentally is lower than 1. This can be an indication of some unknown channel of charge recombination or additional quenching of excitations by either the survived ions or through biexcitonic annihilation of triplets [194] (see Section XIII.C).

When the electron transfer is reversible, the simple expression (3.411) is nevertheless valid at low concentrations but only at fast diffusion, when $k_0/k_D \ll 1$ and $\kappa_b = k_b$. At $\tilde{D} = 10^{-6}\ \mathrm{cm^2/s}$, the results obtained numerically

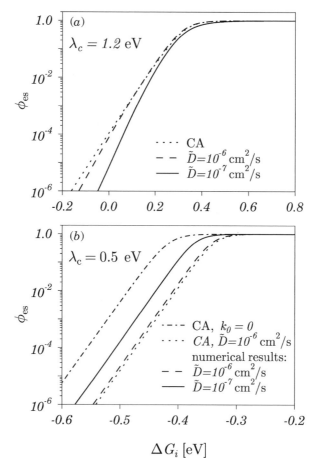

Figure 3.54. The quantum yield of excited states as a function of the ionization free energy at the large (a) and small (b) reorganization energy of electron transfer. The numerically calculated curves for remote transfer are shown by solid lines for slow diffusion ($\tilde{D} = 10^{-7}$ cm^2/s) and by the long dashed lines for the faster one ($\tilde{D} = 10^{-6}$ cm^2/s). Contact calculations (CAs) of the same curves are made with (dotted lines) and without taking into account the reverse electron transfer to the ionized state (dashed–dotted line). (From Ref. 189.)

for the remote transfer are practically the same as follow from the contact formula (3.410) [upper curves in Fig. 3.54(a) and lower curves in Fig. 3.54(b)]. However, at slower diffusion, $\tilde{D} = 10^{-7}$ cm^2/s, the exact curves obtained numerically deviate significantly from the contact ones and in opposite directions in cases (a) and (b).

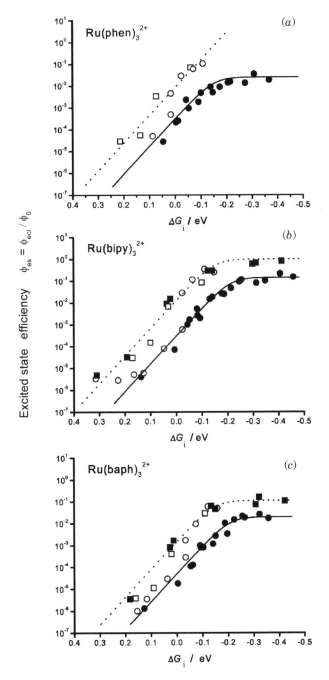

Figure 3.55. Plot of ϕ_{es} versus ΔG_i for the ECL systems involving Ru(phen)$_3^{2+}$ (a), Ru(bipy)$_3^{2+}$ (b), and Ru(baph)$_3^{2+}$ (c) ions in 0.1 M (C$_2$H$_5$)$_4$NPF$_6$ acetonitrile solutions. Data for the ECL systems with nitrocompounds (●), quinones (○), N-methylpyridinium cations (□), and aromatic amines or 2,3,7,8-tetramethoxythianthrene (■). (From Ref. 193.)

According to Eq. (3.413), the competition between recombination and excitation breaks down at

$$\Delta G_i^* = \lambda_c - \frac{\mathscr{E}}{2} \qquad (3.414)$$

Below this boundary the recombination to the ground state is preferable; above it, the excitation dominates. If $\mathscr{E} = 1.6\,\mathrm{eV}$, then at large $\lambda_c = 1.2\,\mathrm{eV}$ the border free energy, $\Delta G_i^* = 0.4\,\mathrm{eV}$, is positive, while at small $\lambda_c = 0.5\,\mathrm{eV}$ the border free energy is negative: $\Delta G_i^* = -0.3\,\mathrm{eV}$. These are cases (a) and (b) compared in Figure 3.54. Beyond the contact approximation the relative positions of the excitation and ionization layers in these cases are the opposite.

In case (a) the excitation is exergonic and therefore strong and remote unlike the endergonic ionization, which is weak and near-contact. The latter plays a minor role because the excitations born in the outer layer can freely separate. However, the excitation layer is reached only by ions that crossed the most remote reaction layer, where they recombine to the ground state. With decreasing diffusion the recombination layer becomes more "black," less transparent for ions, thus decreasing the fraction of them participating in the excitation. As a result, their quantum yield decreases with lowering diffusion [solid line in Fig. 3.54(a)].

The effect is the opposite in case (b), where the situation is inverted. The excitation is weakest and closest to the contact while the much stronger ionization occurs around it, screening excited molecules from coming out. Roughly speaking, they are captured in a narrow strip near the contact and the longer the time, the slower the diffusion. If ionization is artificially switched off, setting $k_0 = 0$, then the doors open and the larger fraction of excitation survives (dashed–dotted curve). If they are closed but diffusion decreases, lengthening the lifetime of excitation in prison, then the yield (3.407) proportional to $\int_0^\infty N^*(t)\,dt$ also increases although not as much [the solid line in Fig. 3.54(b)].

IX. IRREVERSIBLE TRANSFER AT A SHORTAGE OF QUENCHERS

So far the bimolecular ionization of photoexcited molecules was theoretically investigated assuming a huge excess of quenchers and the weakest light pumping. The latter was usually represented as either a short pulse modeled as a δ function or quasistationary excitation that is switched on as a ζ pulse:

$$I = \gamma\delta(t) \qquad\qquad \text{short pulse} \qquad (3.415a)$$

$$I = I_0\zeta(t) = \begin{cases} 0 & \text{at} \quad t < 0 \\ I_0 & \text{at} \quad t \geq 0 \end{cases} \qquad \text{long pulse} \qquad (3.415b)$$

Here $\gamma = N^*(0)/\mathscr{N}$ is a fraction of the excited molecules immediately after the pulse and $\mathscr{N} = N^* + N^\pm + N$ is the total concentration of donors participating in the simplest irreversible reaction (3.257), which we are addressing. We will also restrict our consideration to moderate light intensities when

$$I_0 \ll \frac{1}{\tau_D} \lesssim \frac{1}{\tau_d} \tag{3.416}$$

This assumption, which ensures that $N^* \ll N \approx \mathscr{N}$, allows us to avoid the light saturation that will be considered in the next section.

At the same time we do not have to assume that $N^* \ll c$ as we did previously. Under this condition the acceptor concentration $A \equiv [\mathbf{A}]$ remained almost constant, approximately equal to its initial value c. In what follows we will eliminate this restriction and account for the expendable neutral acceptors whose concentration $A(t)$ decreases in the course of ionization. When there is a shortage of acceptors, the theory becomes nonlinear in the concentration, even in absence of bulk recombination. Under such conditions only general encounter theories are appropriate for a full timescale (non-Markovian) description of the system relaxation. We will compare them against each other and with the properly generalized Markovian and model theories of the same phenomena.

A. Encounter Theories for Moderate Pumping

Instead of Eqs. (3.265), one can obtain the following set of non-linearized IET equations from the general theory

$$\dot{N}^* = -\int_0^t R^*(t')N^*(t-t')A(t-t')dt' - \frac{N^*}{\tau_D} + IN \tag{3.417a}$$

$$\dot{N}^\pm = \int_0^t R^\dagger(t')N^*(t-t')A(t-t')dt' - \int_0^t R^\ddagger(t')[N^\pm(t-t')]^2 dt' \tag{3.417b}$$

where $A(t) = c - N^\pm(t)$ but $N \approx \mathscr{N} = \text{const}$. At light intensity limited by inequality (3.416), the kernels are not affected by the light pumping; they remain the same as in Eqs. (3.266). The auxiliary equations (3.267) for pair distributions, as well as their initial and boundary conditions, also remain unchanged.

The bimolecular ionization producing the energy quenching and primary charge separation is represented by the first integral terms in Eqs. (3.417a) and (3.417b), containing kernels R^* and $R^\dagger$ correspondingly. The latter accounts for the accumulation of ion pairs and their geminate recombination. The final bimolecular recombination of free ions in the bulk is represented by a kernel $R^\ddagger$ from the last term in (3.417b), which is quadratic in ion concentration.

If one produces the excitation with a δ pulse that is so weak that the initial concentration of excited donors satisfies the inequality $N^*(0) = \gamma \mathcal{N} \ll c$, then the acceptors are in great excess from the very beginning. Under such a condition their concentration $A(t) = c$ remains constant at all times and Eqs. (3.417) can be subjected to linearization leading to Eqs. (3.265). These IET equations can be further reduced to UT equations (3.277), which are also linear in c [178]. However, this weakness was overcome in Ref. 195, where the following set of equations was proposed instead:

$$\frac{dN^*(t)}{dt} = -k_I(t)N^*(t)A(t) - \frac{N^*(t)}{\tau_D} \tag{3.418a}$$

$$\frac{dN^{\pm}(t)}{dt} = \int \frac{\partial}{\partial t} p(r,t) d^3 r - k_R(t)[N^{\pm}(t)]^2 \tag{3.418b}$$

The time-dependent ionization and recombination constants are defined by Eqs. (3.276) and (3.277) as before, but $cm(r,t)$ is substituted by $p(r,t)$, which obeys the following equation:

$$\left[\frac{\partial}{\partial t} + W_R(r) - \hat{L}^+\right] p(r,t) = W_I(r)n(r,t)N^*(t)A(t) \tag{3.419}$$

By substituting $(\partial/\partial t)p(r,t)$ from this equation into Eq. (3.418) and integrating the latter over r, we arrive at the final form of the nonlinear UT equations:

$$\frac{dN^*(t)}{dt} = -k_I(t)N^*(t)A(t) - \frac{N^*(t)}{\tau_D} \tag{3.420a}$$

$$\frac{dN^{\pm}(t)}{dt} = k_I(t)N^*(t)A(t) - \int W_R(r)p(r,t)d^3 r - k_R(t)[N^{\pm}(t)]^2 \tag{3.420b}$$

Equations (3.420)–(3.421) constitute the formal basis of the extended unified theory, which accounts for the nonlinear effects resulting from the bimolecular production and recombination of ions. Only one step remains to be taken to go from UT to its simplified, Markovian version, which describes the asymptotic relaxation of the system long after the pulse. Using the Green function of Eq. (3.419), one can represent the space integral of its solution in the following form:

$$\int p(r,t)d^3 r = \int_0^t \int d^3 r' \int G_R(r,r',t-t')d^3 r W_I(r')n(r',t')N^*(t')A(t')dt' \tag{3.421}$$

In the limit of the large times we have

$$n(r,t) \approx n(r,\infty) = n_s, \qquad \int G_R(r,r',\infty)d^3r = \varphi(r')$$

Coming to this extreme in Eq. (3.421), we get

$$\int p(r,t)d^3r = k_i \bar{\varphi}_m \int_0^t N^*(t')A(t')dt' \qquad (3.422)$$

Here

$$k_i = \lim_{t\to\infty} k_I(t) = \int W_I(r)n_s(r)d^3r, \quad \text{and} \quad \bar{\varphi}_m = \int \varphi(r')f_m(r')d^3r' \qquad (3.423)$$

is the Markovian initial distribution of ions, which follows from Eq. (3.312) as $\tau_D \to \infty$:

$$f_m(r) = \lim_{\tau_D\to\infty} f_0^{iet}(r) = \frac{W_I(r)n_s(r)}{\int W_I(r')n_s(r')d^3r'} \qquad (3.424)$$

After the differentiation expression (3.422) with respect to time, one can substitute the result into Eq. (3.418b). Then in the Markovian limit $t \to \infty$ the whole set (3.418) is transformed to

$$\frac{dN^*(t)}{dt} = -k_i N^*(t)A(t) - \frac{N^*(t)}{\tau_D} \qquad (3.425a)$$

$$\frac{dN^\pm(t)}{dt} = k^\pm N^*(t)A(t) - k_r[N^\pm(t)]^2 \qquad (3.425b)$$

where

$$k_r = \lim_{t\to\infty} k_R(t), \quad \text{and} \quad k^\pm = k_i \bar{\varphi}_m \qquad (3.426)$$

Here an important definition for the rate constant of free carrier production, $k^\pm$, is given. The latter differs from the ionization rate constant by a multiplier equal to the charge separation quantum yield $\bar{\varphi}_m$, obtained in the Markovian approximation. This difference indicates that the number of photogenerated ions that avoid geminate recombination and become free is less than their total amount, $\bar{\varphi}_m < 1$.

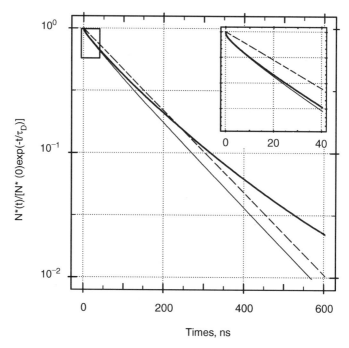

Figure 3.56. Semilogarithmic plot of the irreversible quenching kinetics with a large excess of acceptors $[N^*(0) = 10^{-4}\,M \ll 10^{-2}\,M = c]$ calculated with IET (thick line), UT (thin line), and Markovian theory (dashed line). The remaining parameters are $\sigma = 6\,\text{Å}$ the preexponent for the ionization rate $w_i = 1000\,\text{ns}^{-1}, D = 1.2 \times 10^{-6}\,\text{cm}^2/\text{s}, l = 1.0\,\text{Å}, \; k_i = 1271\,\text{Å}^3/\text{ns}$. The initial nonstationary quenching is shown in the insert. (From Ref. 195.)

Unfortunately, the Markovian theory has a number of drawbacks when applied to an essentially non-Markovian problem like that under consideration. The reduction of UT to its Markovian version deprives UT, misrepresenting the kinetics of energy quenching and charge accumulation at short τ_D. In Figure 3.56 we compare the kinetics of the excited state quenching in IET, UT, and the Markovian theory using the exponential model for the transfer rate, Eq. (3.53). The UT of irreversible transfer, known to be exact in the case of the "target problem" (immobile $\mathbf{D}^*$ and an independently moving point particles $\mathbf{A}$), is taken as a primary standard. IET reproduces well the initial quenching, including its nonstationary stage, but deviates from the long-time asymptote of UT. This is due to a false IET tail that has been already demonstrated in Figure 3.53 and cannot be removed unless MET is used instead of IET [39]. On the contrary, the long-time asymptotic behavior in the Markovian theory is correct but the initial nonstationary quenching is lost completely.

B. Contact and Exponential Models

Using the contact transfer rates (7.16) in the rate constant definitions (3.278), we reduce Eq. (3.420b) to its contact analog:

$$\frac{dN^{\pm}(t)}{dt} = k_0 n(\sigma, t) N^*(t) A(t) - k_c p(\sigma, t) - k_c f(\sigma, t) [N^{\pm}(t)]^2 \qquad (3.427)$$

Here $p(r,t)$ obeys the contact variant of Eq. (3.419) with a proper boundary condition

$$\frac{\partial p(r,t)}{\partial t} - \hat{L}^+ p(r,t) = \frac{k_0}{4\pi\sigma^2} \delta(r-\sigma) n(r,t) N^*(t) A(t), \quad 4\pi\sigma^2 \frac{\partial p}{\partial r}\bigg|_{r=\sigma} = k_c p(\sigma, t)$$

$$(3.428)$$

while $n(r,t)$ and $f(r,t)$ obey the contact variants of Eqs. (3.279) identical to Eqs. (3.19) and (3.20).

Contrary to this approximation, the exponential model, considered in Section V.A, does not assume recombination to be contact, but suggests that it takes place with a uniform backward transfer rate k_{-et} within the reaction sphere of the volume $v = 4\pi\sigma^3/3$. As a result, Eq. (3.419) is replaced by the following one:

$$\left[\frac{\partial}{\partial t} + k_{-et} + k_{sep}\right] p(r,t) = W_I(r) n(r,t) N^*(t) A(t) \quad r \le \sigma \qquad (3.429)$$

where k_{sep} was introduced in Eq. (3.189). In addition, EM assumes that ionization also takes place only within the same reaction sphere as recombination. Therefore, integrating Eq. (3.429) over space, we obtain a similar equation for the total number of ions inside the reaction sphere, $\Omega_{in} = \int_0^\sigma p(r,t) d^3r$:

$$\left[\frac{\partial}{\partial t} + k_{-et} + k_{sep}\right] \Omega_{in} = k_I(t) N^*(t) A(t) \qquad (3.430)$$

where k_I acquires the usual definition (3.34). Contrary to the original equation (3.190), the nonlinear pumping term on the right-hand side of this equation accounts for the bimolecular ionization that populates the reaction sphere. In the same way the bimolecular recombination may be accounted for in the equation for the total amount of ions $N^{\pm} \equiv \Omega = \Omega_{in} + \Omega_{out}$:

$$\frac{dN^{\pm}(t)}{dt} = k_I(t) N^*(t) A(t) - k_{-et} \Omega_{in}(t) - k_R(t) [N^{\pm}(t)]^2 \qquad (3.431)$$

At a great excess of acceptors, equations of this sort with $A = c$ were used a number of times [109,185,187,196], but in their Markovian version; that is, the time-dependent rate constants $k_I(t)$ and $k_R(t)$ were substituted by their asymptotic (Markovian) values k_i and k_r given in Eqs. (3.423) and (3.426). Still they represent the remote electron transfer and can be essentially larger their contact analogs [195].

C. Charge Accumulation and Recombination after δ Pulse

Let us concentrate on the initial geminate stage of the reaction considered in Section VI.D with UT at a large excess of acceptors. When the light excitation is instantaneous and weak, the bimolecular charge recombination in the bulk can be ignored during a limited time interval. Nonetheless, this initial interval can exceed the larger of the two times: the lifetime of the excited donor τ_D and the diffusional encounter time τ_d as in Figure 3.34(b).

1. Acceptors in Excess

If in addition acceptors are present in great excess, then both the accumulation and recombination of the geminate ion pairs can be considered, using instead of the particle densities the survival probabilities of the excited state, $R(t) = N^*(t)/N^*(0)$, and ions, $R^\pm = N^\pm/N^*(0)$. In particular, they obey the original UT equations that follow from the general set (3.418) at $k_R = 0$ and $A = c$

$$\dot{R} = -ck_I(t)R - \frac{R}{\tau_D} \qquad R(0) = 1 \qquad (3.432a)$$

$$\dot{R}^\pm = \int \frac{\partial p(r,t)}{\partial t} d^3r \qquad R^\pm(0) = 0 \qquad (3.432b)$$

where

$$\left[\frac{\partial}{\partial t} + W_R(r) - \hat{L}^+\right] p(r,t) = cW_I(r)n(r,t)N^*(t) \qquad (3.433)$$

Under the same conditions the full set of EM equations is

$$\dot{R} = -ck_I(t)R - \frac{R}{\tau_D}, \qquad \dot{R}^\pm = ck_I(t)R - k_{-et}\Omega_{in} \qquad (3.434)$$

where

$$\frac{d\Omega_{in}}{dt} + (k_{-et} + k_{sep})\Omega_{in} = ck_I(t)R \qquad (3.435)$$

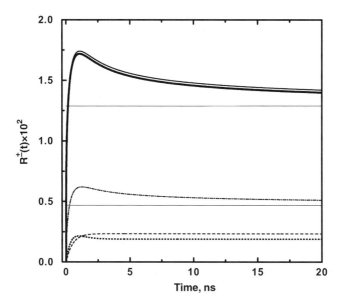

Figure 3.57. The ion survival probability as a function of time at $\tau_D = 0.5$ ns with a great excess of acceptors. In line with UT and IET (above) and Markovian theory (below) (dashed curve), the contact approximation (dashed–dotted line in the middle) and exponential model with $k_{sep} = k_{-et} = 1.0$ ns^{-1} (dotted line) are also shown. The horizontal thick lines indicate the free-ion quantum yield ϕ. The concentrations and ionization parameters are the same as in Figure 3.56, while $w_r = 3.4$ ns^{-1}, $\tilde{D} = D = 1.2 \times 10^{-6}$ cm^2/s, $k^{\pm} = 784$ Å^3/ns, and $k_r = 486$ Å^3/ns. (From Ref. 195.)

These or similar sets were sometimes used for the rough interpretation of the experimental data [109,187], but in principle EM is much worse than other approximations. The escape from the reaction zone and even more so from the Coulomb well does not proceed by a single jump described as an exponential (rate) process even if k_{sep} is given a reasonable estimate as in Eq. (3.91). This simplification ignores all subsequent re-contacts and an essential nonexponentiality of the whole geminate process [20].

In Figure 3.57 we show almost the full identity of the results obtained with IET and UT. They demonstrate the well pronounced maximum appearing when τ_D is shorter than the characteristic time of the subsequent geminate recombination [22]. In the contact approximation the results are qualitatively the same but the ionization quantum yield ψ is half as much as in distant theories (see Table III). This was expected because at such a short τ_D a significant fraction of ions are produced during the initial static ionization that is missed in the contact approximation.

TABLE III
Quantum Yields of Ionization (ψ), Charge Separation ($\bar{\varphi}$), and
Free Ion Production ($\phi = \psi\bar{\varphi}$)

Theories	$\psi \times 10^3$	$\bar{\varphi}$	$\phi \times 10^3$
UT	21.7	0.59	12.8
IET	21.5	0.59	12.7
Contact	9.3	0.50	4.6
Markovian	3.7	0.61	2.3
EM	3.7	0.50	1.9

Neither the maximum nor the descending branches of the upper curves, representing geminate recombination, are reproduced in the Markovian theory. It predicts the monotonous ion accumulation and still further decrease in the ionization quantum yield ψ. This is because the Markovian theory does not account for either static or subsequent nonstationary electron transfer. When ionization is under diffusional control, both these are faster than the final (Markovian) transfer. EM is a bit better in this respect. As a non-Markovian theory, it accounts at least for static ionization and qualitatively reproduces the maximum in the charge accumulation kinetics. However, the subsequent geminate recombination develops exponentially in EM because the kinematics of ion separation is oversimplified in this model. It roughly contradicts an actual diffusional separation of ions, characterized by numerous recontacts and the power dependence of long-time separation kinetics studied in a number of works [20,21,187].

It is interesting that the Markovian charge separation quantum yield $\bar{\varphi}$ is the largest one. As was shown in Section VI.C, in the low-concentration limit the initial charge distributions (5.55) is the same for UT and IET, but differs significantly from the Markovian one, Eq. (3.424), when the lifetime τ_D is short enough. Due to the complete neglection of nonstationary ionization, the Markovian density is smaller near the contact and larger out of it, so that the ions have greater survival probability. In contrast, in EM the ions are initially entirely inside the reaction zone, where they have minimum chance to escape. Correspondingly, their separation yield is smaller than in the Markovian theory despite the fact that ψ is equal in both theories (see Table III).

2. Shortage of Acceptors

The deficiency of acceptors leads to electron transfer saturation, illustrated by Figure 3.58. In the limit of infinite τ_D shown in this figure, all the theories discussed here, Markovian theory included, become equivalent. The saturation takes place from the very beginning and lasts as long as $N^*(t)$ (solid line) is greater than the acceptor concentration $A(t)$ (dotted line). As soon as the

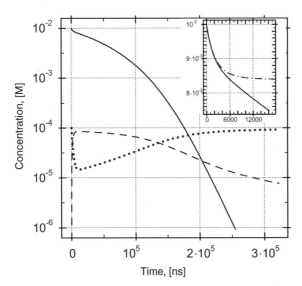

Figure 3.58. Decay of the excited donor concentration (solid line) accompanied by ion accumulation/recombination (dashed line) and depletion of neutral acceptors (dotted line) at $w_i = w_r = 1000\,\text{ns}^{-1}$ and $\tau_D = \infty$ (the remaining parameters are the same as in Figs. 3.57 and 3.58). The shortest stage of excited donor decay is shown in the insert in comparison to the excitation decay without bimolecular recombination in the bulk (dashed–dotted line). The charge separation quantum yield $\bar{\varphi} = 6.2\%$, $N^*(0) = 10^{-2}\,\text{M}$, $c = 10^{-4}\,\text{M}$. (From Ref. 195.)

situation changes with time, the saturation is removed and $A(t)$ approaches its upper limit, $c = 10^{-4}$ M. Within the saturation region there are two stages: the sharp initial one, when ions are accumulated; and the subsequent quasistationary stage, when their concentration remains approximately constant (dashed line).

Initially, the concentration of excited molecules goes down approximately to the level shown in the insert by a dashed–dotted line, which is obtained by ignoring the bimolecular recombination. This level does not equal $N^*(0) - c$ as one might expect, but is much lower. The reason is that only a $\bar{\varphi}$ fraction of c ions produced at first transfer are separated. The rest of them recombine in the geminate pair, restoring the $c(1 - \bar{\varphi})$ neutral particles that are ready to accept electrons once again. Hence, in the absence of bimolecular recombination, the concentration of excitations approaches the following limit

$$N^*(\infty) = N^*(0) - c - c(1 - \bar{\varphi}) - c(1 - \bar{\varphi})^2 - \cdots = N^*(0) - \frac{c}{\bar{\varphi}} \quad (3.436)$$

where each term of the expansion accounts for first, second, and further electron transfer to neutral acceptors in the same pair. In the next stage the bimolecular recombination is important and maintains a quasistationary concentration of charged and neutral acceptors. This stage continues until the concentration of excited donors and charged acceptors become approximately equal. Then the saturation is removed and the excitations are totally quenched, after which the remaining free ions are discharged in the bulk.

D. Evolution During ζ Pulse

1. Acceptors in Excess

If the electron acceptors are in great excess, one can express the relaxation of the excited state population $N^*(t)$ during and after arbitrary light excitation, through the survival probability of the excited donors after δ pulse, $R(t)$ as was done in Eq. (3.5). By substituting Eq. (3.415) into Eq. (3.5), we obtain the following for ζ pulse:

$$N^*(t) = I_0 N \int_0^t R(\tau)d\tau \tag{3.437}$$

Extension of this procedure to the ion relaxation $N^{\pm}(t)$ is not as straightforward. The rate of charge pumping (3.432) is maximal immediately after δ-pulse excitation, but changes and turns to zero very soon. Since the stationary pumping consists of the infinite sequence of such events, the source term is actually the sum of them, which can be obtained by application of the same convolution procedure as in Eq. (3.437):

$$\Pi(t) = I_0 N \int_0^t \dot{R}^{\pm}(\tau)d\tau = I_0 N R^{\pm}(t) \tag{3.438}$$

Summing this with the recombination term, we have the following equation for the free ions accumulation/recombination:

$$\dot{N}^{\pm} = I_0 N R^{\pm}(t) - k_R(t)\left[N^{\pm}\right]^2 \tag{3.439}$$

Equations (3.437) and (3.439) together with the original UT equations (3.432) constitute the formal basis of the generalized unified theory (GUT) [195]. The latter can be used to find the system response to the ζ pulse, provided the acceptor concentration is sufficiently large. In this way one can obtain the accumulation kinetics of excitations and free ions and their stationary concentrations:

$$N_s^* = I_0 N \tilde{R}^*(0), \qquad N_s^{\pm} = \sqrt{I_0 N \phi / k_r} \tag{3.440}$$

where the free ion quantum yield $\phi = R^{\pm}(\infty)$ is the same as in Eq. (3.274). Substituting N_s^* into Eq. (3.8), we confirm the conventional definition of the fluorescence quantum yield (3.11). On the other hand, while substitution of $N_s^{\pm}$ into the conventional definition of conductivity, $euN_s^{\pm}$ (u is the free carrier mobility), specifies it as a function of light strength and the electron transfer parameters.

Figure 3.59 (left) illustrates the kinetics of excitation and ion accumulation obtained from the numerical solution of the GUT equations (3.437) and (3.439). For comparison, the same result was obtained from the IET equations (3.417), with a large excess of electron acceptors. The difference between the results is not essential but becomes larger when c increases. This difference is in favor of

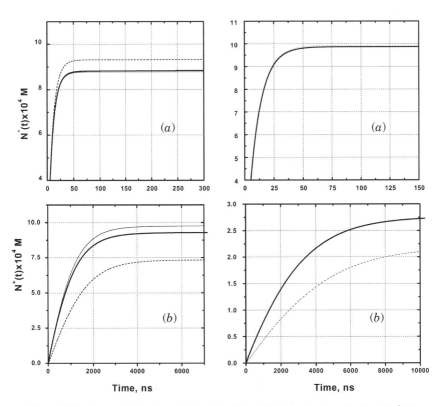

Figure 3.59. Accumulation of excitations (a) and ions (b) during ζ pulse at $I_0 \mathcal{N} = 10^{-4}\mathrm{M/ns}$. Left: kinetics at great excess of acceptors ($c = 10^{-2}\,\mathrm{M}$). Right: the same but at a shortage of acceptors ($c = 10^{-3}\,\mathrm{M}$). Thick line—IET; thin line—GUT; dashed line—Markovian theory. The parameters are $\tau_D = 10\,\mathrm{ns}, w_i = w_r = 1000\,\mathrm{ns}^{-1}$; $k_i = k_r = 1271\,\text{Å}^3/\mathrm{ns}, k^{\pm} = 78.8\,\text{Å}^3/\mathrm{ns}$, $\tilde{D} = D = 1.2 \times 10^{-6}\,\mathrm{cm}^2/\mathrm{s}, l = 1.0\,\text{Å}$. (From Ref. 195.)

GUT because this theory as well as UT account for the higher-order corrections in c which are absent in IET.

These non-Markovian results can be compared with those obtained with the Markovian theory. Contrary to UT, the light pumping can be incorporated in the Markovian equations (3.425) as an additive term, as well as in IET

$$\frac{dN^*}{dt} = -k_i N^* A - \frac{N^*}{\tau_D} + I_0 N \qquad (3.441a)$$

$$\frac{dN^\pm}{dt} = k_i \bar{\varphi}_m N^* A - k_r \left[N^\pm\right]^2 \qquad (3.441b)$$

where $A = c - N^\pm$ as usual. The difference seen in Figure 3.59 (left) between the Markovian and other results is due to the usage of the stationary parameters k_i and $\bar{\varphi}_m$ which do not account for the initial nonstationary development of the process. This difference exists at any finite lifetime and is larger when τ_D is shorter.

2. Shortage of Acceptors

There is also a significant difference at c smaller than N_s^*, when electron transfer is saturated. Since in this situation $\dot{R}^\pm \propto A(t)N * (t)$ is essentially non-linear in the reactant concentrations, the convolution (8.24) is impossible. Thus GUT does not hold, and only IET provides a proper non-Markovian solution. However, the nonlinear effect of acceptor depletion is accounted for in the Markovian theory as well. Using these two theories, we estimated the difference between the non-Markovian and Markovian description of accumulation kinetics under saturation conditions (Fig. 3.59, right). As before, it is larger for the charged products than for excitations and stronger with shorter lifetime τ_D.

The significance and extent of the saturation effect can be easily estimated in the stationary regime available for analytical investigation. The stationary solution of the Markovian equations (3.441) is trivial

$$N_s^* = \frac{I_0 N \tau_D}{1 + k_i A_s \tau_D}, \qquad N_s^\pm = \sqrt{\frac{I_0 N \bar{\varphi}_m}{k_r}}\, \psi \qquad (3.442)$$

where

$$\psi = \frac{k_i A_s \tau_D}{1 + k_i A_s \tau_D} = 1 - \eta \qquad (3.443)$$

At a great excess of acceptors, when $A_s = c$, Eqs. (3.442) are identical to the earlier obtained equations (3.440). However, in general the stationary concentration of neutral acceptors $A_s \neq c$, but obeys the cubic equation, which follows

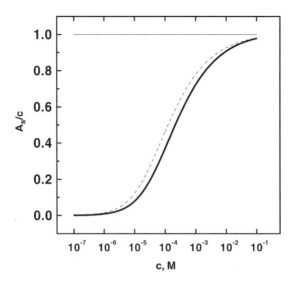

Figure 3.60. The stationary fraction of neutral acceptors as a function of their total density c in IET (thick line) and Markovian theory (dashed line). The parameters are $I_0 N_0 = 10^{-4}\,\mathrm{M/ns}$, $\tau_D = 10\,\mathrm{ns}$, $w_i = w_r = 1000\,\mathrm{ns^{-1}}$; $\bar{\varphi}_m = 0.062$, $k_i = 1271\,\mathrm{\mathring{A}^3/ns}$, $k_r = 1271\,\mathrm{\mathring{A}^3/ns}$. (From Ref. 195.)

from the stationary solution of Eqs. (3.441):

$$k_r(c - A_s)^2(1 + k_i\tau_D A_s) = I_0 N\tau_D k_i \bar{\varphi}_m A_s \qquad (3.444)$$

Because of such a complex dependence $A_s(c)$, shown in Figure 3.60, the inverse quantum yield from Eq. (3.443), $1/\eta = 1 + k_i A_s \tau_D$, is linear in A_s but not in c. As can be seen from Eq. (3.444), the condition of linearity in c is given by the following inequality:

$$k_i \bar{\varphi}_m N_s^* \ll k_r c \qquad (3.445)$$

Otherwise, the depletion of neutral acceptors results in the quadratic dependence of $1/\eta$ on c at small concentrations. This violation of the Stern–Volmer law cannot be reproduced with GUT, which is valid only at large c when the inequality (3.445) holds.

Unlike GUT, IET has no restrictions of validity at small c. The equations for the stationary concentrations follow from the general IET equations (3.417) as $t \to \infty$ are

$$0 = -\tilde{R}^*(0)N_s^* A_s - \frac{N_s^*}{\tau_D} + I_0 N \qquad (3.446a)$$

$$0 = \tilde{R}^\dagger(0)N_s^* A_s - \tilde{R}^\ddagger(0)\left[N_s^\pm\right]^2 \qquad (3.446b)$$

Taking into account the definitions given for the Laplace transforms of the kernels in Eqs. (3.382), (3.386), and (3.389), we can put the stationary IET equations in exactly the same form as in the Markovian theory:

$$0 = -\kappa_g N_s^* A_s - \frac{N_s^*}{\tau_D} + I_0 N \qquad (3.447a)$$

$$0 = \kappa_g \bar\varphi N_s^* A_s - k_r \left[N_s^\pm\right]^2 \qquad (3.447b)$$

The irreversibility of ionization leads to

$$\kappa_g = \tilde R^*(0) = Q = \frac{1}{\tau_D} \int W_I(r)\tilde v(r;0)\, d^3 r \qquad (3.448)$$

and

$$\bar\varphi = \bar\varphi_\beta = \int \varphi(r_0) f_0^{iet}(r_0) d^3 r_0 \qquad (3.449)$$

Here $f_0^{iet}(r_0)$ was as defined in Eq. (3.309) while $\varphi(r_0)$ is given by Eq. (3.388) with $W_B = 0$. The solutions of Eqs. (3.447) coincide with those obtained from the Markovian theory, (3.442)–(3.444), but with κ_g substituted for k_i and $\bar\varphi$ for $\bar\varphi_m$:

$$\eta = \frac{N_s^*}{I_0 N \tau_D} = \frac{1}{1 + \kappa_g A_s \tau_D}, \qquad N_s^\pm = \sqrt{\frac{I_0 N \bar\varphi}{k_r}}\,(1 - \eta) \qquad (3.450)$$

where A_s is the solution to the cubic equation similar to the Markovian equation (3.444):

$$k_r (c - A_s)^2 (1 + \kappa_g \tau_D A_s) = I_0 N \tau_D \kappa_g \bar\varphi A_s \qquad (3.451)$$

The substitutions of k_i and $\bar\varphi_m$ by κ_g and $\bar\varphi$ account for the non-Markovian alterations incorporated by IET.

E. Markovian versus Non-Markovian Theories

In Figure 3.61 we illustrate the difference between the two non-Markovian theories, UT and IET, and their Markovian analog. Inspecting the Stern–Volmer plot, one can see from the insert that the IET curve is nonlinear (quadratic) at low concentrations where the saturation of electron transfer takes place. The Markovian theory in principle reproduces this effect, although less accurately. At higher c both of them reproduce the linear Stern–Volmer dependence η^{-1}

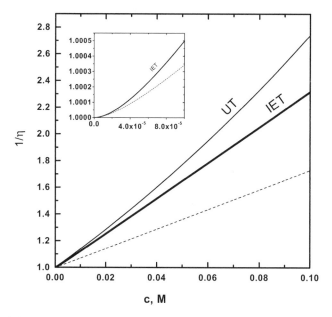

Figure 3.61. The quantum yield of fluorescence as a function of acceptor concentration c in IET (thick line), UT (thin line), and Markovian theory (dashed line). The nonlinearity of IET and the Markovian results at very low concentrations, which are shown in the insert, is caused by the saturation of electron transfer in this region. The parameters are the same as in Fig. 3.60.

versus c, but the tangent (Stern–Volmer constant) is underestimated in the Markovian theory.

On the contrary, UT is not valid at low concentrations, where it misses the saturation effect. On the other hand, it is better than the two others at high concentrations, where it reproduces the nonlinear increase of η^{-1} with c, originating from the nonstationary and static energy quenching. Such a superlinearity is usually attributed to the nonlinear corrections to IET available with DET, UT, and a number of other theories (see Section XII) [54,197,198,136]. Fortunately, this effect is very easy to distinguish from the saturation phenomenon; at higher concentrations of acceptors their depletion by electron transfer is removed, while the non-linearity related to nonstationary quenching is enhanced.

Hence, both the IET and Markovian theory provide the lowest-order approximation for the fluorescence quantum yield with respect to acceptor concentration. This approximation is the only limitation of the validity of IET. Because of this limitation, IET is unable to describe properly the long-time asymptote of the system response to instantaneous excitation (Fig. 3.56) and the nonlinearity of the Stern–Volmer law at high concentrations (Fig. 3.61). On the other hand, it accounts for the effect of electron acceptors depletion and all non-Markovian effects. They are seen not only in the system response to either δ or ζ pulses, but also in the principal characteristics of the stationary regime, κ_g and $\bar{\varphi}$, which are different from their Markovian analogs, k_i and $\bar{\varphi}_m$ [199]. Moreover, there is a non-Markovian effect arising at higher intensities of light, which affects the kernels of the IET equations as well as the Stern–Volmer constants. It will be considered in the next section.

X. STATIONARY PHOTOCHEMISTRY IN STRONG LIGHT

In strong light there is an exhaustion of the ground-state population that we ignored in Eq. (3.5) and later on assuming that $N \approx \mathcal{N} = \text{const}$. Now is the time to admit that the stationary population of the ground state is in fact $N_s(c)$ in the presence of quenchers and $N_s(0)$ in their absence, but at strong illumination either of these quantities is not equal to the total number of donors.

1. *Quantum Yield Definitions.* Having in mind that the population of the excited state is $N_s^*(c)$ or $N_s^*(0)$, correspondingly, we can define the absolute values of the quantum yields for these cases as follows:

$$\eta_{abs}(c) = \frac{AN_s^*(c)}{I_0 N_s(c)} \quad \text{and} \quad \eta_{abs}(0) = \frac{AN_s^*(0)}{I_0 N_s(0)} \tag{3.452}$$

In the absence of quenchers the saturation of state populations can be described by the universal equation:

$$\dot{N}^* = I_0 N - \frac{N^*}{\tau} \tag{3.453}$$

From its stationary solution we get

$$\frac{N_s^*(0)}{N_s(0)} = I_0 \tau \tag{3.454}$$

Using these results for the definition of the relative quantum yield, we obtain the following formula used in Refs. 125 and 200:

$$\eta(c) = \frac{\eta_{abs}(c)}{\eta_{abs}(0)} = \frac{N_s^*(c)}{I_0 \tau N_s(c)} \tag{3.455}$$

This differs from definition (3.8) as much as $N_s(c)$ differs from N that was previously identified with a total number of D molecules, $\mathcal{N}$. There is also an alternative definition of the relative quantum yield. It relates to one another the emitted light in the presence and absence of quenchers, $F(c) = AN_s^*(c)$ and $F(0) = AN_s^*(0)$ [63,142,201]:

$$\eta'(c) = \frac{F(c)}{F(0)} = \frac{N_s^*(c)}{N_s^*(0)} = \frac{N_s^*(c)}{I_0 \tau N_s(0)} \tag{3.456}$$

Generally speaking, this definition is different from that given in Eq. (3.455):

$$\eta' = \eta \frac{N_s(c)}{N_s(0)} \tag{3.457}$$

The difference disappears only at weak pumping, when $N_s(c) \approx N_s(0) \approx \mathcal{N} = \text{const}$, but can be essential when the saturation takes place. Which definition is actually appropriate in this case depends on what is actually measured.

2. Harnessing of Convolution. To make use of the definition (3.455), one should know the stationary concentration of excitations N_s^*. It is available from the convolution recipe (3.6) that was extended to arbitrary light intensity by Naumann and Szabo [201].

$$N_s^* = I_0 \mathcal{N} \int_0^\infty e^{-(I_0 + 1/\tau)t'} P(t') dt' \tag{3.458}$$

where

$$P(t) = \frac{N^*(t)}{N^*(0)} \bigg|_{I_0 = 1/\tau = 0} \tag{3.459}$$

The DET expression, $P(t) = \exp\left(-c \int_0^t k(t') dt'\right)$, provides actually an exact solution for the target problem [138] used as a test for other approximations.

Since the total number of fluorescent molecules $\mathcal{N}$ remains constant at any I_0

$$N^* + N = \mathcal{N} \tag{3.460}$$

their stationary concentrations in the excited and ground states are

$$N_s^*(c) = I_0 \mathcal{N} \tilde{P}\left(I_0 + \frac{1}{\tau}\right), \qquad N_s(c) = \mathcal{N}\left[1 - I_0 \tilde{P}\left(I_0 + \frac{1}{\tau}\right)\right] \tag{3.461}$$

Substituting Eq. (3.461) into Eq. (3.455), we obtain the most general definition of the relative quantum yield resulting from the convolution recipe:

$$\frac{1}{\eta} = \frac{\tau}{\tilde{P}(I_0 + 1/\tau)} - I_0\tau \tag{3.462}$$

If there is no quencher in the sample, then

$$\tilde{P}\left(I_0 + \frac{1}{\tau}\right) = \left[I_0 + \frac{1}{\tau}\right]^{-1} \quad \text{and} \quad N_s^*(0) = \frac{I_0\mathcal{N}}{I_0 + 1/\tau}$$

Using this result together with $N^*(c)$ from Eq. (3.461) in Eq. (3.456), we obtain

$$\eta' = [I_0 + 1/\tau] \cdot \tilde{P}\left(I_0 + \frac{1}{\tau}\right) \tag{3.463}$$

As was expected, this yield is essentially different from that defined in Eq. (3.462):

$$\frac{1}{\eta'} - 1 = \frac{1}{1 + I_0\tau} \cdot \left(\frac{1}{\eta} - 1\right) \tag{3.464}$$

Using the Stern–Volmer representation for both quantum yields

$$1/\eta' = 1 + cK_{SV}^0, \qquad 1/\eta = 1 + c\kappa_0\tau \tag{3.465}$$

we obtain the following relationship between κ_0 favored here and the Stern–Volmer constant at low concentrations K_{SV}^0, arising from the alternative yield definition (3.456) [142]:

$$K_{SV}^0 = \frac{\kappa_0}{I_0 + 1/\tau} \tag{3.466}$$

Note that κ_0 is identical to k_Q^s from Ref. 142 and k_Q^{ss} from Ref. 201.

A. Impurity Induction of Excitation Decay

First let us consider the simplest reaction, (3.1), which is actually the irreversible transfer (3.161), which occurs at $W_B = 0$ and $1/\tau_B = \infty$:

$$D^* + Q \overset{W_q}{\Longrightarrow} D + Q \tag{3.467}$$
$$I_0 \updownarrow \tau$$

After δ-pulse excitation there is no more pumping and the quenching kinetics can be described by the IET equations (3.162) with $W_B = \langle S \rangle = 0$ and the proper simplification of $\langle R \rangle$. However, in presence of permanent pumping, $\langle S \rangle \neq 0$, and both kernels have to be redefined starting from the general IET equation (3.103a). It should be adopted for intramolecular relaxation catalyzed by inert particles whose concentration $c = [Q]$ remains invariable. Substituting c for $\mathbf{N}_B$ and

$$\mathbf{N}_D = \begin{pmatrix} N^* \\ N \end{pmatrix} \tag{3.468}$$

for $\mathbf{N}_A$, we obtain the following from Eq. (3.103a):

$$\frac{d}{dt}\mathbf{N}_D = \hat{\mathbf{Q}}\mathbf{N}_D(t) + c \int_0^t \mathbf{M}(\tau)\mathbf{N}_D(t - \tau)d\tau \tag{3.469}$$

The kernel of this equation was defined in Eq. (3.104). All the matrices involved are of the second rank because the basis consists of only two donor states, excited and ground:

$$\mathbf{M} = \begin{pmatrix} -\Sigma_1 & -\Sigma_2 \\ \Sigma_1 & \Sigma_2 \end{pmatrix}, \quad \hat{\mathbf{W}} = \begin{pmatrix} -W_q(r) & 0 \\ W_q(r) & 0 \end{pmatrix}, \quad \mathbf{F} = \begin{pmatrix} v_1 & \mu_2 \\ \mu_1 & v_2 \end{pmatrix} \tag{3.470}$$

The latter one is governed by the auxiliary matrix equation (3.106) where

$$\hat{\mathbf{Q}} = \begin{pmatrix} -1/\tau & I_0 \\ 1/\tau & -I_0 \end{pmatrix} \quad \text{and} \quad \hat{\mathbf{L}} = \begin{pmatrix} \hat{L}_1 & 0 \\ 0 & \hat{L}_2 \end{pmatrix} \tag{3.471}$$

The inclusion of the rate of light transitions into the relaxation operator $\mathbf{Q}$ in line with the decay rate $1/\tau$, makes the kernel light-dependent, a fact that was completely ignored until now. This rate affects the pair correlation functions that obey two sets of equations that follow from (3.106)

$$\left(\partial_t - \hat{L}_1 + W_q(r) + \frac{1}{\tau}\right)v_1(r,t) = I_0\mu_1(r,t) \tag{3.472a}$$

$$(\partial_t - \hat{L}_2 + I_0\mu_1(r,t)) = \left(W_q(r) + \frac{1}{\tau}\right)v_1 \tag{3.472b}$$

and

$$(\partial_t - \hat{L}_2 + I_0)v_2(r,t) = \left(W_q(r) + \frac{1}{\tau}\right)\mu_2 \tag{3.473a}$$

$$\left(\partial_t - \hat{L}_1 + W_q(r) + \frac{1}{\tau}\right)\mu_2(r,t) = I_0v_2(r,t) \tag{3.473b}$$

with the initial conditions

$$v_1(r,0) = v_2(r,0) = 1, \qquad \mu_1(r,0) = \mu_2(r,0) = 0$$

1.   Zero Kernel

It is not easy to foresee that one of the elements of $\mathbf{M}$ in (3.470) equals zero:

$$\tilde{\Sigma}_2(s) = 0 \tag{3.474}$$

It follows from the important relationship derived from Eqs. (3.472) and (3.473) in Ref. 200 that

$$\tilde{\mu}_1(r,s) = \frac{I_0}{s + I_0}\tilde{v}_1(r,s) \tag{3.475}$$

As a result of the same relationship, another element of this matrix kernel is:

$$\tilde{\Sigma}_1(s) = s\left\{\frac{s + I_0 + 1/\tau}{s + I_0}\right\} \int W_q(r)\tilde{v}_1(r,s)\, d^3r \tag{3.476}$$

Bearing in mind that only this element of $\mathbf{M}$ differs from zero, we obtain from the matrix IET equation (3.469) and the particle conservation law (3.460) a single integral equation:

$$\frac{d}{dt}N^* = -\frac{N^*}{\tau} - I_0 N^* + I_0 \mathscr{N} - c\int_0^t dt'\Sigma_1(t')N^*(t-t'), = -\frac{d}{dt}N \tag{3.477}$$

The general solution of this equation is given by its Laplace transformation:

$$\tilde{N}^*(s) = \frac{I_0\mathscr{N}/s + N^*(0)}{s + 1/\tau + I_0 + c\tilde{\Sigma}_1(s)} \tag{3.478}$$

Hence, the problem is reduced to solely pinpointing the IET kernel, $\Sigma_1(s)$.

2.   Verification of the Convolution Recipe

This problem can be essentially simplified assuming the motion in pairs D*B and DB is the same. In highly polar solvents the force interactions in these pairs are either the same or negligible. If in the latter case the diffusion coefficients are equal, then the corresponding operators are

$$\hat{L}_1 = \hat{L}_2 = D\Delta \tag{3.479}$$

Under this condition $v_1 + \mu_1 = 1$, and a single auxiliary equation follows from the set (3.472) for either $\tilde{v}_1$ or $\tilde{v} = s\tilde{v}_1/(s + I_0)$ [200]. The original of the latter obeys the equation

$$\left[\partial_t - \hat{L}_1 + W_q(r) + \frac{1}{\tau} + I_0\right] v(r, t) = 0 \qquad (3.480)$$

with the initial condition $v(r, 0) = 1$. Expressed through $\tilde{v}$ the kernel (9.25) looks simpler:

$$\tilde{\Sigma}_1(s) = \left\{s + I_0 + \frac{1}{\tau}\right\} \int d^3 r W_q(r) \tilde{v}(r, s) = \tilde{\Sigma}_0\left(s + I_0 + \frac{1}{\tau}\right) \qquad (3.481)$$

but even more simple is

$$\tilde{\Sigma}_0(s) = s \int d^3 r W_q(r) \tilde{n}(r, s) \qquad (3.482)$$

which is expressed through the conventional pair density $n(r, t)$ commonly used in DET (Section II.B). The Laplace transformation of the latter simply relates to that of $v(r, t)$:

$$\tilde{n}\left(r, s + I_0 + \frac{1}{\tau}\right) = \tilde{v}(r, s) \qquad (3.483)$$

and its original, $n(r, t)$, obeys Eq. (3.35), which follows from Eq. (3.480) at $I_0 = 1/\tau = 0$.

The Laplace transformation of expression (3.459) can be easily obtained from Eqs. (3.478) and (3.481):

$$\tilde{P}(s) = \frac{1}{s + c\tilde{\Sigma}_0(s)} \qquad (3.484)$$

Substituting Eq. (3.481) into Eq. (3.478), we come to the following relationship between $\tilde{N}^*(s)$ and $\tilde{P}(s)$ under the action of ζ pulse:

$$\tilde{N}^*(s) = I_0 \mathcal{N} \frac{\tilde{P}(s + I_0 + 1/\tau)}{s} \qquad (3.485)$$

The inverse Laplace transformation of this relationship provides the convolution recipe for calculation $N^*(t)$:

$$N^*(t) = I_0 \mathcal{N} \int_0^t e^{-(I_0 + 1/\tau)t'} P(t') dt' \qquad (3.486)$$

This expression determines the kinetics of the excitation accumulation from $N^*(0) = 0$ up to the stationary value N_s^*, after instantaneous switching the permanent illumination at $t = 0$. This is a particular case of the more general convolution recipe derived in Ref. 201 for pulses of arbitrary shape. Hence, for equal diffusion coefficients, the convolution formula follows from IET as well as from the many-particle theory employed in Ref. 201. The generality of the latter allows us to use in the convolution formula the system response $P(t)$, calculated with any available theory. The same is valid for the stationary equation (3.458), used above. Although $P(t)$ obtained with different theories is different, as well as $N^*(t)$, the relationship between N^* and P remains the same.

3.   Light-Dependent Stern–Volmer Constant

To obtain the quantum yield of fluorescence from Eq. (3.455), we need the stationary concentrations of D molecules in their excited and ground states, which are available from the solution of Eq. (3.477) at $\dot{N}^* = \dot{N} = 0$:

$$N_s^* = \mathcal{N}\frac{I_0\tau}{1 + I_0\tau + c\tilde{\Sigma}_1(0)\tau} = \mathcal{N} - N_s \qquad (3.487)$$

Using this result in (3.455), we not only confirm the ideal (linear in c) Stern–Volmer law, (3.11) or (3.271), but also define the IET constant for irreversible quenching at strong light, $\kappa_0 = \tilde{\Sigma}_1(0) = \tilde{\Sigma}_0(I_0 + 1/\tau)$. According to Eqs. (3.482) and (3.483), this is

$$\kappa_0(I_0) = \left(I_0 + \frac{1}{\tau}\right)\int W_q(r)\tilde{v}(r,0)d^3r \qquad (3.488)$$

In the contact approximation

$$W_q(r) = k_0\frac{\delta(r-\sigma)}{4\pi r^2} \qquad \text{where} \qquad k_0 = \int W_q(r)d^3r \qquad (3.489)$$

Substituting this rate and diffusional operator (3.479) into Eq. (3.480), one can easily solve it and, using the result in Eq. (3.488), find the contact solution of the problem:

$$\kappa_0 = \frac{k_0 k_D}{k_D + \alpha_0 k_0} \qquad (3.490)$$

The important correction factor

$$\alpha_0 = \frac{1}{1 + \sqrt{\tau_d(I_0 + 1/\tau)}} \qquad (3.491)$$

includes the dependence on not only diffusional encounter time $\tau_d = \sigma^2/D$ and τ but also the light intensity I_0. Equation (3.490) with α_0 from (3.491) represents the straightforward generalization of the well-known result of IET, [Eq. (3.273)] and DET [Eq. (3.27)]. On the other hand, the substitution of κ_0 from (3.490) into the relationship (3.466) gives exactly the same K_{SV}^0 that was obtained in the low-concentration limit [see Eq. (3.23) of Ref. 142] by the superposition approximation. However, the latter has an advantage as well as DET. Both of them are suitable for the prediction of the Stern–Volmer constant concentration dependence, which is absent in IET. The accuracy of these predictions, investigated in Refs. 133 and 201, will be a subject for discussed further in Section XII.

In the kinetic control limit ($k_D \to \infty$) there is no dependence of the Stern–Volmer constant $\kappa_0 = k_0$ on either $1/\tau$ or I_0. However, it builds up with both of them in the alternative, diffusion control limit ($k_0 \to \infty$):

$$\kappa_0(I_0) = k_D\left(1 + \sqrt{\tau_d(I_0 + 1/\tau)}\right) \qquad (3.492)$$

This effect originates from the nonstationary quenching (3.21), which is the more pronounced with smaller k_D compared to k_0. From the general definition of κ_0 in Eq. (3.13) it follows that the decay of the excited state confines the fluorescence quenching to a limited time interval $t < \tau$, where the quenching rate is larger than k_D. The earlier the excitation decays, the closer is $\kappa_0 \approx k(\tau)$ to its maximum value k_0. This is because at earliest times only those excitations are quenched that have quenchers in their near vicinity. Then the time is ripe for more isolated excitations, and so on. However, the strong pumping soon restores those that were quenched first and are ready to absorb the new photons. The favorable situation is preserved until there are quenchers in the near surroundings, that is, within a limited encounter time τ_d. Therefore the reexcited molecules are quenched faster than the rest, facilitating the total quenching whose rate is characterized by the field-dependent Stern–Volmer constant (3.492).

To make more general conclusions one should give up the contact approximation and return back to Eqs. (3.472a) and (3.472b), which operate with remote quenching $W_q(r)$ and different diffusional operators for the excited- and ground-state donors. By resolving them with respect to v_1 and using the result in Eq. (3.476), one can find $\kappa_0 = \tilde{\Sigma}_1(0)$. We are able to perform these numerical calculations, having at our disposal a special program developed in Ref. 162. The results shown in Figure 3.62 demonstrate that the difference between the exponential quenching rate and its contact model (3.489) is significant. It significantly affects the field dependence of the Stern–Volmer constant: compare the solid curve (b) and its contact analog (dashed line), both obtained for equal diffusion coefficients.

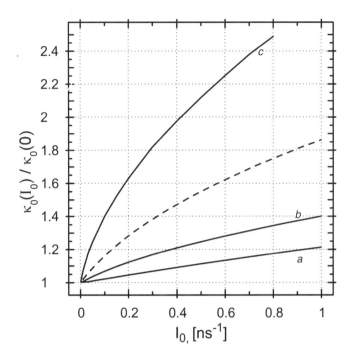

Figure 3.62. The light dependence of the Stern–Volmer constant $\kappa_0(I_0)$ for diffusional quenching with given exponential rate $\left(W_q = 10^3 \exp[-2(r - \sigma)/L]\,\text{ns}^{-1}\right)$, but at different diffusion in pairs containing the excited molecules: (a) $D^* = 0.1D_D$, (b) $D^* = D = 10^{-5}\,\text{cm}^2/\text{s}$ (dashed line—the same, but in contact approximation), (c) $D^* = 10D$. Other parameters: $\sigma = 5\,\text{Å}$, $L = 1.0\,\text{Å}$, $\tau = 10\,\text{ns}$, $k_0 = \int W_q(r)d^3r = 1.9 \times 10^5\,\text{Å}^3/\text{ns}$. (From Ref. 200.)

In the same figure we also demonstrate the sensitivity of the results to the ratio of the diffusion coefficients in the stable and excited pairs, $D \cdots Q$ and $D^* \cdots Q$, which increases from (a) to (c). The difference in the diffusion coefficients ignored in the convolution recipe affects the field dependence no less than the contact approximation. This is the unique advantage of IET, that it is free of any limitations of this sort and can be used for quantitative investigations of field effects, at any space dispersion of quenching rate and at arbitrary diffusion coefficients. In fact, this is diffusion of the excited molecule that affects the field dependence of κ_0. By slowing down this diffusion, one increases the encounter time τ_d and thus enhances the field acceleration of quenching.

This field dependence is actually the qualitative test of whether the quenching is under kinetic or diffusional control. These mechanisms can be distinguished experimentally without heavy measurements of the diffusion coefficients. If there is no field dependence of κ_0, it is kinetic-quenching; if κ_0 rises with I_0, the quenching is diffusional. However, when K_{SV} is actually measured, then the factor $1/(I_0 + 1/\tau)$ competing in Eq. (3.466) with $\kappa_0(I_0)$ can change this dependence to the opposite [142]. To eliminate this trivial factor, Eq. (3.466) should be used to extract κ_0 from K_{SV}.

B. Quenching by Electron Transfer

According to the reaction scheme (3.257) of the irreversible quenching by electron transfer, the electron donor has three electronic states involved in the interaction with light and electron acceptor, while the latter has only two states participating in the charge transfer:

$$\mathbf{N}_D = \begin{pmatrix} N^* \\ N^+ \\ N \end{pmatrix} \quad \text{and} \quad \mathbf{N}_A = \begin{pmatrix} A \\ A^- \end{pmatrix} \tag{3.493}$$

A few conservation laws should be added to these definitions:

$$N + N^* + N^+ = \mathcal{N}, \qquad A + A^- = c, \qquad N^+ = A^- \tag{3.494}$$

Due to strong light pumping, the vector $\mathbf{N}_D$ has to have a higher rank here than in Eq. (3.258). The light-induced and radiationless transitions $D^* \rightleftharpoons D$ should be accounted for in the intramolecular relaxation operator

$$\hat{\mathbf{Q}}_D = \begin{pmatrix} -1/\tau & 0 & I_0 \\ 0 & 0 & 0 \\ 1/\tau & 0 & -I_0 \end{pmatrix} \tag{3.495}$$

defined in the three-state basis of the donor molecule.

Vectors (3.493) obey the following general equations of integral encounter theory:

$$\frac{d}{dt}\mathbf{N}_D = \text{Tr}_A \int_0^t \mathbf{M}(\tau)\mathbf{N}_D(t-\tau) \otimes \mathbf{N}_A(t-\tau)d\tau + \mathbf{Q}_D\mathbf{N}_D(\tau) \tag{3.496}$$

$$\frac{d}{dt}\mathbf{N}_A = \text{Tr}_D \int_0^t \mathbf{M}(\tau)\mathbf{N}_D(t-\tau) \otimes \mathbf{N}_A(t-\tau)d\tau \tag{3.497}$$

Here $\mathbf{N}_D \times \mathbf{N}_A$ is a direct vector product in the basis of collective states of the donor–acceptor pair. The latter is a direct product of the three-state D basis with the two-state A basis. To account for the bimolecular reaction of electron transfer, one must define the reactivity and relaxation matrices on this basis:

$$
\hat{\mathbf{W}} = \begin{pmatrix} -W_I & 0 & 0 & 0 & 0 & 0 \\ 0 & 0 & 0 & 0 & 0 & 0 \\ 0 & 0 & 0 & 0 & W_R & 0 \\ 0 & 0 & 0 & 0 & 0 & 0 \\ W_I & 0 & 0 & 0 & -W_R & 0 \\ 0 & 0 & 0 & 0 & 0 & 0 \end{pmatrix} \qquad \hat{\mathbf{Q}} = \mathbf{I}_2 \otimes \hat{\mathbf{Q}}_D \qquad (3.498)
$$

Here $\mathbf{I}_2$ denotes the 2×2 identity matrix.

The main IET equations (3.104) and (3.106) remain valid:

$$
\tilde{\mathbf{M}}(s) = \int \hat{\mathbf{W}}(r)\tilde{\mathbf{F}}(r,s)[s\hat{\mathbf{I}} - \hat{\mathbf{Q}}]d^3r \qquad (3.499a)
$$

$$
\partial_t \mathbf{F} = \hat{\mathbf{L}}\mathbf{F} + \hat{\mathbf{W}}\mathbf{F} + \hat{\mathbf{Q}}\mathbf{F} \qquad (3.499b)
$$

Written in components, these matrix equations compose the following set [200]:

$$
\partial_t N^*(t) = -\int_0^t \Sigma_1(t-t')N^*(t')A(t')dt' + \int_0^t \Sigma_2(t-t')N(t')A(t')dt'
$$

$$
-\int_0^t \Sigma_3(t-t')N^+(t')A^-(t')dt' - \frac{N^*}{\tau} + I_0N \qquad (3.500a)
$$

$$
\partial_t N(t) = \int_0^t \Sigma_4(t-t')N^*(t')A(t')dt' - \int_0^t \Sigma_5(t-t')N(t')A(t')dt'
$$

$$
+\int_0^t \Sigma_6(t-t')N^+(t')A^-(t')dt' + \frac{N^*}{\tau} - I_0N \qquad (3.500b)
$$

Here Σ_i are elements of the matrix kernel $\mathbf{M}$. The latter is similar to that in Eq. (3.470) but of a higher rank: (6×6). The particle balance provides all other equations:

$$
\partial_t N^+ = \partial_t A^- = -\partial_t A = -\partial_t(N + N^*) \qquad (3.501)
$$

1. New and Zero Kernels

IET provides the definitions for all six kernels figuring in Eqs. (3.500) [200]. They are expressed through inter- and intramolecular transfer rates and nine pair densities: ν_1, ν_2, ν_3 and $\mu_1, \mu_2, \mu_3, \mu_4, \mu_5, \mu_6$, which obey equations similar to

Eqs. (3.472) and (3.473). As before, not all those from the nine pair densities are independent. They share two relationships that are similar to (3.475) that were established in the Appendix to Ref. 200:

$$I_0 \tilde{v}_1(r, s) - (s + I_0)\tilde{\mu}_1(r, s) = \tilde{\Sigma}_2 = 0 \qquad (3.502)$$

$$I_0 \tilde{\mu}_3(r, s) - (s + I_0)\tilde{\mu}_4(r, s) = \tilde{\Sigma}_5 = 0 \qquad (3.503)$$

These relationships reduce the number of auxiliary equations to seven, turn to zero two of the six kernels, and simplify the definitions of the rest:

$$\tilde{\Sigma}_1(s) = \frac{s(s + I_0 + 1/\tau)}{s + I_0} \int d^3r W_I(r)\tilde{v}_1(r, s) \qquad (3.504a)$$

$$\tilde{\Sigma}_3(s) = s \int d^3r W_I(r)\tilde{\mu}_2(r, s) \qquad (3.504b)$$

$$\tilde{\Sigma}_4(s) = \frac{s(s + I_0 + 1/\tau)}{s + I_0} \int d^3r W_R(r)\tilde{\mu}_3(r, s) \qquad (3.504c)$$

$$\tilde{\Sigma}_6(s) = s \int d^3r W_R(r)\tilde{v}_2(r, s) \qquad (3.504d)$$

After such a reduction, the IET equations (3.500) take the simpler form:

$$\partial_t N^* = -\int_0^t \Sigma_1(t - t')N^*(t')A(t')dt'$$

$$-\int_0^t \Sigma_3(t - t')N^+(t')A^-(t')dt' - \frac{N^*}{\tau} + I_0 N \qquad (3.505a)$$

$$\partial_t N = \int_0^t \Sigma_4(t - t')N^*(t')A(t')dt'$$

$$+\int_0^t \Sigma_6(t - t')N^+(t')A^-(t')dt' + \frac{N^*}{\tau} - I_0 N \qquad (3.505b)$$

Substituting these equations into Eq. (3.501), we obtain an equation for ion concentration:

$$\partial_t N^+ = \int_0^t [\Sigma_1(t - t') - \Sigma_4(t - t')]N^*(t')A(t')dt'$$

$$-\int_0^t [\Sigma_6(t - t') - \Sigma_3(t - t')]N^+(t')A^-(t')dt' = \partial_t A^- \qquad (3.506)$$

In the IET of moderate pumping considered in the previous section, N^* was linear in I_0 as in Eqs. (3.5) and (3.4) so that there was no need in the equation for N, because $N \approx \mathcal{N} = \text{const}$. As a result, in Eqs. (3.417) there were only three integral terms and corresponding kernels: one, $R^* \equiv \Sigma_1$, for N^* and two others, $R^\dagger \equiv \Sigma_1 - \Sigma_4 = R^* - \Sigma_4$ and $R^\ddagger \equiv \Sigma_6$, for N^+. All kernels had a clear physical sense. Σ_1 and Σ_6 related to the bimolecular quenching of excitation and ion recombination, respectively. The kernel $R^\dagger \equiv \Sigma_1 - \Sigma_4$ was responsible for primary ionization (Σ_1) followed by geminate recombination of the ion pair (Σ_4) [38,202].

Only Σ_3 is new and rather unusual. It describes the overpopulation of ion pairs $[D^+ \cdots A^-]$, produced by electron transfer in closely situated products of ion recombination in the bulk, excited before their separation. Since the density of the correlated pairs is enhanced through bulk recombination, the ionization becomes faster than at the entirely homogeneous distribution of the reactants. The terms involving Σ_3 account for this effect. However, unlike other kernels, Σ_3 vanishes in the lowest-order approximation with respect to I_0. Therefore, this kernel and all the integral terms involving it are absent in the theory of the previous section, as well as in other theories, developed in the same approximation.

2. Stationary Solution

Let us introduce the stationary constants related to the kernels discussed above:

$$\kappa_g = \tilde{\Sigma}_1(0), \quad \kappa_i = \tilde{\Sigma}_1(0) - \tilde{\Sigma}_4(0) \tag{3.507}$$

$$\kappa_r = \tilde{\Sigma}_6(0), \quad \kappa' = \tilde{\Sigma}_3(0) \tag{3.508}$$

Unlike standard Markovian rate constants these constants depend on the excitation lifetime and light intensity. Using them, we can obtain a set of algebraic equations for the stationary concentrations of reactants and reaction products passing to the limit $t \to \infty$ in Eqs. (3.505):

$$0 = -\kappa_g N_s^* A_s - \kappa'[N_s^+]^2 - \frac{N_s^*}{\tau} + I_0 N_s \tag{3.509a}$$

$$0 = \kappa_i N_s^* A_s + (\kappa' - \kappa_r)[N_s^+]^2 \tag{3.509b}$$

$$A_s + A_s^- = A_s + N_s^+ = c, \quad N_s + N_s^+ + N_s^* = \mathcal{N} \tag{3.509c}$$

From Eqs. (3.509a) and (3.509b), one can easily find the ratio of stationary concentrations of excited and nonexcited fluorophores:

$$\frac{N_s^*}{N_s} = \frac{I_0 \tau}{1 + (\kappa_g + \beta \kappa') \tau A_s} \tag{3.510}$$

where $\beta = \kappa_i/(\kappa_r - \kappa')$. Substituting this ratio into the most general definition of the fluorescence quantum yield (3.455), we can prove that it obeys the modified Stern–Volmer law from Eq. (3.450)

$$\frac{1}{\eta} = 1 + \kappa_0 \tau A_s \tag{3.511}$$

but with the ab initio–defined light-dependent constant:

$$\kappa_0 = \kappa_g + \frac{\kappa_i \kappa'}{\kappa_r - \kappa'} \tag{3.512}$$

Since A_s is not a total concentration of electron acceptors c but a neutral fraction of them, the relationship between A_s and $c = [A] + [A^-]$, as well as the $\eta^{-1}(c)$ dependence, are complex and non-linear at large I_0. In this sense the original Stern–Volmer law breaks down with an increase in light intensity as well as with a decrease in the total quencher concentration c. In both cases one has to find first the $A_s(c)$ dependence before using it in Eqs. (3.511). As follows from Eq. (3.509b) and the conservation law for acceptors ($N_s^+ = c - A_s$), this dependence is given by the following relationship:

$$A_s = c - N_s^+ = c - \frac{1}{2}\left[\sqrt{\beta N_s^*(\beta N_s^* + 4c)} - \beta N_s^*\right] \tag{3.513}$$

Excluding A_s and N_s^+ from Eqs. (3.509), one can solve them numerically getting $N^*(c)$ and hence $A_s(c)$. However, the depletion of acceptors is a rather simple effect already considered in the previous section. It can be eliminated by taking a large excess of acceptors. Then $A_s \approx c$ and the single effect worthy of discussion is the light dependence of the Stern–Volmer constant κ_0 essentially affected by the strong pumping especially due to the second term in Eq. (3.512).

3.   Contact Approximation

The simplest solution of the problem can be obtained if both the ionization and recombination rates, $W_I(r)$ and $W_R(r)$, are assumed to be contact as in Eq. (3.368). According to the analysis presented in Section VII.A, this is possible only in case (a) of Figure 3.36, within the NN subregion, where both the forward and backward electron transfers occur in the normal Marcus regions. For the sake of simplicity, we also neglect the force interaction between reactants and assume diffusion to be the same in all pairs as in Eq. (3.479):

$$\hat{L}_1 = \hat{L}_2 = \hat{L}_3 = D\Delta \tag{3.514}$$

Under such conditions a few identities appear: $\nu_1 + \mu_5 + \mu_3 = 1$; $\nu_2 + \mu_2 + \mu_6 = 1$; $\nu_3 + \mu_1 + \mu_4 = 1$, which reduce the number of the independent auxiliary equations that were solved in Appendix C to Ref. 200. From there we also borrow the final results for all κ values:

$$\kappa_g = \frac{k_0 k_D (k_c + k_D)(1 + I_0\tau)}{k_D[k_c + k_D + I_0\tau(k_0 + k_c + k_D)] + k_0(k_c + k_D + I_0\tau k_c)\alpha_0} \qquad (3.515)$$

$$\kappa' = \frac{k_0 k_c k_D I_0\tau(1 - \alpha_0)}{k_D[k_c + k_D + I_0\tau(k_0 + k_c + k_D)] + k_0(k_c + k_D + I_0\tau k_c]\alpha_0} \qquad (3.516)$$

$$\kappa_r = \frac{k_c k_D[k_D(1 + I_0\tau) + k_0(I_0\tau + \alpha_0)]}{k_D[k_c + k_D + I_0\tau(k_0 + k_c + k_D)] + k_0(k_c + k_D + I_0 k_c\tau)\alpha_0} \qquad (3.517)$$

Here α_0 is the same as in Eq. (3.491) and similarly to $\kappa_g\bar{\varphi}$ in Eq. (3.447b) there is

$$\kappa_i = \kappa_g\varphi_c, \qquad (3.518)$$

where φ_c is also the charge separation quantum yield but in the contact approximation, (3.211).

Using these results the Stern–Volmer constant (3.512) can be presented as follows:

$$\kappa_0 = \kappa_g\left[1 + \varphi_c\frac{\kappa'}{\kappa_r - \kappa'}\right] = \kappa_g\Theta \qquad (3.519)$$

where

$$\kappa_g = \frac{k_0 k_D(1 + I_0\tau)}{k_D(1 + I_0\tau) + k_0\left[\alpha_0 + I_0\tau\frac{k_D + k_c\alpha_0}{k_D + k_c}\right]} \qquad (3.520)$$

and

$$\Theta = 1 + \frac{k_D}{k_D + k_c} \cdot \frac{k_0 I_0\tau(1 - \alpha_0)}{(1 + I_0\tau)(k_D + k_0\alpha_0)} \qquad (3.521)$$

Setting $I_0\tau = 0$ here, we return to the linear theory where $\Theta = 1$, $\alpha_0^{-1} = 1 + \sqrt{\tau_d/\tau}$ and κ_g turns to κ_0 from Eq. (3.273). In strong light, the reactions under kinetic and diffusional control should be discriminated since they are affected differently by light.

If recombination is under kinetic control, $k_D \gg k_c > k_c\alpha_0$, then

$$\kappa_g = \frac{k_0 k_D(1 + I_0\tau)}{k_D(1 + I_0\tau) + k_0[\alpha_0 + I_0\tau]}, \qquad \Theta = 1 + \frac{k_0 I_0\tau(1 - \alpha_0)}{(1 + I_0\tau)(k_D + k_0\alpha_0)} \qquad (3.522)$$

If ionization is also under kinetic control, $k_D \gg k_0 > k_0\alpha_0$, then $\Theta \approx 1$, $\alpha_0 \approx 1$ and

$$\kappa_0 = \kappa_g = k_0$$

does not depend on light at all.

In the opposite limit when the reaction is strongly controlled by diffusion, $k_D \ll k_c\alpha_0 < k_c$ and $k_D \ll k_0\alpha_0 < k_0$, we obtain

$$\kappa_g \approx k_D\left(1 + \sqrt{\tau_d(I_0 + 1/\tau)}\right), \qquad \Theta \approx 1 + \frac{k_D}{k_c}\frac{I_0\tau}{1 + I_0\tau}\sqrt{\tau_d(I_0 + 1/\tau)}$$

$$(3.533)$$

Although κ_g is exactly the same as in Eq. (3.492), the Stern–Volmer constant (3.529) differs from it because of the multiplier Θ. Both κ_0 and κ_g, shown by solid and dashed lines in Figure 3.63, increase with the light intensity and the difference between them is attributed to Θ.

The insertion of the $A_s(c)$ dependence into Eq. (3.521) breaks the linearity of the Stern–Volmer law [200]. The resulting curves $\eta^{-1}(c)$ are shown in Figure 3.64, taking into account the field dependence of the Stern–Volmer

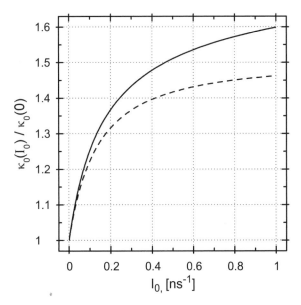

Figure 3.63. The field dependence of the Stern–Volmer constant $\kappa_0(I_0)$ with (solid line) and without (dashed line) accounting for the correction proportional to κ'. (From Ref. 200.)

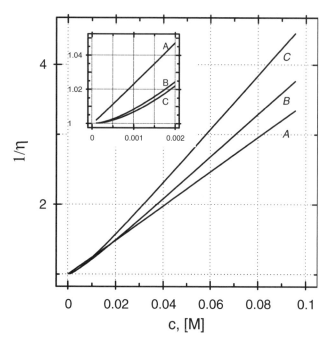

Figure 3.64. The bend in concentration dependence of $1/\eta$ at different light strength $I_0 = 10^{-3}\,\text{ns}^{-1}$ (a); $10^{-2}\,\text{ns}^{-1}(b)$; $10^{-1}ns^{-1}(c)$. The remaining parameters are $k_D = 6000\,\text{Å}^3/\text{ns}$, $k_0 = k_c = 10^4\text{Å}^3/\text{ns}$, $\tau_d = 0.25\,\text{ns}$, $\tau = 10\,\text{ns}$. The parabolic distortion of the high-field curves shown in the insert originates from the electron acceptor depletion at low concentrations. (From Ref. 200.)

constant $\kappa_0(I_0)$. Except for curve a for the lowest field, which reproduces the classical linear law, all the rest are nonlinear at low concentrations and even intersect. This is due to electron acceptor depletion ($A_s \ll c$), which is more pronounced with a stronger field. As a result, the signs of the field effect are diametrically opposite at low and high concentrations. To avoid this complication, one should measure the photoconductivity in line with the fluorescence. If the former allows one to find A_s^- and hence $A_s = c - A_s^-$, then the linearity of Eq. (3.521) can be established experimentally and used to find the true κ_0 defined in Eq. (3.522).

C. Intermolecular Energy Transfer

The quenching of fluorescence by energy transfer from the excited fluorophores to other impurities was described in Sections II and III but only in the absence of light (after pulse excitation). There we studied the simplest quenching (3.43), which is irreversible (due to the high exergonicity of transfer), as well as quasiresonant and therefore reversible transfer, represented by the reaction scheme (3.92). Here we will address the former (and simpler) reaction, but taking into account the permanent pumping of the arbitrary light intensity:

$$D^* + B \xrightarrow{W} D + B^* \qquad (3.534)$$

$$I_0 \uparrow\downarrow \tau_D \qquad\qquad \downarrow \tau_B$$

To account for the high intensity effects, the IET has to be used here instead of the DET employed in Section II.

Using the matrix form of IET, we have to introduce the following vectors of states:

$$\mathbf{N}_D = \begin{pmatrix} N^* \\ N \end{pmatrix} \qquad \text{and} \qquad \mathbf{N}_B = \begin{pmatrix} B^* \\ B \end{pmatrix} \qquad (3.535)$$

where B^* and B are concentrations of the excited and ground states of the quencher. These vectors obey the following matrix IET equations:

$$\frac{d}{dt}\mathbf{N}_D = \mathrm{Tr}_B \int_0^t \mathbf{M}(\tau)\mathbf{N}_B(t-\tau) \otimes \mathbf{N}_D(t-\tau)d\tau + \hat{\mathbf{Q}}_D\mathbf{N}_D \qquad (3.536a)$$

$$\frac{d}{dt}\mathbf{N}_B = \mathrm{Tr}_D \int_0^t \mathbf{M}(\tau)\mathbf{N}_B(t-\tau) \otimes \mathbf{N}_D(t-\tau)d\tau + \hat{\mathbf{Q}}_B\mathbf{N}_B \qquad (3.536b)$$

where $\hat{\mathbf{Q}}_D$ and $\hat{\mathbf{Q}}_B$ are matrices representing not only the decay of both excited molecules as in Eqs. (3.102) but also the light excitation of one of them:

$$\hat{\mathbf{Q}}_D = \begin{bmatrix} -1/\tau_D & I_0 \\ 1/\tau_D & -I_0 \end{bmatrix} \qquad \hat{\mathbf{Q}}_B = \begin{bmatrix} -1/\tau_B & 0 \\ 1/\tau_B & 0 \end{bmatrix} \qquad (3.537)$$

The integral parts of Eqs. (3.536) account for the bimolecular reactions such as the inter-molecular energy transfer in the present case. In the binary integral theories the kernels of the integral terms, $\hat{\mathbf{M}}$, must be defined through the solution of the pair problem bounded to the encounter of single-donor and single-acceptor molecules. However, there are a few ways of doing this using IET. This is the time to find the unique one that avoids the emergence of extra but zero kernels as in Eqs. (3.474) or (3.502) and (3.503).

1. Radical Reformulation of IET

Generally speaking, the IET kernel is just the space average of the **T** matrix. For the simplest irreversible reactions, an equivalence of IET and the average

T-matrix approximation developed in Ref. 190 was shown in Refs. 39 and 42. The matrix generalization of this approach was obtained in Refs. 49, 134, and 203. Here we are using its modification proposed in Ref. 125.

For the reactions in highly polar solvents ($r_c = 0$), the kernel was given an alternative form, much more convenient for further analytical treatment:

$$\tilde{\mathbf{M}}(s) = \int d^3r \hat{\mathbf{W}}(r)\left(\hat{\mathbf{I}} + \tilde{\mathbf{\Phi}}(r,s)\right) \tag{3.538}$$

The newly introduced matrix pair density $\mathbf{\Phi}(r,t)$ obeys the following equation:

$$\partial_t \mathbf{\Phi}(r,t) = \hat{\mathbf{L}}\mathbf{\Phi} + \hat{\mathbf{W}}\mathbf{\Phi} + \hat{\mathbf{Q}}\mathbf{\Phi}, \qquad \mathbf{\Phi}(r,0) = \hat{\mathbf{W}}(r) \tag{3.539}$$

There is the relationship between the new ($\mathbf{\Phi}$), and old ($\mathbf{F}$), pair distributions matrices:

$$\tilde{\mathbf{\Phi}}(r,s) = \tilde{\mathbf{F}}(r,s)\left[s\hat{\mathbf{I}} - \hat{\mathbf{Q}}\right] - \hat{\mathbf{I}} \tag{3.540}$$

The old one obeys the equation (3.106):

$$\partial_t \mathbf{F}(r,t) = \hat{\mathbf{L}}\mathbf{F} + \hat{\mathbf{W}}\mathbf{F} + \hat{\mathbf{Q}}\mathbf{F}, \qquad \mathbf{F}(r,0) = \hat{\mathbf{I}} \tag{3.541}$$

It is easy to see that the Laplace transformations of Eqs. (3.539) and (3.541) are identical in view of the relationship (3.540).

The principal difference is in the initial conditions to Eq. (3.541) and Eq. (3.539). The initial condition to Eq. (3.539) is much more convenient because it automatically reveals the zero elements of the kernel, thus allowing us to decrease the rank of the matrix equations. The reaction under consideration illustrates this point looking at the matrix form of main operators in the basis of the four collective states of reacting pairs: (D^*B^*, DB^*, D^*B, DB). They are

$$\hat{\mathbf{W}} = \begin{cases} 0 & 0 & 0 & 0 \\ 0 & 0 & W(r) & 0 \\ 0 & 0 & -W(r) & 0 \\ 0 & 0 & 0 & 0 \end{cases} \qquad \hat{\mathbf{Q}} = \hat{\mathbf{I}}_D \otimes \hat{\mathbf{Q}}_B + \hat{\mathbf{Q}}_D \otimes \hat{\mathbf{I}}_B \tag{3.542}$$

where $\hat{\mathbf{I}}_D$ and $\hat{\mathbf{I}}_B$ are identity matrices in the basis of individual particles. Since $\mathbf{\Phi}(r,s) = \int \tilde{\mathbf{G}}(r,r_0,s)\hat{\mathbf{W}}(r_0)d^3r_0$, only those elements of $\mathbf{\Phi}(r,s)$ should be different from zero that belong to the column containing nonzero elements of $\hat{\mathbf{W}} = \mathbf{\Phi}(r,0)$:

$$\mathbf{\Phi}(r,s) = \begin{Bmatrix} 0 & 0 & \tilde{f}_2(r,s) & 0 \\ 0 & 0 & \tilde{f}_4(r,s) & 0 \\ 0 & 0 & \tilde{f}_1(r,s) & 0 \\ 0 & 0 & \tilde{f}_3(r,s) & 0 \end{Bmatrix}, \qquad \tilde{\mathbf{M}}(s) = \begin{Bmatrix} 0 & 0 & 0 & 0 \\ 0 & 0 & \tilde{\Sigma}(s) & 0 \\ 0 & 0 & -\tilde{\Sigma}(s) & 0 \\ 0 & 0 & 0 & 0 \end{Bmatrix}$$

(3.543)

The kinetic equations for the concentrations of excited particles, can be obtained by writing Eqs. (3.536) in components:

$$\frac{dN^*(t)}{dt} = -\int_0^t dt' \Sigma(t') N^*(t-t') B(t-t') - \frac{N^*(t)}{\tau_D} + I_0 N(t) \qquad (3.544a)$$

$$\frac{dB^*(t)}{dt} = \int_0^t dt' \Sigma(t') N^*(t-t') B(t-t') - \frac{B^*(t)}{\tau_B} \qquad (3.544b)$$

To close the set (3.544), one has to take into account the particle balance

$$B^*(t) + B(t) = c, \qquad N^*(t) + N(t) = \mathcal{N} \qquad (3.545)$$

where c and $\mathcal{N}$ are the overall concentrations of each species correspondingly. They remain constant over time.

2. Reduction of Auxiliary Equations

Only 2 non-zero elements remain in matrix $\tilde{\mathbf{M}}$ defined in Eq. (3.538). However, this is not the case if one starts with the old approach formulated in Eqs. (3.104) and (3.106) or (3.499). Some cumbersome calculations were necessary to show that the kernel element (3.474) equals zero as well as $\tilde{\Sigma}_2$ and $\tilde{\Sigma}_5$ in Eqs. (3.502) and (3.503). The new approach does not require such additional work.

The single nonzero element of $\tilde{\mathbf{M}}(s)$ from Eq. (3.543)

$$\tilde{\Sigma}(s) = \int d^3r W(r)(1 + \tilde{f}_1(r,s)), \qquad (3.546)$$

is expressed via the Laplace transformation of $f_1(r,t)$, which is the diagonal element of matrix $\mathbf{\Phi}(r,t)$. This element is coupled with three off-diagonal elements of the same matrix, but if diffusion is the same for the excited and nonexcited particles, then $\hat{\mathbf{L}} = D\Delta \cdot \hat{\mathbf{I}}$ is diagonal and

$$f_1 + f_2 + f_3 + f_4 = 0 \qquad (3.547)$$

Taking into account this relationship, we can retain only three independent components in the vector that obeys the matrix equation similar to the Laplace transformation of Eq. (3.539) [125]

$$(s - D\Delta + \hat{\mathbf{Q}}_0) \begin{pmatrix} \tilde{f}_1 \\ \tilde{f}_2 \\ \tilde{f}_3 \end{pmatrix} = -W(r) \begin{pmatrix} 1 + \tilde{f}_1 \\ 0 \\ 0 \end{pmatrix} \tag{3.548}$$

where

$$\hat{\mathbf{Q}}_0 = \begin{pmatrix} -\frac{1}{\tau_D} & \frac{1}{\tau_B} & I_0 \\ -I_0 & -\left(I_0 + \frac{1}{\tau_D} + \frac{1}{\tau_B}\right) & -I_0 \\ \frac{1}{\tau_D} - \frac{1}{\tau_B} & -\frac{1}{\tau_B} & -I_0 - \frac{1}{\tau_B} \end{pmatrix} \tag{3.549}$$

3.   Contact Approximation

As before, the analytical solution becomes possible only in the contact approximation (3.40). It is inappropriate for the dipole–dipole transfer rate (3.44) but can be considered as a reasonable model for the exponential "exchange" quenching carried out by the dipole-forbidden triplet–triplet energy transfer [204].

In the contact approximation the general matrix solution of Eq. (3.548) can be expressed through its Green function $\hat{\mathbf{G}}_0(\sigma, \sigma, s)$ [125]:

$$\begin{pmatrix} \tilde{f}_1(\sigma, s) \\ \tilde{f}_2(\sigma, s) \\ \tilde{f}_3(\sigma, s) \end{pmatrix} = -(\hat{\mathbf{I}}_0 + \hat{\mathbf{G}}_0(\sigma, \sigma, s)\hat{\mathbf{K}}_r)^{-1}\hat{\mathbf{G}}_0(\sigma, \sigma, s) \begin{pmatrix} k_0 \\ 0 \\ 0 \end{pmatrix} \tag{3.550}$$

Here $\hat{\mathbf{I}}_0$ is a 3×3 identity matrix and $\hat{\mathbf{K}}_r$ is a matrix of the same rank but with a single nonzero element: $(\hat{\mathbf{K}}_r)_{1,1} = k_0$. In order to find $\hat{\mathbf{G}}_0(\sigma, \sigma, s)$, one has to specify a transformation that renders matrix $\hat{\mathbf{Q}}_0$ diagonal and apply it to Eq. (3.548), taking into account the well-known expression (3.240). After a cumbersome but straightforward matrix calculation, one can obtain from the contact analog of Eq. (3.546) the sought-after kernel

$$\tilde{\Sigma}(s) = k_0\left[1 + \tilde{f}_1(\sigma, s)\right] = k_0 k_D \frac{1/\tau_D + I_0}{[k_D + k_0\alpha(s)][1/\tau_D + I_0] + I_0 k_0[\beta(s) - \gamma(s)]} \tag{3.551}$$

where

$$\alpha(s) = \left(1 + \sqrt{\tau_d(s + I_0 + 1/\tau_D)}\right)^{-1} \tag{3.552a}$$

$$\beta(s) = \left(1 + \sqrt{\tau_d(s + 1/\tau_B)}\right)^{-1} \tag{3.552b}$$

$$\gamma(s) = \left(1 + \sqrt{\tau_d(s + I_0 + 1/\tau_D + 1/\tau_B)^{-1}}\right)^{-1} \tag{3.552c}$$

4. Stationary Solution

As $t \to \infty$, $\dot{N}^* = \dot{B}^* \to 0$ and Eqs. (3.544) and (3.555) turn to the set of algebraic equations regarding N_s^*, B_s^* and N_s, B_s. One can easily obtain from their solution the generalized Stern–Volmer law similar in form to (3.521):

$$\frac{1}{\eta} = \frac{I_0 N_s \tau_D}{N_s^*} = 1 + \kappa_0 \tau_D B_s \tag{3.553}$$

as well as the relationship between the nonexcited energy acceptors B_s and their total concentration, c [125]:

$$2B_s = c - I_0 \mathcal{N} \tau_B - \frac{I_0 + 1/\tau_D}{\kappa_0} + \sqrt{\left[c - I_0 \mathcal{N} \tau_B - \frac{I_0 + 1/\tau_D}{\kappa_0}\right]^2 + 4c\,\frac{I_0 + 1/\tau_D}{\kappa_0}} \tag{3.554}$$

Here the Stern–Volmer constant is

$$\kappa_0 = \tilde{\Sigma}(0) = \frac{k_0 k_D}{k_D + \alpha_0 k_0 + \frac{I_0 k_0 \tau_D}{1 + I_0 \tau_D}[\beta_0 - \gamma_0]} \tag{3.555}$$

where $\alpha_0 = \alpha(0)$, $\beta_0 = \beta(0)$, and $\gamma_0 = \gamma(0)$ are the corresponding values of α, β, and γ, respectively, introduced in Eqs. (3.552). In the limit $I_0 \to 0$ this expression turns to the simplest one, (3.27) or (3.273), obtained with DET or IET correspondingly.

As is obvious from Eq. (3.554), if the excitation accepted by **B** decays immediately, $(\tau_B = 0)$, the density of the acceptors in their ground state is preserved: $B_s = c$. At any finite τ_B the fraction of energy acceptors resides in the excited state **B*** so that in general $B_s < c$ and may be much less than c at strong pumping (see Fig. 3.65). The simple analysis shows that $B_s \simeq c$ until the light intensity is limited by the following condition:

$$I_0 \tau_B \ll \frac{c}{\mathcal{N}} \tag{3.556}$$

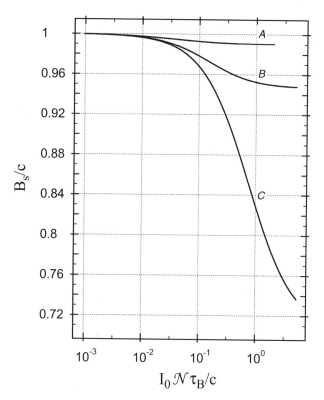

Figure 3.65. Depletion of the ground-state population of energy acceptor B_s with an increase in the dimensionless light intensity $I_0 \mathcal{N} \tau_B / c$ at different τ_B equal to $1\,\mathrm{ns}(A), 10\,\mathrm{ns}(B),$ $100\,\mathrm{ns}(C)$. $\mathcal{N}/c = 0.1, c = 10^{-2}\,\mathrm{M}, \tau_D = 10\,\mathrm{ns}, k_0 = 10^5\,\text{Å}^3/\mathrm{ns}, k_D = 6 \times 10^3\,\text{Å}^3/\mathrm{ns}, \tau_d = 0.25\,\mathrm{ns}.$ (From Ref. 125.)

Under this condition the light pumping affects only the Stern–Volmer constant (3.555).

As $\tau_B \to 0$, the energy acceptors are very effective quenchers with $\beta_0 = \gamma_0 = 0$, so that the general expression (3.555) reduces to its simpler form (3.490). This field dependence is represented by the upper curve in Figure 3.66, which relates to the shortest τ_B. The dissimilarity of our present reaction (3.544), from the simplest one, (3.467), reveals itself in Figure 3.66 at larger τ_B when the dependence $\kappa_0(I_0)$ changes qualitatively . At very large τ_B the horizontal transitions shown in Figure 3.67 are switched off and the density of the double excited pairs, $\mathbf{D}^*\mathbf{B}^*$, is enhanced by light pumping. The decay of these states contributes only to the fluorescence intensity because $\mathbf{D}^*$ surrounded by $\mathbf{B}^*$ cannot be quenched. In other words, being excited, $\mathbf{B}s$ become inoperative. The closest neighbors of $\mathbf{D}^*$ are excited in first turn. The energy

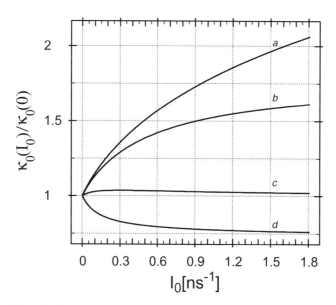

Figure 3.66. The Stern–Volmer constant dependence on the light strength at different lifetimes of the excited acceptor of energy, $\tau_B = 0.1, 1, 10, 100 \, \text{ns}$ (from top to bottom: a,b,c,d), and $\tau_D = 10 \, \text{ns}$. The remaining parameters (k_0, k_D, τ_d) are the same as in the previous Figure 3.65. (From Ref. 125.)

transfer disables this neighborhood, thus diminishing the quenching constant κ_0. The disposition of reaction partners is accounted for in the memory functions (kernels) of IET.

If the energy acceptors are in great excess $(c \gg \mathcal{N})$, then the condition (3.556) is not too rigid; the interval of relatively weak fields where $B_s \simeq c$ is rather wide. Within this interval the conventional Stern–Volmer law is valid and its constant is given by Eq. (3.555). This statement relates to the three upper curves shown in Figure 3.68, which are almost linear in c. However, at a much higher density of fluorophores the inequality (3.556) is inverted at small c and the concentration dependence of the quantum yield becomes curvilinear similar

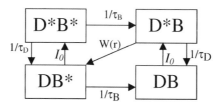

Figure 3.67. The schematic representation of light-induced and radiationless transitions between four states of the collective basis of the reactant pair. (From Ref. 125.)

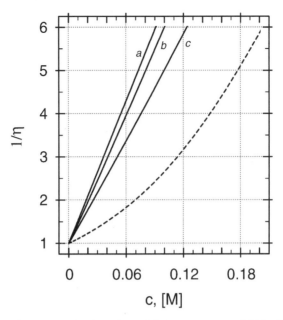

Figure 3.68. The concentration dependence of the inverse quantum yield for different strengths of light pumping $I_0 = 0.01\,\text{ns}^{-1}$ (a), $0.1\,\text{ns}^{-1}$ (b), $1\,\text{ns}^{-1}$ (c) with a great excess of energy acceptors ($\mathcal{N} = 10^{-3}\,\text{M}$ is almost everywhere smaller than c). Dashed curve—the same as in case (b) but for $\mathcal{N} = 3 \times 10^{-2}\,\text{M}$. The parameters k_0, k_D, τ_d are the same as in Figure 3.65, whereas $\tau_D = 10\,\text{ns}$, $\tau_B = 50\,\text{ns}$. (From Ref. 125.)

to the dashed line in Figure 3.68. As always, the pronounced light intensity effects are seen when the transfer reaction is under diffusional control but disappear in the kinetic control limit, when $k_0 \ll k_D$. Hence, if light affects κ_0, this is an indication of diffusion-controlled energy transfer.

XI. SPIN CONVERSION

The RIP $[D^+ \cdots A^-]$ arising in the spinless reaction scheme (3.257) has actually 4 spin sub-levels: one for the singlet state, $^1[D^+ \cdots A^-]$, and three for the triplet one, $^3[D^+ \cdots A^-]$. In low magnetic fields they are practically degenerate and therefore the equilibrium relationship of singlet to triplet populations is $1 : 3$. However, the forward electron transfer generates either a singlet or triplet RIP depending on what is a precursor (excited reactant). The further relaxation to equilibrium is carried out by spin conversion, which is either a coherent or incoherent process. The latter is just an approximation although convenient and popular [205–209] and well justified for RIPs composed of transition metal complexes with strong spin-orbital coupling and anisotropic g factor [210]. It was implemented in UT [28,32,149,150] and an IET [107,123] using the only parameter: the spin conversion rate k_s.

For the Δg mechanism of spin conversion in moderate magnetic fields $(\omega_- T_2 \ll 1)$ [29,32]

$$k_s = \left[\frac{1}{2T} + \frac{\omega_-^2 T}{12} \right] \tag{3.557}$$

Here $T = T_1 = T_2$ is the relaxation time, longitudinal and transversal, and the frequency detunning

$$\omega_- = \frac{(g_+ - g_-)\beta}{\hbar} H \tag{3.558}$$

where β is the Bohr magneton, g_+ and g_- are g factors of radical ions in the pair, and H is the external magnetic field. The relaxation of the singlet and triplet populations to their equilibrium values is governed by a set of four equations that in the lowest-order approximation with respect to the small parameter $\omega_-^2 T_2^2$ reduces to the following two [32]

$$\dot{m}_S = k_s m_T - 3k_s m_S \tag{3.559a}$$
$$\dot{m}_T = -k_s m_T + 3k_s m_S \tag{3.559b}$$

where m_S is the population of the singlet state $|S\rangle$ and m_T is the sum of the populations of triplet substates, $|T_+\rangle, |T_0\rangle, |T_-\rangle$.

In the coherent (Hamiltonian) approach to the four-state spin system, the state populations are just the diagonal elements of the corresponding density matrix $\hat{\rho}(r,t)$, which obeys the Bloch or Redfield equation [211]:

$$\frac{\partial}{\partial t}\hat{\rho}(t) = -i[\mathbf{H}, \hat{\rho}] - +\hat{\mathbf{R}}\hat{\rho} \tag{3.560}$$

Here $\mathbf{H}$ is the spin Hamiltonian and $\hat{\mathbf{R}}$ is the relaxation superoperator, accounting for the population and phase relaxation in the collective states basis of the RIP: $|T_+\rangle, |T_-\rangle, |T_0\rangle$ and $|S\rangle$.

Here we will confine ourselves mainly to the incoherent (rate) description of spin conversion, which provides the simplest way to account for spin-forbidden recombination of RIPs in complex photochemical reactions. Since k_s from Eq. (3.557) depends on H through $\omega_-(H)$, the charge separation yield is also affected by the magnetic field. Therefore, the ratio

$$M(H) = \frac{\bar{\varphi}(H) - \bar{\varphi}(0)}{\bar{\varphi}(0)} \tag{3.561}$$

is used to characterize the magnetic field effect (MFE) in the charge separation. The coherent description of spin conversion was also used to calculate the MFE by means of UT [149,150], and the results were reviewed in Ref. 32.

A. Triplet Precursor of RIP

The irreversible quenching of the triplet Ruthenium complex by methylviologen [148] represented by Eq. (3.329) was considered within the spinless theory in Section VII. The more detailed reaction scheme accounting for the spin states and their incoherent conversion is as follows:

$$
\begin{array}{ccc}
(^3Ru^{2+})^* + MV^{2+} \xrightarrow{W_I} \; ^3[\dot{R}u^{3+} \cdots \dot{M}V^+] & \longrightarrow & ^2\dot{R}u^{3+} + \, ^2\dot{M}V^+ \\
k_s \downarrow\uparrow 3k_s & & \\
^1[\dot{R}u^{3+} \cdots \dot{M}V^+] & \longrightarrow & ^2\dot{R}u^{3+} + \, ^2\dot{M}V^+ \qquad (3.562) \\
\Downarrow W_S & & \\
^1[Ru^{2+} \cdots MV^{2+}] & &
\end{array}
$$

Instead of the universal recombination rate W_R there is a sequence of two vertical transitions. The triplet–singlet conversion precedes the backward electron transfer from the singlet RIP to the ground state. Symbolizing Ru^{2+} by D and MV^{2+} by A, we can formalize this reaction scheme in the following way:

$$
\begin{array}{ccccc}
& & ^2\dot{D}^+ & + & ^2\dot{A}^- \\
& & \nearrow \quad \bar{\varphi} \quad \searrow & & \\
^3D^* + A \xrightarrow{W_I} \; ^3[D^+ \cdots A^-] & \underset{3k_s}{\overset{k_s}{\rightleftharpoons}} & ^1[D^+ \cdots A^-] & & (3.563) \\
\downarrow \tau & & \Downarrow W_S & & \\
D & & [D \cdots A] & &
\end{array}
$$

The UT equation (3.289a) continues to describe the kinetics of bimolecular charge separation, but the total survival probability of charged products is the sum of fractions originating from the triplet and singlet RIPs:

$$
P(t) = c \int_\sigma^\infty [m_T(r,t) + m_S(r,t)] 4\pi r^2 dr \qquad (3.564)
$$

These fractions obey the set of extended equations (3.559) substituted for Eq. (3.290):

$$
\frac{\partial}{\partial t} m_T(r,t) = \hat{L}^+ m_T(r,t) + 3k_s m_S(r,t) - k_s m_T(r,t) + W_I(r) n(r,t) N(t) \qquad (3.565a)
$$

$$
\frac{\partial}{\partial t} m_S(r,t) = \hat{L}^+ m_S(r,t) - 3k_s m_S(r,t) + k_s m_T(r,t) - W_S(r) m_T(r,t) \qquad (3.565b)
$$

with the initial conditions $m_T(r,0) = m_S(r,0) = 0$ and the reflecting boundary conditions.

1. Charge Separation Yield

The general solution to Eqs. (3.565) has the following form:

$$\begin{pmatrix} m_T(t) \\ m_S(t) \end{pmatrix} = \int_0^t dt' \int d^3r' \begin{pmatrix} p_{TT}(t-t') & p_{ST}(t-t') \\ p_{TS}(t-t') & p_{SS}(t-t') \end{pmatrix} \begin{pmatrix} W_I(r')n(r',t')N(t') \\ 0 \end{pmatrix}$$

$$(3.566)$$

We need only $p_{TT}(r,r',t)$ and $p_{TS}(r,r',t)$, which are the Green functions for the populations of triplet and singlet RIPs originating from the initially excited triplet state. Another pair, $p_{ST}(r,r',t)$ and $p_{SS}(r,r',t)$, are the similar Green functions of Eqs. (3.565) but for RIPs produced from a singlet excitation. From Eqs. (3.564) and (3.566) one can easily obtain

$$P(t) = c \int d^3r \int d^3r' W_I(r') \int_0^t p(r,r',t-t')n(r',t')N(t')dt' \qquad (3.567)$$

Here $p = p_{TT} + p_{TS}$ is composed of triplet and singlet components that yield the set of homogeneous differential equations following from Eqs. (3.565):

$$\frac{\partial}{\partial t}p_{TT} = -k_s p_{TT} + 3k_s p_{TS} + \tilde{D}\frac{1}{r^2}\frac{\partial}{\partial r}r^2 e^{r_c/r}\frac{\partial}{\partial r}e^{-r_c/r}p_{TT} \qquad (3.568a)$$

$$\frac{\partial}{\partial t}p_{TS} = k_s p_{TT} - 3k_s p_{TS} + \tilde{D}\frac{1}{r^2}\frac{\partial}{\partial r}r^2 e^{r_c/r}\frac{\partial}{\partial r}e^{-r_c/r}p_{TS} - W_S(r)p_{TS} \qquad (3.568b)$$

They should be solved with the same reflecting boundary conditions as before, but with other initial conditions peculiar to the Green functions:

$$p_{TT}(r,r',0) = \frac{\delta(r-r')}{4\pi r^2}, \qquad p_{TS}(r,r',0) = 0 \qquad (3.569)$$

The free-ion quantum yield $\phi = P(\infty)$ resulting from Eqs. (3.567) and (3.294), is

$$\phi = c \int d^3r \int d^3r' p(r,r',\infty)m_0(r') = \psi\bar{\phi} \qquad (3.570)$$

As previously in Eq. (3.274), it is the product of the total ionization quantum yield ψ, given by Eq. (3.295), and the charge separation quantum yield

$$\bar{\phi} = \int \varphi(r')f_0(r')d^3r' = 1 - \bar{\phi}_s \qquad (3.571)$$

where $\bar{\phi}_s$ is the quantum yield of the recombination products.

The averaging in Eq. (3.571) is performed over a normalized initial distribution of RIPs, (3.299). If $f_0(r') = \delta(r' - r_0)/4\pi r_0^2$, then $\bar{\varphi}$ is identical to the partial charge separation yield:

$$\varphi(r_0) = \int_\sigma^\infty p(r, r_0, \infty)d^3 r = \frac{1}{1 + Z(r_0)/\tilde{D}} \tag{3.572}$$

From now on the recombination efficiency Z depends not only on the starting point r_0 and encounter diffusion coefficient $\tilde{D}$ but also on the spin conversion rate k_s.

2. Contact Approximation

To proceed further with the analytic investigation, we have to confine ourself to the contact recombination, which is a reasonable approximation for the remote start and narrow reaction layer adjacent to the contact. In this approximation the reaction term in Eqs. (3.568) can be omitted

$$\frac{\partial}{\partial t}p_{TT} = -k_s p_{TT} + 3k_s p_{TS} + \tilde{D}\frac{1}{r^2}\frac{\partial}{\partial r}r^2 e^{r_c/r}\frac{\partial}{\partial r}e^{-r_c/r}p_{TT} \tag{3.573a}$$

$$\frac{\partial}{\partial t}p_{TS} = k_s p_{TT} - 3k_s p_{TS} + \tilde{D}\frac{1}{r^2}\frac{\partial}{\partial r}r^2 e^{r_c/r}\frac{\partial}{\partial r}e^{-r_c/r}p_{TS} \tag{3.573b}$$

but the reaction should be accounted for in the boundary conditions to this set:

$$\left(\frac{\partial p_{TT}}{\partial r} + \frac{r_c}{r^2}p_{TT}\right)\Bigg|_{r=\sigma} = 0, \qquad 4\pi\tilde{D}\sigma^2\left(\frac{\partial p_{TS}}{\partial r} + \frac{r_c}{r^2}p_{TS}\right)\Bigg|_{r=\sigma} = k_c^S p_{TS}(\sigma, t)$$
$$\tag{3.574}$$

The contact reaction constant $k_c^S = \int W_S(r)d^3 r$ is the main parameter of the theory, and the yield of recombination products is $k_c^S \tilde{p}_{TS}(\sigma, r_0, 0) = 1 - \varphi(r_0)$.

In this approximation the rigorous solution can be expressed through the Green function of free diffusion, $G_0(\sigma, r', s)$ [28]:

$$\varphi(r_0) = 1 - \frac{\frac{1}{4}k_c^S[\tilde{G}_0(\sigma, r_0, 0) - \tilde{G}_0(\sigma, r_0, 4k_s)]}{1 + \frac{3}{4}k_c^S\tilde{G}_0(\sigma, \sigma, 4k_s) + \frac{1}{4}k_c^S\tilde{G}_0(\sigma, \sigma, 0)} \tag{3.575}$$

For highly polar solvents ($r_c = 0$), there is an analytic expression for the Green function of the free encounter diffusion without reaction [20,28]:

$$\tilde{G}_0(\sigma, r', s) = \frac{1}{4\pi r'\tilde{D}}\frac{e^{-(r'-\sigma)\sqrt{s/\tilde{D}}}}{1 + \sigma\sqrt{s/\tilde{D}}} \tag{3.576}$$

Using this in Eq. (3.575), we obtain

$$\varphi(r_0) = 1 - k_c^S \tilde{p}_{TS}(\sigma, r_0, 0) = 1 - \frac{\sigma \frac{1}{4} k_c^S \left[1 + x - \exp\left\{ -x\left(\frac{r_0 - \sigma}{\sigma} \right) \right\} \right]}{r_0 \quad k_D(1+x) + k_c^S \left[1 + \frac{1}{4} x \right]} \quad (3.577)$$

where the spin conversion is represented by the dimensionless parameter:

$$x = \sqrt{\frac{4k_s \sigma^2}{\tilde{D}}} = \sqrt{\alpha \frac{k_c^S}{k_{\tilde{D}}}}, \quad \text{or} \quad \alpha = \frac{16\pi\sigma^3 k_s}{k_c^S} \quad (3.578)$$

Evidently, $x^2 = 4k_s \tau_d$ is a degree of conversion during encounter time.
   It follows from Eqs. (3.572) and (3.577) that

$$\frac{Z(r_0)}{\tilde{D}} = \frac{1 + x - \exp\left\{ -x\left(\frac{r_0 - \sigma}{\sigma} \right) \right\}}{\frac{4r_0}{\sigma} \left[1 + x/4 + (1+x)k_D/k_c^S \right] - \left[1 + x - \exp\left\{ -x\left(\frac{r_0 - \sigma}{\sigma} \right) \right\} \right]} = \frac{k_{isc}\sigma^2}{3\tilde{D}} \quad (3.579)$$

where the last expression establishes the interrelationship between UT and the exponential model [212–216]. The latter introduces the intersystem crossing rate k_{isc} as an exact analog of k_{-et}, but for spin-forbidden recombination. In highly polar solvents ($r_c \to 0$) it relates to the recombination efficiency Z in the same very way as in Eq. (3.194). However, Z represented in Eq. (3.579) is not a constant but exhibits the complex dependence on diffusion and the spin conversion rate.

3. Spin Conversion Control

If there is no spin conversion ($x = k_s = 0$), the recombination is entirely absent: $Z \equiv 0$ and $\varphi \equiv 1$. Otherwise we have the following for the slow and fast spin conversions:

$$\frac{Z(r_0)}{\tilde{D}} = \begin{cases} \frac{x}{4} \cdot \frac{k_c^S}{k_c^S + k_D} & \text{at} \quad x \to 0 \\ \frac{k_c^S}{k_c^S(r_0 - \sigma)/\sigma + 16\pi r_0 \tilde{D}} & \text{at} \quad x \to \infty \end{cases} \quad (3.580)$$

Only the result for the fast conversion ($x \to \infty$) is identical to the spinless theory equation (3.208), provided $k_c^S/4$ is substituted for k_c in Eq. (3.205). Such a substitution conditioned by a full equilibration of the spin state populations results in a proper redefinition of the EM parameter z in highly polar solvents ($r_c = 0$):

$$z = \frac{k_c^S}{16\pi\sigma} \quad (3.581)$$

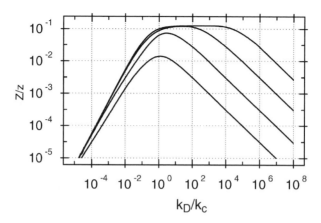

Figure 3.69. The diffusional dependence of Z for RIPs that start from the initial separation $r_0 = 8\sigma$ ($\sigma = 5\,\text{Å}$) and recombine at contact with the rate $k_c^S = 628\,\text{Å}^3/\text{ns}$, at different $\alpha = 10^{-3}, 0.1, 10, 10^3$ (from bottom to top) [32].

In the spinless theory the diffusional dependence of Z at a distant start is composed of the diffusional and kinetic branches of the geminate recombination; see the lower curves in Figure 3.26. When diffusion increases, the ascending (diffusional) branch gives way to the kinetic plateau $Z = (\sigma/r_0)z$. However, with a decrease in the spin conversion rate, both these branches are split into two subregions; one of them is affected while the other is indifferent to spin effects [150]. The unaffected diffusional branch has the same declination as before, while the affected one follows it with a smaller slope (Fig. 3.69) The lower is α, the earlier the former branch gives way to the latter. Similarly, the kinetic plateau gives way to the descending branch as soon as recombination falls under the combined spin kinetic control. This also occurs at slower diffusion with smaller α. At the smallest α represented in Figure 3.69 by the lowest curve, there is no plateau at all and the reaction is under spin conversion control everywhere.

The increase of Z due to the acceleration of the spin conversion with the magnetic field was in fact detected experimentally [Fig. 3.70(b)], although in the rather restricted diffusion (viscosity) range. This effect resembles the field enhancement of diffusional recombination obtained earlier in the low (natural) magnetic field (Fig. 3.23). The data are in qualitative agreement with the theoretical prediction that the ascending branch of the recombination efficiency shifts up with the magnetic field [Fig. 3.70(a)]. Accordingly, at any given diffusion the RIP separation quantum yield $\bar{\varphi}(H)$ decreases with H as well as

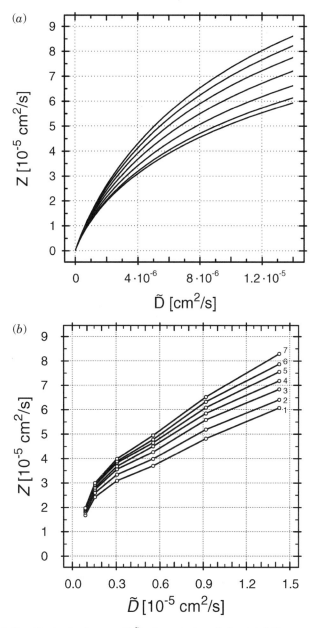

Figure 3.70. The band of curves $Z(\tilde{D})$: theoretical prediction (a) falling approximately in the region where the $^3[Ru(bpy)_3]^{2+}/MV^{2+}$ reaction was experimentally studied (b). Curves 1–7 corresponds to magnetic fields of 0, 0.5, 1.0, 1.5, 2.0, 2.5, and 3.0 T. (From Refs. 150 and 210.)

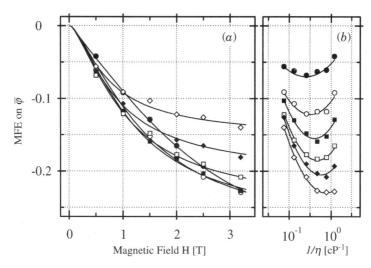

Figure 3.71. (a) Magnetic field dependence of MFE on $\bar{\varphi}$, $M = [\bar{\varphi}(H)/\bar{\varphi}(0) - 1]$ versus H, observed for solvent mixtures (H_2O/ACN 1 : 1 with ethylene glycol) of different viscosities [in centipoise (cP)]: $\diamond$ 13.40, $\blacklozenge$ 7.68, $\square$ 3.88, $\blacksquare$ 2.15, $\circ$ 1.29, $\bullet$ 0.83. (b) Data of (a) represented as the viscosity dependence of the MFE at various magnetic fields in teslas (T): $\diamond$ 3.2, $\blacklozenge$ 2.5, $\square$ 2.0, $\blacksquare$ 1.5, $\circ$ 1.0, $\bullet$ 0.5. The solid lines represent second order polynomial approximations drawn to aid the viewer. (From Ref. 149.)

$M(H)$ from Eq. (3.561) shown in Figure 3.71(a). Following from these data, the extremal viscosity dependence of MFE at any given field [Fig. 3.71(b)] has a plausible explanation within the same unified theory but with the coherent spin conversion in the reduced (two spin states) basis [32,149]. The latter simplification was removed in the subsequent work where all four states were considered [150].

However, the applicability of the contact approximation to the system in question is doubtful. At the lowest viscosities the forward transfer is under kinetic control and the initial distribution of RIPs is adjacent to the contact [149]. Hence most of the starts are not far from contact. Sometimes the optimal r_0 is 10.5 Å at $\sigma = 10$ Å and $L = 0.75$ Å [150]. Under such conditions the RIPs are starting from the inside recombination zone and their recombination should be hindered by diffusion. Hence the diffusional dependence of Z must be the opposite, similar to that inherent in the upper curves in Figure 3.26. As was shown in Section VII.D, this is not the case, only due to the significant change of Pekar's factor γ in line with the viscosity variation (Tables I and II). The acceleration of the RIP recombination with diffusion was given an alternative explanation there, namely, that this is due to the increase in the reorganization energy caused by growing γ (Fig. 3.46). With this explanation L appears to be twice as large (1.35 Å), making the contact approximation even less pertinent. Nonetheless, the contact approximation remains very attractive in view of its simplicity. For electron transfer this is just a model allowing for an analytical solution, but for the proton or atom transfer ($L < 0.1$ Å) it provides a reasonable description of the phenomena.

B. Singlet Precursor of RIP

An example of the bimolecular photoionization of the excited singlet molecule given by Eq. (3.330) can be represented by the following comprehensive scheme:

$$
\begin{array}{ccc}
& A^- \quad + \quad D^+ & \\
& \nearrow \quad \bar{\varphi} \quad \searrow & \\
{}^1A^* + D \; \overset{W_I}{\Longrightarrow} \; {}^1[A^- \cdots D^+] & \underset{k_s}{\overset{3k_s}{\rightleftharpoons}} & {}^3[A^- \cdots D^+] \\
\downarrow \tau \qquad\qquad \Downarrow W_S & & \Downarrow W_T \\
A \qquad\qquad [A \cdots D] & & [{}^3A^* \cdots D]
\end{array}
\qquad (3.582)
$$

Unlike reaction (3.563), this one proceeds through two parallel channels of geminate recombination: to the singlet (ground) and triplet (excited) states. Therefore, one should discriminate between the partial efficiencies, Z_S and Z_T, which contribute to the total one:

$$
Z = Z_S + Z_T \qquad (3.583)
$$

Correspondingly, there are three terms in the quantum yield balance instead of only two of them as in Eq. (3.571):

$$
\bar{\varphi}_t + \bar{\varphi}_s + \bar{\varphi} = \frac{Z_T}{\tilde{D} + Z} + \frac{Z_S}{\tilde{D} + Z} + \frac{\tilde{D}}{\tilde{D} + Z} = 1 \qquad (3.584)
$$

The very existence of the triplet channel of geminate recombination was first recognized and experimentally proved by Weller et al., who studied the quenching of excited pyrene (Py) by dimethylaniline (DMA) and a few other electron donors [217,219]. However, the quenching of perylene (Per) fluorescence by DMA or o-DMT (N,N-dimethyl-o-toluidine) studied besides others in a later work was given a quantitative description within the spinless exponential model [109]. Since EM fails to describe the geminate kinetics correctly (see Section V), the problem was revised by means of noncontact UT [187]. The pseudoexponential separation kinetics of radical ions in the Per/ o-DMT system was explained assuming that after the spin conversion in RIPs they recombine into the excited triplet products. The magnetic field effect was also predicted in this work assuming that the singlet–triplet transitions in the zero field are different from those in the high magnetic field H:

$$^1[A^- \cdots D^+] \underset{k_s}{\overset{3k_s}{\rightleftharpoons}} {}^3[A^- \cdots D^+] \quad (H = 0) \qquad (3.585a)$$

$$^1[A^- \cdots D^+] \underset{k_s}{\overset{k_s}{\rightleftharpoons}} {}^3[A^- \cdots D^+] \quad (H \to \infty) \qquad (3.585b)$$

The equalization of the rates in both directions at high H results from the splitting of the triplet that disables the $|T_+\rangle$ and $|T_-\rangle$ states when the spin-conversion is carried out by a weak hyperfine coupling. A similar effect was predicted and backed experimentally in the classical works with Py [217–219].

The quenching of Per through the triplet channel was confirmed in 2003 by a straightforward observation of the triplet products of geminate recombination [98]. Moreover, the value of their quantum yield $\bar{\varphi}_t$ was detected and the viscosity dependence of this quantity was also specified. Having on hand two independent components of Eq. (3.584), $\bar{\varphi}_t$ and $\bar{\varphi}$, one can easily obtain two corresponding recombination parameters

$$Z_T = \tilde{D}\,\frac{\varphi_t}{\bar{\varphi}} \quad \text{and} \quad Z = \tilde{D}\left(\frac{1}{\bar{\varphi}} - 1\right) \qquad (3.586)$$

as well as the third one, $Z_S = Z - Z_T$. The main goal of the theory is to specify the diffusional dependence of all of them and compare it with the experimental data.

It is instructive to compare the general theory with the EM description of the two-channel reaction (II) [213,214]. The yields of triplets and separated ions are represented as follows:

$$\varphi_t = \frac{k_{\text{isc}}}{k_{\text{sep}} + k_{-et} + k_{\text{isc}}}, \qquad \varphi = \frac{k_{\text{sep}}}{k_{\text{sep}} + k_{-et} + k_{\text{isc}}} \qquad (3.587)$$

The only difference from the single-channel EM outlined above (Section V.A) is the substitution of k_{-et} by the sum of the spin-allowed and spin-forbidden transfer rates $k_{-et} + k_{\text{isc}}$, to the ground and triplet states, respectively. Like k_{-et}, the intersystem crossing rate k_{isc} does not depend on viscosity. Moreover, EM does not separate the two different steps of the forbidden transition: spin conversion to the triplet RIP and subsequent allowed electron transfer into the triplet product [212–216]. However, as has been shown in Section XI.A, even in the case of a single channel but spin-forbidden reaction (I), one should discriminate between the spin conversion and subsequent recombination through electron transfer. The qualitative difference between the spin-allowed and spin-forbidden recombination will be confirmed here once again as well as the crucial dependence of both on the rate of spin conversion.

Rearranging Eqs. (3.586), we can give the uniform definitions for the singlet and triplet recombination efficiencies

$$Z_T = \tilde{D}\frac{\bar{\varphi}_t}{\bar{\varphi}} \qquad \text{and} \qquad Z_S = \tilde{D}\frac{\bar{\varphi}_s}{\bar{\varphi}} \tag{3.588}$$

where all the quantum yields are averaged over the normalized initial distribution of RIPs produced by the bimolecular ionization:

$$\bar{\varphi} = \int \varphi(r)f_0(r)d^3r, \qquad \bar{\varphi}_s = \int \varphi_s(r)f_0(r)d^3r, \qquad \bar{\varphi}_t = \int \varphi_t(r)f_0(r)d^3r \tag{3.589}$$

1. Contact Approximation

Assuming the contact recombination through both reaction channels

$$W_S = \frac{k_c}{4\pi\sigma^2}\,\delta(r-\sigma), \qquad W_T = \frac{k_t}{4\pi\sigma^2}\,\delta(r-\sigma) \tag{3.590}$$

we get the following definitions for all the partial quantum yields:

$$\varphi_s(r) = k_c\tilde{p}_{SS}(\sigma,r,0), \qquad \varphi_t(r) = k_t\tilde{p}_{ST}(\sigma,r,0), \qquad \varphi(r) = 1 - \varphi_s(r) - \varphi_t(r) \tag{3.591}$$

Here p_{SS} and p_{ST} are the Green functions that enter equation (3.566) but with the pumping term shifted down, because of the singlet precursor. For the same reason their initial conditions are different from those in Eq. (3.569):

$$p_{ST}(r,r',0) = 0, \qquad p_{SS}(r,r',0) = \frac{\delta(r-r')}{4\pi r^2} \tag{3.592}$$

In the contact approximation these Green functions obey equations similar to Eqs. (3.573)

$$\frac{\partial}{\partial t}p_{ST} = -k_s\,p_{ST} + 3k_s\,p_{SS} + \tilde{D}\frac{1}{r^2}\frac{\partial}{\partial r}r^2 e^{r_c/r}\frac{\partial}{\partial r}e^{-r_c/r}p_{ST} \tag{3.593a}$$

$$\frac{\partial}{\partial t}p_{SS} = k_s p_{ST} - 3k_s p_{SS} + \tilde{D}\frac{1}{r^2}\frac{\partial}{\partial r}r^2 e^{r_c/r}\frac{\partial}{\partial r}e^{-r_c/r}p_{SS} \tag{3.593b}$$

but with the boundary conditions accounting for recombination through both reaction channels:

$$\left(\frac{\partial p_{ST}}{\partial r} + \frac{r_c}{r^2} p_{ST}\right)\bigg|_{r=\sigma} = k_t p_{ST}(\sigma, t), \quad 4\pi \tilde{D}\sigma^2 \left(\frac{\partial p_{SS}}{\partial r} + \frac{r_c}{r^2} p_{SS}\right)\bigg|_{r=\sigma} = k_c p_{SS}(\sigma, t)$$

$$(3.594)$$

In a highly polar solvent ($r_c = 0$) the solution to these equations is known [220]

$$\tilde{p}_{ST}(\sigma, r, 0) = \frac{3\sigma}{4r} \frac{xk_D + \left[1 - e^{-x(r-\sigma)/\sigma}\right](k_D + k_c)}{[k_D(1+x) + k_t](k_D + k_c) + \frac{3}{4}xk_D(k_t - k_c)}$$

$$(3.595)$$

and all quantum yields of the recombination and separation are expressed through this single quantity [220]:

$$\varphi_s(r) = \frac{k_c}{k_c + k_D} \left\{\frac{\sigma}{r} - \tilde{p}_{ST}(\sigma, r, 0)(k_t + k_D)\right\}$$

$$(3.596a)$$

$$\varphi_t(r) = k_t \tilde{p}_{ST}(\sigma, r, 0)$$

$$(3.596b)$$

$$\varphi(r) = 1 - \frac{k_c/4\pi r\tilde{D} + (k_t - k_c)\tilde{p}_{ST}(\sigma, r, 0)}{1 + k_c/k_D}$$

$$(3.596c)$$

Here $x = \sqrt{4k_s\sigma^2/\tilde{D}}$ is exactly the same as in Eq. (3.578) and all the results crucially depend on this parameter.

2. Only Singlet Recombination

If there is no recombination of the triplet RIPs, as was assumed in Ref. 221, then $k_t = 0$ and $\varphi_t(r) = 0$ as well. This is a single-channel reaction but the yield of its products

$$\varphi_s(r) = 1 - \varphi(r) = k_c \frac{\sigma}{r} \frac{1 + x/4 + (3/4)\left[1 - \exp\left(-x\frac{r-\sigma}{\sigma}\right)\right]}{k_D(1+x) + k_c(1 + x/4)}$$

$$(3.597)$$

is different from that obtained with the spinless theory, Eq. (3.203). The latter is reproduced only in the absence of spin conversion, at $x = k_s = 0$:

$$\varphi(r) = \begin{cases} 1 - \dfrac{\sigma}{r_0} \dfrac{k_c}{k_D + k_c} & \text{at } k_s = 0 \\[3mm] 1 - \dfrac{\sigma}{r_0} \dfrac{(1/4)k_c}{k_D + (1/4)k_c} & \text{at } k_s = \infty \end{cases}$$

$$(3.598)$$

In the opposite limit of infinitely fast spin conversion the recombination through the singlet channel is 4 times slower because of the full equilibration of the spin state populations of RIPs.

A similar result was expected in Ref. 221 for $\varphi_s(\sigma) = 1 - \varphi(\sigma)$, but the obtained difference between the recombination rates in the opposite limits was half as much: k_c for the slowest conversion and $k_c/2$ for the fastest one. This is because the isotropic Δg mechanism determining the spin conversion in Ref. 221 mixes the singlet with the $|T_0\rangle$ sublevel only. In the rate approximation one can easily get the same, assuming that the spin transitions between the singlet and triplet RIPs occurs with equal rates in the forward and backward directions as in Eq. (3.585b). However, the transition from the slow to the fast conversion limit resulting from the rate approximation differs somehow from that obtained with the Hamiltonian approach in Ref. 221.

3. Only Triplet Recombination

In principle, the RIP recombination to the excited triplet product is less exergonic than to the ground state, so that one may expect $k_t > k_c$. Coming to the extreme, let us assume that $k_c = 0$, that is, $\varphi_s(r) = 0$ as well. This is again the single-channel recombination, but unlike the previous one, it is spin-forbidden. As follows from Eqs. (3.596a) and (3.595) for this particular case, we have

$$\varphi(r) = 1 - \varphi_s(r) = 1 - \frac{\sigma}{r} \frac{(3/4)k_t\left[1 + x - \exp\left\{-x\left(\frac{r-\sigma}{\sigma}\right)\right\}\right]}{k_D(1+x) + k_t[1 + (3/4)x]} \qquad (3.599)$$

This is exactly the same result as for the spin-forbidden recombination from the triplet precursor, but with k_t substituted for k_c in Eq. (3.577) and $\frac{3}{4}$ for $\frac{1}{4}$, according to the rates and the weights of the reacting states. As a result, $\varphi(r) = 1/(1 + Z/\tilde{D})$, and $Z(r)$ has the same properties as its analog studied in the previous subsection:

$$\frac{Z(r)}{\tilde{D}} = \begin{cases} \frac{3x}{4} \cdot \frac{k_t}{k_t + k_D} & \text{at} \quad x \to 0 \\ \frac{k_t}{k_t(r-\sigma)/\sigma + 16\pi r \tilde{D}/3} & \text{at} \quad x \to \infty \end{cases} \qquad (3.600)$$

Only the last result coincides with that of the contact model provided that there is a contact start, no Coulomb interaction, and $z = 3k_t/16\pi\sigma$.

4. Contact Recombination with Equal Rates

The dissimilarity of the double-channel recombination from either of the single-channel reactions considered above is seen from the particular example of equal recombination rates [220,222]:

$$k_c = k_t \qquad (3.601)$$

From Eqs. (3.596b), (3.596c) and (3.595), we obtain for this case

$$\varphi_t(r) = \frac{3\sigma}{4r} \frac{k_c/k_D}{1+x+k_c/k_D} \left[\frac{x}{1+k_c/k_D} + 1 - e^{-x(r-\sigma)/\sigma} \right], \qquad \varphi(r) = 1 - \frac{\sigma}{r} \frac{k_c/k_D}{1+k_c/k_D}$$
$$(3.602)$$

With the constraint (3.601) at any spin conversion $\varphi(r)$ is exactly the same as in the equation of the spinless theory, (3.211). This is because the transitions between the singlet and triplet RIPs do not modulate their recombination rates, leaving $z = k_c/4\pi\sigma$ as well as $\varphi(r)$ unchanged. However, the partition of the product yields, $1 - \varphi(r)$, between singlet and triplet fractions, crucially depends on the conversion rate and diffusion.

It is often assumed that not only recombination takes place in the contact but the RIPs start from there also [222]. This assumption essentially simplifies the problem by sacrificing its relevancy. Setting $r = \sigma$ also, we get from Eq. (3.602) the simpler relationships:

$$\varphi_t(\sigma) = \frac{3}{4} x k_D \frac{k_c}{(k_c + k_D)[k_c + k_D(1+x)]}, \qquad \varphi(\sigma) = \frac{1}{1 + k_c/k_D} \qquad (3.603)$$

Here the separation yield $\varphi(\sigma)$ is exactly the same as in the exponential model, which also implies that RIPs start from the recombination zone. At the same time the yield of triplet products is proportional to the spin conversion degree x or more specifically to $x k_D = 8\pi\sigma^2 \sqrt{k_s \tilde{D}}$. To make this proportion evident, we use $\sqrt{\tilde{D}}$ as an argument in Figure 3.72. Unlike φ and φ_s, which are monotonous although opposite functions of $\sqrt{\tilde{D}}$, the triplet yield φ_t tends to zero at either small or large $\tilde{D}$, having a maximum in between. The extremal behavior of this quantity was first discovered in the work of the Schultens [222]. Although they used the coherent description, and not the rate, of the spin conversion, Figure 3.5 of their work is in essence the same as our Figure 3.72.

However, the assumption that RIPs start only from contact is very artificial. The starting distance is the closest to contact when the normal ionization is under kinetic control (Section VI.C). Even in this case the distribution of starting distances, which reproduces the normalized ionization rate $W_I(r)$, is not a δ function but a quasiexponential smooth function with the width equal to $L \sim 1 \div 2$ Å.

The problem is aggravated by the fact that in the contact approximation the quantum yields and recombination efficiencies Z_T and Z depend essentially on

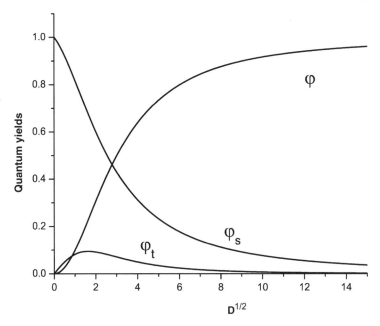

Figure 3.72.   The quantum yields at the contact start and equal recombination rates as functions of the diffusion coefficient given in units $\mathring{A}^2/ns$ ($k_c = k_t = 800\,\mathring{A}^3/ns$ and $k_s = 0.01\,ns^{-1}$). (From Ref. 220.)

the starting point of the RIPs (Fig. 3.73). Particularly striking is the $Z(\tilde{D})$ sensitivity to very small deviations of r from σ; in the contact approximation the slope of the linear increase of Z tends to infinity when the starting point approaches contact.

5.   *Noncontact Double-Channel Recombination*

This is the chief drawback of the contact approximation discussed in Section V.C and illustrated by Figures 3.25 and 3.26. At any $r > \sigma$ we have the following in the contact approximation: $\lim_{\tilde{D}\to 0} Z(r) = 0$ and $\lim_{\tilde{D}\to 0} \varphi(r) \neq 0$. For remote electron transfer, everything is quite the reverse: $\lim_{\tilde{D}\to 0} Z(r) = const$ and $\lim_{\tilde{D}\to 0} \varphi(r) = 0$.

Since $Z \to 0$ at small $\tilde{D}$ in the contact approximation, $\varphi(r)$ obtained with this approximation has a plateau in this region, while $\varphi_t(r) = Z_T/(Z + \tilde{D})$ strongly deviates there from $\varphi_t(\sigma)$ increasing as $\tilde{D} \to 0$ faster, when $r - \sigma$ is smaller (Fig. 3.74). Since the experimental data obtained in Ref. 98 is out of this region, they were fitted rather well in the contact approximation [220]. The starting distance $r = 7.51\,\mathring{A}$ chosen from the best fit exceeds insignificantly $\sigma = 7.5\,\mathring{A}$. Although this difference is much less than any real L, the solution is so sensitive to its value that it offers considerable scope for speculation.

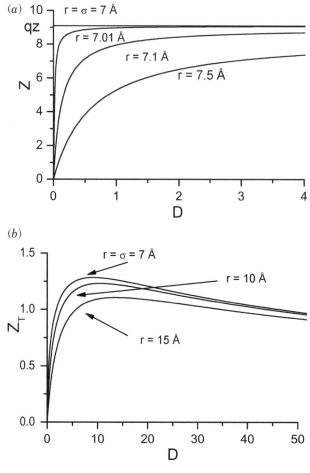

Figure 3.73. The diffusional dependence of total (*a*) and triplet (*b*) recombination efficiencies at the equal recombination rates as in the previous figure but from a different starting separation of radical ions. (From Ref. 220.)

The instability of the contact description of geminate recombination stimulated another attempt to fit the same data using the exponential transfer rates in the general unified theory of remote reactions (3.53) [223]. UT not only enables eliminating the artificial sensitivity of the results to the initial separation but also avoids the very necessity of choosing the starting distance r. The existing numerical program for solving UT equations [224] provides us with a true distribution over r and the quantum yields of all products at any given r. The averaging of partial quantum yields over such distributions is requested by Eqs. (3.589).

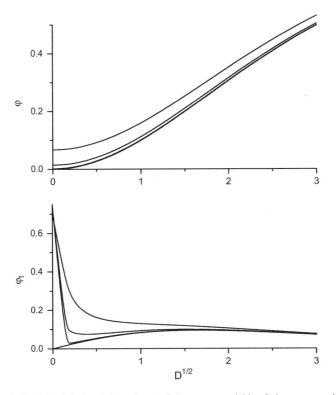

Figure 3.74. The diffusional dependence of the quantum yields of charge separation φ and triplet products φ_t, corresponding to the efficiencies shown in the previous figure at starting distance $r = 7, 7.01, 7.1, 7.5\,\text{Å}$ (from bottom to top). (From Ref. 220.)

Using the averaged values of all the yields in Eqs. (3.588), one obtains the recombination efficiencies Z_T and Z_S as well as $Z = Z_T + Z_S$. The only parameters varying in such calculations are the preexponents and space decrements of the ionization rate $W_c \exp[-2(r - \sigma)/l_I]$ and the recombination rate to singlet $W_s \exp[-2(r - \sigma)/l_R]$. The recombination rate to the triplet is assumed to be the same as the latter but with another preexponent, W_t, and the ratio W_t/W_s is subject to fitting as well. The results obtained from the best fit are shown in Figure 3.75.

Since the recombination rate was found to be more extended than the ionization one ($l_R > l_I$), there is no surprise that the curve $Z(\tilde{D})$ goes through the maximum as the upper curve in Figure 3.41. This effect was attributed in Section VII.B to extension of the recombination layer due to the high exergonicity of recombination. Although this factor is absent when the model of exponential rates is used, l_R can be stretched instead to reach the same effect:

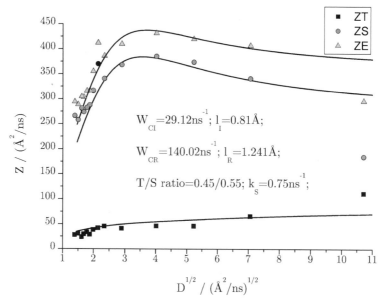

Figure 3.75. The experimental quantum yields of singlet (•) and triplet (■) recombination products as well as the charge separation yield (▲). Solid lines—their fitting with UT using the exponential rates for ionization, $W_I = W_c \exp[-2(r - \sigma)/l_I]$, and for recombination to either the ground state, $W_S = W_s \exp[-2(r - \sigma)/l_R]$, or the triplet excited state, $W_T = W_t \exp[-2(r - \sigma)/l_R]$. Parameters obtained from the best fit are the following: $W_c = 29.12 \, \text{ns}^{-1}$, $W_s = 77 \, \text{ns}^{-1} = 1.2 \, W_t$, $l_I = 0.81 \, \text{Å}, l_R = 1.24 \, \text{Å}, k_s = 0.75 \, \text{ns}^{-1}$, $\sigma = 7.5 \, \text{Å}$. (From Ref. 223.)

creation of ions deep inside the recombination layer at fast diffusion. The latter helps them escape from there, causing Z to decrease with $\tilde{D}$.

However, stretching of the tunneling length may be an artifact. If the true Marcus rates were substitute for exponents then the effect would be the same or even more pronounced without significant variation of l. If this makes the decrease of Z sharp enough to pass through the last experimental point, then the same will happen to Z_S, which decreases with $\tilde{D}$ in a similar way. As soon as the last Z point becomes smaller, both $\bar{\varphi} = \tilde{D}/[\tilde{D} + Z]$ and $\bar{\varphi}_t = Z^T/[\tilde{D} + Z]$ have to increase correspondingly as well as Z_T. Hence, the transition from the exponential to the true Marcus rates could improve the total agreement with the experiment even more.

C. Exciplex Precursor of RIP

According to scheme (3.231), the singlet RIPs can transform into exciplexes or appear from them with the rates k_a and k_d, respectively. Taking into account the spin states, the total reaction of photoionization/recombination through exciplexes becomes a bit more complex:

$$^1A^* \; + \; D \qquad\qquad A^- \; + \; D^+$$

$$_I\swarrow \qquad \searrow^{II} \qquad \nearrow \quad_{\bar\varphi}\quad\searrow$$

$$^1[A^-D^+] \; \underset{k_a}{\overset{k_d}{\rightleftharpoons}} \; ^1[A^-\cdots D^+] \; \underset{k_s}{\overset{3k_s}{\rightleftharpoons}} \; ^3[A^-\cdots D^+] \qquad\qquad (3.604)$$

$$\Big\downarrow \tau_{exc} \qquad\qquad \Big\Downarrow w_S \qquad\qquad \Big\Downarrow w_T$$

$$[AD] \qquad\qquad [A\cdots D] \qquad\qquad [^3A^*\cdots D]$$

There are two parallel channels of energy quenching: by either contact formation of exciplex with subsequent dissociation to RIP [Eq. (3.604), scheme I] or by remote formation of RIP with subsequent association (transformation) to exciplex [Eq. (3.604), scheme II]. The last one was considered first by means of unified theory [29], which was extended later to account for both schemes together [30]. Since the results were comprehensively reviewed in Chapter IX of Ref. 32, there is no need to do the same here. It should only be noted that the theory of scheme I has been generalized to account exciplex formation, not only by encounters of excited reactants but also by a straightforward light excitation of existing complexes of the same particles [31].

1. Unified Theory of Scheme I

The quantum yield of exciplex fluorescence is

$$\phi_e = W_f \int_0^\infty P_e(t)dt = \eta_0\varphi_E \qquad\qquad (3.605)$$

where W_f is the rate of light emission from exciplex, $\eta_0 = W_f\tau_{exc}$ is the quantum yield of fluorescence, provided that the exciplex does not dissociate, and

$$\varphi_E = \frac{\tilde{P}_e(0)}{\tau_{exc}} \qquad\qquad (3.606)$$

is the exciplex quantum yield expressed through its integral population, $\int_0^\infty P_e(t)dt = \tilde{P}_e(0)$. The latter in turn can be found from the general UT equations of the problem [31]

$$\tilde{P}_e(0) = \frac{\tau_E\psi_e}{1 - \tau_E k_a k_d \tilde{p}_{SS}(\sigma,0)} \qquad\qquad (3.607)$$

where $1/\tau_E = 1/\tau_{exc} + k_d$. If exciplexes are created by a straightforward complex excitation, ψ_e should be substituted by 1. Alternatively, if they are the products of the bimolecular contact association of $^1A^* + D$ with the rate constant k_e then $\psi_e = ck_e \int_0^\infty n(\sigma, t)N(t)dt$, where $N = [^1A^*]$ as in Eq. (3.295). Equation (3.607) is in fact the Laplace transformation of the solution to Eqs. (3.233) extended for the case of spin assisted exciplex formation in Ref. 31.

The quantum yield of exciplexes produced in the bimolecular reaction is

$$\varphi_E = \psi_e \left[\frac{1}{1 + k_d\tau_{exc}} + \frac{k_d\tau_{exc}}{1 + k_d\tau_{exc}} \chi_E \right] \qquad (3.608)$$

where

$$\chi_E = \frac{k_a \tilde{p}_{SS}(\sigma, 0)}{1 + k_d\tau_{exc}(1 - k_a\tilde{p}_{SS}(\sigma, 0))} \qquad (3.609)$$

Here $\tilde{p}_{SS}$ is the Laplace transformation of the solution to the coupled equations (3.593), where $k_a + k_c$ is substituted for k_c in the boundary conditions. In the spinless theory they reduce to a single equation that is the contact analog of Eq. (3.234) for $G \equiv p_{SS}$.

2. Spin-Affected Quantum Yields

However, the theory of exciplex dissociation cannot be made spinless like that for photoacids (Section V.D). The dissociation products are radical ions and the spin conversion in RIPs essentially affects φ_E as well as other quantities listed in Eq. (3.589). To illustrate this phenomenon, let us concentrate on the fluorescence yield, which is affected through $\chi_E(k_s)$ and the charge separation quantum yield $\varphi(\sigma)$. We will consider the general solution obtained for these two quantities in Ref. 31 only in the simplest case of highly polar solvents for which the Green functions are well known.

It is easier to compare the quantum yields in the two opposite limiting cases: (1) $k_t = 0$ and (2) $k_t = \infty$. For case 1

$$\chi_E = \frac{k_{ass}}{k_{ass} + k_c + 4k_Dq}, \qquad \varphi(\sigma) = \frac{4k_Dq}{k_{ass} + k_c + 4k_Dq} \qquad (3.610)$$

where the overall association constant $k_{ass} = k_a/(1 + k_d\tau_{exc})$ and $q = (1 + x)/(4 + x)$. For case 2

$$\chi_E = \frac{k_{ass}}{k_{ass} + k_c + k_D\beta}, \qquad \varphi(\sigma) = \frac{k_D}{k_{ass} + k_c + k_D\beta} \qquad (3.611)$$

where $\beta = 1 + \frac{3}{4}x$. Formulas (3.610) and (3.611) establish the upper and lower limits for the x dependence of χ_E and $\varphi(\sigma)$ (Fig. 3.76). The curves for the

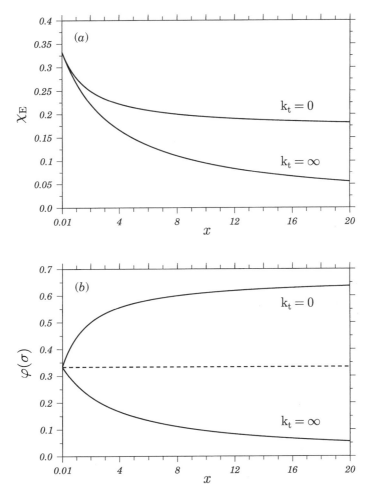

Figure 3.76. The effect of spin conversion on χ_E (a) and $\varphi(\sigma)$ (b) at different rates of RIP recombination to triplet products, k_t. The horizontal dashed line relates to $k_t = k_a + k_c$. (From Ref. 31.)

intermediate values of k_t can be obtained from the general expressions for these quantities obtained in Ref. 31. In particular, it was shown that for $k_t = k_a + k_c$ the charge separation quantum yield does not depend on the spin conversion, which makes no difference between the singlet and triplet RIPs if they recombine at the same rate.

Since in both cases (1 and 2) χ_E decreases with $x = \sqrt{4k_s\sigma^2/\tilde{D}}$, which is a measure of the spin conversion, the same is true for the exciplex yield and its fluorescence. However, the quantum yield of free ions $\varphi(\sigma)$ exhibits the opposite behavior in these two limits: it increases with x in case 1 but decreases in case 2. This means that only in case 2 do both χ_E and $\varphi(\sigma)$ decrease with the spin conversion rate k_s. If the spin conversion is faster in the hydrogen-substituted species than in the deuterated ones, both the fluorescence and free-ion yields should be sensitive to this difference. As was indicated in Ref. 131, in the H-containing species relative to deuterated ones the quantum yield of free ions measured by the ion conductivity is lower. Along with this, the lifetime of the exciplex is shorter, which makes the fluorescence quantum yield smaller. The simultaneous reduction of both these quantities in the RIPs composed of 9,10-dicyanoanthracene (DCA) and durene (DUR), as well as in other pairs [131], points out that the ion recombination through the triplet channel is faster than through the singlet one. More particularly, $k_t > k_a + k_c$ and the situation is closer to case 2 than to case 1.

Even more straightforward evidence in favor of recombination to the triplet state was obtained by heavy-atom substitution into the fluorescence quencher, which also enhances the rate of spin conversion. By measuring the transient absorption of both ion radicals and the triplet products of their recombination, it was demonstrated that the quantum yield of triplets increases when the charge separation yield $\varphi(\sigma)$ decreases as a result of heavy-atom substitution [225]. As was shown in Figure 3.76, $\varphi(\sigma)$ decreases with k_s if recombination to the triplet state is faster than to the singlet state, while the quantum yield of triplets produced from the singlet precursor only increases with k_s. Hence these data also indicate that the triplet channel of recombination is the most efficient.

The spinless variant of the present theory was already discussed in Section V.D and its interrelationship with IET and a number of other theories of exciplexes or stable complexes was disclosed. In the next Section XI.D we also consider not an excited-state but a ground-state particle. It is subjected to thermal dissociation to radicals followed by their geminate and subsequent bimolecular recombination into the fluorescent product.

D. Chemiluminescence

The mechanism of chemically initiated electron exchange luminescence (CIEEL) has been specified and studied in solvents of different viscosities [226–230]. The intramolecular electron transfer in the oxyaryl-substituted adamantyldioxetane anion D^- [**2** (the boldfaced numbers refer to the numbered structures 1–5 in Fig. 3.77)], followed by decomposition of the latter into two radicals $\dot{A}^-$ and $\dot{M}$ (**5**), allows them to diffuse freely in and out of the cage until the backward electron transfer (BET) excites the methyl-m-oxybenzoate anion M^- (**4**), as shown in Figure 3.77. In general, the intermolecular BET mechanism of excitation (right branch) competes with a direct chemiexcitation in the course

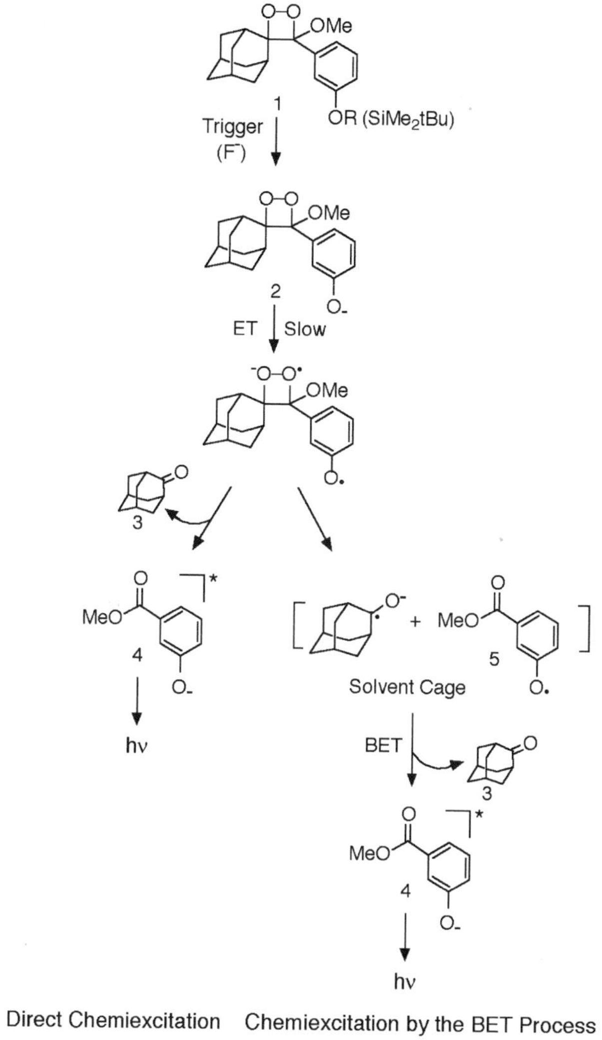

Figure 3.77. Two-channel mechanism of chemically initiated luminescence triggered from a spiroadamantyl-substituted dioxetane (from Ref. 228). Here we discuss only the right branch (CIEEL channel) of the reaction $2 \rightarrow 5 \rightarrow 4 \rightarrow h\nu$.

of D^- decomposition into **3** and **4** (left branch), but the latter was recognized as an inefficient channel [227]. Therefore, we restrict ourselves to only the intermolecular BET channel, but will take into account the spin conversion in the radical pair constituted from adamantanone anion radical $\dot{A}^-$ and methyl-*m*-oxybenzoate radical $\dot{M}$.

1. Detailed Reaction Schemes

As a result of spin conversion, the initial singlet state of this pair changes to a triplet one and thus opens the way for the electron transfer to a triplet product as well. This becomes possible in parallel with the allowed RIP recombination through a singlet channel, to either the excited or ground states, $^1M^-$ or M^- as in Section XI.C:

$$
\begin{array}{c}
[A \cdots M^-] \quad \varphi_{ss} \quad [A \cdots {}^1M^-] \implies A + M^- + h\nu \\[4pt]
k_c \diagup \quad k_r^S \diagup \quad k^* \diagup \\[4pt]
D^- \implies {}^1[\dot{A}^- \cdots \dot{M}] \; \overset{3k_s}{\underset{k_s}{\rightleftharpoons}} \; {}^3[\dot{A}^- \cdots \dot{M}] \overset{k_t}{\longrightarrow} [A \cdots {}^3M^-] \quad \varphi_{st} \\[4pt]
\varphi \\[4pt]
{}^2\dot{A}^- + {}^2\dot{M}
\end{array}
\tag{3.612}
$$

The total rate constant for singlet RIP recombination is the sum of the partial constants for excited- and ground-state production:

$$
k_r^S = k^* + k_c \tag{3.613}
$$

Since electron transfer straight to the ground state of anion, M^-, is much more exergonic than recombination to the excited singlet state, the relative yield of the latter is not expected to be much less than 1:

$$
\xi = \frac{k^*}{k^* + k_c} \lesssim 1 \tag{3.614}
$$

Since the singlet and triplet products of the geminate reaction (3.612) have a singlet precursor (ground-state D^-), their yields are denoted as φ_{ss} and φ_{st}, respectively. Because the initial singlet RIP produced by D^- decomposition is born at the contact distance, no averaging over the initial distribution is necessary and the argument of all yields is σ. They obey a conservation law similar to (3.584):

$$
\varphi_{st} + \varphi_{ss} + \varphi = 1 \tag{3.615}
$$

Here the first index indicates the state of the precursor while the second one marks the state of the product.

In the reaction under study only the singlet S_1 state $^1M^-$ is light-emitting:

$$
{}^1M^- \to M^- + h\nu
$$

Consequently, the chemiluminescence quantum yield is

$$
\Phi^{CIEEL} = \Phi^{fl}\Phi \tag{3.616}
$$

where Φ^{fl} is the fluorescence quantum yield of **4**, which was found to be viscosity-independent and equal to 0.23 for this anion [228]. In contrast, the quantum yield of the excited singlet anions produced by the radical ion recombination Φ changes essentially with solvent viscosity. This fundamental fact, established by Adam et al. [227–230], was given a theoretical justification in our work [231] outlined below.

In general, the total quantum yield of excited singlets Φ is composed of both geminate (φ_{ss}) and bimolecular (ψ^s) contributions:

$$\Phi = \xi[\varphi_{ss} + \psi^s] \tag{3.617}$$

The geminate recombination produces formation of excited singlet anions, some nonluminescent ground-state anions, and triplet excited anions. But it also yields the separated radicals that start to recombine in the bulk to the same singlet, triplet, and ground states of anions:

$$[A \cdots M^-] \xleftarrow{k_c} {}^1[\dot{A}^- \cdots \dot{M}] \xrightarrow{k^*} [A \cdots {}^1M^-] \Longrightarrow A + M^- + h\nu$$

$$2\dot{A}^- + {}^2\dot{M} \qquad k_s \uparrow\downarrow 3k_s \tag{3.618}$$

$$^3[\dot{A}^- \cdots \dot{M}] \xrightarrow{k_t} [A \cdots {}^3M^-]$$

2. Bimolecular Recombination

At the beginning, the number of separated radicals was $N(0) = N_{RIP} \cdot \varphi$, but they disappear with time as a result of bimolecular recombination in the bulk. The current number of free radicals $N(t)$ obeys the conventional rate equation:

$$\dot{N} = -k_r N^2 \tag{3.619}$$

with the rate constant

$$k_r = \left[\frac{1}{4}(\varphi_{ss} + \varphi_{st}) + \frac{3}{4}(\varphi_{ts} + \varphi_{tt})\right] k_D \tag{3.620}$$

The expression in the square brackets is the efficiency of encounters. Since the spins of free radicals are uncorrelated, their encounter forms either a singlet or triplet RIP with the probabilities $\frac{1}{4}$ and $\frac{3}{4}$, respectively [232]. According to the scheme (3.618), each RIP can recombine into singlet or triplet product, but the corresponding yields depend on what is the precursor. In particular, φ_{ss} and φ_{st} are the quantum yields of the singlet and triplet products of the RIP that was initially in the singlet state, while φ_{ts} and φ_{tt} are the yields of the same products but from a triplet precursor.

The rate of singlet product formation from any precursors is

$$\dot{S} = \left(\frac{1}{4}\varphi_{ss} + \frac{3}{4}\varphi_{ts}\right)k_D N^2 \qquad (3.621)$$

while the yield of these products accumulated over the whole time is $\int_0^\infty \dot{S}\,dt$. Referring to a single RIP, this is

$$\psi^s = \int_0^\infty \frac{\dot{S}\,dt}{N_{RIP}} = wk_r \int_0^\infty \frac{N^2 dt}{N_{RIP}} \qquad (3.622)$$

where the efficiency of singlet production per encounter is

$$w = \frac{\varphi_{ss} + 3\varphi_{ts}}{\varphi_{ss} + \varphi_{st} + 3(\varphi_{ts} + \varphi_{tt})} \qquad (3.623)$$

Using the solution of Eq. (3.619) in Eq. (3.622), we obtain, after integration

$$\psi^s = w\varphi \qquad (3.624)$$

3. Singlets and Triplets Yielded from Different RIPs

To analyze the viscosity dependence of the effect, one needs to specify the quantum yields of singlets and triplets produced from any initial spin state of a RIP. For a contact-born singlet RIP, one has to set $r = \sigma$ in Eqs. (3.596) and substitute k_r^S from (3.613) for k_c:

$$\varphi_{ss} = \frac{k_r^S}{k_D + k_r^S}[1 - (k_D + k_t)\tilde{p}_{ST}(\sigma,\sigma,0)] \qquad (3.625a)$$

$$\varphi_{st} = k_t \tilde{p}_{ST}(\sigma,\sigma,0) \qquad (3.625b)$$

These formulas account for the singlet products in either the excited or ground states. Making the same substitution, k_r^S for k_c, we obtain from Eq. (3.595) the contact value of $\tilde{p}_{ST}$:

$$\tilde{p}_{ST}(\sigma,\sigma,0) = \frac{\frac{3}{4}xk_D}{[k_D(1+x) + k_t](k_D + k_r^S) + \frac{3}{4}xk_D(k_t - k_r^S)} \qquad (3.626)$$

Similar formulas were obtained for the contact born triplet RIP as well [231]:

$$\varphi_{ts} = k_r^S \tilde{p}_{TS}(\sigma, \sigma, 0) \tag{3.627a}$$

$$\varphi_{tt} = \frac{k_t}{k_D + k_t}\left[1 - (k_D + k_r^S)\tilde{p}_{TS}(\sigma, \sigma, 0)\right] \tag{3.627b}$$

where $\tilde{p}_{TS}(\sigma, \sigma, 0)$ can be obtained from (3.626) by simply a permutation of k_r^S and k_t and substitution of $\frac{1}{4}$ for $\frac{3}{4}$:

$$\tilde{p}_{TS}(\sigma, \sigma, 0) = \frac{\frac{1}{4}xk_D}{[k_D(1 + x) + k_r^S](k_D + k_t) + \frac{1}{4}xk_D(k_r^S - k_t)} \tag{3.628}$$

If the electron transfer in the triplet RIP is switched off ($k_t = \varphi_{tt} = 0$) and in the singlet RIP it is possible only to the ground state ($k_r^S = k_c$), then φ_{ts} given by Eqs. (3.627a) and (3.628) becomes identical to a single-channel quantum yield, which follows from Eq. (3.577) at $r_0 = \sigma$.

4. Viscosity Dependence

According to Eqs. (3.617) and (3.624) the total yield of excited singlets jointly produced by geminate and bimolecular recombination is

$$\Phi = \xi[\varphi_{ss} + w\varphi] \tag{3.629}$$

The experimental data of Adam et al. (points in Figs. 3.78 and 3.79) provides evidence that this quantity is inversely proportional to the solvent viscosity η

$$\frac{1}{\Phi} = \frac{1}{\Phi_0} + \frac{A}{\eta} \tag{3.630}$$

where $\Phi_0 = 0.5$, and $A = 5.64\,\text{cP}$. The qualitative interpretation of this fact given by the model of the exponentially decaying solvent cage [227–230] or the stochastic model of "discrete tests (collisions)" [233], completely ignored the bimolecular contribution into chemiluminescence.

The reaction in the bulk can be really neglected if the free radicals are unstable as a result of fast decomposition, discharging, or other reasons. But even in this case neither of the abovementioned models is an appropriate tool for the explanation of the phenomenon. The contact diffusional theory provides the alternative interpretation of the effect originating from either geminate recombination alone or together with the reaction in the bulk.

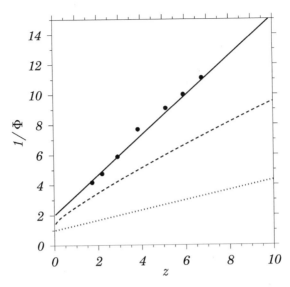

Figure 3.78. The inverse quantum yield of geminate singlet production as a function of $z = k_D/k_t = 4.2/\eta$ at different spin-conversion rates measured by $\gamma = \pi\sigma^3 k_s/k_t$ (from Ref. 231). The upper straight line corresponds to $\gamma = 10^5$, the intermediate dashed line to $\gamma = 10$, and the lower dotted line to $\gamma = 0$. The experimental points are borrowed from Ref. 227.

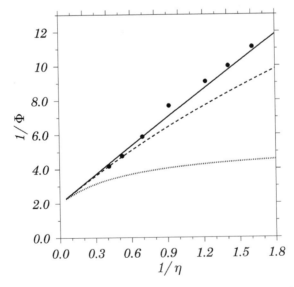

Figure 3.79. The viscosity dependence of the singlet excitation quantum yield Φ given by Eq. (3.646) at different rate constants of recombination to the triplet state: $k_t = 10^{11}$, $10^{10}, 10^9, M^{-1}s^{-1}$ (from top to bottom). The points are the same as in Figure 3.78 and the excitation rate constant $k^* = 8.8 \times 10^8 \, M^{-1}s^{-1}$. (From Ref. 231.)

5. Chemiluminescence of Geminate Origin

If there is really only geminate production of the fluorescent particles, then

$$\frac{1}{\Phi} = \frac{1}{\xi \varphi_{ss}} \tag{3.631}$$

To be identical with the experimental dependence (3.630), this expression should be linear in $k_D = 4\pi\sigma\tilde{D}$, which is related to the solvent viscosity in the usual way [100]:

$$k_D = \frac{8\,RT}{3000\,\eta\zeta} \qquad \zeta \approx 1 \tag{3.632}$$

In general, the linearity $\Phi^{-1}(k_D)$ does not follow from the expression resulting from Eqs. (3.625a) and (3.626):

$$\frac{1}{\varphi_{ss}} = 1 + \frac{k_D}{k_r^S}\frac{k_t(1 + 3x/4) + k_D(1 + x)}{k_t + k_D(1 + x/4)} \tag{3.633}$$

This quantity is approximately linear only in the bounded region of the diffusion variation:

$$\frac{x}{4} \gg 1, \frac{k_t}{k_D} \qquad \text{or} \qquad \frac{k_t}{2\pi\sigma^2\sqrt{k_s}} \ll \sqrt{\tilde{D}} \ll \frac{\sigma}{2}\sqrt{k_s} \tag{3.634}$$

This region is wider when the spin conversion is more rapid, but only in the limit of $k_s \to \infty$ does the linearity hold everywhere. Coming to this extreme, we are able to find from Eqs. (3.631) and (3.633) an unrestricted linear relationship:

$$\frac{1}{\Phi} = \frac{k_r^S + 3k_t}{\xi k_r^S} + \frac{4k_D}{\xi k_r^S} = \frac{k_r^S + 3k_t}{k^*} + \frac{32 \cdot RT}{3000k^*} \cdot \frac{1}{\eta} \tag{3.635}$$

This fits well the experimental equation (3.630), provided

$$k^* = 4.7 \times 10^9\,\mathrm{M}^{-1}\mathrm{s}^{-1} \qquad \text{and} \qquad k_r^S = k^* = 3k_t \qquad (\xi = 1) \tag{3.636}$$

However, if the spin conversion measured in the dimensionless parameter $\gamma = \pi\sigma^3 k_s/k_t$ is not as fast, there are essential deviations from the linearity (Fig. 3.78). Only at $\gamma = 10^5$ is the general dependence (3.631)–(3.633) quasilinear, approaching all the experimental points.

6. *Joint Geminate–Bulk Luminescence*

A completely different type of situation occurs when the singlet excitations are the products of not only geminate but the bulk recombination as well. The efficiency of the excitation in the bulk is given by Eq. (3.623). By substitution there Eqs. (3.625) and (3.627) as well as (3.626) and (3.628), the general expression for w can be put into the following form

$$w = \frac{k_r^S(k_D + k_t)[1 - L]}{4k_r^S k_t + k_D(k_r^S + 3k_t) + k_D(k_t - k_r^S)L} \tag{3.637}$$

where

$$L = \frac{3}{4} \frac{xk_D(k_t - k_r^S)}{(k_D + k_r^S)(k_D + k_t) + xk_D\left[k_D + \frac{1}{4}k_r^S + \frac{3}{4}k_t\right]} \tag{3.638}$$

Besides, the fraction of radicals that escape geminate recombination, $\varphi = 1 - \varphi_{ss} - \varphi_{st}$, also can be specified using Eqs. (3.625) and (3.626):

$$\varphi = \frac{k_D}{k_D + k_r^S}\left[1 - \frac{3xk_D(k_t - k_r^S)}{3xk_D(k_t - k_r^S) + 4(k_D + k_r^S)[k_t + k_D(1 + x)]}\right] \tag{3.639}$$

It is instructive to note that neither of these quantities depends on x if $k_r^S = k_t$. In this particular case the spin conversion does not modulate the rate of recombination so that $\varphi = (1 + k_r^S/k_D)^{-1}$ as in the spinless theory and $w = \frac{1}{4}$ is the weight of the singlet in the uncorrelated pairs composed in the bulk.

Another interesting situation arises if $k_t = \infty$ so that

$$w = \frac{k_r^S(1 - L)}{4k_r^S + k_D(3 + L)} \qquad \text{and} \qquad L = \frac{(3/4)xk_D}{k_D + k_r^S + (3/4)xk_D} \tag{3.640}$$

Coming to the same limit in Eqs. (3.639) and (3.633), we obtain from (3.629):

$$\Phi = \xi \frac{k_r^S + wk_D}{k_r^S + k_D[1 + (3/4)x]} \qquad \text{at} \qquad k_t = \infty \tag{3.641}$$

The geminate contribution to singlet production in this limit is not linear because the inequalities (3.634) do not hold. Regarding the bulk efficiency w, it decreases with the spin conversion and in the limit $x \to \infty$ turns to zero ($L \to 1, w \to 0$). However, this limit is hardly attainable, and the opposite one is the more realistic one.

7. Zero Spin Conversion

In the limit, $x \ll 1$, the spin conversion during the encounter time is negligible, and coming to the extreme, we can set

$$x = L = 0 \qquad (3.642)$$

For this particular case it follows from Eq. (3.637) that

$$w = \frac{k_r^S(k_D + k_t)}{k_r^S(k_D + k_t) + 3k_t(k_D + k_r^S)} = \begin{cases} \frac{(1/4)k_r^S}{(1/4)k_r^S + (3/4)k_t} & \text{at} \quad \tilde{D} \to \infty \\ \frac{1}{4} & \text{at} \quad \tilde{D} \to 0 \end{cases} \qquad (3.643)$$

Besides, under condition (3.642) we have the following from Eqs. (3.633) and (3.639):

$$\frac{1}{\varphi_{ss}} = 1 + \frac{k_D}{k_r^S}, \qquad \varphi = \frac{k_D}{k_D + k_r^S} \qquad (3.644)$$

These quantities are found exactly the same as in EM or contact spinless theory. Using them together with w from Eq. (3.643) in the general definition (3.629), we have

$$\Phi = \xi \frac{k_r^S + wk_D}{k_r^S + k_D} \qquad \text{at} \qquad x = k_s = 0 \qquad (3.645)$$

From the complex formulas (3.645) and (3.643), we obtain the following relatively simple but general relationship:

$$\frac{1}{\Phi} = \frac{1}{\xi} + \frac{3k_D}{4\xi k_r^S(1 + k_D/4k_t)} \qquad (3.646)$$

8. Fitting to Experiment

The dependence (3.646) is linear over a wide range of diffusion provided k_t is high enough:

$$\frac{1}{\Phi} = \frac{1}{\xi} + \frac{3k_D}{4k^*} \qquad \text{at} \qquad 4k_t \gg k_D \qquad (3.647)$$

Absolutely the same result follows from Eqs. (3.629) and (3.640) under conditions (3.642). By identification of the theoretical relationship (3.647) with the experimental dependence (3.630), we gain $\Phi_0 = \xi = 0.5$ and

$A = RT/500\,k^* = 5.64\,\mathrm{cP}$, that is, $k_r^S = 2k^*$ and $k^* = 8.8 \times 10^8\,\mathrm{M}^{-1}\mathrm{s}^{-1}$. This parameterization is different from (3.636) and its inclusion in Eq. (3.646) transforms it to the following one:

$$\frac{1}{\Phi} = 2 + \frac{5.64}{\eta + 1.65 \cdot 10^9/k_t} \tag{3.648}$$

At $k_t = 10^{11}\,\mathrm{M}^{-1}\mathrm{s}^{-1}$, the theoretical line is almost as straight as in the original experimental works (Fig. 3.79). However, at lower rates of triplet recombination $1/\Phi$ levels off earlier and at a lower level, making the studied dependence curvilinear.

As has been shown, the fitting of the linear viscosity dependence of chemiluminescence is completely different if the geminate recombination is considered alone or accompanied by the bulk reaction. In the former case the faster the spin conversion, the better, while in the latter case it can be set to zero provided the rate of electron transfer through the triplet channel is high enough. A similar alternative will be presented in the next section. There the combination of geminate and bulk reaction appears more preferable, especially because the spin conversion carried out by the hyperfine interaction is usually weak.

E. Multiple Rehm–Weller Plot

The reaction scheme for the reversible ionization (3.354) should be essentially extended when accounting for the spin states of all species and their possible conversion in the radical ion pairs:

$$
{}^1D^* + A \underset{W_B}{\overset{W_I}{\rightleftharpoons}} {}^{1,3}[\dot{D}^+ \cdots \dot{A}^-] \overset{\bar{\varphi}}{\longrightarrow} {}^2\dot{D}^+ + {}^2\dot{A}^- \overset{W_B}{\longrightarrow} {}^1D^* + A \tag{3.649}
$$

$$I_0 \uparrow\downarrow \tau_S \qquad\qquad W_S \nearrow\searrow W_T \qquad W_T \nearrow\searrow W_S \qquad \nearrow \tau_S$$

$$D \qquad\qquad\qquad [D \cdots A] \qquad {}^3D^* + A \qquad D + A$$

The conversion of $[\dot{D}^+ \cdots \dot{A}^-]$ from the singlet to triplet state allows the geminate pair to recombine not only in the ground state but also into the excited triplet product, ${}^3D^*$ (Fig. 3.80). However, the quantum yield of the latter inspected in Section XI.B is never as large as necessary to make the ionization fully irreversible. This can be possible only if one takes into account the free ion recombination in the bulk. Since these ions meet with uncorrelated spins, 75% of the newly born RIPs appear in the triplet state from where they recombine fast and irreversibly into the same triplet product, ${}^3D^*$. This mechanism allows extending the Rehm–Weller diffusional plateau up to the border between exergonic and endergonic electron transfer at $\Delta G_i = 0$.

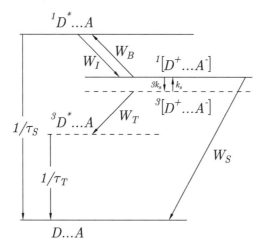

Figure 3.80. The scheme of the energy levels and electronic transitions between collective levels of ionized and neutral reactants. Solid lines are singlet levels and dashed lines are triplet levels of the same pair.

1. Triplet Recombination in Contact IET

To account for the triplet channel of recombination, one should simply add to the set of IET equations (3.357) one more equation [123]:

$$\dot{N}^T = c \int_0^t R^\heartsuit(\tau) N^*(t-\tau) d\tau + \int_0^t R^\spadesuit(\tau) [N^\pm(t-\tau)]^2 d\tau - \frac{N^T}{\tau_T} \qquad (3.650)$$

Two supplementary kernels, $\tilde{R}^\heartsuit$ and $\tilde{R}^\spadesuit$, should be added to four others defined in the contact approximation in Eqs. (3.369). In an extreme case of negligible spin conversion during the geminate stage (i.e., at $\sqrt{4k_s\tau_d} \to 0$), these four kernels are

$$\tilde{R}^*(s) = \frac{k_0(1 + k_c\tilde{g}_2)}{Z}, \qquad \tilde{R}^\dagger(s) = \frac{k_0}{Z}$$

$$\tilde{R}^\#(s) = \frac{1}{4}\frac{k_b}{Z}, \qquad \tilde{R}^\ddagger(s) = \frac{1}{4}\frac{k_b + k_c + k_0 k_c \tilde{g}_1}{Z} + \frac{3}{4}\frac{k_t}{Z_t} \qquad (3.651)$$

Here $Z = (1 + k_0\tilde{g}_1)(1 + k_c\tilde{g}_2) + k_b\tilde{g}_2$ as before and $Z_t = 1 + \tilde{g}_2 k_t$ is expressed through the rate of triplet RIP recombination to the neutral triplet product, k_t.

The shape of the kernels (3.651) is rather obvious from the physical point of view. The kernels R^* and $R^\dagger$ are exactly the same as in Eq. (3.369) of the spinless theory. When spin conversion during encounter is negligible, the backward electron transfer to the ground state remains the single channel of geminate ion recombination. Therefore, the kernel $R^\#$ is 4 times smaller than the similar kernel in the spinless case; only a quarter of the RIPs, originating from random encounters in the bulk are formed in the singlet state. The last kernel, $R^{\ddagger}$, is represented as the sum of two contributions taken with the equilibrium weights of the singlet and triplet states. The first of them, $(k_b + k_c + k_f k_c g_s)/X_0$, is exactly the same as in the spinless case, but weighted with the coefficient $\frac{1}{4}$ assigned by the spin statistics. Another one, k_t/Z_t, is responsible for the irreversible recombination of the triplet RIPs into the excited triplet state of neutral product, $^3D^* \cdots A$. Again, the weight $\frac{3}{4}$ accounts for a probability of forming triplet RIPs during the random encounters of counterions in solution.

As long as triplets are not formed in the course of geminate ion recombination, but all of them owe their origin entirely to bimolecular ion recombination in the bulk, we have

$$\tilde{R}^\heartsuit = 0, \qquad \text{but} \qquad \tilde{R}^\spadesuit \propto \frac{3}{4} k_t \qquad (3.652)$$

At fast bimolecular recombination of RIPs to the triplet products, the fluorescence quenching by electron transfer becomes irreversible.

2. Stationary Quenching of Fluorescence

Under permanent illumination ($I_0 = \text{const}$) the stationary concentrations $N_s^*, N_s^\pm$, and N_s^T obey the set of equations that follow from Eqs. (3.357) and (3.650):

$$0 = -c \int_0^\infty R^*(\tau) d\tau N_s^* + \int_0^\infty R^\#(\tau) d\tau (N_s^\pm)^2 - \frac{N_s^*}{\tau_D} + I_0 N \qquad (3.653a)$$

$$0 = c \int_0^\infty R^\dagger(\tau) d\tau N_s^* - \int_0^\infty R^\ddagger(\tau) d\tau (N_s^\pm)^2 \qquad (3.653b)$$

$$0 = c \int_0^\infty R^\heartsuit(\tau) d\tau N_s^* + \int_0^\infty R^\spadesuit(\tau) d\tau (N_s^\pm)^2 - \frac{N_s^T}{\tau_T} \qquad (3.653c)$$

The solution of these equations provides us with the stationary values for all the concentrations:

$$N_s^* = \frac{I_0 N \tau_D}{1 + c\tau_D \left[\tilde{R}^*(0) - \tilde{R}^\#(0) \tilde{R}^\dagger(0) / \tilde{R}^\ddagger(0) \right]} \qquad (3.654a)$$

$$N_s^\pm = \sqrt{c N_s^* \tilde{R}^\dagger(0) / \tilde{R}^\ddagger(0)} \qquad (3.654b)$$

$$N_s^T = c\tau_T \left[\tilde{R}^\heartsuit(0) + \tilde{R}^\spadesuit(0) \frac{\tilde{R}^\dagger(0)}{\tilde{R}^\ddagger(0)} \right] N_s^* \qquad (3.654c)$$

The relative fluorescence quantum yield defined in Eq. (3.8) is the ratio of the stationary singlet excitation concentration in the presence of quenchers to the same concentration in their absence. By substituting into this definition N_s^* from Eq. (3.654a), we confirm that the fluorescence quantum yield obeys the Stern–Volmer law (3.363) with the same constant as in Eq. (3.364), but with the contact φ_c substituted for $\bar{\varphi}$:

$$\kappa_0 = \kappa_g[1 - \chi\varphi_c] \qquad (3.655)$$

All parameters present in this equation are expressed through the IET kernels [123,188]:

$$\kappa_g = \tilde{R}^*(0), \qquad \chi = \frac{\tilde{R}^{\#}(0)}{\tilde{R}^{\ddagger}(0)}, \qquad \varphi_c = \frac{\tilde{R}^{\dagger}(0)}{\tilde{R}^*(0)} \qquad (3.656)$$

As always, κ_g is the rate constant of the excitation quenching by electron transfer during the primary geminate process, but only a limited fraction of singlet excitations, $1 - \varphi_c$, is actually quenched forever at this stage. All the rest, separated by a yield φ_c, encounter later in the bulk, restoring the singlet excitation of the donor with a probability χ. The forward electron transfer is truly irreversible when $\chi = 0$, but in the opposite limit, $\chi = 1$, all separated ions regenerate the excited donors and $\kappa_0 = \kappa_g(1 - \varphi_c)$ is much less than the geminate one (see Section VIII.D). To be more specific, we can calculate χ using the contact estimates of all the kernels given in Eqs. (3.651) and (3.652):

$$\chi = \frac{k_b(k_D + k_t)}{k_b(k_D + 4k_t) + [k_c(k_D + 4k_t) + 3k_D k_t]\left(1 + \frac{k_0}{k_D(1+\sqrt{\tau_d/\tau_D})}\right)} \qquad (3.657)$$

If there is no recombination to the triplet ($k_t = 0$), then ions can recombine to either the ground or the excited state of the donor with the corresponding rate constants, $k_c(\Delta G_i)$ and $k_b(\Delta G_i)$. Assuming that their maximal values are the same, we represent them in Figure 3.81(a) as normalized bell-shaped curves marked by k_c and k_b and split by the value of the excitation energy of donor molecule, $\mathscr{E}$. At the point halfway between them should be a border vertical line. To the right of it recombination to the excited state dominates, to the left of it ions presumably recombine to the ground state. Quantitatively this statement is expressed by the $\chi(\Delta G_i)$ dependence, which follows from Eq. (3.657):

$$\chi = \frac{k_b}{k_b + k_c\left(1 + \frac{k_0}{k_D(1+\sqrt{\tau_d/\tau_D})}\right)} \qquad \text{at} \qquad k_t = 0 \qquad (3.658)$$

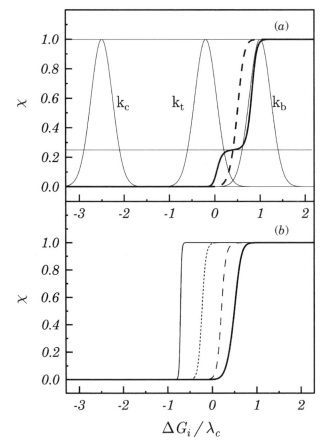

Figure 3.81. The probability of reverse excitation as a function of the ionization free energy ΔG_i at $\mathscr{E} = 3.5\lambda_c, \mathscr{E}_T = 2.3\lambda_c, \lambda_c = 35\,T$: (a) $k_t^m = k_c^m = k_b^m = 2000k_D$ (thick line); $k_t^m = k_c^m = k_b^m = 20k_D$ (dashed line); (b) the family of the curves with different $k_t^m/k_D = 20; 0.1; 10^{-5}; 0$ (from right to left), at $k_c^m = k_b^m = 20k_D$. (index m denotes maximal values of the rate constants).

This is the left stepwise curve depicted in Figure 3.81(*b*). To the right of this step the ionization is reversible; to the left it is irreversible.

The situation changes qualitatively if we add the triplet channel of recombination, which is also represented by the normalized bell-shaped curve marked as k_t in Figure 3.81(*a*). It is shifted to the right from the curve k_c by the value of the triplet energy $\mathscr{E}_T$ and therefore competes more successfully with recombination to excited state. Within this curve triplet production dominates over singlet excitation and the corresponding $\chi(\Delta G_i)$ is given by another formula, following from Eq. (3.657):

$$\chi = \frac{k_b}{4k_b + (3k_D + 4k_c)\left[1 + \frac{k_0}{k_D(1+\sqrt{\tau_d/\tau_D})}\right]} \le \frac{1}{4} \quad \text{at} \quad k_t \to \infty \quad (3.659)$$

This expression is represented by a small intermediate step with height 0.25 seen on the thick line. However, with a further increase in ΔG_i the triplet recombination constant is not as high any more, but falls down to zero, giving way to the alternative recombination channel: backward to the excited singlet state. Therefore χ rises further on with ΔG_i up to $\chi = 1$.

Such a complex behavior of $\chi(\Delta G_i)$ in the presence of the triplet channel is smoothed at faster diffusion when k_D is closer to the maximal value of the recombination constants. In this case there is a single step border [dashed line in Fig. 3.81(a)] between the irreversible (to the left) and reversible (to the right) reactions. The position of this border essentially depends on the true height of k_t. In Figure 3.81(b) one can see how the border marked by this line is shifted to the right with the increasing rate of triplet recombination.

As a result, the free-energy dependence of the Stern–Volmer constant (3.655) is also not a unique one. Its right edge is shifted correspondingly from left to right when the triplet recombination speeds up. However, the tops of all the curves are cut off by one and the same diffusional plateau [Fig. 3.82(a)]. If the rate of triplet RIPs recombination in different families of reactants is different, it can be responsible for the multiple Rehm–Weller plots in the quasiresonant region [Fig. 3.82(b)]. It was discovered by Jacques and Allonas [234] but given another explanation. From our point of view, expressed in Ref. 123, the descending branches of the Stern–Volmer constants calculated from Eqs. (3.657) and (3.655) are positioned at different places depending on k_t, as in Figure 3.82(a). For a comparison in Figure 3.82(b) we represented three different Rehm–Weller plots in acetonitrile redrawn from Figure 2 of the original experimental work [234]. The striking similarity of the plots is in favor of our explanation. Since the same picture was obtained later in another system [235], the multiple Rehm–Weller plot seems to be a general feature of the phenomenon in the quasiresonant region. It has the same origin as the difference between the reversible and irreversible ionization shown in Figure 3.16 and discussed in detail in Section II.F.

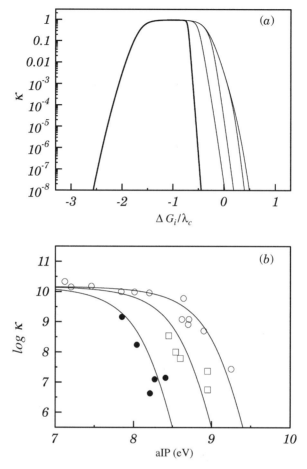

Figure 3.82. Rehm–Weller plot for a few systems which differ by the rate of triplet recombination: (*a*) the theoretical curves for $k_t^m/k_D = 20; 0.1; 10^{-5}; 10^{-8}; 0$ (from right to left)—the remaining parameters are the same as in Figure 3.81(*b*); (*b*) interpolation through the experimental points from Figure 3.2 of Ref. 234.

XII.   HIGH-CONCENTRATION EFFECTS

Besides a number of advantages demonstrated above, IET was also shown to have a few disadvantages that revealed themselves at high concentrations. This is the false asymptote $t^{-3/2}$ intimated in Sections VIII.E (Fig. 3.53) and IX.A (Fig. 3.56) and the absence of the static contribution to the normalized initial distribution of transfer products proportional to the quencher concentration c (Fig. 3.33). In a similar way, IET looses the concentration dependence of the

Stern–Volmer constant responsible for the nonlinearity of the Stern–Volmer law predicted by DET and UT [see Eqs. (3.30)–(3.33) and Fig. 3.61]. All these drawbacks arise because of the limited validity of IET, which is merely the lowest-order approximation with respect to particle concentration c.

To get rid of all these disadvantages, IET was modified using the intuitive binary conception [40] based on the exactly solvable models [137,236] and diagrammatic summation [40]. Later on the kinetic equations of MET were rederived in the same group adapting the methods of nonequilibrium statistical mechanics to the reacting systems [134]. A universal method of constructing the hierarchy for correlation patterns has been developed. The new derivation of the binary non-Markovian kinetic equations is applicable to any reacting system [134]. It was ultimately proved that MET gives more accurate estimates of the quantum yield of fluorescence [133] and describes the quenching kinetics in a wider time interval than does IET [41].

Finally, MET was imparted the matrix form similar to that of IET. A newly developed original method based on the many-particle master equation led to an infinite hierarchy for vector correlation patterns (VCPs) that can be truncated in two different ways [43,44]. The simplest one reproduces the conventional IET, while the other allows a general modification of the kernel, resulting in the matrix formulation of MET applicable to complex multistage reactions.

However, MET is not a unique theory accounting for the higher-order concentration corrections. Similar results were obtained within the fully renormalized YLS (Yang–Lee–Shin) theory [132], which is also integrodifferential and employs the kernels containing concentration corrections as compared to those in IET. It was shown in Ref. 41 that both these theories, MET and YLS, provide the correct asymptotic expressions for binary kinetics, but differ slightly in the nonlinear terms of the concentration expansion. There were also a number of other attempts to overcome the concentration limitations of the theory made by the same Korean group earlier (in superposition approximation [139,141,142]) and later [numerous multiparticle kernel (MPK) theories [51,126]].

These and other theories are compared below and, where possible, also with the exact solution [138] provided by DET for the target problem. The common feature of almost all these theories is the contact approximation for the transfer reaction. Therefore the same approximation will be used in IET and MET to compare them with competing theories.

A. Reversible Intermolecular Transfer

At first we should return to the problem of reversible transfer between metastable reactants considered in Section III. Using MET developed in Ref. 44, we can confirm the previously obtained results but eliminate the abovementioned demerits of IET. MET corrects the long-time kinetics of transfer and accounts for the concentration dependence of the Stern–Volmer constant.

1. MET versus IET

The general formalism developed in Ref. 43 provides the following definition of the matrix kernel (3.104) through the Green function of the pair evolution during encounter:

$$\mathbf{M}(t) = \int d^3 r \hat{\mathbf{W}}(r) \left[\delta(t) + \int d^3 r_0 \mathbf{G}(r|r_0, t) \hat{\mathbf{W}}(r_0) \right] \tag{3.660}$$

The auxiliary equation for the Green function is as follows

$$(\partial_t - \hat{\mathbf{L}} - \hat{\mathbf{W}}(r) - \hat{\mathbf{Q}})\mathbf{G}(r|r_0, t - t_0) = \delta(t - t_0)\delta(r - r_0)\hat{\mathbf{I}} \tag{3.661}$$

where $\hat{\mathbf{Q}} = \hat{\mathbf{Q}}_A + \hat{\mathbf{Q}}_B$ and $\hat{\mathbf{I}}$ is the unity operator.

The kernel subjected to modification in MET takes a form similar to (3.660)

$$\Sigma(t) = \int d^3 r \hat{\mathbf{W}}(r) \left[\delta(t) + \int d^3 r_0 \mathbf{G}^{\text{eff}}(r|r_0, t) \hat{\mathbf{W}}(r_0) \right] \tag{3.662}$$

but the effective Green function obeys another auxiliary equation:

$$(\partial_t - \hat{\mathbf{L}} - \hat{\mathbf{W}}(r) - \hat{\mathbf{Q}} + \mathscr{R})\mathbf{G}^{\text{eff}}(r|r_0, t - t_0) = \delta(t - t_0)\delta(r - r_0)\hat{\mathbf{I}} \tag{3.663}$$

An additional operator in the brackets, $\mathscr{R}$, accounts for the triplet events when the strange particle ($\mathbf{B}'$ or $\mathbf{A}'$) joins the couple $[\mathbf{A} \ldots \mathbf{B}]$ during their encounter. This operator acts on the effective Green function as follows:

$$\begin{aligned}
\mathscr{R}\mathbf{G}^{\text{eff}} = \text{Tr}_{\mathbf{B}'} \int_0^t [\mathbf{M}_{\mathbf{AB}'}(t - t') \otimes \exp\{\hat{\mathbf{Q}}_B(t - t')\}][\mathbf{N}_{\mathbf{B}'}(t') \otimes \mathbf{G}_{\mathbf{AB}}^{\text{eff}}(r, t'|r_0, t_0)]dt' \\
+ \text{Tr}_{\mathbf{A}'} \int_0^t [\mathbf{M}_{\mathbf{A'B}}(t - t') \otimes \exp\{\hat{\mathbf{Q}}_A(t - t')\}][\mathbf{N}_{\mathbf{A}'}(t') \otimes \mathbf{G}_{\mathbf{AB}}^{\text{eff}}(r, t'|r_0, t_0)]dt'
\end{aligned}$$
$$\tag{3.664}$$

where $\mathbf{M}_{\mathbf{AB}} \equiv \mathbf{M}$ has been introduced in Eq. (3.660). Since the last expression contains the particle vectors, $\mathbf{N}_{\mathbf{A}'}$ and $\mathbf{N}_{\mathbf{B}'}$, the Green function as well as the kernel itself are also expressed through them. This means that in general the kinetic equations (3.103) with Σ substituted for $\mathbf{M}$ and the auxiliary equations (3.105) and (3.106) should be solved jointly. However, if the densities of the excited states are negligible compared to the ground-state densities, then the Green function as well as the kernel depend only on the total concentrations of particles, which remain constant.

There was also an interesting and important reformulation of MET as well as IET made in the same work [43]. In the spatially uniform case, when the single-particle VCPs become coordinate independent and the two-particle vector depends on the distance between the particles, the integral equations can be transformed to the extended set of partial differential equations: Eqs. (3.259) for IET and Eqs. (3.348) for MET in Ref. 43. The similar sets were simultaneously but independently also obtained in Ref. 203, although by other means. Moreover, the authors of the latter work developed a general computer code (available at *http://www.fh.huji.ac.il/krissinel/programs/tegro.html*) for numerical solution of these equations. One of the illustrations presented in the same work will be demonstrated in the following paragraphs.

2. Reversible Contact Transfer

Let us directly address the MET application to the reversible reaction between reactants, represented by vectors (3.100). With a large excess of nonexcited particles, the effects of depletion and saturation considered in Sections IX and X can be neglected to proceed further. After rather complex and long calculations made in Ref. 44, Eqs. (3.127) were reproduced, but with a properly modified and concentration-dependent kernel $\Phi(t)$:

$$\dot{N}_A^* = k_b N_A \int_0^t \Phi(\tau) N_B^*(t-\tau)d\tau - k_a N_B \int_0^t \Phi(\tau) N_A^*(t-\tau)d\tau - \frac{N_A^*}{\tau_A} \qquad (3.665a)$$

$$\dot{N}_B^* = -k_b N_A \int_0^t \Phi(\tau) N_B^*(t-\tau)d\tau + k_a N_B \int_0^t \Phi(\tau) N_A^*(t-\tau)d\tau - \frac{N_B^*}{\tau_B} \qquad (3.665b)$$

The Laplace transformation of Φ is given by the expression

$$\tilde{\Phi}(s) = \left[1 + \frac{k_a}{k_D}\beta\left(s + \frac{1}{\tau_A^*}\right) + \frac{k_b}{k_D}\beta\left(s + \frac{1}{\tau_B^*}\right)\right]^{-1} \qquad (3.666)$$

where $\beta(x) = [1 + \sqrt{x\tau_d}]^{-1}$. The modification is concerned only with the lifetimes whose values are corrected by adding the concentration dependent terms:

$$\frac{1}{\tau_A^*} = \frac{1}{\tau_A} + k_a N_B \tilde{\mathscr{F}}(s), \qquad \frac{1}{\tau_B^*} = \frac{1}{\tau_B} + k_b N_A \tilde{\mathscr{F}}(s) \qquad (3.667)$$

where

$$\tilde{\mathscr{F}}(s) = \left[1 + \frac{k_a}{k_D}\beta\left(s + \frac{1}{\tau_A}\right) + \frac{k_b}{k_D}\beta\left(s + \frac{1}{\tau_B}\right)\right]^{-1} \qquad (3.668)$$

is the IET analog of $\tilde{\Phi}(s)$ introduced in Eq. (3.126).

At $s = 0$ the concentration corrections in Eq. (3.667) become the rates of excitation quenching by any partner that does not belong to a given couple (reactant pair). These "bachelors" compete for an excitation with the reactants in a couple when they move apart for a while between successive recontacts. Similar results were obtained with the many-particle theory of diffusion-influenced reactions based on the revised superposition approximation and became known as MPK1 [51]. The authors were the first who managed to obtain concentration corrections to the IET result for the kinetics of reversible energy transfer. In a subsequent modification of their theory, named MPK3 [126], the same authors reached the full correspondence with MET.

3. Stern–Volmer Constant

Making a Laplace transformation of Eqs. (3.665) and using $\tilde{N}_A^*(0)$ in Eq. (3.128), one can easily confirm the Stern–Volmer law (3.30), but with the corrected constant:

$$k_q^{MET} = \frac{k_a}{1/\tilde{\Phi}(0) + k_b N_A \tau_B} \tag{3.669}$$

Presenting this expression as Eq. (3.129), we can generalize the definitions of the forward and backward transfer constants:

$$\frac{K_{AB}}{k_a} = \tilde{\Phi}(0) = \frac{K_{BA}}{k_b} \tag{3.670}$$

where

$$\frac{1}{\tilde{\Phi}(0)} = 1 + \frac{k_a/k_D}{1 + \sqrt{\frac{\tau_d}{\tau_A}\left(1 + k_a\tau_A N_B \tilde{\mathscr{F}}(0)\right)}} + \frac{k_b/k_D}{1 + \sqrt{\frac{\tau_d}{\tau_B}\left(1 + k_b\tau_B N_A \tilde{\mathscr{F}}(0)\right)}} \tag{3.671}$$

where

$$\frac{1}{\tilde{\mathscr{F}}(0)} = 1 + \frac{k_a/k_D}{1 + \sqrt{\tau_d/\tau_A}} + \frac{k_b/k_D}{1 + \sqrt{\tau_d/\tau_B}} \tag{3.672}$$

Neglecting the concentration corrections in the square roots of Eq. (3.671) turns $\tilde{\Phi}(0)$ into $\tilde{\mathscr{F}}(0)$ and the MET equation (3.670) into the previously obtained IET Eq. (3.130).

In the lowest-order approximation with respect to the particle concentrations it follows from Eq. (3.671) that

$$\frac{1}{\tilde{\Phi}(0)} = \frac{1}{\tilde{\mathscr{F}}(0)} - \alpha k_a \tau_A N_B - \beta k_b \tau_B N_A \tag{3.673}$$

where

$$\alpha = \frac{k_a\sqrt{\tau_A\tau_d}}{2k_D(\sqrt{\tau_A}+\sqrt{\tau_d})^2}\tilde{\mathscr{F}}(0) \qquad \beta = \frac{k_b\sqrt{\tau_B\tau_d}}{2k_D(\sqrt{\tau_B}+\sqrt{\tau_d})^2}\tilde{\mathscr{F}}(0) \qquad (3.674)$$

Similar results were also obtained with MPK1 in Ref. 51, but there $\alpha = \beta = 1$. This is the price for the inaccurate low-concentration expansion. The price is especially high in the kinetic control limit ($D \to \infty$) when, according to Eqs. (3.674) $\alpha = \beta = 0$, contrary to Ref. 51, where they remain the same at any D. Since the kinetic quenching is essentially Markovian, it is reasonable that the true concentration corrections to IET should vanish in this limit.

From here on we should discriminate between qualitatively different situations:

- *Energy Quenching.* In this case $\tau_B < \tau_A$ and there is a limit $k_b\tau_B \to 0$ when nothing can be transferred backward to A^* before B^* decays.
- *Equal Times.* This is a border case, $\tau_B = \tau_A = \tau$, when the decay of excited states and the redistribution between them proceed independently.
- *Energy Storing.* At $\tau_B > \tau_A$ the initial quenching of A^* by B results in energy storing there and the subsequent backward transfer to A^* backing their delayed fluorescence.

4. Irreversible Limit ($k_b\tau_B \to 0$)

If the decay time of the partner is too short, then the transfer reaction becomes really irreversible. To be more specific, one can obtain from Eq. (3.669):

$$k_q^{MET} = k_a\tilde{\Phi}(0) = K_{AB}, \qquad \text{at} \qquad k_b\tilde{\Phi}(0)N_A\tau_B \ll 1 \qquad (3.675)$$

Passing to the extreme, we find from Eqs. (3.675) and (3.671):

$$\lim_{k_b\tau_B \to 0} K_{AB} = k_a\left[1 + \frac{k_a/k_D}{1 + \sqrt{\frac{\tau_d}{\tau_A}}\left(1 + k_a\tau_A N_B\tilde{\mathscr{F}}(0)\right)}\right]^{-1} = \kappa \qquad (3.676)$$

where

$$\tilde{\mathscr{F}}(0) = \frac{1}{1 + (k_a/k_D)/(1 + \sqrt{\tau_d/\tau_A})}$$

The concentration dependence of the MET Stern–Volmer constant $\kappa(N_B)$ will be considered in more detail in the following paragraphs, in comparison to other estimates of $\kappa(c)$.

5. Border Case ($\tau_B = \tau_A$)

If the times are equal, one can separate the state decay from the energy redistribution, introducing a change in the variables in Eqs. (3.665): $N_A^*(t) = N_A^*(0)e^{-t/\tau}P_A^*(t)$ and $N_B^*(t) = N_B^*(0)e^{-t/\tau}P_B^*(t)$. In the course of time the energy redistribution turns the initial values $P_A^*(0) = 1$ and $P_B^*(0)$ into the stationary fractions of the excited states, P_A^s and P_B^s, whose ratio is

$$Q = \frac{P_B^s}{P_A^s} = K\frac{N_B}{N_A} \tag{3.677}$$

where $K = k_a/k_b$ is the equilibrium constant. Using this parameter, the conventional Stern–Volmer law (3.30), with constant (3.669), can be rewritten for equal times as follows

$$\eta = \frac{1 + Q\eta_0 - \eta_0}{1 + Q - \eta_0} \tag{3.678}$$

where

$$\eta_0 = \frac{1}{1 + K_{AB}(\tau)N_B\tau} \tag{3.679}$$

Equations (3.678) and (3.679) are the intermolecular analogs of Eqs. (3.178) and (3.179) obtained for the intramolecular transfer between the states with equal decay rates.

The kinetics of energy redistribution is governed by the equation

$$P_A^*(t) = P_A^s + \frac{F(t)}{k_a N_B}$$

where $P_A^s = 1/(1 + Q)$. The quantity $F(t)$ calculated in the framework of MPK1 is exponential at long times [51]

$$F(t) = k_a N_B e^{-k_A(\infty)N_B t} + k_b N_A e^{-k_B(\infty)N_A t} \tag{3.680}$$

where $k_A(\infty) = k_a k_D/(k_a + k_D)$ and $k_B(\infty) = k_b k_D/(k_b + k_D)$ are the stationary rate constants of the forward and backward transfers. The asymptotic dependence (3.680) differs qualitatively from its IET analog obtained in Ref. 49 and confirmed in Ref. 51:

$$F(t) \propto \left(\frac{\tau_d}{t}\right)^{3/2} \tag{3.681}$$

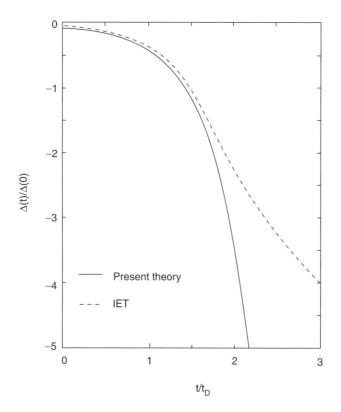

Figure 3.83. The population redistribution between the excited states at $\tau_A = \tau_B$, $4\pi\sigma^3 N_A = 4\pi\sigma^3 N_B = 0.1$, $k_a/2 = k_b = k_D$. "Present theory" is MPK1 developed in Ref. 51, from which this figure is taken.

This is again the false IET asymptote that was already indicated in Eq. (3.404) as well as in Figs. 3.53 and 3.56. The MPK1 kinetics of equilibration represented by $\Delta(t)/\Delta(0) = (P_A^*(t) - P_A^s)/(1 - P_A^s)$ is shown in Fig. 3.83 in comparison to its IET analog.

A similar correction to IET is inherent in MET as well. For irreversible transfer $(k_b = P_A^s = 0)$, the quenching kinetics represented by $P_A^*(t)$ was obtained in Ref. 203 with the original program designed to solve the differential form of IET and MET equations. As seen from Figure 3.84, the difference between the curves representing these solutions is insignificant within the validity limits for IET established in Ref. 39

$$t \leq t_p = \frac{\ln(\xi^{-1})}{kN_B} \qquad (3.682)$$

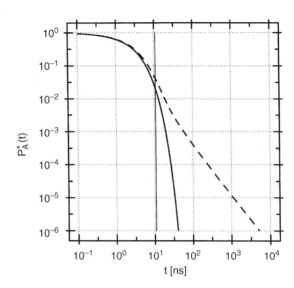

Figure 3.84. The quenching kinetics obtained with IET (dashed line) and MET (solid line). The vertical strip denotes the upper boundary of the region of IET validity. Parameters of exponential transfer rate: $W_c = 100\,\text{ns}^{-1}$, $l = 1\,\text{Å}$, $\sigma = 5\,\text{Å}$. The remaining parameters are $N_B = 0.1\,\text{M}$, $D = 10^{-5}\,\text{cm}^2/\text{s}$. (From Ref. 203.)

where $\xi = \frac{4}{3}\pi\sigma^3 N_B$ and k is stationary (Markovian) rate constant. As stated by the same authors, the validity region for MET covers a much wider time interval:

$$t \ll \frac{\xi^{-1}}{kN_B} \tag{3.683}$$

This was confirmed in Ref. 203 by comparing the MET solution with that given in Eq. (3.177) and considered as an exact solution for the target problem: $P_A^* = \exp[-N_B \int_0^t k(t')dt']$. The difference between this result and the solid line in Figure 3.84 was found to be less than the precision of numerical calculations and therefore invisible. On the contrary, the dashed curve representing the IET significantly deviates from the exact result when t exceeds t_p, indicated by the vertical line in this figure.

6.   Energy Storing ($\tau_B > \tau_A$)

The IET kinetics of reversible transfer of excitation, which lives longer on the acceptor than on the donor, was discussed in full detail in Sections IV.A and IV.B. In Ref. 44 the same was done and confirmed with MET. Here we have restricted ourselves to the concentration dependence of the Stern–Volmer constant, which is most important in the extreme case $\tau_B = \infty$ and under diffusion control.

In this particular case the power expansion in concentrations (3.673) is not possible, and one should address the general equations (3.670) and (3.671) from, which it follows that

$$K_{AB} = \frac{k_a}{1 + k_a/k_D(1 + \sqrt{x}) + k_b/k_D(1 + \sqrt{y})} \tag{3.684}$$

where

$$x = \frac{\tau_d}{\tau_A} + k_a N_B \tau_d \tilde{\mathscr{F}}(0) \quad \text{and} \quad y = k_b N_A \tau_d \tilde{\mathscr{F}}(0) \quad \text{at} \quad \tau_B = \infty \tag{3.685}$$

Since x and y are concentration-dependent, K_{AB} is also a nonlinear function of concentration.

In the case of irreversible energy transfer, such a dependence is given by Eq. (3.676). An increase of x, linear in quencher concentration N_B, accelerates the rate of excitation decay, thus enhancing the role of nonstationary quenching. This well-known effect exists only in the limit of diffusion-controlled transfer when

$$\kappa = \frac{k_a}{1 + k_a/k_D(1 + \sqrt{x})} \approx k_D(1 + \sqrt{x}) \quad \text{at} \quad k_D \ll k_a \tag{3.686}$$

In the zero-concentration limit, κ turns to its IET value $\kappa_0 = \lim_{N_B \to 0} \kappa$ given by Eq. (3.27). Under diffusional control κ increases with a concentration from this minimal value, (3.79), up to k_0 (see Section XII.B).

When the backward transfer is taken into account, K_{AB} becomes smaller than κ as a result of partial restoration of the excited state. This effect is greater the larger is the kinetic rate constant k_b, or the smaller is the diffusional constant k_D. It is the most pronounced at minimal concentrations of A ($y = 0$). As this concentration increases, the effect is hindered as shown in Figure 3.85. According to MET

$$\frac{K_{AB}}{\kappa} = \left[1 + \frac{k_b(1 + \sqrt{x})}{k_a(1 + \sqrt{y})}\right]^{-1} \quad \text{at} \quad k_D \ll k_a, k_b \tag{3.687}$$

where

$$x = \frac{\tau_d}{\tau_A} + \tau_d z k_a N_B, \quad y = \tau_d z k_b N_A, \quad z = \frac{1}{1 + k_a/[k_D(1 + \sqrt{\tau_d/\tau_A})]} \tag{3.688}$$

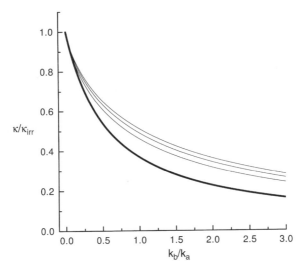

Figure 3.85. The Stern–Volmer constant of reversible energy quenching at $\tau_B = \infty$, K_{AB}, related to its irreversible analog κ as a function of backward energy transfer rate constant k_b related to the forward one, k_a. The thick line is an IET result, while the thin lines are obtained with MET at different concentrations of A: $4\pi\sigma^3 N_A/3 = 0.05, 0.15, 0.3$ (from bottom to top). The remaining parameters are $4\pi\sigma^3 N_B/3 = 0.15$, $\tau_A = 2\tau_d$, and $k_D \ll k_a, k_b$. (From Ref. 44.)

The shortage of B^* lifetime produced by "bachelors" makes the forward energy transfer less reversible, thus enhancing its rate. However, the acceleration of the diffusional transfer in both directions in a more concentrated solution does not produce the fluorescence quenching in the case under consideration. The quenching constant (3.669) differs from its IET analog (3.129) only by the definitions of K_{AB} and K_{BA}. Whatever they are, $k_q^{MET} = k_q = 0$ when $\tau_B = \infty$. The energy exchange with stable particle does not result in any quenching but only in delayed fluorescence.

B. Irreversible Quenching

As was shown when the transfer reaction is completely irreversible $k_q \equiv \kappa$ is the only quenching constant and it is a function of the single concentration $c \equiv N_B$. Here we report the $\kappa(c)$ dependence as predicted by numerous multiparticle theories. It is premised by a brief comparison of the initial product distributions at different concentrations.

1. Initial Distributions

From a comparison of the initial distributions at a single concentration $c = 10^{-3}$M (Fig. 3.33), we have already concluded that IET loses the contribution of nonstationary quenching to the near-contact region, while MET

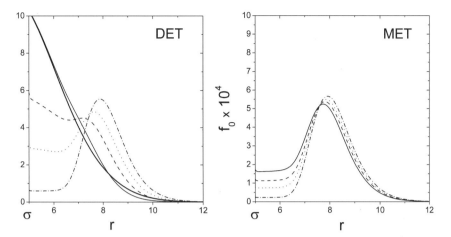

Figure 3.86. DET and MET initial distributions resulting from diffusional ionization at different $c = 0.1, 0.5, 1.0, 2.0$ (from bottom to top) and $D = 10^{-7} \, \text{cm}^2/\text{s}$. The thick solid line reflects the static distribution obtained with DET at the same parameters of the exponential transfer rate: $W_c = 1000 \, \text{ns}^{-1}, l = 1 \, \text{Å}, \sigma = 5 \, \text{Å}$.

accounts for it only partially as compared to DET. It follows from Figure 3.86 that the difference between the MET and DET distributions obtained for the exponential transfer rate (3.53) increases with concentration and finally becomes qualitative. Since DET can be considered as an exact theory of the target problem (for point particles), this is an indication that the inaccuracy of MET grows with c.

From the physical point of view at largest c, the initial distribution should take the form of a static one. This follows from the definition of $f_0(r, c)$ in Eq. (3.310) if we account for the fact that

$$\int_0^\infty n(r,t)N(t)dt = \int_0^\infty n(r,t) \exp\left(-\frac{t}{\tau} - c \int_0^t dt' \int W_I(r)n(r,t')d^3r\right)dt \quad (3.689)$$

where τ is the lifetime of excitation. At higher c the exponent decays faster and cuts $n(r, t)$ at earlier times. But at small times an approximate expression for

$$n(r,t) \approx e^{-W_I(r)t} \quad (3.690)$$

coincides in fact with the static solution of Eq. (3.291) valid at $\hat{L} = D = 0$. Using it in expression (3.689), we transform this quantity to its static value as well as the whole distribution $f_0(r, c)$.

This is the reason why the low-concentration distribution, having the well-pronounced maximum in the case of diffusional ionization, transforms with c

into the monotonous distribution, approaching the shape of the static one. Such a transformation experiences only the DET distribution, while that obtained with MET does not change qualitatively (Fig. 3.86). This is an indication that MET is just the next-order approximation with respect to c, which is better than IET but not ideal. In what follows we will demonstrate this using the example of the concentration dependence of the Stern–Volmer constant $\kappa(c)$.

2. Moderate Concentrations of Quenchers

In the lowest approximation with respect to c, the Stern–Volmer constant in DET, Eq. (3.27), is the same as in IET, Eq. (3.273). Under diffusional control the difference between κ and κ_0 is expected to be the largest, but not in this approximation when

$$\kappa = \kappa_0 = k_D\left(1 + \sqrt{\tau_d/\tau}\right) \qquad \text{at} \qquad k_D \ll k_0 \qquad \text{and} \qquad c \to 0 \quad (3.691)$$

However, at any finite c the IET value κ_0 is just the lower limit of $\kappa(c)$. It was obtained long ago by means of the contact DET [65]

$$\kappa(\xi) = k_D \frac{\sqrt{3\xi + \tau_d/\tau} + \Phi\,\tau_d/\tau}{\sqrt{3\xi + \tau_d/\tau} - 3\xi\,\Phi} \qquad \text{at} \qquad k_D \ll k_0 \qquad (3.692)$$

where

$$\Phi = \exp\left(\frac{9\xi^2}{\pi(3\xi + \frac{\tau_d}{\tau})}\right) \text{erfc}\, \frac{3\xi}{\sqrt{\pi(3\xi + \frac{\tau_d}{\tau})}}$$

and the dimensionless concentration $\xi = 4\pi\sigma^3 c/3$ is a binary parameter. This parameter is assumed to be small in all theories except DET, which was proved to be exact (valid at any ξ) for a particular case of the target problem (immobile excitation-quenched independently moving point quenchers). The DET Stern–Volmer constant $\kappa(\xi)$ is an increasing function of ξ (Fig. 3.87). Only its initial value coincides with that obtained with IET: $\kappa(0) = \kappa_0$.

This well-known shortage of IET is partially eliminated in MET, which accounts for the higher-order terms in the expansion $\kappa(\xi) = \kappa_0 + \beta'\xi + \cdots$ by substituting the corrected kernel $\Sigma(t) = R^*(t)\exp(-ckt)$ for the IET kernel $R^*(t)$ in Eq. (3.269a):

$$\dot{N} = -c \int_0^t \Sigma(t - t')N(t')dt' - \frac{N}{\tau} \qquad (3.693)$$

As a result, the generalized definition of

$$\kappa = \tilde{\Sigma}(0) \qquad (3.694)$$

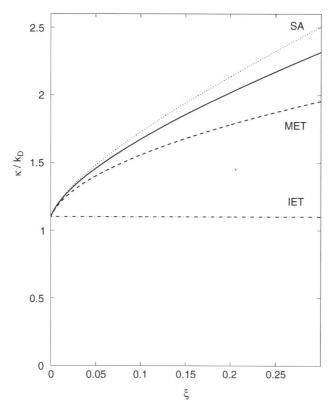

Figure 3.87. The Stern–Volmer constants as functions of the dimensionless concentration $\xi = 4\pi\sigma c^3/3$ obtained in the contact approximation and under diffusional control at $\tau_d/\tau_D = 0.01$. The thick line represents DET, which is exact for immobile donors and independently moving acceptors. The rest of the curves are obtained with the approximate methods, SA, MET, and IET. (From Ref. 133.)

provides the better estimate of the Stern–Volmer constant than does κ_0 defined in Eq. (3.272). For diffusional quenching [133]

$$\kappa^{MET}(\xi) = \tilde{R}^*(ck) = k_D\left(1 + \sqrt{\tau_d/\tau + 3\xi}\right) \tag{3.695}$$

A slightly different result was first obtained using a "mean field" approach [63] and then confirmed by some other methods [237,238], including the super-position approximation (SA) [201]:

$$\kappa^{SA}(\xi) = k_D\left(1 + 3\xi/2 + \sqrt{\tau_d/\tau + 3\xi + (9/4)\xi^2}\right) \tag{3.696}$$

TABLE IV
Linear Concentration Correction to Stern-Volmer Constant

Theories	A
DET	$\dfrac{3}{2}\left[\sqrt{\dfrac{\tau}{\tau_d}}+2-\dfrac{4}{\pi}\right]$
MET	$\dfrac{3}{2}\left[\sqrt{\dfrac{\tau}{\tau_d}}\right]$
SA	$\dfrac{3}{2}\left[\sqrt{\dfrac{\tau}{\tau_d}}+1\right]$

This result is easy to generalize for arbitrary light intensity by a simple substitution of $1/\tau + I_0$ for $1/\tau$ [201].

The difference between these and other estimates of κ can be quantitatively characterized by the first (linear in concentration) correction to κ_0:

$$\kappa = \kappa_0 + Ak_D\xi \qquad (3.697)$$

The values of A arising from different theories are collected in Table IV [133].

From these data and Figure 3.87 representing the concentration dependence of $\kappa(\xi)$, it is clear that the SA curve is closer to that of DET, although it slightly overestimates κ while MET underestimates this value considered as exact. It can also be seen that the linear part of the $\kappa(\xi)$ dependence is very short even in the concentration domain of Figure 3.87. The same dependence is much more instructive when considered in the full concentration range as shown below.

3. Arbitrary Quenching Concentrations

It is remarkable that the integral equation (3.693) proved to be formally exact [51], so that all theories differ only in their definition of the kernel, given as $\Sigma(t)$ or $\tilde{\Sigma}(s)$. Sometimes the kernel is explicitly defined in the original works, but more often the reduction to the integral form and extraction of the kernel is a separate problem solved in Ref. 46. In a few cases this procedure was nontrivial and required rather long and sophisticated calculations that are of no interest except for the final results represented by the Laplace transformed kernels $\tilde{\Sigma}$ in Table V.

In Table V we used the Laplace transformation of the general solution of Eq. (3.693):

$$\tilde{N}(s) = \int_0^\infty e^{-(s+u)t}P(t)dt = \tilde{P}(s+u) \qquad (3.698)$$

where $P(t)$ is the usual DET expression for the quenching kinetics (3.177) and $u = 1/\tau$. Besides, the Green function of the free diffusion equation, $G_0(r, r', t)$,

TABLE V
Integral Kernels in Different Theories

Theory	$k_0/\tilde{\Sigma}(s)$
LSA(= IET)	$1 + k_0 g(s + u)$
LESA	$1 + k_0 g(s + u + ck_0)$
CA, MPK1 (= DET)	$\left[ck_0\tilde{P}(s+u)\right]/\left[1 - (s+u)\tilde{P}(s+u)\right]$
MPK2	$1 + k_0 g\left(s + u + ck_0\tilde{\Sigma}(s)\right)$
MPK3 (= MET)	$1 + k_0 g\left(s + u + ck_0/[1 + k_0 g(s+u)]\right)$
SCRTA	$1 + k_0 g(s + u + ck_f), \qquad k_f = k_0/[1 + k_0 g(ck_f)]$

was used to define the auxiliary function

$$g(s) = \tilde{G}_0(\sigma, \sigma, s) = \frac{1}{k_D\left(1 + \sqrt{s\tau_d}\right)} \qquad (3.699)$$

This function is easily related to the "ideal" Stern–Volmer constant

$$\kappa_0 = \kappa(\xi = 0) = \frac{k_0}{1 + k_0 g(u)} \qquad (3.700)$$

which is the same for all theories listed in Table V.

The first group of theories was based on the superposition approximation (SA) used for truncation of an infinite hierarchy of equations for a reduced distribution function at the pair distribution level [239,240]. It was generalized and applied to the reversible reactions by Lee and Karplus [139] and their successors [145,174,176]. After linearization over deviations from equilibrium [52,63], the theory became simpler and finally was recognized as one identical to IET, provided the reduction to the phenomenological equations is not done [31,175]. This is why the linearized superposition approach (LSA) provides exactly the same kernel as IET.

SA was also applied to the reversible reactions considered in Section XII.C.4. Since the results were not satisfactory, an extended superposition approach (ESA) was developed, then linearized, and later known as LESA [241]. Independently, a similar linearization over deviations from equilibrium was also made in Ref. 242. Although the asymptotic description of the quenching kinetics is improved, it was recognized [242] that LESA is not valid with a large equilibrium constant K because the superposition approach worsens when K increases [241]. This is especially true at earlier times when the deviations from equilibrium are not small. However, the authors who constructed LESA claimed that it is "applicable at all times" [241]. Therefore, it was taken for comparison with other approximations. In the irreversible limit ($K \to \infty$), the kernels obtained in both works [241,242] coincide with that listed as LESA in Table V.

The largest body of research carried out by different authors with different methods [120,243–245], on closer inspection, are noted to be identical to the

DET of contact irreversible reactions, that is, to the classical Collins–Kimball theory [2]. This is equally true for a few works [63,170] based on the convolution approximation (CA) and that published by Lee et. al. as MPK1 [51]. The latter was an extension to reaction $A + B \rightleftharpoons C + D$, the earliest association/dissociation theory of the same name [246] (see Section XII.B.4). In fac ;, a kernel for the reversible bimolecular reaction based on four-particle decoupling of the hierarchical chain was artificially composed in MPK1 from two fragments originating from the irreversible forward and backward DET kernels. Although it is doubtful whether such a compilation is well grounded in the case of irreversible quenching, MPK1 is entirely identical to DET.

However, upgrading of the theory was not stopped. After the first attempt the MPK2 [247] and MPK3 [126] have arisen whose kernels are different. As can be seen from Table V, the kernel for MPK2 is actually defined as a solution of the transcendent equation. Another one, for MPK3, appears to be identical to that provided by MET. More recently a new theory named the self-consistent relaxation time approximation (SCRTA), was developed by Gopich and Szabo [180]. In the irreversible case the inverse relaxation time of SCRTA, $k_f c$, is defined by the transcendent equation for k_f shown in Table V. It is equivalent to the one expressed by the Laplace transformation of $k(t)$: $c\tilde{k}(k_f) = 1$ [177]. In spite of this complication, SCRTA is competitive in the concentration corrections with other theories.

As has been proved in Eq. (3.700), the lowest limit of the Stern–Volmer constant $\kappa_0 = \kappa(0)$ is the same for all contact theories. However, there is also the upper limit of $\kappa(c)$ reached at largest c. At the very beginning the quenching kinetics is always exponential

$$R(t) = \exp(-t/\tau - ck_0 t + \cdots) \tag{3.701}$$

but it remains the same until almost the end, if the concentration of quenchers c is high enough. With the exponential $R(t)$, we obtain from Eq. (3.10) the classical Stern–Volmer law but with the maximal constant

$$k_0 = \lim_{c \to \infty} \kappa(c)$$

Now we have to specify the Stern–Volmer constant behavior between these two limits, in the interval $\kappa_0 \le \kappa(c) \le k_0$.

In the semilogarithm plot of Figure 3.88 the concentration dependence $\kappa(c)$ is represented by the S-like curves related to different theories. The main one is that of DET, which is expected to be exact for the target problem for independently moving point quenchers. This is also true for all equivalent theories of irreversible transfer (CA, MPK1, Vogelsang theory [243,244]).

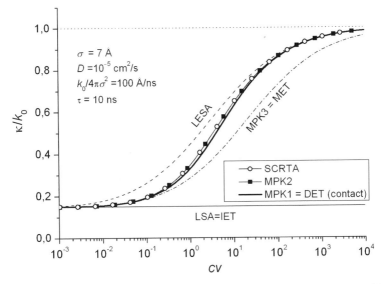

Figure 3.88. The concentration dependence of the Stern–Volmer constant κ in units of k_0 for a number of contact theories, provided $k_0 = 3.7 \times 10^{10} \, \text{M}^{-1}\text{s}^{-1}$ is the same for all of them. (From Ref. 46.)

From the comparison of this result with others, it is clear that LESA is just poor interpolation between border limits. This is partially true also for MET/MPK3, although it is rather good for small concentrations as was actually expected [39,40,44,45]. At least it is better than IET and LSA, which reproduce the ideal Stern–Volmer law and are suitable for calculation of only concentration-independent constants (horizontal dashed line). Two remaining theories, MPK2 and SCRTA, give results that are very close to each other and to the exact one. This comparison is instructive to those who like to employ one of the existing contact theories to the particular problem of interest.

Even more important is the comparison of contact and remote electron transfer shown in Figure 3.89. The calculations were performed by DET for the exponential transfer rates with different tunneling lengths l (dashed lines). The smaller l is, the closer are the results obtained to those gained in the contact approximation. The latter is actually justified for only the smallest $l = 0.1 \, \text{Å}$, which is reasonable for proton transfer but rather unfavorable for electron transfer with $l = 1 \div 2 \, \text{Å}$.

4. Fitting to Real Data

To make the last point clear, a straightforward fitting of the remote transfer theory to the data presented by Stevens and Biver III [248] was undertaken in

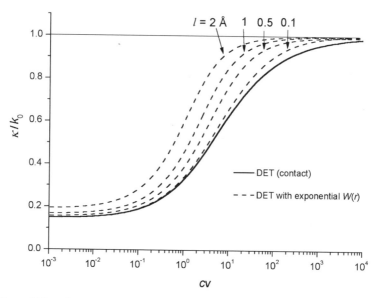

Figure 3.89. The same as in Figure 3.88 but for the exponential transfer rates at different tunneling lengths l indicated in the figure. The contact approximation ($l \to 0$) is represented by a solid line. (From Ref. 46.)

Ref. 46. Neither a contact nor exponential approximation of W is good for real fitting. Instead, one should use the true Marcus rate of electron transfer accounting for the energy balance and the properties of the solution. In the simplest case of a single-channel reaction in a highly polar solvent this rate is given by Eq. (3.47). The highly exergonic fluorescence quenching of 9,10-dicyanoanthracene (DCNA) by N,N,N',N'-tetramethyl-p-phenylenediamine (TMPD) was studied in Ref. 248 in the broad range of TMPD concentrations at three different temperatures ($T = 0°C, 22°C, 50°C$) in acetonitrile solutions. The corresponding diffusion coefficients are listed in Table VI together with other data taken from the original article [248] and the contact reorganization energy λ_0 borrowed from another work [18].

In Figure 3.90 the results of the best fit to the experimental points from Ref. 248 are shown as functions of the dimensionless concentration cv, where

TABLE VI
Parameters of DCA Quenching by TMPD in Acetonitrile

D (0°C)	D (22°C)	D (50°C)	σ	τ	$-\Delta G$	λ_0
$2.4 \times 10^{-5}\ \mathrm{cm^2/s}$	$3.2 \times 10^{-5}\ \mathrm{cm^2/s}$	$4.3 \times 10^{-5}\ \mathrm{cm^2/s}$	$7.5\ \mathring{A}$	12.6 ns	1.81 eV	1.3 eV

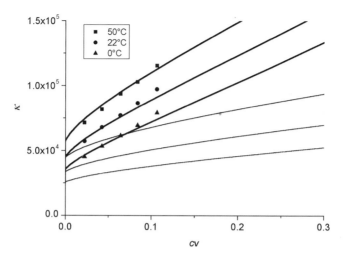

Figure 3.90. The experimental concentration dependence of the Stern–Volmer constant for three different temperatures (points) fitted by DET with the single-channel Marcus transfer rate (thick lines). The thin lines represent the contact analogs of the curves above for the same temperatures (decreasing from top to bottom) and the diffusional control of quenching ($k_0 = \infty$). (From Ref. 46.)

$v = \frac{4}{3}\pi\sigma^3$. The agreement is rather satisfactory and could even be better if a number of additional factors were taken into account. Instead of a single-channel transfer rate (3.47), one could better use its multichannel analog (3.51). The reorganization energy attributed to the quantum channels comprises an additional λ_i, which is sometimes around 0.3 eV [225]. Besides, the liquid structure near the contact and the spacial dependence of the diffusional coefficients as well as the chemical anisotropy that is essential sometimes were ignored [249]. Because of these simplifications, there are only two varying parameters: the tunneling length L and the matrix element V_0. Found from the best fit, they are [46]

$$L = 1.6\,\text{Å}, \qquad V_0 = 4.07 \times 10^{-2}\,\text{eV} \qquad (3.702)$$

With these parameters known, the kinetic rate constant (3.37) and the contact value of the rate, $W_c = W(\sigma)$, have definite although different values for all three temperatures studied.

All the k_0 values listed in Table VII are much larger than the contact diffusional constants $k_D = 4\pi\sigma D$, so that the reactions are strongly in the diffusional limit, as was expected [248]. The same is true for the diffusional $\kappa(c)$, which is larger everywhere than the Stern–Volmer constant obtained in the contact approximation. The family of such curves compared in Figure 3.90

TABLE VII
Kinetic Constants and Contact Rates

T (°C)	0	22	50
$k_0 \times 10^{-6}$ (Å³/ns)	0.979	1.004	1.03
W_c (ns⁻¹)	484	546	619

indicates that the experimental points closely fitted by the remote DET are far above their contact estimates.

However, not only the contact lines themselves are beneath all the points but their slope is also too small to explain the true concentration dependence. Perhaps these difficulties stimulated Stevens to propose the original "finite-sink approximation" instead of the existing theory [103,250]. The inconsistency of this model is seen from the very fact that it does not discriminate between the Stern–Volmer and the steady-state constants, κ and $k = \lim_{t\to\infty} k(t)$, and ascribes to the latter the concentration dependence, which is not inherent to k in principle. At the same time the model predicts the linear dependence of the results in coordinates $1/\kappa$ versus $c^{1/3}$, and this linearity was confirmed experimentally a number of times [6,248,250,251]. According to their finite-sink model [248], the straight lines representing this dependence for different temperatures should intersect at a common point where "the static quenching limit is unambiguously located."

To inspect this statement, the previous figure was redrawn in Ref. 46 in similar but dimensionless coordinates: k_0/κ versus $(cv)^{1/3}$ (Fig. 3.91). One more curve representing the Stern–Volmer constant for the static quenching (dashed–dotted line) was obtained from the static kinetics (3.45) integrated in the general definition of κ, Eq. (3.10). At the maximum concentrations where k_0/κ approaches 1, all the curves merge, reaching the true static limit. But the experimental points are far from this region. Occasionally, they are near the bend of the S-like curves, where the straight line with the viscosity dependent slope is rather a good approximation. The intersection of these lines at a common point that mimics the "finite-sink approximation" is only artificial, and the point itself has nothing in common with the static quenching. This artifact would be easily removed if the concentration region studied experimentally were extended a little bit either to the right or to the left.

However, even the existing data are unique. It was already demonstrated in Figure 3.90 that their fitting is inaccessible for all contact theories. From Figure 3.91 we see that this is equally correct for MET whose results cannot be advanced so far in the high-concentration region without losing accuracy. Only the old classical DET is good enough for the appropriate fitting and interpretation of these data.

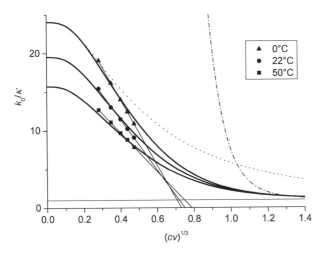

Figure 3.91. The same DET curves (solid lines) and experimental data (points) as in Figure 3.90 but with the Stevens coordinates. The static quenching (dashed–dotted line) and the MET analog of DET with the same parameterization (dotted line) are shown for only 0°C temperature. The straight lines intersecting at the common point are the Stevens interpolation of the experimental results. (From Ref. 46.)

C. Reversible Dissociation in Dense Solution

Let us now direct our attention to the reversible dissociation of the exciplex, as exemplified in Eq. (3.231). It can be formally represented by the following reaction scheme:

$$
\begin{array}{ccc}
\mathrm{A}^* + \mathrm{B} & \underset{k_d}{\overset{k_a}{\rightleftharpoons}} & \mathrm{C}^* \\
\downarrow u_A & & \downarrow u_C
\end{array}
\qquad (3.703)
$$

where $u_A = 1/\tau_A$ and $u_C = 1/\tau_C$ are the decay rates of excitations that may be either equal or different. The principal difference in what was considered in Section V.D is the large concentration of B particles in solution. This makes possible their bimolecular association with A and violates the unimolecular character of exciplex dissociation inherent in diluted solutions. The reversible reaction (3.703) develops in one or the opposite directions depending on what was initially created by a light pulse, A^* or C^*.

The vast majority of competing theories of the phenomenon were aimed at specifying the kinetics of the system approaching the equilibrium at $u_A = u_C = 0$. It is usually assumed that B particles are present in great excess, so that their concentration $c = [B]$ remains constant. In the particular case of reaction (3.231), the desirable concentration of H^+ in solution is provided by the addition of an inert acid (e.g., $HClO_4$). The added protons compete with a geminate proton for the restoration of C^*.

The kinetics of approaching equilibrium is universally described by the relaxation function $R(t)$ defined as follows:

$$\frac{P_C(t) - \mathscr{P}_C}{P_C(0) - \mathscr{P}_C} = R(t) = \frac{\mathscr{P}_A - P_A(t)}{\mathscr{P}_A - P_A(0)} \qquad (3.704)$$

Here $P_C(t) = N_C(t)/[N_C(0) + N_A(0)]$ and $P_A(t) = N_A(t)/[N_C(0) + N_A(0)]$ are the fractions of C and A molecules approaching with time their equilibrium values

$$\mathscr{P}_C = \frac{cK_{eq}}{1 + cK_{eq}} = 1 - \mathscr{P}_A \qquad (3.705)$$

where $K_{eq} = k_a/k_d$ is the equilibrium constant introduced in Eq. (3.84).

In dilute solutions, $\mathscr{P}_C = 0 = 1 - \mathscr{P}_A$, the only possible reaction is the irreversible dissociation of C whose kinetics $R(t) \equiv P_e(t)$ was studied in Section V.D. As was shown there, the long-time behavior of this quantity obeys the power law (3.249). This asymptotic expression represents the very end of the kinetics when, for instance, $P_C(t)$ starting from $P_C(0) = 1$ is already three or four orders of magnitude smaller. This tail is hardly available for detection even with contemporary single-photon counting. Besides, the question arises as to what the difference is in the precursor time evolution of $P_C(t)$ or $P_A(t)$ predicted using a number of different theoretical methods.

Here we switch our attention from the asymptote to the total time evolution of $P_C(t)$ or $P_A(t)$, which entirely contribute to the relative quantum yields of the fluorescence:

$$\eta_A = u_A \int_0^\infty P_A(t)dt, \qquad \eta_C = u_C \int_0^\infty P_C(t)dt \qquad (3.706)$$

Both of these equations can be universally expressed through the Stern–Volmer constant κ. The latter one has been used already in Section XII.B as a standard for comparison between the different theories of irreversible quenching. In the case of the reversible reaction (3.703), the problem is more difficult, especially at high concentrations of B molecules. After the dissociation, other B molecules can be involved in the reaction with A. This many-particle competition for the partner couples the motion of the molecules, making the problem unsolvable analytically. Thus only approximate solutions were obtained by means of different methods and assumptions whose validity very often remains unclear.

The repeating binding–unbinding processes change the spatial distribution of B particles around A. But the movement of A also affects the concentration profile of the B particles. Therefore, the motion of both A and B should have influence on the reaction. Here we consider only two opposite limits considered in Ref. 47. One is known as the "target problem," when A and C are static $(D_A = D_C = 0)$ while the pointlike B molecules move independently $(D_B \neq 0)$. The opposite one is the "trap problem," in which a single particle A accomplishing diffusion with $D_A \neq 0$ between immovable B particles can be trapped by one of them and form an immovable C particle $(D_B = D_C = 0)$. In both cases it is assumed that the reaction takes place only at the contact distance. This is definitely the case for the proton transfer (3.231). The vast majority of existing theories also deal only with contact reactions, although a few of them can be extended to the distant (electron or energy) transfer as well.

1.   Contact Dissociation/Association

When the B particles are present in great excess, the following integrodifferential equations linear in c describe the reversible contact reaction [47]:

$$\frac{dP_A}{dt} = -u_A P_A - ck_a \int_0^t \Sigma(t-\tau)P_A(\tau)d\tau + k_d \int_0^t \Sigma(t-\tau)P_C(\tau)d\tau \quad (3.707a)$$

$$\frac{dP_C}{dt} = -u_C P_C + ck_a \int_0^t \Sigma(t-\tau)P_A(\tau)d\tau - k_d \int_0^t \Sigma(t-\tau)P_C(\tau)d\tau \quad (3.707b)$$

They may be obtained by means of the matrix IET but only together with the kernel $\Sigma(t) = F(t)$ specified by its Laplace transformation (3.244), which is concentration-independent. However, from the more general point of view, Eqs. (3.707) are an implementation of the memory function formalism in chemical kinetics. The form of these equations shows the essentially non-Markovian character of the reversible reactions in solution; the kernel holds the memory effect, and the convolution integrals entail the prehistoric evolution of the process. In the framework of ordinary chemical kinetics $\Sigma(t) = \delta(t)$, so that the system (3.707) acquires the purely differential form. In fact, this is possible only in the limit when the reaction is entirely under kinetic control.

As was shown in Section XII.B, all the theories of irreversible transfer are also reducible to a similar but single-integral equation (3.693) and differ only by the definitions they give to its kernel. The theories of reversible geminate dissociation proposed by Berg [120], Burshtein [31], and Agmon and Szabo [170,177] were also represented by a single equation (3.242) that follows from Eq. (3.707b) at $c = 0$. Now we extend this conclusion for the reversible reaction at $c \neq 0$. In many sophisticated theories, which will be compared, the results are presented differently. However, all of them after some manipulation can be shown to obey Eqs. (3.707), although the kernels are specific for each of them. Omitting these manipulations, we concentrate on the results: the difference between the concentration dependencies of the kernels obtained from different theories.

Equations (3.707) can be formally solved in the Laplace domain. The quantities $\tilde{P}_{A,C}(s) = \int_0^\infty P_{A,C}(t)\exp(-st)dt$, which are the Laplace images of $P_A(t)$ and $P_C(t)$, can be expressed through the Laplace transformation $\tilde{\Sigma}(s)$ of the kernel $\Sigma(t)$, in the following way:

$$\tilde{P}_A(s) = \frac{(s + u_C)P_A(0) + k_d\tilde{\Sigma}(s)}{Q(s)} \tag{3.708a}$$

$$\tilde{P}_C(s) = \frac{(s + u_A)P_C(0) + ck_a\tilde{\Sigma}(s)}{Q(s)} \tag{3.708b}$$

where the denominator

$$Q(s) = (s + u_C)(s + u_A) + [ck_a(s + u_C) + k_d(s + u_A)]\tilde{\Sigma}(s) \tag{3.709}$$

and the sum of the initial probabilities $P_A(0) + P_C(0) = 1$.

2. Approaching Equilibrium

For stable particles ($u_A = u_C = 0$), one can easily deduce from the last formula the Laplace transformation of the relaxation function:

$$\tilde{R}(s) = \frac{\tilde{P}_C(s) - \mathscr{P}_C/s}{1 - \mathscr{P}_C} = \frac{1}{s + (k_d + k_ac)\tilde{\Sigma}(s)} \tag{3.710}$$

The concentration-independent kernel of IET, denoted later as $\tilde{\Sigma}_0(s)$, is simply $\tilde{F}(s)\big|_{k_c=0}$, where $\tilde{F}(s)$ is borrowed from Eq. (3.244):

$$\tilde{\Sigma}_0(s) = \frac{1}{1 + k_ag(s)} \tag{3.711}$$

Here $g(s)$ from Eq. (3.699) is the Laplace transformation of the Green function of the free diffusion around a hard sphere with the reflecting boundary condition at contact. Gopich and Doktorov also obtained a result similar to Eqs. (3.710)–(3.711) except that they used the Green function of penetrable spheres [252]. Even earlier the same result was derived in the framework of a statistical nonequilibrium thermodynamics, but instead of using contact $\tilde{G}_0(\sigma, \sigma, s)$, using the free-space Green function [174]: $\tilde{G}_0(\sigma, 0, s) = \exp\left(-\sqrt{s\sigma^2/D}\right)/k_D$. At small $s \ll D/\sigma^2$ the difference between the Green functions disappears since all

of them are approximated as $g(s) \approx [1 - \sqrt{s\,\sigma^2/D}]/k_D$. Using this approximation in Eq. (3.711) and then making the inverse Laplace transformation of
Eq. (3.710), one obtains the asymptotic relaxation to the equilibrium [174]:

$$R(t) = \frac{K_{eq}}{(1 + cK_{eq})(4\pi Dt)^{3/2}} \qquad \text{at} \qquad t \to \infty \qquad (3.712)$$

This power-law dependence on time is the same as in Eq. (3.249), where Z
should be set to 1 for a stable complex ($\tau_{exc} = \infty$). The peculiarity of the dense
solution is the concentration-dependent multiplier $(1 + cK_{eq})^{-1}$ arising in
(3.712). However, this concentration dependence is inherent in IET only, as it
differs in other theories compared below. Using more adequate methods for
studying the target [180,241,242,253,254] and trap problems [180,242,252], it
was proved that

$$R(t) = \begin{cases} \dfrac{K_{eq}}{(1+cK_{eq})^2(4\pi Dt)^{3/2}} & \text{target problem} \\[3ex] \dfrac{K_{eq}}{\sqrt{1+cK_{eq}}(4\pi Dt)^{3/2}} & \text{trap problem} \end{cases} \qquad (3.713)$$

The most general theory valid at the arbitrary diffusion coefficients, D_A, D_B, and
D_C, also corrects only the concentration dependence in the denominator of
Eq. (3.712), but in full accordance with both the target and trap limits [255].

3. Quantum Yields and Quenching Constant

If $c = 0$, only the dissociation of C makes any sense. At $c \neq 0$ there is an
alternative: either (1) A can be initially excited [$P_A(0) = 1, P_C(0) = 0$] or (2) C
[$P_A(0) = 0, P_C(0) = 1$]. The former case is typical impurity-quenching (by B
molecules) whose relative quantum yield $\eta_{AA} = \eta_A|_{P_A(0)=1}$ can be obtained from
Eqs. (3.706) and (3.708a):

$$\frac{1}{\eta_{AA}} = 1 + c\kappa \qquad (3.714)$$

This is the conventional Stern–Volmer law but with the constant defined as
follows:

$$\kappa = \frac{k_a\tilde{\Sigma}(0)/u_A}{1 + k_d\tilde{\Sigma}(0)/u_C} \qquad (3.715)$$

If the fluorescence of C is registered at the same initial conditions, then its
quantum yield η_{CA} can be found from Eqs. (3.706) and (3.708b):

$$\eta_{CA} = \eta_C|_{P_A(0)=1} = 1 - \eta_{AA} \qquad (3.716)$$

This conservation law for excitations holds, since there is no recombination to the ground state accompanying the transfer in either direction.

It is instructive to note that κ from Eq. (3.715) can be represented in a form similar to its electron transfer analog (3.655) but with $\varphi_c = 1$, because the recombination to the ground state is absent here ($k_c = 0$):

$$\kappa = \kappa_g[1 - \chi] \tag{3.717}$$

Here

$$\kappa_g = k_a\tilde{\Sigma}(0)\tau_A \quad \text{and} \quad \chi = \frac{k_d\tilde{\Sigma}(0)}{1/\tau_C + k_d\tilde{\Sigma}(0)} \tag{3.718}$$

have the sense of the geminate Stern–Volmer constant and the dissociation efficiency of the unstable C particle.

If this is not an A but a C particle initially excited, then the quantum yields of fluorescence through both channels, η_{CC} and η_{AC}, can be found from the same formulas (3.706) and (3.708) but with the opposite initial conditions: $P_A(0) = 0, P_C(0) = 1$. The conservation law takes place in this case as well

$$\eta_{CC} = \eta_C|_{P_C(0)=1} = 1 - \eta_A|_{P_C(0)=1} = 1 - \eta_{AC} \tag{3.719}$$

but instead of the Stern–Volmer we have another law:

$$\frac{1}{1 - \eta_{CC}} = K_{eq}\frac{u_C}{u_A}\frac{1 + c\kappa}{\kappa} = \frac{1}{\eta_{AC}} \tag{3.720}$$

From this linear relationship two essential parameters of the problem can be obtained by fitting the experimental data: $(u_C/u_A)K_{eq}$ and κ. The latter is an invariant of the problem that is available experimentally in two different ways: from either the usual Stern–Volmer law (3.714) or its analog for the reverse process, (3.720). Unlike κ_0 from IET Eq. (3.655) or (3.364), the present κ as well as $\Sigma(0)$ are concentration-dependent.

The concentration-independent IET quantity $\kappa_0 = \lim_{c \to 0} \kappa(c)$, considered as an "ideal" Stern–Volmer constant, can be derived from (3.715) with $\tilde{\Sigma}(0) = \tilde{\Sigma}_0(u_A)$ from (3.711)

$$\kappa_0 = \frac{k_a'/u_A}{1 + k_a'g(u_A)} \tag{3.721}$$

where $k_a' = k_a u_C/(u_C + k_d)$ is the overall association rate constant renormalized as a result of the reverse reaction. This is the relationship that also follows from the equations of ordinary chemical kinetics applied to reaction (3.703).

If the association is irreversible ($k_d = 0$), the Stern–Volmer constant (3.715) is

$$\kappa_{irr} = \frac{k_a \tilde{\Sigma}_{irr}(0; c)}{u_A} \qquad (3.722)$$

where

$$\tilde{\Sigma}_{irr}(s; c) = \frac{1}{ck_a}\left[\tilde{R}^{-1}(s; c) - (s + u_A)\right] \qquad (3.723)$$

and $\tilde{R}$ is the Laplace transformation of expression (3.4) with $1/\tau = u_A$. The concentration dependence of such a κ was investigated in Section XII.B.

For reversible reactions, κ is constructed from two parts that are responsible for the forward ($k_a \tilde{\Sigma}(0)/u_A$) and backward ($k_d \tilde{\Sigma}(0)/u_C$) processes. Both of them are proportional to $\tilde{\Sigma}(0)$, which can also serve as a standard for comparison of the different theoretical approaches. The majority of them provide $\tilde{\Sigma}(s)$ when two lifetimes are equal, $u_A = u_C$. In this case the formal solution is the same as that for the reversible reaction between the ground-state particles but multiplied by $\exp(-u_A t)$. In the Laplace space it leads to a shift in the Laplace variable, $s \rightarrow s + u_A$.

4. Collapsed Theories

The concentration dependence of the Stern–Volmer constant has been used already as a test for the validity of the theories of intermolecular (Section XII.B) and intramolecular electron transfer (Section IV.E). However, the statistical theory of Felderhof and Jones [256], which addressed the latter case but did not withstand the test [257], was excluded from the abovementioned comparison. The same should be done with a few theories of dissociation/association. These are the theory of Vogelsang and Hauser [243] and the so-called phenomenological rate equations proposed by Lee and Karplus [139] and used in a number of other works [145,174,176].

In Vogelsang's approach the ideal Stern–Volmer constant takes the form

$$\kappa_0 = \frac{k_a'/u_A}{1 + k_a g(u_A)}$$

which is different from the exact one, Eq. (3.721). The main point of this theory is that at the moment just after the dissociation, all B particles (including the released one) form the equilibrium distribution around A. It is clear that only the fast diffusion or fast relaxation of C can validate the theory, but it is irrelevant in the majority of other cases. Moreover, Vogelsang's theory gives an incorrect result for the equilibrium probabilities (3.705), thus violating the mass action law. For all these reasons we exclude this theory from further consideration.

The modified rate equation (MRE) approximation [177] was also constructed to describe the non-Markovian character of diffusion-controlled reversible reactions. The forward reaction rate $k(t)$ is the same as in contact DET, Eq. (3.21), but with k_a substituted for k_0. As for the backward reaction rate, it was modified to be proportional to $k(t)$:

$$\frac{d}{dt}P_A = -ck(t)P_A + \frac{k_d}{k_a}k(t)P_C - u_A P_A \qquad (3.724a)$$

$$\frac{d}{dt}P_C = ck(t)P_A - \frac{k_d}{k_a}k(t)P_C - u_C P_C \qquad (3.724b)$$

This was done to "predict the correct equilibrium solution at long times" if $u_A = u_C = 0$. In the case of equal lifetimes, $u_A = u_C \neq 0$, the ideal Stern–Volmer constant found from Eqs. (3.724) was shown to be [47]

$$\kappa_0 = K_{eq}\left[1 - u_A \tilde{R}\left(0; K_{eq}^{-1}\right)\right]$$

This result does not coincide with the IET value (3.721); that is, MRE does not hold the geminate limit. MRE equations can be justified in the kinetic control limit only. Moreover, as was shown in Section V.D at $u_A \neq u_C$, MRE loses the phenomenon of the delayed fluorescence through the particle with a shorter lifetime (see Fig. 3.28). This also put it out of comparison.

5.   Encounter Theories

The contact IET provides the following Laplace transformation of the kernel [258]:

$$\tilde{\Sigma}(s) = \frac{k_D[1 + \sqrt{(s + u_A)\tau_d}]}{k_a + k_D[1 + \sqrt{(s + u_A)\tau_d}]} = \tilde{\Sigma}_0(s + u_A) \qquad (3.725)$$

This is not surprising since IET represents just the small concentration limit, when $\Sigma(c, t) \to \Sigma_0(t)$.

The unified theory (UT) considers the bimolecular generation of exciplexes, or their light-induced generation in line with their subsequent reversible dissociation (Section V.D). Under the condition of instantaneous exciplex formation, this theory reduces to IET.

The modified encounter theory is the only one that gives different forms for $\tilde{\Sigma}(s)$ in the trap (Table VIII) and target (Table IX) limits. In the case of the

TABLE VIII
$\tilde{\Sigma}(s)$ for the Target Problem

Theory	$\tilde{\Sigma}(s)^{-1}$
CA	$\tilde{\Sigma}_{irr}^{-1}(s;c)$
LSA = IET	$\tilde{\Sigma}_0^{-1}(s+u_A)$
LESA, $u_A=u_C$	$\mathscr{P}_A\tilde{\Sigma}_0^{-1}(s+u_A)+\mathscr{P}_C\tilde{\Sigma}_0^{-1}(s+u_A+ck_a+k_d)$
MPK1, $u_A=u_C$	$\mathscr{P}_A\tilde{\Sigma}_0^{-1}(s+u_A)+\mathscr{P}_C\tilde{\Sigma}_{irr}^{-1}(s;c+K_{eq}^{-1})$
MPK2,	$(1-\nu(s))\tilde{\Sigma}_0^{-1}(s+\alpha_-(s))+\nu(s)\tilde{\Sigma}_0^{-1}(s+\alpha_+(s))$
$u_A=u_C$	$\mathscr{P}_A\tilde{\Sigma}_0^{-1}(s+u_A)+\mathscr{P}_C\tilde{\Sigma}_0^{-1}\left(s+u_A+(ck_a+k_d)\tilde{\Sigma}(s)\right)$
MPK3 =	$(1-\mu(s))\tilde{\Sigma}_0^{-1}(s+\beta_-(s))+\mu(s)\tilde{\Sigma}_0^{-1}(s+\beta_+(s))$
MET(at $u_A=u_C$)	$\mathscr{P}_A\tilde{\Sigma}_0^{-1}(s+u_A)+\mathscr{P}_C\tilde{\Sigma}_0^{-1}\left(s+u_A+(ck_a+k_d)\tilde{\Sigma}_0(s+u_A)\right)$
SCRTA, $u_A=u_C$	$\mathscr{P}_A\tilde{\Sigma}_0^{-1}(s+u_A)+\mathscr{P}_C\tilde{\Sigma}_0^{-1}(s+u_A+\lambda)$
	$\lambda=\frac{ck_a+k_d}{1+k_ag(\lambda)/\left[k_Dg\left(\mathscr{P}_A^2\lambda\right)\right]}$

Auxiliary functions

$$\alpha_\pm(s)=\tfrac{1}{2}\left(Y(s)\pm\sqrt{Y(s)^2-4Z(s)}\right)$$

$$Y(s)=u_A+u_C+(ck_a+k_d)\tilde{\Sigma}(s)$$

$$Z(s)=u_Au_C+(ck_au_C+k_du_A)\tilde{\Sigma}(s)$$

$$\nu(s)=\frac{\alpha_+(s)-u_C-k_d\tilde{\Sigma}(s)}{\alpha_+(s)-\alpha_-(s)}$$

$$\beta_\pm(s)=\tfrac{1}{2}\left(y(s)\pm\sqrt{y(s)^2-4z(s)}\right)$$

$$y(s)=u_A+u_C+(ck_a+k_d)\tilde{\Sigma}_0(s+u_A),$$

$$z(s)=u_Au_C+(ck_au_C+k_du_A)\tilde{\Sigma}_0(s+u_A)$$

$$\mu(s)=\frac{\beta_+(s)-u_C-k_d\tilde{\Sigma}_0(s+u_A)}{\beta_+(s)-\beta_-(s)}$$

contact reaction with $u_A=u_C$, MET was shown to be identical with MPK3 [126].

6. Convolution Approaches

Berg's work [120] was chronologically the first theory of a diffusion-influenced reversible reaction that was later on reproduced and analyzed by Agmon and

TABLE IX
$\tilde{\Sigma}(s)$ for the Trap Problem

Theory	$\tilde{\Sigma}(s)^{-1}$
MPK3 = MET(at $u_A=u_C$)	$\tilde{\Sigma}_0^{-1}\left(s+u_A+ck_a\left/\left[\frac{k_d}{s+u_C}+\tilde{\Sigma}_0^{-1}(s+u_A)\right]\right.\right)$
SCRTA, $u_A=u_C$	$\tilde{\Sigma}_0^{-1}\left(s+u_A+ck_a\left/\left[\frac{k_d}{s+u_A}+\tilde{\Sigma}_0^{-1}(0)\right]\right.\right)$

Szabo [170]. Berg assumed that at any moment of dissociation a newborn A–B pair is surrounded by the equilibrium distribution of B particles. As a result, this theory works well near the limit, when the dissociation events are rare and the trap is mostly bound ($cK_{eq} \gg 1$). At the same time, it correctly reduces to the opposite, geminate limit, when $c = 0$ and an isolated A–B pair is formed as a result of the decay of C (Section V.C). There are also other authors who started later from the same assumption and obtained the same result [170,177,245]. In Ref. 170 it was called the *convolution approximation* (CA).

7. Superposition Approach

The *superposition approximation* (SA) suggested in Refs. 177 and 259 is essentially a nonlinear theory that cannot be represented in the form of Eqs. (3.707). The same is true for the extended version of SA [260]. For this reason, we focus on two derivatives of these theories linearized near the equilibrium: the linearized superposition approximation (LSA) and the linearized extended superposition approximation (LESA). It was found that LSA developed in a number of works [139,175,255,260] is in fact identical to IET (see Table VIII). They both have the same concentration-independent kernel $\tilde{\Sigma}(s)$. As for LESA, it was, strictly speaking, created for the reactions in the ground state [241,242], but can be easily extended to the case of equal lifetimes, $u_A = u_C$.

8. Multiparticle Kernel Approximation

In a series of works [126,132,246,247] a set of approximate solutions for the contact reactions was suggested. These solutions are based on a hierarchical system of diffusional equations for n-particle probabilities. The truncation of this system at the second order has led to the so-called multiparticle kernel 3 (MPK3) approximation [126]; the third order has given MPK2 theory [132,247], but well before the effort was mounted to truncate this system at the fourth order [246]. This earliest attempt, known as MPK1, turned out to be the best for the reversible dissociation/association reaction. It correctly reduces to the limits available for strict investigation:

- To the irreversible case, $k_d = 0$, where it coincides with DET
- To the geminate case, $c = 0$, where it reproduces IET

A comparison with the Brownian dynamics simulations of the reaction (3.703) in the target limit for infinite lifetimes [261] showed an excellent agreement even at the unusually large value of c. Probably, MPK1 was not generalized for the case of different lifetimes because of the complexity of the derivation.

However, MPK1 as well as MPK2 describe only the target problem, while MPK3 is able to treat the trap problem as well (Table IX). At equal lifetimes MPK3 exactly reproduces MET in both these limits [126,252].

9. The Self-Consistent Relaxation Time Approximation (SCRTA)

The SCRTA of Szabo and Gopich [180] considers two pair correlation functions, $p_{AB}(r,t)$ and $p_{CB}(r,t)$, that describe the density of B particles around the unbound and bound traps, respectively. These functions are bounded by the set of equations taking into account the fact that a given couple, A and B, disappear not only because of their own association in C but also because of a similar alliance of A with some other B (a "bachelor" coming from the bulk). These functions become equal on reaching equilibrium.

SCRTA deals with equal lifetimes, but can account for the arbitrary diffusion coefficients of the reactants. It can also afford the usage of the distance-dependent reaction rates, but breaks down near the irreversible limit [180].

10. Concentration Dependence of κ

Since only some of the theories deal with the case when $u_A \neq u_C$, we examine here only the case of equal lifetimes $u_A = u_C = 1/\tau$. The genuine discrimination between the theories can be either made by numerical simulations or achieved by a comparison of theoretical $\kappa(c)$ dependencies (Fig. 3.92) with those obtained experimentally.

According to Eq. (3.715), κ differs from $\tilde{\Sigma}(0)$ especially at large c where it approaches the upper limit:

$$\kappa_m = \frac{k_a \tau}{1 + k_d \tau} \to K_{eq} \qquad \text{at} \qquad \tau \to \infty$$

All theories compared in Figure 3.92 give the same minimal and maximal values of κ, at $c = 0$ and $c = \infty$, respectively. However, they are very different in between. Two of them, LESA and MPK3, essentially deviate from all the rest describing the target problem [Fig. 3.92(a)]. This is not a surprise because we have already detected the same studying the irreversible quenching (Fig. 3.98). Some discrimination between others can be made, judging by the right asymptotic behavior over long times, Eq. (3.713). There are only two in between LESA and MPK3 that are capable of reproducing the target problem: MPK1 and SCRTA. A difference between them is noticeable but greatly reduces with increasing τ. Still MPK1 has the advantage of excellent correspondence with the Brownian dynamic simulations of the whole time behavior of $R(t)$ obtained in Ref. 261. The experimental investigation of the same dependencies is possible in only a limited range of concentrations where the results should be compared with those presented in Figure 3.92(b).

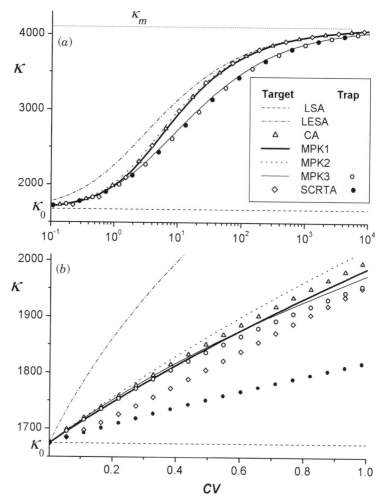

Figure 3.92. The Stern–Volmer constant κ at $u_A = u_C = 1/\tau$ as a function of the dimensionless concentration cv ($v = 4\pi\sigma^3/3$, $\sigma = 7$Å, $k_a/v = 43\,\text{ns}^{-1}$, $k_d = 5\,\text{ns}^{-1}$, $D = 100\,\text{Å}^2/$ ns $= 10^{-5}\,\text{cm}^2/\text{s}$): (a) the entire concentration dependence from the minimal (ideal) value κ_0 up to the maximal one, κ_m; (b) the low-concentration region. (From Ref. 47.)

XIII. BIEXCITONIC REACTIONS

In Section IX we considered the reactions whose rates are nonlinear in the reactant concentrations because of a shortage of one of them. In Section X a similar effect arises as a result of the light exhaustion of the population of reactive states. Here we address reactions that are bilinear in concentration of excited reactants just because the reaction occurs only at their encounters in liquid or solid media. In the latter case these are the excitons that are mobile and interact with each other.

A. Anti-Smoluchowski Transient Kinetics

The delayed fluorescence of an excited aromatic molecule, $^1A^*$, may be sustained by annihilation of the triplet excitations at their encounters in solution:

$$^3A^* + {}^3A^* \overset{W_{TT}}{\Longrightarrow} A + {}^1A^* \to 2A + h\nu \qquad (3.726)$$

In the conventional rate theory as well as in DET, the kinetics of triplet annihilation and subsequent singlet decay is described by a set of differential equations:

$$\dot{T} = -\frac{1}{\tau_T}T - 2\,k(t)T^2(t) \qquad T(0) = T_0 \qquad (3.727a)$$

$$\dot{S} = -\frac{1}{\tau_S}S + k(t)T^2(t) \qquad S(0) = 0 \qquad (3.727b)$$

where $S = [{}^1A^*]$ and $T = [{}^3A^*]$ are concentrations of singlets and triplets. Since the triplet lifetime $\tau_T \gg \tau_S$ and annihilation are also slower than the singlet decay rate $1/\tau_S$, we come to approximate the quasistationary solution of equations (3.727):

$$S \approx \tau_S k(t)T^2, \qquad \text{where} \qquad T \approx T_0 e^{-t/\tau_T} \qquad (3.728)$$

At earlier times, $t \ll \tau_T$, the intensity of the delayed fluorescence is proportional to its rate, k_f, and initial concentration of triplets in square:

$$I(t) = k_f S(t) \approx k_f \tau_S k(t)T_0^2 \qquad (3.729)$$

This transient kinetics is determined solely by the time-dependent rate constant of annihilation $k(t)$.

According to the Smoluchowski theory of diffusion-controlled bimolecular reactions in solutions, this "constant" decreases with time from its kinetic value, k_0 to a stationary (Markovian) value, which is k_D under diffusional control. In the contact approximation it is given by Eq. (3.21), but for remote annihilation with the rate $W_{TT}(r)$ its behavior is qualitatively the same as far as $k(t)$ is defined by Eq. (3.34)

$$k(t) = \int W_{TT}(r)n(r,t)d^3r \qquad (3.730)$$

where, according to Eqs. (3.35) and (3.36)

$$\dot{n} = -W_{TT}(r)n + \hat{L}n \quad \text{and} \quad n(r,0) = 1 \qquad (3.731)$$

This is due to the uniform initial condition $k(t)$, which starts from its maximal value $k_0 = \int W_{TT}d^3r$ and drops with time, approaching the stationary limit from above. Contrary to this famous result, the experimental study of delayed fluorescence of anthracene in viscous solution [262] showed quite the opposite time behavior of $k(t)$. As it is initially much less than k_D, the rate constant increases with time, approaching the long-time asymptote (3.56) from below. The authors of Ref. 262 called this anomaly "the anti-Smoluchowski time behavior of the delayed fluorescence." They properly attributed it to a nonuniform distribution of triplets generated by the intersystem conversion from singlets that are preliminary quenched by the resonant energy transfer.

From now on we should be more specific, taking into account the preparation of the system. In this particular experiment the triplets are produced not by the light excitation in their own (dipole-forbidden) band, but indirectly, by the intersystem conversion, $^1A^* \rightarrow {}^3A^*$, accompanying the singlets' decay after their instantaneous creation. In fact, $1/\tau_S = k_f + k_{isc}$, where k_{isc} is the monomolecular rate of intersystem conversion. Besides these intramolecular processes, there is also biexcitonic self-quenching of singlets, causing one of them to disappear

$$A + {}^1A^* \longleftarrow A + {}^1A^{**} \overset{W_{SS}}{\Longleftarrow} {}^1A^* + {}^1A^* \overset{W_{SS}}{\Longrightarrow} {}^1A^{**} + A \longrightarrow {}^1A^* + A \qquad (3.732)$$

and bimolecular quenching of singlets by triplets:

$$^1A^* + {}^3A^* \overset{W_{ST}}{\Longrightarrow} A + {}^3A^{**} \longrightarrow A + {}^3A^* \qquad (3.733)$$

It was assumed that the quasiresonant transfer of excitation with the rates W_{SS} or W_{ST} is due to the multipolar interaction between the excited molecules and the decay of the doubly excited molecules $^{1,3}A^{**}$ is practically instantaneous [263].

By taking into account one or both of these mechanisms, one can calculate the total number of triplets T_0 accumulated at the end of singlet decay, as well as their pair distribution at that time $g(r,0)$ [262,263]. The latter shows the shortage of the nearest neighbors quenched first, due to the strongest interaction. As a result

$$k(0) = \int W_{TT}(r)g(r,0)d^3r \qquad (3.734)$$

is found to be smaller than k_0 and even smaller than $k_D = \lim_{t \to \infty} k(t)$. Although $k(t)$ later increases it nevertheless remains smaller than k_D for as long as the near-contact dip in triplet–triplet distribution is washed out by the encounter diffusion.

Considering a reasonable physical model, the authors of Refs. 262 and 263 obtained the non-uniform initial pair distribution $g(r,0)$, assuming that the biexcitonic quenching is static and too slow to significantly affect the singlet and triplet evolution during excitation. The acceleration of quenching by diffusion or stronger excitation resulting in deformation of $g(r,0)$ cannot be accounted for within their intuitive rate model, as well as within DET, which also operates with the time-dependent rate constants. The rigorous solution to this problem was obtained later [264] in the framework of the memory function formalism instead of the rate concept. So far it is the most complex and impressive application of IET to a particular problem.

1.   IET of Biexcitonic Annihilation

The integral kinetic equation is bilinear with respect to vectors $\mathbf{N_A}$ and $\mathbf{N_B}$ of two interacting particles, but their components (the concentrations of reactant and product states) are the same

$$\mathbf{N_A} = \begin{pmatrix} S \\ T \\ S_0 \end{pmatrix} = \mathbf{N_B} \tag{3.735}$$

where S and T are the densities of the excited states as before, while S_0 is the density of the ground state. As long as we are interested in the transient effect prevailing at $t \ll \tau_T$, we can set $\tau_T = \infty$ and get the simplest operator of intramolecular relaxation:

$$\hat{\mathbf{Q}}_A = \begin{pmatrix} -k_f - k_{\text{isc}} & 0 & 0 \\ k_{\text{isc}} & 0 & 0 \\ k_f & 0 & 0 \end{pmatrix} = \hat{\mathbf{Q}}_B \tag{3.736}$$

In the collective basis of states, where the straight product $\mathbf{N}_A \times \mathbf{N}_B$ is a vector, the corresponding operator $\hat{\mathbf{Q}}$ is a matrix 9x9. The same is true regarding the reaction operator $\hat{\mathbf{W}}(r)$. These two operators, defined in Ref. 264, determine the kernel (3.104), which enters the integral equation:

$$\dot{\mathbf{N}}_A = Sp_B \int_0^\infty \hat{\mathbf{M}}(\tau) \mathbf{N}_B(t - \tau) \times \mathbf{N}_A(t - \tau) d\tau + \hat{\mathbf{Q}}_A \mathbf{N}_A \tag{3.737}$$

A single equation instead of the pair (3.103) is sufficient because $\mathbf{N_A}$ and $\mathbf{N_B}$ are identical. Written in components, this equation takes the following form:

$$\dot{S} = I(t) - \frac{S}{\tau_S} + \int_0^t M_{SS}(\tau)S^2(t-\tau)d\tau + \int_0^t M_{ST}(\tau)S(t-\tau)T(t-\tau)d\tau$$

$$+ \int_0^t M_{TT}(\tau)T^2(t-\tau)d\tau \qquad\qquad (3.738a)$$

$$\dot{T} = k_{\rm isc}S + \int_0^t R_{SS}(\tau)S^2(t-\tau)d\tau + \int_0^t R_{ST}(\tau)S(t-\tau)T(t-\tau)d\tau$$

$$- 2\int_0^t M_{TT}(\tau)T^2(t-\tau)d\tau \qquad\qquad (3.738b)$$

The pumping term $I(t)$ reproduces the shape of the light pulse, and the terms linear in S describe the natural singlet decay and interconversion, while all the others are bilinear in the exciton concentration and integral in principle.

2. Triplet Phantoms

The IET expressions of all the kernels through the reaction rates and pair distributions as well as the auxiliary equations for all these distributions were derived and presented in Ref. 264. Moreover, Eq. (3.738b) was transformed there to a form having a nontrivial physical interpretation:

$$\dot{T} = k_{\rm isc}S - 2\int W_{TT}(r)m_{TT}(r,t)d^3r - 2\int_0^t M_{TT}(\tau)T^2(t-\tau)d\tau \qquad (3.739)$$

The last term in this equation is just the IET rate of triplet annihilation, similar to its DET analog in Eq. (3.727a). It is defined by the kernel

$$M_{TT}(t) = s\int W_{TT}(r)[\delta(t) + \dot{v}_4(r,t)]d^3r \qquad\qquad (3.740)$$

where

$$\dot{v}_4 = \hat{L}v_4 - 2W_{TT}(r)v_4, \qquad v_4(r,0) = 1 \qquad\qquad (3.741)$$

Another integral term in (3.739) is positive and large at the beginning. It reduces the total rate significantly, setting it to increase at a later time in contrast to the normal Smoluchowski behavior. The quantity m_{TT} is a pair density of triplet "phantoms," which are the triplet pairs that were not born because of preliminary quenching of parent singlet–triplet and singlet–singlet pairs. This density is negative but yields the usual diffusional equation

$$\dot{m}_{TT}(r,t) = \hat{L}m_{TT}(r,t) - 2W_{TT}(r)m_{TT}(r,t) + j_{TT}(r,t) \qquad (3.742)$$

where $m_{TT}(r,0) = 0$ and

$$\tilde{j}_{TT}(r,s) = 2k_{\text{isc}}[(s+k)\tilde{v}_2(r,s) - 1]\widetilde{ST}(s) + 2k_{\text{isc}}\tilde{m}_{ST}(r,s) \qquad (3.743)$$

Here $\widetilde{ST}(s)$ is the Laplace transformation of the product $S(t)T(t)$ and $m_{ST}(t)$ is the density of pairs composed from the singlet and triplet phantoms. This density is also negative and yields the corresponding diffusional equation, which accounts for the singlet–triplet quenching, decay of singlet phantoms, and pumping due to precursor singlet–singlet quenching, $j_{ST}(r,t)$

$$\dot{m}_{ST}(r,t) = \hat{L}m_{ST}(r,t) - W_{ST}(r)m_{ST}(r,t) - \frac{1}{\tau_S}m_{ST}(r,t) + j_{ST}(r,t) \qquad (3.744)$$

where $m_{ST}(r,0) = 0$. The pumping term is given by its Laplace transformation as the previous one

$$\tilde{j}_{ST}(r,s) = k_{\text{isc}}\left[\left(s+\frac{2}{\tau_S}\right)\tilde{v}_1(r,s) - 1\right]\widetilde{SS}(s) \qquad (3.745)$$

where again $\widetilde{SS}(s)$ is the Laplace transformation of $S^2(t)$. The functions $v_1(r,t)$ and $v_2(r,t)$ obey the usual auxiliary equations

$$\dot{v}_1 = \hat{L}v_1 - 2W_{SS}(r)v_1 - \frac{2}{\tau_S}v_1, \qquad \dot{v}_2 = \hat{L}v_2 - W_{ST}(r)v_2 - \frac{1}{\tau_S}v_2 \qquad (3.746)$$

with initial conditions $v_1(r,0) = v_2(r,0) = 1$ and reflecting boundary conditions.

3. Initial Annihilation Kinetics

For times longer than τ_S but shorter than the characteristic time of triplet annihilation, $T \approx T_0$ and the first term in Eq. (3.739) turns to zero, reducing this equation to

$$\dot{T} = -2\int W_{TT}(r)m_{TT}(r,t)d^3r - 2\int_0^t M_{TT}(\tau)d\tau T_0^2 \qquad (3.747)$$

where T_0 is the maximal triplet concentration accumulated during the pulse. Taking into account the IET definition of M_{TT}, one can reduce this equation to [264]

$$\dot{T} = -2k(t)T_0^2 \qquad (3.748)$$

where

$$k(t) = \int W_{TT}(r)g(r,t)d^3r \qquad (3.749)$$

is expressed through

$$g(r,t) = m_{TT}(r,t)/T_0^2 + v_4(r,t) \qquad (3.750)$$

which is the total pair distribution including triplets and their phantoms.

When phantom production is accomplished ($j_{TT} = 0$), Eq. (3.742) for $m_{TT}(r,t)$ becomes identical to the auxiliary equation for $v_4(r,t)$, Eq. (3.741). Therefore the same equation is also valid for $g(r,t)$:

$$\dot{g}(r,t) = \hat{L}g(r,t) - 2W_{TT}(r)g(r,t) \qquad (3.751)$$

although with the nontrivial initial condition $g(r,0)$ that has to be specified.

4. Initial Conditions

This condition was intuitively derived for a rectangular pulse of duration τ assuming that the quenching of singlets is static and there is no triplet annihilation during the pumping [262]

$$g(r,0) = \frac{p}{1 + W_{ST}\tau_S} + \frac{(1-p)}{(1 + W_{ST}\tau_S)(1 + W_{SS}\tau_S)} \qquad (3.752)$$

where

$$1 - p = 2\frac{\tau_S^2}{\tau^2}\left(\frac{\tau}{\tau_S} - 1 + e^{-\tau/\tau_S}\right) \qquad (3.753)$$

For dipole–dipole quenching

$$W_{SS} = \frac{1}{\tau_S}\frac{R_{SS}^6}{r^6} \qquad \text{and} \qquad W_{ST} = \frac{1}{\tau_S}\frac{R_{ST}^6}{r^6}$$

Using these definitions in Eq. (3.752), one obtains

$$g(r,0) \approx \begin{cases} \frac{r^{12}}{(r^6+R_{ST}^6)(r^6+R_{SS}^6)} & \text{after short pulse} & \tau \ll \tau_S \quad (p=0) \\ \frac{r^6}{r^6+R_{ST}^6} = g_0(r) & \text{after long pulse} & \tau \gg \tau_S \quad (p=1) \end{cases} \tag{3.754}$$

The initial distribution for short pulse ($p = 0$) was first derived in Ref. 263, but then generalized for arbitrary p in Refs. 262 and 265. All distributions differ from $n(r,0) = 1$ in that the broad near-contact gap is present in each of them. At long times it is burned by only singlet–triplet quenching because the singlet–singlet quenching is already over.

Initial conditions (3.754) were obtained assuming that the biexcitonic quenching finishes before the triplet annihilation starts. However, during pumping one cannot ignore the triplet annihilation during the acting pulse. This annihilation is carried out by a short-range "exchange interaction" so that its rate is a sharp exponential function of the distance:

$$W_{TT}(r) = W_c e^{-2(r-\sigma)/L} = W_e e^{-2r/L} \tag{3.755}$$

The MET employed in Ref. 264 allows us to find the distribution of triplets at the end of a long pulse, which accounts for their static annihilation:

$$g(r,0) = g_0(r) \frac{[2W_{TT}\tau - 1 + \exp(-2W_{TT}\tau)]}{2W_{TT}^2\tau^2} \tag{3.756}$$

where

$$g_0(r) = \frac{1}{1 + (R_{ST}/r)^6}$$

was defined in Eq. (3.754).

The integrands from (3.734), $W_{TT}(r)g(r,0)$, are always smaller than their DET analog, $W_{TT}(r)n(r,0) = W_{TT}(r)$, shown by the upper straight line in Figure 3.93(a). In particular, $W_{TT}(r)g_0(r)$ is smaller than $W_{TT}(r)$ since the preliminary singlet quenching pushes curve A down, burning the broad gap in $g_0(r)$. With $g(r,0)$ from (3.756), the integrands are even smaller because of the additional short-range gap produced by the triplet annihilation during illumination. Comparison of lines B and C in Figure 3.93(b) shows that this gap is deeper when the acting pulse is longer.

5. Transient Kinetics

Equations (3.768) and (3.769) are in fact the same as in DET, as well as Eq. (3.751), which should be solved with initial condition (3.752). This was done

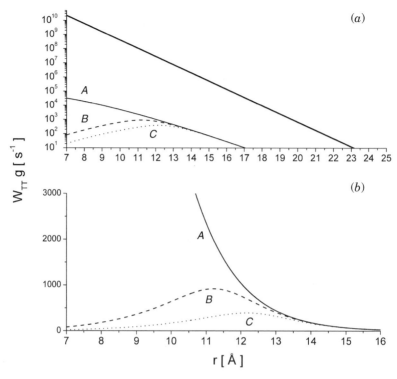

Figure 3.93. The space dependence of integrands $W_{TT}(r)g(r,0)$ for $g(r,0) = 1$ [thick line in plot (*a*)], for $g(r,0) = g_0(r)$ (thin line, *A*), and for the finite pulse duration: $\tau = 35$ ns (dashed line, *B*), and $\tau = 140$ ns (dotted line, (*C*); $W_e = 2.56 \times 10^{14}$ s^{-1}, $L = 1.5$ Å, $R_{ST} = 66$ Å, $\tau_S = 7$ ns.

in Ref. 262 for a long pulse setting $g(r,0) = g_0(r)$ and taking $W_{TT}(r)$ from (3.755). With this solution used in Eq. (3.769), the authors obtained $k(t)$, which appears not only in Eq. (3.768) but also in the equation for singlets generated by triplet annihilation:

$$\dot{S} = -\frac{S}{\tau_S} + k(t)T_0^2 \tag{3.757}$$

Since the lifetime of singlets is very short, it has a quasi-stationary solution

$$S(t) = k(t)\tau_S T_0^2 \tag{3.758}$$

which determines the intensity of the delayed fluorescence at the very beginning of the triplet decay:

$$I_{DF}(t) = k_f S(t) = k(t)k_f \tau_S T_0^2, \qquad \text{at} \qquad t \ll \tau_T \tag{3.759}$$

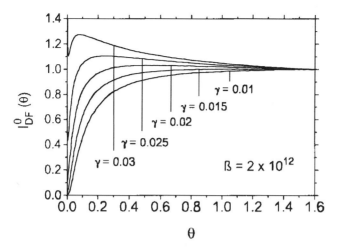

Figure 3.94. The normalized theoretical time dependence of the intensity of the delayed fluorescence for fixed $\beta = W_e R_{ST}^2/D$ and variable $\gamma = L/2R_{ST}$. (From Ref. 262.)

The transient effect is accomplished within this interval when $k(t)$ reaches its stationary value $k = \lim_{t\to\infty} k(t)$, which is approximately k_D. Relating the fluorescence intensity to its asymptotic value, Nickel et al. [262] introduced the normalized value which serves as a characteristic of the transient effect:

$$I_{DF}^0(t) = \frac{I_{DF}(t)}{I_{DF}(\infty)} = \frac{k(t)}{k} \qquad (3.760)$$

In Figure 3.94 they plotted this value as a function of dimensionless time $\theta = Dt/R_{ST}^2$. Keeping $\beta = W_e R_{ST}^2/D$ constant, they changed $\gamma = L/2R_{ST}$, that is, L. If L were larger than the size of the initial gap R_{ST}, the latter would not significantly affect the classical Smoluchowski behavior of $k(t)/k$ approaching 1 from above. But in reality $L \ll R_{ST}$, and such behavior is seen only at the end of the process and for the largest $\gamma = 0.03$. At earlier times the sharp increase in $I_{DF}^0(\theta)$ indicates the anti-Smoluchowski time behavior of $k(t)/k$, which is more pronounced when the tunneling parameter shorter is ($L \sim 1.5\,\text{Å}$) regarding the characteristic dimension of the multipolar interaction, measured by $R_{ST} \sim 70\,\text{Å}$.

The dramatic experimental findings of Nickel et al. strengthened their theory. The time-resolved measurements of the delayed fluorescence of anthracene in a 1:1 mixture of methylcyclohexane (MCH) and methylcyclopentane (MCP) unambiguously indicated that the initial rise in $I_{DF}^0(t)$ can be attributed solely to the expected anti-Smoluchowski behavior of $k(t)$ (Fig. 3.95). The reasonable fit of these data with $\gamma = 0.125$ allowed estimation of the value of $L = 1.58\,\text{Å}$,

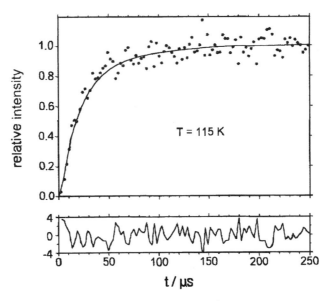

Figure 3.95. The normalized initial time dependence I^0_{DF} of the delayed fluorescence from anthracene $(3 \times 10^{-5} \, \text{M})$ in MCH/MCP at 115 K (●). The theoretical curve (smooth line) was calculated with $\beta = 2 \times 10^{10}, \gamma = 0.0125, D/R^2_{ST} = 6.4 \times 10^3 \, \text{s}^{-1}$. The weighted residuals are shown below. (From Ref. 262.)

provided that $R_{ST} = 63.3 \, \text{Å}$ is in good agreement with the independent spectroscopic data. Although L is slightly dependent on the temperature of measurement, it is always in the framework of the expected values for this parameter, which is analogous to the tunneling length studied above.

Nonetheless, the true picture is more complex. There is a sharp decrease in $k(t)$ at the shortest times noticed theoretically in Ref. 262. It precedes the anti-Smoluchowski stage, which starts just after $k(t)$ passes through the minimum. This pseudonormal drop in $k(t)$ was ascribed by the authors to the static triplet annihilation known from studies of solid solution [263]. However, the static and subsequent nonstationary annihilation of triplets started not after the pulse but simultaneously with switching on the light. At the end of the pulse the annihilation has already burned a short-range dip, seen in profiles B and C in Figure 3.93. Immediately after the pulse the triplet annihilation deepens this gap, forcing $k(t)$ to decrease further on. With the original initial distribution (3.756) obtained in Ref. 264, this effect was shown to be weaker than in the Nickel theory and soon gave way to the anti-Smoluchowski acceleration of the triplet annihilation (Fig. 3.96). It was stressed in Ref. 262 that the "initial fast decay of I^0_{DF} would be observable, but in practice it would be extremely weak"

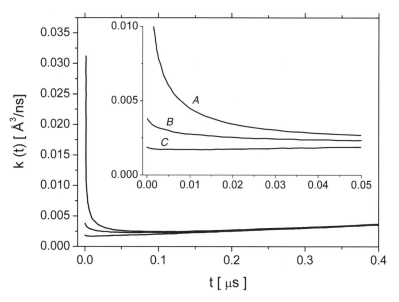

Figure 3.96. The rate constant immediately after excitation without (*A*) and with (*B,C*) the correction for triplet annihilation during pumping. The pulse duration is $\tau = 35$ ns (*B*) or $\tau = 140$ ns (*C*), the encounter diffusion coefficient $D = 2.56 \times 10^{-9}$ cm^2/s; the remaining parameters are the same as in Figure 3.93. (From Ref. 264.)

and its experimental detection "would be impossible." This is the more so in view of the hindering of this effect with τ shown in Figure 3.96. The problem is aggravated by the encounter diffusion of excitation, which facilitates all reactions between them during the pulse. So far it was ignored for simplicity, but MET allows us to account for it, at least numerically.

B. Self-Quenching and Photoconductivity

Apart from the excitonic self-quenching (3.732) there is also a channel of exciton annihilation resulting in charge separation. It was established long ago that in solid anthracene this is the only reaction channel producing the free carriers responsible for the crystal conductivity [266–268]. This process is represented by the following scheme of biexcitonic ionization followed by charge recombination:

$$2A \leftarrow \begin{array}{c} A^- + A^+ \\ A \cdots A \end{array} \begin{array}{c} \nwarrow \\ \searrow \end{array} [A^- \cdots A^+] \overset{W_I}{\Longleftarrow} [A^* + A^*] \overset{W_I}{\Longrightarrow} [A^+ \cdots A^-] \begin{array}{c} \nearrow A^+ + A^- \to 2A \\ \searrow A \cdots A \end{array}$$

This scheme is generally accepted and verified in a number of crystals (see, e.g., the book by Pope and Swenberg [269] and references cited therein). The same mechanism is responsible for the ionic photoconductivity of liquid solutions of porphyrins and metalloporphyrins [270,271].

The photoinduced ionization of benzophenon in acetonitrile has also been reported to proceed via triplet–triplet annihilation at very low laser pulse intensities [272]. The biexcitonic ionization in this system has been studied by applying the transient photoconductivity technique and described with the conventional (Markovian) rate equations, with the time-independent rate constants [273]. Such equations can be represented as follows

$$\dot{N}^* = -2k_i[N^*]^2 - \frac{N^*}{\tau_D} + I(t)N, \qquad N^*(0) = 0 \qquad (3.761a)$$

$$\dot{N}^\pm = k_+[N^*]^2 - k_r[N^\pm]^2, \qquad N^\pm(0) = 0 \qquad (3.761b)$$

where $I(t)$ is the rate of exciton generation by either δ or ζ pulses (3.415). As to the rate constants of ionization, free-carrier production and recombination (k_i, k_+, and k_r), they usually remain unknown fitting parameters.

It is likely that there is only one exception to this rule given in the earliest DET [274]. There the non-Markovian kinetic equation for nonlinear (biexciton) annihilation after δ-pulse excitation was derived from first principles:

$$\dot{N}^* = -2k_I(t)[N^*]^2 - \frac{N^*}{\tau_D} \qquad N^*(0) = N_0 \qquad (3.762)$$

The initial condition for N^* is prepared by instantaneous excitation, after which the annihilation "rate constant" $k_I(t)$ decreases with time, approaching its stationary (Markovian) value k_i as $t \to \infty$. The non-Markovian generalization of another equation, (3.761), became possible only in the framework of the unified theory, where it takes the integral form. Unfortunately, the system response to the light pulses of finite duration or permanent illumination remains a problem for either UT or DET. The convolution recipes such as (3.5) or (3.437) are inapplicable to annihilation, which is bilinear in N^*. Therefore we will start from IET, which is solely capable of consistent consideration of stationary absorbtion and conductivity [199]. Then we will turn to UT and the Markovian theories applied to the relaxation of the instantaneously excited system described in Ref. 275.

1. Integral Theory of the Phenomenon

According to the general theory of Section III, one should first introduce the vector of state populations analogous to (3.735):

$$\mathbf{N_A} = \begin{pmatrix} N^* \\ N^+ \\ N^- \end{pmatrix} = \mathbf{N_B} \qquad (3.763)$$

Its time evolution after instantaneous photoexcitation satisfies the integral kinetic equation (3.737) with the initial condition

$$\mathbf{N_A}(0) = \begin{pmatrix} N_0 \\ 0 \\ 0 \end{pmatrix} = \mathbf{N_B}(0)$$

In the collective state basis a direct product of the population vectors is

$$\mathbf{N}_A \times \mathbf{N}_B = \begin{pmatrix} N^*N^* \\ N^+N^- \\ N^-N^+ \end{pmatrix} \tag{3.764}$$

and the relaxation matrices are

$$\hat{\mathbf{Q}} = \begin{pmatrix} -2/\tau_D & 0 & 0 \\ 0 & 0 & 0 \\ 0 & 0 & 0 \end{pmatrix} \quad \text{and} \quad \hat{\mathbf{W}}(r) = \begin{pmatrix} -2W_I(r) & 0 & 0 \\ W_I(r) & -W_R(r) & 0 \\ W_I(r) & 0 & -W_R(r) \end{pmatrix} \tag{3.765}$$

where $W_I(r)$ and $W_R(r)$ are the rates of the forward and backward electron transfers.

The kernel of the integral equation (3.737)

$$\hat{\mathbf{M}}(t) = \int \hat{\mathbf{W}}(r) \left[\delta(t) + \frac{d}{dt}\hat{\mathbf{F}}(r,t) - \hat{\mathbf{F}}(r,t)\hat{\mathbf{Q}} \right] d^3r \tag{3.766}$$

is defined via a matrix of pair correlation functions of reactants and products that obeys the auxiliary equation

$$\frac{\partial \hat{\mathbf{F}}}{\partial t} = \hat{\mathbf{L}}\hat{\mathbf{F}} + \hat{\mathbf{W}}\hat{\mathbf{F}} + \hat{\mathbf{Q}}\hat{\mathbf{F}}, \qquad \hat{\mathbf{F}}(\mathbf{r},0) = \hat{\mathbf{I}} \tag{3.767}$$

where

$$\hat{\mathbf{L}} = \begin{pmatrix} \hat{L}_I & 0 & 0 \\ 0 & \hat{L}_R & 0 \\ 0 & 0 & \hat{L}_R \end{pmatrix} \quad \text{and} \quad \hat{\mathbf{F}} = \begin{pmatrix} v_1 & 0 & 0 \\ \mu & v_2 & 0 \\ \mu & 0 & v_2 \end{pmatrix} \tag{3.768}$$

Here $\hat{L}_I$ and $\hat{L}_R$ are the operators of encounter diffusion of excitons and counterions while v_i and μ obey the auxiliary diffusional equations with reflecting boundary conditions and the following initial ones:

$$v_1(r,0) = v_2(r,0) = 1, \qquad \mu(r,0) = 0$$

Incorporating the matrix integral equation into the components of the vector (3.763), we get the following set of equations:

$$\dot{N}^* = -\frac{N^*}{\tau_D} - \int_0^t M(\tau)[N^*(t-\tau)]^2 d\tau \tag{3.769a}$$

$$\dot{N}^{\pm} = \int_0^t R^*(\tau)[N^*(t-\tau)]^2 d\tau - \int_0^t R(\tau)[N^{\pm}(t-\tau)]^2 d\tau \tag{3.769b}$$

where

$$M(t) = 2 \int d^3 r W_I(r) \left[\delta(t) + \dot{v}_1 + \frac{2v_1}{\tau_D} \right] \tag{3.770a}$$

$$R^*(t) = \int d^3 r \left[W_I(r) \left(\delta(t) + \dot{v}_1 + \frac{2v_1}{\tau_D} \right) - W_R(r) \left(\dot{\mu} + \frac{2\mu}{\tau_D} \right) \right] \tag{3.770b}$$

$$R(t) = \int d^3 r W_r [\delta(t) + \dot{v}_2]. \tag{3.770c}$$

The first kernel represents the biexciton ionization, the second one controls the geminate recombination of charged products, and the last one governs the bimolecular recombination of separated ions.

Let us rewrite the equation for $N^{\pm}$ in the form

$$\dot{N}^{\pm} = \frac{d}{dt} \int d^3 r \, p(r,t) - \int_0^t R(\tau)[N^{\pm}(t-\tau)]^2 d\tau \tag{3.771}$$

where the pair density $p(r,t)$ obeys the following equation:

$$\dot{p} = \hat{L}_{RP} p - W_R(r)p + J(r,t) \tag{3.772}$$

This equation follows from a comparison of Eqs. (3.769b) and (3.771), taking into account Eq. (3.770b). It contains the pumping term introduced in Ref. 199:

$$J(r,t) = W_I(r) \int_0^t d\tau \left\{ \delta(\tau) + \dot{v}_1(r,\tau) + \frac{2v_1}{\tau_D} \right\} [N^*(t-\tau)]^2 \tag{3.773}$$

where

$$\dot{v}_1 = \hat{L}_I v_1 - 2W_I(r)v_1 - 2v_1/\tau_D , \qquad v(r,0) = 1 \tag{3.774}$$

Integration of the pair density $p(r,t)$ over r gives the total concentration of geminate ion pairs that have survived by time t after δ excitation:

$$P(t) = \int p(r,t)d^3r \tag{3.775}$$

In the absence of bulk recombination, $N^\pm \equiv P$ remains finite as $t \to \infty$. Coming to the limit $N_0 \to 0$, one can exclude the bulk recombination and define the free-ion quantum yield as the ratio of ions that escaped geminate recombination $2P(\infty)$ to the total initial concentration of excitations generated by pulse N_0:

$$\phi = \frac{2P(\infty)}{N_0} \tag{3.776}$$

The factor 2 accounts for the fact that one-half of the ionized excitations transforms to positive ions and the other half to negative ions.

2. Stationary Fluorescence and Free-Ion Production

An important advantage of the integral theory is its capacity to account for arbitrary light excitation by additive inclusion of the pumping term into the IET kinetic equations (3.769):

$$\dot{N}^* = I(t)N - \frac{N^*}{\tau_D} - \int_0^t M(\tau)[N^*(t-\tau)]^2 d\tau \tag{3.777a}$$

$$\dot{N}^\pm = \dot{P} - \int_0^t R(\tau)[N^\pm(t-\tau)]^2 d\tau \tag{3.777b}$$

where $P(t)$ was defined in Eq. (3.775). If one is interested in the stationary regime only, when $I(t) = I_0$, then setting $\dot{N}^* = \dot{N}^\pm = 0$, the set (3.777) is reduced to the algebraic equations

$$I_0 N = \frac{N_s^*}{\tau_D} + 2\kappa_0[N_s^*]^2 \tag{3.778a}$$

$$\dot{P}(\infty) = k_r[N_s^\pm]^2 \tag{3.778b}$$

where N_s^* and $N_s^\pm$ are stationary concentrations of excitations and free carriers and

$$\kappa_0 = \frac{1}{2}\int_0^\infty M(\tau)d\tau \quad \text{and} \quad k_r = \int_0^\infty R(\tau)d\tau \tag{3.779}$$

Using here Eq. (3.770), we obtain the Stern–Volmer constant

$$\kappa_0 = \frac{2}{\tau_D} \int W_I(r)\tilde{n}\left(r, \frac{2}{\tau_D}\right)d^3r \tag{3.780}$$

where $n(r,t) = \nu_1 \exp(2t/\tau_D)$ is the distribution of the excitation that obeys the conventional equation of differential encounter theory (3.35) with $2W_I$ substituted for W.

The generation rate of free carriers may be represented as follows [264]:

$$\dot{P}(\infty) = \int J(r')\varphi(r')d^3r' \tag{3.781}$$

Here $\varphi(r') = \int G_R(r,r',\infty)d^3r$ is the quantum yield of charge separation from the initial distance r'. It is expressed through the Green function $G_R(r,r',t)$ of the homogeneous equation (3.772) and integrated with the flux (3.773), which stabilizes in the long run, reaching the value

$$J(r) = \lim_{t\to\infty} J(r,t) = \frac{2W_I(r)}{\tau_D}\tilde{n}\left(r, \frac{2}{\tau_D}\right)[N_s^*]^2 \tag{3.782}$$

where $\tilde{n}(r,t)$ is the same as in (3.780).

From Eqs. (3.781) and (3.782) we finally obtain the generation rate of free charges:

$$\dot{P}(\infty) = \kappa_0\bar{\varphi}[N_s^*]^2 \tag{3.783}$$

Here the quantum yield of free carriers

$$\bar{\varphi} = \int f_0(r)\varphi(r)d^3r \tag{3.784}$$

is averaged over the normalized initial distribution of charged products of biexcitonic ionization:

$$f_0(r) = \frac{J(r)}{\int J(r)d^3r} \tag{3.785}$$

Using Eq. (3.783) in the set of equations (3.778), one can easily solve them, obtaining the stationary concentration of excitations and free carriers:

$$N_s^* = \frac{\sqrt{1 + 8\kappa_0\tau_D^2 I_0 N} - 1}{4\kappa_0\tau_D} \quad \text{and} \quad N_s^{\pm} = \sqrt{\frac{\kappa_0\bar{\varphi}}{k_r}}N_s^* \tag{3.786}$$

The product of the stationary concentration of carriers $N_s^{\pm}$ and their mobility u determines the stationary conductivity of the sample $\sigma = euN_s^{\pm}$. It is proportional to the exciton concentration $N_s^*(I_0)$, which is in general the nonlinear function of light. However, at moderate light pumping, when $8\kappa_0\tau_D^2 I_0 N \ll 1$, we obtain

$$N_s^* = I_0 \tau_D N \qquad \text{and} \qquad \sigma \sim N_s^{\pm} \sim N_s^* \sim I_0 \qquad (3.787)$$

The experimentally detected linearity of $\sigma(I_0)$ dependence was imperatively attributed in the pioneering works to the biexcitonic origin of the photocurrent [266,267].

The relative quantum yield of self-quenched fluorescence can reasonably be defined as the ratio of high concentration of excitons at strong pumping to their concentration at the weakest pumping:

$$\eta = \frac{N_s^*}{N_s^*|_{I_0 \to 0}} = \frac{N_s^*}{I_0 N \tau_D} = \frac{\sqrt{1 + 8\kappa_0\tau_D^2 I_0 N} - 1}{4\kappa_0\tau_D^2 I_0 N} \qquad (3.788)$$

At weak pumping this equation acquires the form of the Stern–Volmer law (3.12):

$$\eta \approx 1 - 2I_0 N \kappa_0 \tau_D^2 = 1 - 2N_s^* \kappa_0 \tau_D \qquad \text{at} \qquad I_0 \to 0 \qquad (3.789)$$

where κ_0 is the corresponding self-quenching constant. Since nonstationary quenching precedes the stationary stage, this Stern–Volmer constant defined in Eq. (3.779) significantly exceeds the Markovian self-quenching constant (3.14):

$$k_i = \lim_{\tau_D \to \infty} \kappa_0 = \int W_I(r) n_s(r) d^3 r \qquad (3.790)$$

where $n_s(r) = \lim_{t \to \infty} n(r,t) = \lim_{s \to 0} s\tilde{n}(r,s)$. They are identical only at $\tau_D = \infty$ (Fig. 3.97). Note that both these constants are 2 times larger than their analogs for impurity quenching, which results in the quenching of only one excitation per encounter.

Since for every two excitations only one ion of a given sign is produced, the rate constant of their generation in Eq. (3.783) is only $\kappa_0\bar{\varphi}$. It is even less than κ_0, because the fraction of separated ions that escape geminate recombination and contribute to the stationary photoconductivity, $\bar{\varphi} \leq 1$. This quantity strongly depends on the distribution of charged products of biexciton ionization, $f_0(r)$. At short τ_D they are generated in the course of the initial nonstationary ionization, mainly at the contact distance, where recombination is the fastest and the charge separation yield $\varphi(r \approx \sigma)$ is the lowest. When τ_D increases, the

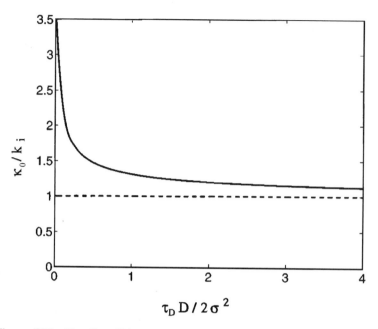

Figure 3.97. The Stern–Volmer constant of biexcitonic quenching at $\tau_D = 0.5\,\text{ns}$, $\tilde{D} = 10^{-5}\,\text{cm}^2\text{s}^{-1}$, $\sigma = 5\,\text{Å}$, $r_c = 10\,\text{Å}$, $N_0 = 0.2\,\text{M}$, and the exponential rates of the forward and backward electron transfer with $W_I(\sigma) = 500\,\text{ns}^{-1}$, $W_R(\sigma) = 500\,\text{ns}^{-1}$ and $L = 1\,\text{Å}$. The dashed line shows the Markovian result, $k_i = 4.84\,\text{M/ns}$. (From Ref. 199.)

distribution becomes broader and shifts from the contact, if diffusion is slow. As a result, the free-carrier quantum yield (3.784) should increase with τ_D, approaching the upper limit at $\tau_D \to \infty$ (Fig. 3.98). However, one should keep in mind that the contribution from the nonstationary quenching into the ion distribution and hence in $\bar{\varphi}$ is different in IET and UT. Since the latter is preferable at high concentrations (Section XII.B), we should now turn our attention to the unified theory.

3.   Unified Theory of the Phenomenon

Following the regular procedure described in Section IV.A and employed in Section VI.B, the IET equations (3.777) were transformed to their UT analogs, valid when $I = 0$ [275]:

$$\dot{N}^* = -2k_I(t)[N^*]^2 - \frac{N^*}{\tau_D} \tag{3.791a}$$

$$\dot{N}^{\pm} = \int \dot{p}(r,t)d^3r - k_R(t)[N^{\pm}]^2 \tag{3.791b}$$

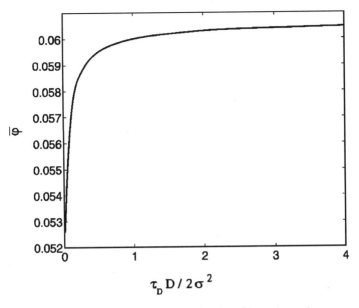

Figure 3.98.   The free-carrier quantum yield as a function of decay time at the same parameters as in Figure 3.97. (From Ref. 199.)

Here

$$k_I(t) = \int d^3 r W_I(r) n(r,t), \qquad k_R(t) = \int d^3 r W_R(r) v_2(r,t) \qquad (3.792)$$

where n and v_2 obey the auxiliary equations similar to (3.279):

$$\dot{n} = \hat{L}_I n - 2 W_I(r) n, \qquad \dot{v}_2 = \hat{L}_R v_2 - W_R(r) v_2, \qquad n(r,0) = v_2(r,0) = 1 \qquad (3.793)$$

If Eq. (3.773) is also subjected to the same transformation, the pumping rate $J(r,t)$ takes the following form [275]:

$$J(r,t) = W_I(r) n(r,t) [N^*(t)]^2 \qquad (3.794)$$

This is the biexcitonic analog of what we obtained for impurity quenching in Eq. (3.288). It determines the initial distribution of ions that can be obtained from Eq. (3.772), setting $\hat{L}_R = W_R = 0$:

$$p_0(r) = p(r,\infty)|_{W_R = \tilde{D} = 0} = \int_0^\infty J(r,t) dt = \tilde{J}(r,0) \qquad (3.795)$$

According to Eqs. (3.775) and (3.776), the free-ion quantum yield can be represented as

$$\phi = \frac{2}{N_0} \int p(r, \infty) d^3 r = \frac{2}{N_0} \int \lim_{s \to 0} s\tilde{p}(r, s) d^3 r \qquad (3.796)$$

where

$$\tilde{p}(r, s) = \int d^3 r_0 \tilde{G}_R(r, r_0, s) \tilde{J}(r_0, s) \qquad (3.797)$$

is expressed through the Green function of Eq. (3.772), $G_R(r, r_0, t)$. The survival probability of the RIP initially separated by distance r_0 is also expressed via the same Green function:

$$\varphi(r_0) = \int d^3 r \, G_R(r, r_0, t) = \lim_{s \to 0} s \int d^3 r \, \tilde{G}_R(r, r_0, s) \qquad (3.798)$$

From Eqs. (3.796)–(3.798) we obtain the total yield of free ions

$$\phi = \int d^3 r_0 \varphi(r_0) g(r_0) \qquad (3.799)$$

where

$$g(r_0) = \frac{2 p_0(r_0)}{N_0} = \frac{2}{N_0} \int_0^\infty J(r, t) dt \qquad (3.800)$$

is the distribution of RIPs over the initial separations. From Eqs. (3.795), (3.792), and (3.791a), we see that

$$\int p_0(r) d^3 r = \frac{1}{2} \left[N_0 - \frac{\tilde{N}(0)}{\tau_D} \right] = \frac{N_0}{2}(1 - \eta)$$

where $\eta = \tilde{N}(0)/N_0 \tau_D$ is the relative fluorescence quantum yield. Hence, the total ionization quantum yield ψ is given by the following expression:

$$\psi = 1 - \eta = \int \frac{2 p_0(r)}{N_0} d^3 r = \int g(r) d^3 r \qquad (3.801)$$

Using ψ, we can introduce the normalized initial distribution of RIPs

$$f_0(r) = \frac{g(r)}{\psi} = \frac{\tilde{p}_0(r)}{\int \tilde{p}_0(r) \, d^3 r} = \frac{\tilde{J}(r, 0)}{\int \tilde{J}(r, 0) \, d^3 r} \qquad (3.802)$$

where the last equality was obtained taking into account Eq. (3.795).

With this distribution, the free-carrier quantum yield (3.799) can be brought to the same form it had in the theory of impurity quenching

$$\phi = \psi\bar{\varphi} \qquad (3.803)$$

where the charge separation quantum yield

$$\bar{\varphi} = \int \varphi(r)f_0(r)\, d^3r \qquad (3.804)$$

Both these yields depend on the shape of the initial RIP distribution $f_0(r)$ defined in Eq. (3.802). Taking into account that $J(r) = \lim_{s\to 0} s\tilde{J}(s)$, it is easy to see that the IET [Eq. (3.802)] and the UT [Eq. (3.785)] are in fact identical. Nonetheless, the initial distributions in these theories are different because the pumping rates (3.773) and (3.794) are not the same. The latter is preferable at high concentrations.

4.   Markovian Theory

In the long run all the rate constants in Eqs. (3.791) approach their stationary values and UT reduces to the Markovian theory. As for ionization and recombination, the corresponding rate constants are evident

$$k_i = \lim_{t\to\infty} k_I(t) = \int W_I(r)n_s(r)d^3r, \qquad k_r = k_R(\infty) = \int d^3r W_R(r)v_s(r)$$

$$(3.805)$$

where k_i is the same as in Eq. (3.790) and $v_s(r) = v_2(r,\infty)$. As for the source term in Eq. (3.791), its Laplace transformation can be found using Eq. (3.797):

$$\int \tilde{p}(r,s)\, d^3r = \int s\tilde{p}(r,s)d^3r = \int d^3r_0 \left[[s\tilde{G}_R(r,r_0,s)]d^3r\, \tilde{J}(r_0,s) \right.$$

As $s \to 0$, Eq. (3.798) might be used to replace a term in the square brackets by the survival probability of the pair $\varphi(r_0)$:

$$\int_0^\infty e^{-st}dt \int \dot{p}(r,t)\, d^3r = \int d^3r_0\varphi(r_0) \int_0^\infty e^{-st}dtJ(r_0,t) \qquad s\tau_d \ll 1$$

Therefore, in the time domain we have

$$\int \dot{p}(r,t)d^3r = \int d^3r_0\varphi(r_0)J(r_0,t) = \int d^3r\varphi(r)W_I(r)n_s(r)[N^*(t)]^2, \qquad t \gg \tau_d$$

$$(3.806)$$

In the last estimate we used Eq. (3.794) and replaced the pair density $n(r, t)$ by its stationary value $n_s(r) = \lim_{t \to \infty} n(r, t)$.

Substituting Eqs. (3.805) and (3.806) into the UT equations (3.791), we get their Markovian analog valid at $t \gg \tau_d$:

$$\dot{N}^* = -2k_i[N^*]^2 - N^*/\tau_D \tag{3.807a}$$

$$\dot{N}^{\pm} = k_i \bar{\varphi}_m[N^*]^2 - k_r[N^{\pm}]^2 \tag{3.807b}$$

Here $\bar{\varphi}_m$ is the Markovian charge separation yield

$$\bar{\varphi}_m = \int \varphi(r_0) f_m(r_0) d^3 r_0 \tag{3.808}$$

which is averaged over the normalized Markovian distribution of annihilation products:

$$f_m(r_0) = \frac{W_I(r_0) n_s(r_0)}{\int W_I(r_0) n_s(r_0) d^3 r_0} \tag{3.809}$$

This is exactly the same distribution that results from the IET analog (3.312) in the limit $\tau_D \to \infty$. In this limit the general UT/DET distribution (3.802) also reduces to the Markovian one provided the pumping is low enough. Under this condition one can set $N^*(t) \approx N_0$ at $N_0 k_i t \ll 1$ and obtain the following from Eq. (3.794): $\tilde{J}(r, s) \approx W_I(r) \tilde{n}(r, s) N_0^2$. Inserting this result into Eq. (3.802), we have as $N_0 k_i \tau_d \ll 1$ and $\tau_D = \infty$:

$$f_0(r) = \lim_{s \to 0} \frac{\tilde{J}(r, s)}{\int \tilde{J}(r, s) d^3 r} \approx \lim_{s \to 0} \frac{W_I(r, s) s \tilde{n}(r, s)}{\int W_I(r, s) s \tilde{n}(r, s) d^3 r} = f_m(r) \tag{3.810}$$

Hence the Markovian distribution is common for all non-Markovian ones but only in the limit $\tau_D \to \infty$. This distribution completely ignores the nonstationary annihilation and therefore does not depend on the exciton concentration and lifetime. The difference between $f_0(r)$ and $f_m(r)$ becomes more pronounced when τ_D is reduced (Fig. 3.99). Under diffusion control both of them have a well-pronounced maximum near the effective ionization radius. However, the near-contact contribution of the nonstationary annihilation increases with shortening τ_D, on account of the main maximum. Finally (as $\tau_D \to 0$), the UT distribution tends to become exponential, as $W_I(r)$, while the Markovian one remains unchanged.

Using the same input data, we can now compare the kinetics of the charge accumulation/recombination obtained with the non-Markovian UT equations

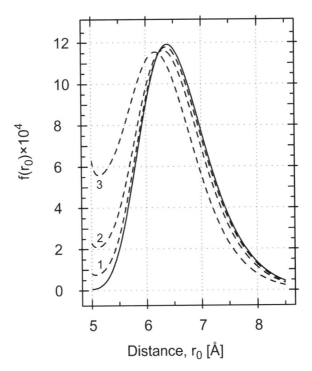

Figure 3.99. The normalized distribution of ions produced by the biexcitonic ionization, according to Markovian theory (thick line) and non-Markovian UT distributions at different excitation lifetimes (dashed lines): (1) $\tau_D = 10^6$ ns, (2) $\tau_D = 100$ ns, (3) $\tau_D = 10$ ns. (From Ref. 275.)

(3.791) and their Markovian analogs (3.807). In Figure 3.100(a) this comparison is made for the "fast ionization" interrupted by the rapid decay of the excitations. The initial charge accumulation is performed by static annihilation seen as a shoulder on the ascending branch of the UT curve. This stage is followed by diffusional ionization interrupted by excitation decay. After the charge accumulation is accomplished, the geminate recombination results in the partial decrease in charge concentration, which levels off at $N^{\pm} = \phi$, provided there is no bulk recombination (dashed line). Otherwise ions fade away up to 0 with a relatively slow bulk recombination rate. This is the typical behavior similar to that observed for impurity quenching in Section VIII.C (Fig. 3.29).

In contrast, the Markovian theory leads to a much simpler picture—it cuts off the maximum related to the geminate process and underestimates essentially the charge separation quantum yield. This is a consequence of the full ignorance of

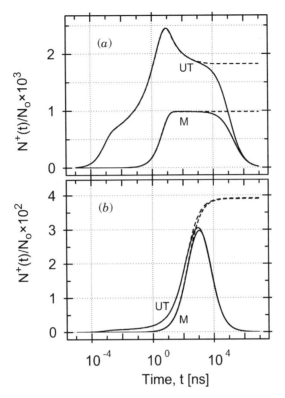

Figure 3.100. The kinetics of charge accumulation without bulk recombination (dashed lines) and taking into account the bulk recombination (solid lines) calculated with the unified theory (UT) and its Markovian version (M). (a) fast ionization ($\tau_D = 10$ ns); (b) slow ionization ($\tau_D = 10^6$ ns) at $N_0 = 0.01$ M, $D = \tilde{D} = 10^{-6}$ cm^2/s, $r_c = 0$. (From Ref. 275.)

the initial nonstationary events, static and diffusional. However, at much longer τ_D (the slow ionization limit), these events are less important; the accumulation of charges becomes monotonous and the results of both theories are semiquantitatively the same [Fig. 3.100(b)]. Hence, for the long-living triplet excitations, the Markovian theory is almost as good as general UT.

The same is true regarding stationary luminescence and conductivity. Unlike UT, the Markovian theory allows additive inclusion of the light-pumping term in its equations (3.761). From their comparison with microscopically derived Eqs. (3.807), we see that

$$k_+ = k_i \bar{\varphi}_m \tag{3.811}$$

With this notification we can find the following from the stationary solution of Eqs. (3.761):

$$N_s^* = \frac{\sqrt{1 + 8k_i\tau_D^2 I_0 N} - 1}{4k_i\tau_D} \quad \text{and} \quad N_s^{\pm} = \sqrt{\frac{k_i\bar{\varphi}_m}{k_r}}N^* \qquad (3.812)$$

These are exactly the same expressions as obtained with IET in Eq. (3.786) but with substitution of k_i for κ_0 and $\bar{\varphi}$ for $\bar{\varphi}_m$. In view of Eqs. (3.790) and (3.312), the Markovian results are valid only at $\tau_D = \infty$. To account for the excitons decay, one should address the non-Markovian IET and UT.

However, it is important to bear in mind that any encounter theory implies that the reactants execute a diffusional motion localized at a definite coordinate all the time. For excitons in crystals this can be true if they are incoherent. Coherent excitons are actually waves with the given wavevectors, whose annihilation rate was calculated in Ref. 276, with the usual perturbation theory ignoring exciton decay. If the excitons are incoherent, their annihilation rate, depending on the distance between them, was used to investigate the nonlinear fluorescence quenching in molecular crystals considering excitons as freely moving quasiparticles [277]. Alternatively, the resonance transfer of excitons from site to site can be considered as a random-walk motion and their self-quenching as a hopping reaction [72,278]. In both cases the diffusional description of the encounter motion in the course of the reaction becomes possible if either the free path or step length is smaller than the effective reaction radius. This is a rather rigid condition that is most likely met in liquids and not in solids.

C. Electrochemical Luminescence

In Section VIII.F the competition between electron transfer to the excited and ground states during recombination of electrochemically injected ions was studied in the framework of spinless IET. In fact, the primary products of such recombination are usually the low-energy triplet excitations. They are produced from triplet RIPs, while singlet RIPs recombine to the ground state:

$$^2D^+ + {}^2A^- \begin{array}{c} \nearrow \\ 3k_s \uparrow\downarrow k_s \\ \searrow \end{array} \begin{array}{c} {}^3[D^+ \cdots A^-] \underset{W_I}{\overset{W_B}{\rightleftharpoons}} D + {}^3A^* \\ \\ {}^1[D^+ \cdots A^-] \overset{W_R}{\longrightarrow} D + A \end{array} \qquad (3.813)$$

The cage reaction of the RIPs develops according to the scheme in Figure 3.101, provided recombination to the excited singlet is impossible because of deficiency

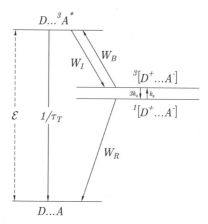

Figure 3.101. The energy scheme for the cage recombination of RIP into the excited triplet and ground states with indication of electron transitions and their rates.

of energy. However, it can be produced via the triplet–triplet annihilation in the bulk as in reaction (3.726):

$$^3A^* + {}^3A^* \overset{W_{TT}}{\Longrightarrow} {}^1A^* + A \tag{3.814}$$

In Section XIII.A this process was responsible for the delayed fluorescence. Here the self-quenching of triplets, following their electrochemical generation, lowers their quantum yield and resulting phosphorescence.

1. Integral Theory of the Phenomenon

First, we have to substitute the spinless theory Eqs. (3.357) by the following set [194]:

$$\dot{N}^{\pm}(t) = c \int_0^t R^{\dagger}(\tau) N_T(t - \tau) d\tau - \int_0^t R^{\ddagger}(\tau) [N^{\pm}]^2 (t - \tau) d\tau \tag{3.815a}$$

$$\dot{N}_T(t) = -c \int_0^t R^*(\tau) N_T(t - \tau) d\tau + \int_0^t R^{\#}(\tau) [N^{\pm}]^2 (t - \tau) d\tau$$

$$- \int_0^t R^{\star}(\tau) N_T^2(t - \tau) d\tau - \frac{N_T(t)}{\tau_T} \tag{3.815b}$$

$$\dot{N}_S(t) = \frac{1}{2} \int_0^t R^{\star}(\tau) N_T^2(t - \tau) d\tau - \frac{N_S(t)}{\tau_S} \tag{3.815c}$$

The quenchers of the excitations are assumed to be present in great excess, so that their concentration $c = [D]$ is practically constant and we have to follow only the concentrations of ions as well as the triplet and singlet excitations, $N_T(t) = [^3A^*]$ and $N_S(t) = [^1A^*]$. The lack of light pumping is compensated for by the electrochemical injection of ions represented by the initial conditions identical to (3.408):

$$N_T(0) = N_S(0) = 0, \qquad N^{\pm}(0) = N_0^{\pm} \tag{3.816}$$

To optimize the specification of all the kernels as elements of the matrix operator $\hat{\mathbf{R}}(t)$, the latter is defined as the original of $\tilde{\mathbf{M}}(s)$ from Eq. (3.528):

$$\hat{\mathbf{R}}(t) = \int \hat{\mathbf{W}}(r)\left[\delta(t) + \hat{\boldsymbol{\Phi}}(r,t)\right]d^3r \tag{3.817}$$

Here $\hat{\boldsymbol{\Phi}}$ is the matrix of pair correlation functions obeying the auxiliary equation (3.529) with helpful initial condition $\hat{\boldsymbol{\Phi}}(r,0) = \hat{\mathbf{W}}(r)$. Still the problem remains rather complex. Solving it in Ref. 194, all the kernels were specified using as input data only the diffusion coefficients and transfer rates: $W_I(r)$, $W_B(r)$, $W_R(r)$, and $W_{TT}(r)$.

2. Low Concentration of Injected Ions

In this limit we can ignore the triplet annihilation neglecting the terms quadratic in N_T together with the last equation in the set (3.815). The remaining equations take the form of Eqs. (3.357) with $I(t) = 0$

$$\dot{N}^{\pm}(t) = c\int_0^t R^{\dagger}(\tau)N_T(t-\tau)d\tau - \int_0^t R^{\ddagger}(\tau)[N^{\pm}]^2(t-\tau)d\tau \tag{3.818a}$$

$$\dot{N}_T(t) = -c\int_0^t R^*(\tau)N_T(t-\tau)d\tau + \int_0^t R^{\#}(\tau)[N^{\pm}]^2(t-\tau)d\tau - \frac{N_T(t)}{\tau_T} \tag{3.818b}$$

but their kernels have different definitions accounting for the spin states.

In the contact approximation the Laplace transformations of these kernels are

$$\tilde{R}^{\dagger}(s) = 4k_0\frac{1 + k_c\tilde{g}_0}{X}, \qquad \tilde{R}^{\ddagger}(s) = \frac{3k_b + k_c(1 + k_0\tilde{g}_1 + 4k_b\tilde{g}_0)}{X} \tag{3.819a}$$

$$\tilde{R}^*(s) = k_0\frac{4 + k_c(\tilde{g}_2 + 3\tilde{g}_0)}{X}, \qquad \tilde{R}^{\#}(s) = 3k_b\frac{1 + k_c\tilde{g}_0}{X} \tag{3.819b}$$

where $k_0 = \int W_I d^3r$, $k_b = \int W_B d^3r$, $k_c = \int W_R d^3r$, and

$$X = [1 + k_0\tilde{g}_1][4 + k_c(\tilde{g}_2 + 3\tilde{g}_0)] + k_b[3\tilde{g}_2 + \tilde{g}_0 + 4k_c\tilde{g}_2\tilde{g}_0] \tag{3.820}$$

Here are three diffusional Green functions

$$\tilde{g}_1(s) = \frac{1/k_D}{1 + \sqrt{s\tau_d + \tau_d/\tau_T}}, \quad \tilde{g}_2(s) = \frac{1/k_D}{1 + \sqrt{s\tau_d}}, \quad \tilde{g}_0(s) = \frac{1/k_D}{1 + \sqrt{s\tau_d + 4k_s\tau_d}}$$

but only the last of them accounts for the spin conversion through $4k_s\tau_d$.

Since the only excited products of the charge recombination are triplets, their quantum yield is given by the definition identical to that of Eq. (7.55):

$$\phi_{es} \equiv \phi_T = \frac{\int_0^\infty N_T(t)dt}{N_0^\pm \tau_T} = \chi_T \eta \qquad (3.821)$$

Making the Laplace transformation of Eqs. (3.818), one can easily resolve them regarding $\tilde{N}_T(0)$ and insert the result in equality (3.821). In this way it was found [194] that

$$\chi_T = \frac{\tilde{R}^\#(0)}{\tilde{R}^\ddagger(0)} \qquad (3.822)$$

and

$$\eta = \frac{1}{1 + c\kappa_0\tau_T} \qquad (3.823)$$

Here χ_T is the quantum yield of triplets produced by the geminate recombination, while η is the quantum yield of their phosphorescence. The latter is quenched by the subsequent bulk ionization, in encounters of triplets with neutral quenchers. The relative phosphorescence yield obeys the usual Stern–Volmer law, whose constant is identical to that of Eq. (3.364)

$$\kappa_0 = \tilde{R}^*(0)\left[1 - \frac{\tilde{R}^\ddagger(0)}{\tilde{R}^*(0)} \times \frac{\tilde{R}^\#(0)}{\tilde{R}^\ddagger(0)}\right] = \kappa_g[1 - \chi_T\varphi(\sigma)] \qquad (3.824)$$

where $\varphi(\sigma)$ is $\bar{\varphi}$ in the contact approximation.

Since spin conversion manifests itself variously as either fast or slow, it makes sense to consider these limits separately.

3. Fast Conversion

At $\sqrt{k_s\tau_d} \gg 1$, the equilibrium relationship between the populations of singlet and triplet RIPs is sustained during the encounter time and discrimination between these states is hardly possible. This is the most favorable case for a spinless theory. Coming to the extreme $k_s \to \infty$, we get $\tilde{g}_0 = 0$ and

$$\frac{1}{4}X = (1 + k_0\tilde{g}_1)(1 + k_c^{eq}\tilde{g}_2) + k_b^{eq}\tilde{g}_2 = Z^{eq} \qquad (3.825)$$

Here the rate constants arise with the equilibrium weights of the singlet and triplet states:

$$k_c^{eq} = \frac{1}{4}k_c \qquad \text{and} \qquad k_b^{eq} = \frac{3}{4}k_b$$

The replacement of k_c and k_b by k_c^{eq} and k_c^{eq} makes the spinless expression Z identical with Z^{eq} and transforms all the spinless kernels (3.369) to their modified analogs, accounting for the fast spin conversion:

$$\tilde{R}^* = \frac{k_0[1 + k_c^{eq}\tilde{g}_2]}{Z^{eq}}; \quad \tilde{R}^{\#} = \frac{k_b^{eq}}{Z^{eq}}; \quad \tilde{R}^{\dagger} = \frac{k_0}{Z^{eq}}; \quad \tilde{R}^{\ddagger} = \frac{k_b^{eq} + k_c^{eq}(1 + k_0 g_1)}{Z^{eq}}$$

All other results obtained previously with the spinless theory are also generalized in the same way. In particular, using the appropriate kernels in Eq. (3.822), we obtain the geminate quantum yield of triplets in the following form:

$$\chi_T = \left[1 + \frac{k_c^{eq}}{k_b^{eq}}\left(1 + \frac{k_0/k_D}{1 + \sqrt{\tau_d/\tau_T}}\right)\right]^{-1} \qquad (3.826)$$

Since quenching of the triplets is assumed negligible as in Refs. 191 and 192, the quantum yield of triplets coincides with the yield of their formation:

$$\phi_{es} = \chi_T, \qquad \text{at} \qquad c\kappa_0\tau_T \ll 1 \qquad (3.827)$$

Moreover, at high exergonicity of ionization one can set $k_0 \approx 0$ and obtain the follownig from Eq. (3.826):

$$\chi_T \approx \frac{1}{1 + k_c^{eq}/k_b^{eq}} = \frac{1}{1 + k_c/3k_b} \qquad (3.828)$$

This result is similar to its spinless analog, Eq. (3.411). The difference between them is only in the numerical coefficient 3, originating from the ratio of the equilibrium weights of the spin states: $\frac{3}{4}$ and $\frac{1}{4}$.

If $k_0 \neq 0$, then Eq. (3.826) must be used instead of (3.828). The concentration reduction of η also should be taken into account, because it

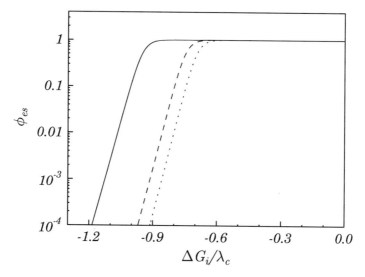

Figure 3.102. The quantum yield of reversible triplet production, $\phi_{es} \equiv \phi_T$, at different concentrations of quenchers. Solid line, $c = 0$; dashed line, $c = 10^{-3}$ M; dotted line, $c = 0.1$ M. Here $\mathscr{E} = 4\lambda_c$, $\lambda_c = 20T$, $\tau_T = 10^{-3}$ s, $k_0^{max} = k_c^{max} = 20k_D$, $k_D = 5 \times 10^9$ M^{-1}s^{-1}, $k_s\tau_d = 1$. (From Ref. 194.)

reduces the total triplet yield (3.821). In view of Eqs. (3.823) and (3.824), the latter is:

$$\phi_T = \frac{\chi_T}{1 + c\kappa_g[1 - \chi_T\varphi(\sigma)]\tau_T} \qquad (3.829)$$

As shown in Fig. 3.102, the parallel shift of the ascending branches of the $\phi_{es}(\Delta G_i)$ curves is due to the different initial concentration of quenchers, c. This concentration was not likely to be controlled because the simplest theory used in all experimental works identified ϕ_T with χ_T. Equation (3.829) shows that the distinction between them is essential and can be responsible for a similar shift seen in Figure 3.55 for different quenchers.

4. Slow Conversion

If the spin conversion is frozen during the encounters ($\sqrt{k_s\tau_d} \ll 1$), the results are slightly different. In the extreme case of no spin conversion $\tilde{g}_0 = \tilde{g}_2$ and the general expression (3.820) reduces to the following one:

$$X = 4(1 + k_c\tilde{g}_2)X_0, \qquad \text{where} \qquad X_0 = 1 + k_b\tilde{g}_2 + k_0\tilde{g}_1 \qquad (3.830)$$

Therefore in this limit the kernels are different:

$$\tilde{R}^*(s) = \frac{k_0}{X_0} = \tilde{R}^\dagger(s); \qquad \tilde{R}^\#(s) = \frac{3}{4}\frac{k_b}{X_0}; \qquad \tilde{R}^\ddagger(s) = \frac{3}{4}\frac{k_b}{X_0} + \frac{1}{4}\frac{k_c}{1 + g_2 k_c}$$

The first two represent the reversible ionization of the triplet excitations and accumulation of triplet RIPs. In the absence of the spin conversion, since there is no geminate recombination of triplet RIPs to the ground state these kernels are equal. $\tilde{R}^\#$ describes the recombination of triplet RIPs to the triplet excited states. The last kernel represents the recombination of ions to either the triplet or ground state, in proportion to the equilibrium weights of competing channels.

The geminate quantum yield of triplets is obtained by substitution of the corresponding kernels into definition (3.822):

$$\chi_T = \frac{1}{1 + \kappa_c/3\kappa_b} \tag{3.831}$$

Here κ_r and κ_b are the Stern–Volmer constants for ion recombination to the ground and triplet states, respectively:

$$\kappa_c = \frac{k_c k_D}{k_D + k_c}, \qquad \kappa_b = \frac{k_b k_D}{k_D + k_b + k_0/\left(1 + \sqrt{\tau_d/\tau_T}\right)} \tag{3.832}$$

The latter differs from the former because the reversible transfer to the unstable triplet state is affected by the triplet decay.

Even at large $\Delta G_i > 0$ where excitation of the triplets is practically irreversible ($k_0 \approx 0$), the difference between Eqs. (3.828) and (3.831) does not disappear. The Stern–Volmer constants $\kappa_{c,b}$ are identical to the kinetic rate constants $k_{c,b}$ only when they both are smaller than k_D. This is true near the point of inflection in the $\chi_T(\Delta G_i)$ dependence, which is equally far from the maxima of both constants. Near this point $k_b \approx k_c \ll k_D$, and there should be no difference between the fast and slow spin conversions. This is the point of intersection of all the curves shown in Figure 3.103. As can be seen from this figure, accounting for the spin state is not an essential factor in electro-chemiluminescence. This is because the electrochemical preparation of ions is not spin-selective.

When recombination to the triplet state becomes highly exergonic, the recombination to the ground state is switched off and all RIPs eventually turn to the triplet excitations. In this region the quantum yield of triplets saturates, approaching a plateau whose height reaches its maximum value $\chi_T = 1$, provided there is no triplet annihilation. What happens when this condition is violated is discussed below on the basis of general equations (3.406).

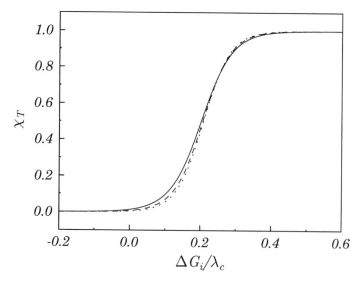

Figure 3.103. The quantum yield of triplet production at zero spin conversion (solid line), medium spin conversion rate ($k_s\tau_d = 1$, dashed line), and an infinitely fast conversion (dotted line). Here $\mathscr{E} = 1.5\lambda_c$, $\lambda_c = 35T$, $k_0^{max} = k_c^{max} = 20k_D$. (From Ref. 194.)

5. High Concentration of Injected Ions

When triplets are generated at sufficiently high concentration, their annihilation produces a number of singlet excitations. In principle, they contribute to the fluorescence of the solution, but it proceeds in different time and frequency ranges and can be hardly detected. However, the quantum yield of singlets

$$\phi_S = \frac{\int_0^\infty N_S(t)dt}{N_0^\pm \tau_S} \tag{3.833}$$

is not negligible. According to reaction (3.814), the birth of each singlet costs the lives of two triplets. Triplet annihilation results in a lowering of their χ_T.

To find both ϕ_S and ϕ_T one should solve the general set of integral equations (3.406) which account for the biexcitonic annihilation of triplets. For this goal an efficient numerical program was developed in Ref. 194. It provides the desired solution in the whole free-energy range and beyond the contact approximation. The rates of backward and forward electron transfer to the excited and ground states were taken in the form of Eq. (3.47) or (3.51), while $W_{TT} = C/r^6$ was assumed to be dipole–dipole in origin.

When ΔG_i is negative, the ion recombination to the triplet state is endergonic and reversible, unlike recombination to the ground state. The latter strongly dominates over triplet production in this case, and the quantum yields of both excited states are very small: $1 \gg \phi_T \gg \phi_S$. These quantum yields sharply increase with ΔG_i along the ascending branches in Figure 3.104

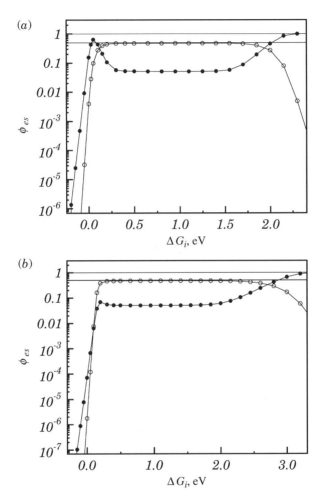

Figure 3.104. The quantum yields of singlets (○) and triplets (●) without (top) and taking into account the vibronic modes (bottom). Here $\lambda_c = 0.8$ eV, $\mathscr{E} = 2.1$ eV, $\tau_T = 10^{-4}$ s, $\tau_S = 10^{-8}$ s, $N_0^\pm = 10^{-4}$ M, $c = 0.1$ M, $\sigma = 7.2$ Å, $L = 1$ Å, $W_I(\sigma) = W_R(\sigma) = 2.3 \times 10^{12}$ s^{-1}, $W_{TT}(\sigma) = 10^{11}$ s^{-1}. For the bottom graph the internal mode reorganization energy is $\lambda_i = 0.3$ eV, and the frequency of this quantum mode $\hbar\omega_i = 0.1$ eV. (From Ref. 194.)

because the situation is changing in favor of the triplet excitation and subsequent production of singlets. At positive and high ΔG_i, the recombination to the ground state is completely switched off and only triplets are produced by charge recombination. However, at a high initial concentration of ions, the triplet yield does not reach the maximum $\phi_T = 1$ immediately as in Figure 3.103, but levels off at a plateau that is an order or two lower than the maximum value. This is an expected effect of the triplet annihilation that contributes to the production of singlet excitations. The singlet quantum yield increases even more sharply. It also reaches the plateau $\phi_S = \frac{1}{2}$ whose value is specified by the reaction scheme (3.814): one singlet per two triplets.

The competition between the annihilation of triplets and their decay with time τ_T determines the height of the triplet plateau. It is lower, the slower is the triplet decay and the faster their annihilation. The length of the plateau is determined by the rate of ion recombination to the triplet state. In the contact approximation this is $k_b k_D / (k_b + k_D)$, where $k_b(\Delta G_i)$ is the bell-shaped Marcus free-energy dependence. Its top is cut by the diffusional plateau, $k_D = 4\pi\sigma D$. Within this plateau the relationship between the triplet decay and annihilation does not depend on (ΔG_i) as well as the triplet quantum yield ϕ_T. At excessively high ΔG_i, the diffusional control of recombination to triplets gives way to the kinetic control with rate constant k_b going down with ΔG_i. There the rate of triplet production slows down and their concentration as well as annihilation tend to diminish with ΔG_i. Finally, only monomolecular decay of triplets is left and their quantum yield approaches 1.

The most unexpected feature seen in Figure 3.104 is the maximum between the ascending branches representing the accumulation of triplets under kinetic control of recombination and the plateau where this reaction is controlled by the encounter diffusion of ions. The maximum is reached soon after recombination to the triplet state overrides recombination to the ground state, but does not yet produce as many triplets as is necessary to make annihilation faster than natural decay. As long as the triplets are not essentially quenched by annihilation, their contribution to the phosphorescence is maximal, but sharply decreases with a further increase of ΔG_i, which facilitates their production and even more annihilation.

This effect is more pronounced when accumulation of triplets starts earlier and the maximum is reached further from the plateau where the annihilation rate is maximal. This is demonstrated in Figure 3.104(a), where $W_B(r)$ is calculated assuming that the electron transfer is assisted only by the solvent mode with reorganization energy $\lambda_c = 0.8$ eV. However, the maximum almost disappears when the internal mode with $\lambda_i = 0.3$ eV is added to assist the same transfer [Fig. 3.104(b)]. They make $W_B(\Delta G_i)$ broader, especially at the highly exergonic wing where triplet production is strongly facilitated, making annihilation more effective. The result looks much better in comparison with the available experimental data, where the mere existence of the maximum is questionable. Nevertheless, one can probably recognize the maximum in some experimental data obtained in Refs. 191 and 192 because there is a significant decrease in ϕ_T of the latest point, in comparison with the maximal one ($3 \div 5$ times).

Unfortunately the straightforward comparison of the present theory with those data is hardly possible because in the real experiments the distribution of reactants is not homogeneous. The counterions are injected sequentially from a common electrode. Their local densities differ significantly from each other and change with time as a result of not only the reaction but also their spatial diffusion out of the electrode. There are also some doubts as to whether the annihilation is the only mechanism of triplet quenching. The latter can be carried out in encounters with ions as well [279,280]. These channels of additional recombination can be taken into account but only for the homogeneous solution [194]. Hence, the problem remains open for further discussion.

XIV. CONCLUSIONS

In the context of a single review, it is difficult to cover all the great diversity of encounter theory applications. Some of the applications that were not covered are listed here:

- The photo-generation of free carriers in p-phenilene vinelene due to electron transfer to oxygen, which suppresses the singlet oxygen generation through energy transfer [202]
- The impressive application of IET/MET to the study of Lotka–Volterra autocatalytic oscillatory reactions [203]
- The IET description of degenerate electron transfer reactions [281]
- The extension of matrix IET to bulk reversible association–dissociation reactions [282] and reversible excitation binding [283]
- The generalization of matrix MET tailored to account for the internal quantum states of reactants [284]

All these applications in conjunction with those included in the present review testify to the fact that encounter theory has been converted to a solo branch of non-Markovian chemical kinetics.

However, encounter theory is subdivided into IET, MET, and DET/UT, and neither of them is perfect. IET is in fact the most universal and general approach to binary reactions in three-dimensional space at low concentration of reactants. Moreover, this is a common limit for all other theories valid at higher concentrations. IET is applied to contact and distant reactions, both reversible and irreversible, and is especially suited for analytic calculations of quantum yields. Matrix IET is expanding to multistage reactions of arbitrary complexity especially in the form imparted to it in Section X.C.

However, IET does not describe well the static and long-time asymptote of diffusional quenching that are not of binary origin. These shortages are partially removed by MET. It retains all the advantages of IET but extends the time limits of the theory as in Eq. (3.683) and provides the first nonlinear concentration corrections to the Stern–Volmer law (Section XII).

However, at still larger concentrations only DET/UT is capable of reaching the kinetic limit of the Stern–Volmer constant and the static limit of the reaction product distribution. On the other hand, this theory is intended for only irreversible reactions and does not have the matrix form adapted for consideration of multistage reactions. The latter is also valid for competing theories based on the superposition approximation or nonequilibrium statistical mechanics. Moreover, most of them address only the contact reactions (either reversible or irreversible). These limitations strongly restrict their application to real transfer reactions, carried out by distant rates, depending on the reactant and solvent parameters. On the other hand, these theories can be applied to reactions in one- and two-dimensional spaces where binary approximation is impossible and encounter theories inapplicable.

There are also two essential simplifications of the reaction mechanism accepted in almost all the theories. They imply that the reactivity of the partners is spherically isotropic and their encounter motion can be considered as a continuous diffusion at any distance. Both these assumptions are simply the approximations that have been overcome already with DET under the pressure of experimental facts.

First, the diffusional radical reactions in solutions whose rates are proportional to D sometimes have rate constants that are much smaller than their contact estimate for the isotropic black sphere, $k_D = 4\pi\sigma D$. It was proved that this is the result of chemical anisotropy of the reactants. Partially averaged by translational and rotational diffusion of reactants, this anisotropy manifests itself via the encounter efficiency $w \leq 1$, which enters the rate constant $k_D = w4\pi\sigma D$ [249]. Even the model of white spheres with black spots is more appropriate for such reactions than the conventional Smoluchowski model.

The random walk with finite step length is the conventional model of motion of incoherent excitons in solids. If this length is smaller than the steepness parameter of the transfer rate, the diffusional description of the encounters is valid. Otherwise the reaction that is switched on by a single jump is qualified as the hopping one [72,167]. The step length in liquid solutions can be compatible with molecular size. At least, the penetration into the first coordination shell should be a jumpwise process substituted for continuous diffusion. If this is a limiting stage of the reaction, it is in fact hopping and can be described with the modified model of diffusion-controlled reactions (MMDCR) [166], which was shown to be a hopping limit of the most general random-walk theory [167].

Both chemical anisotropy and finite jump size are the important features of the pseudocontact reactions of electron and especially proton transfer. They should be incorporated in IET and MET in approximately the same way as in DET.

Acknowledgments

The author is very grateful to Drs. A. Popov, O. Igoshin, and MSc. V. Gladkikh for their valuable help. I also wish to thank the Israeli Science Foundation (Project N 6863) for supporting this work.

References

1. M. V. Smoluchowski, *Z. Phys. Chem.* **92**, 129 (1917).

2. F. C. Collins and G. E. Kimball, *J. Colloid. Sci.* **4**, 425 (1949).

3. N. N. Tunitskii and Kh. S. Bagdasar'yan, *Opt. Spectrosc.* **15**, 303 (1963).

4. S. F. Kilin, M. S. Mikhelashvili, and I. M. Rozman, *Opt. Spectrosc.* **16**, 576 (1964).

5. I. I. Vasil'ev, B. P. Kirsanov, and V. A. Krongaus, *Kinet. Kataliz* **5**, 792 (1964).

6. I. Z. Steinberg and E. Katchalsky, *J. Chem. Phys.* **48**, 2404 (1968).

7. A. B. Doktorov and A. I. Burshtein, *Sov. Phys. JETP* **41**, 671 (1975).

8. G. Wilemski and M. Fixman, *J. Chem. Phys.* **58**, 4009 (1973).

9. A. I. Burshtein and A. G. Kofman, *Opt. Spectrosc.* **40**, 175 (1976).

10. A. B. Doktorov, A. A. Kipriyanov, and A. I. Burshtein, *Sov. Phys. JETP* **47**, 623 (1978); *Opt. Spectrosc.* **45**, 279, 640 (1978).

11. V. M. Agranovich and M. D. Galanin, *Electron Excitation Energy Transfer in Condensed Matter* (North-Holland, Amsterdam, 1982) (Russian ed., Nauka, 1978).

12. M. Tachiya, *Radiat. Phys. Chem.* **21**, 167 (1983).

13. D. E. Eads, B. G. Dismer, and G. R. Fleming, *J. Chem. Phys.* **93**, 1136 (1990).

14. M. J. Philling and S. A. Rice, *J. Chem. Soc. Faraday Trans. 2* **71**, 1563 (1975).

15. Yu. A. Berlin, *Dokl. Akad. Nauk SSSR* **223**, 625 (1975).

16. S. M. B. Costa, A. I. Macanita, and S. J. Formosinho, *J. Phys. Chem.* **88**, 4089 (1984).

17. A. I. Burshtein, E. I. Kapinus, I. Yu. Kucherova, and V. A. Morozov, *J. Lumin.* **43**, 291 (1989).

18. V. S. Gladkikh, A. I. Burshtein, H. L. Tavernier, and M. D. Fayer, *J. Phys. Chem. A* **106**, 6982 (2002).

19. K. M. Hong and J. Noolandi, *J. Chem. Phys.* **68**, 5163 (1978).

20. A. I. Burshtein, A. A. Zharikov, N. V. Shokhirev, O. B. Spirina, and E. B. Krissinel, *J. Chem. Phys.* **95**, 8013 (1991).

21. A. I. Burshtein, A. A. Zharikov, and N. V. Shokhirev, *J. Chem. Phys.* **96** 1951 (1992).

22. A. I. Burshtein, *Chem. Phys. Lett.* **194**, 247 (1992)

23. R. C. Dorfman and M. D. Fayer, *J. Chem. Phys.* **96**, 7410 (1992).

24. A. I. Burshtein, E. Krissinel, and M. S. Mikhelashvili, *J. Phys. Chem.* **98**, 7319 (1994).

25. A. I. Burshtein and A. Yu. Sivachenko, *J. Photochem. Photobiol. A* **109**, 1 (1997).

26. S. Murata and M. Tachiya, *J. Chim. Phys.* **93**, 1577 (1996).

27. S. Murata and M. Tachiya, *J. Phys. Chem.* **100**, 4064 (1996);

28. A. I. Burshtein and E. Krissinel, *J. Phys. Chem. A* **102**, 816 (1998).

29. A. I. Burshtein and E. Krissinel, *J. Phys. Chem. A* **102**, 7541 (1998).

30. A. I. Burshtein, *Chem. Phys.* **247**, 275 (1999); erratum **250**, 111 (1999).

31. A. I. Burshtein, *J. Chem. Phys.* **117**, 7640 (2002).

32. A. I. Burshtein, *Adv. Chem. Phys.* **114**, 419 (2000).

33. V. P. Sakun, *Physica A* **80**, 128 (1975).

34. H. Mori, *Progr. Theor. Phys.* **33**, 423; **34**, 399 (1965).

35. B. J. Berne, J. P. Boon and S. A. Rice, *J. Chem. Phys.* **45**, 1086 (1966).

36. A. B. Doktorov, *Physica A* **90**, 109 (1978).

37. A. A. Kiprianov, A. B. Doktorov, and A. I. Burshtein, *Chem. Phys.* **76**, 149, 163 (1983).

38. A. I. Burshtein and P. A. Frantsuzov, *J. Chem. Phys.* **106**, 3948, (1997).

39. A. A. Kipriyanov, I. V. Gopich, and A. B. Doktorov, *Chem. Phys.* **187**, 241 (1994); **191**, 101 (1995).

40. A. A. Kipriyanov, I. V. Gopich, and A. B. Doktorov, *Physica A* **255**, 347 (1998).

41. I. V. Gopich, A. A. Kipriyanov, and A. B. Doktorov, *J. Chem. Phys.* **110**, 10888 (1999).

42. A. A. Kipriyanov, O. A. Igoshin, and A. B. Doktorov, *Physica A* **275**, 99 (2000).

43. K. L. Ivanov, N. N. Lukzen, A . B. Doktorov, and A. I. Burshtein, *J. Chem. Phys.* **114**(I), 1754 (2001)

44. K. L. Ivanov, N. N. Lukzen, A . B. Doktorov, and A. I. Burshtein, *J. Chem. Phys.* **114**(II), 1763 (2001).

45. K. L. Ivanov, N. N. Lukzen, A . B. Doktorov, and A. I. Burshtein, *J. Chem. Phys.* **114**(III), 5682 (2001).

46. A. V. Popov, V. S. Gladkikh, and A. I. Burshtein, *J. Phys. Chem. A* **107**, 8177 (2003).

47. A. V. Popov and A. I. Burshtein, *J. Phys. Chem. A* **107**, 9688 (2003).

48. N. N. Lukzen, A. B. Doktorov, and A. I. Burshtein, *Chem. Phys.* **102**, 289 (1986).

49. A. I. Burshtein and N. N. Lukzen, *J. Chem. Phys.* **103**, 9631 (1995).

50. A. I. Burshtein and N. N. Lukzen, *J. Chem. Phys.* **105**, 9588 (1996).

51. J. Sung, J. Chi, and S. Lee, *J. Chem. Phys.* **111**, 804 (1999).

52. W. Naumann, *J. Chem. Phys.* **110**, 3926 (1999).

53. D. Rehm and A. Weller, *Israel J. Chem.* **8**, 259 (1970).

54. Th. L. Nemzek and W. R. Ware, *J. Chem. Phys.* **62** 479 (1975).

55. M. Sikorski, E. Krystowiak, and R. P. Steer, *J. Photochem. Photobiol. A* **117**, 1 (1998).

56. S. A. Rice, in *Comprehensive Chemical Kinetics*, Vol. 25, *Diffusion-Limited Reactions*, C. H. Bamford, C. F. H. Tripper and R. G. Kompton, eds. (Elsiever, Amsterdam, 1985).

57. A. Weller, *Z. Electrochem.* **58**, 849 (1954); *Z. Phys. Chem.* **17**, 224 (1958).

58. N. M. Trieff and B. R. Sundheim, *J. Phys. Chem.* **69**, 2044 (1965).

59. W. R. Laws and L. Brand, *J. Phys. Chem.* **83**, 795 (1979).

60. B. Cohen, D. Huppert, and N. Agmon, *J. Phys. Chem. A* **105**, 7165 (2001).

61. R. W. Wijnaendts van Resandt, *Chem. Phys. Lett.* **95**, 205 (1983).

62. M. R. Eftink and C. A. Ghiron, *Anal. Biochem.* **114**, 199 (1981).

63. A. Szabo, *J. Phys. Chem.* **93**, 6929 (1989).

64. A. I. Burshtein and Yu. I. Naberukhin, *Preprint of Institute of Theoretical Physics*, Ukraina Academy, 74-160P Kuiv., 1974.

65. R. W. Noyes, in *Progress of Reaction Kinetics*, G. Porter, ed. (Pergamon Press, 1961), Vol. 1, p. 129.

66. S. Murata, M. Nishimura, S. Y. Matsuzaki, and M. Tachiya, *Chem. Phys. Lett.* **219**, 200 (1994).

67. N. Matsuda, T. Kakitani, T. Denda, and N. Mataga, *Chem. Phys.* **190**, 83 (1995).

68. A. A. Zharikov and N. V. Shokhirev, *Chem. Phys. Lett.* **190**, 423 (1992).

69. A. I. Burshtein, *J. Lumin.* **34**, 162 (1985).

70. M. Yokoto and O. Tanimoto, *J. Phys. Soc. Japan* **22**, 779 (1967).

71. B. Ya. Sveshnikov and V. I. Shirokov, *Opt. Spectrosc.* (USSR) **12**, 320 (1962).

72. A. I. Burshtein, *Sov. Phys. Usp.* **27**, 579 (1984).

73. M. Inokuti and F. Hirayama, *J. Chem. Phys.* **43**, 1978 (1965).

74. D. F. Thomas, J. J. Hopfield, and W. M. Augustyniak, *Phys. Rev.* **140**, 202 (1965).

75. K. I. Zamaraev and R. F. Khairutdinov, *Russ. Chem. Rev.* **47**, 518 (1978).

76. Yu. Georgievskii and A. I. Burshtein, *J. Chem. Phys.* **101**, 10858 (1995).

77. Yu. Georgievskii, A. I. Burshtein, and B. Chernobrod, *J. Chem. Phys.* **105**, 3108 (1996).

78. A. V. Barzykin, P. A. Frantsuzzov, K. Seki, and M. Tachiya, *Adv. Chem. Phys.* **123**, 511 (2002).

79. R. A. Marcus, *J. Chem. Phys.* **24**, 966 (1956); **43**, 679 (1965).

80. S.Efrima and M. Bixon, *Chem. Phys. Lett.* **25**, 34 (1974).

81. J. Jortner and M. Bixon, *J. Chem. Phys.* **88**, 167 (1988).

82. J. R. Miller, L. T. Calcaterra, and G. L. Closs, *J. Am. Chem. Soc.* **106**, 3047 (1984).

83. H. D. Roth, *J. Phys. Chem.* **97**, 13037 (1993).

84. A. I. Burshtein, P. A. Frantsuzov, and A. A. Zharikov, *Chem. Phys.* **155**, 91 (1991).

85. A. I. Burshtein and P. A. Frantsuzov, *J. Lumin.* **51**, 215 (1992).

86. B. Sipp and R. Voltz, *J. Chem. Phys.* **79**, 434 (1983).

87. A. A. Kipriyanov and A. B. Doktorov, *Chem. Phys. Lett.* **246**, 359 (1995).

88. A. B. Doktorov and A. A. Kipriyanov, *Mol. Phys.* **88**, 453 (1996).

89. S. Nishikawa, T. Asahi, T. Okada, N. Mataga, and T. Kakitani, *Chem. Phys. Lett.* **185**, 237 (1991).

90. S. Murata, S. Y. Matsuzaki, and M. Tachiya, *J. Phys. Chem.*, **95**, 5354 (1995).

91. R. C. Dorfman, Y. Lin, and M. D. Fayer, *J. Phys. Chem.* **94**, 8007 (1990).

92. L. Song, R. C. Dorfman, S. F. Swallen, and M. D. Fayer, *J. Phys. Chem.* **95**, 3454 (1991).

93. L. Song, S. F. Swallen, R. C. Dorfman, K. Weidemaier, and M. D. Fayer, *J. Phys. Chem.* **97**, 1374 (1992).

94. K. Kakitani, N. Matsuda, T. Denda, N. Mataga, and Y. Enomoto, *Ultrafast Reaction Dynamics and Solvent Effects*; *AIP Conf. Proc.*, New York, Y. Gauduel and P. J. Rossky, eds., 1993, p. 298.

95. H. L. Tavernier and M. D. Fayer, *J. Chem. Phys.* **114**, 4552 (2001).

96. X. Allonas, P. Jacques, A. Accary, M. Kessler, and F. Heisel, *J. Fluoresc.* **10**, 237 (2000).

97. H. L. Tavernier, M. M. Kalashnikov, and M. D. Fayer, *J. Chem. Phys.* **113**, 10191 (2000).

98. G. Angulo, G. Grampp, A. A. Neufeld, and A. I. Burshtein, *J. Phys. Chem. A* **107**, 6913 (2003) .

99. A. A. Kipriyanov and A. A. Karpushin, *Preprint 23*, Siberian Branch of Academy of Science, Novosibirsk (in Russian), 1988.

100. A. Spernol and K. Z. Wirtz, *Naturforsch* **8a**, 522 (1953).

101. S. I. Vavilov and I. M. Frank, *Z. Phys.* **69**, 100 (1931).

102. J. C. Andre, M. Niclause, and W. R. Ware, *Chem. Phys.* **28**, 371 (1978).

103. B. Stevens, *Chem. Phys. Lett.* **134**, 519 (1987).

104. M. A. R. B. Castanho and M. J. E. Prieto, *Biochim. Biophys. Acta* **1373**, 1 (1998).

105. G. Angulo and A. I. Burshtein (in preparation).

106. S. S. Jayanthi and P. Ramamurthy, *J. Phys. Chem.* **101**, 2016 (1997); **102**, 511 (1998).

107. A. I. Burshtein and K. L. Ivanov, *J. Phys. Chem. A* **105**, 3158 (2001).

108. R. A. Marcus and P. Siders, *J. Phys. Chem.* **86**, 622 (1982)

109. N. Mataga, T. Asahi, J. Kanda, T. Okada, and T. Kakitani, *Chem. Phys.* **127**, 249 (1988).

110. K. Kikuchi, Y. Takahashi, T. Katagiri, T. Niwa, M. Hoshi, and T. Miyashi, *Chem. Phys. Lett.*, **180**, 403 (1991).

111. A. I. Burshtein and E. Krissinel, *J. Phys. Chem.* **100**, 3005 (1996).

112. (a) Y. Enomoto, T. Kakitani, A. Yoshimori, Y. Hatano, and M. Saito, *Chem. Phys. Lett.* **178**, 235 (1991); (b) Y. Enomoto, T. Kakitani, A. Yoshimori, and Y. Hatano, *Chem. Phys. Lett.* **186**, 366 (1991).

113. T. Asahi, M. Ohkohchi, and N. Mataga, *J. Phys. Chem.* **97** 13132 (1993).

114. T. Kakitani, N. Matsuda, A. Yoshimori, and N. Mataga, *Prog. React. Kinet.* **20**, 347 (1995).

115. N. Mataga, J. Kanda, and T. Okada, *J. Phys. Chem.* **90**, 3880 (1986).

116. N. Mataga, Y. Kanda, T. Asahi, H. Miyasaka, T. Okada, and T. Kakitani, *Chem. Phys.* **127**, 239 (1988).

117. T. Kakitani, A. Yoshimori, and N. Mataga, *J. Phys. Chem.* **96**, 5385 (1992).

118. K. Kikuchi, T. Katagiri, T. Niwa, Y. Takahashi, T. Suzzuki, H. Ikeda, and T. Miyashi, *Chem. Phys. Lett.* **193**, 155 (1992).

119. K. Kikuchi, T. Niwa, Y. Takahashi, H. Ikeda, and T. Miyashi, *J. Phys. Chem.* **97**, 5070 (1993).

120. O. G. Berg, *Chem. Phys.* **31**, 47 (1978).

121. S. Yasui, M. Tsujimoto, K. Itoh, and A. Ohno, *J. Org. Chem.* **65**, 4715 (2000).

122. S. Yasui, K. Itoh, M. Tsujimoto, and A. Ohno, *Bull. Chem. Soc. Japan* **75**, 1311 (2002).

123. A. I. Burshtein and K. L. Ivanov, *Phys. Chem. Chem. Phys.* **4**, 4115 (2002).

124. A. A. Kipriyanov, A. B. Doktorov, and A. I. Burshtein, *Chem. Phys.* **76** 149, 163 (1983).

125. O. A. Igoshin and A. I. Burshtein, *J. Lumin.* **92**, 123 (2000).

126. J. Sung and S. Lee, *J. Chem. Phys.* **112**, 2128 (2000).

127. S. Mukamel, *Chem. Phys.* **37**, 33 (1979).

128. R. W. Stoughton and G. K. Rollefson, *J. Am. Chem. Soc.* **62**, 2264 (1940).

129. K. C. Hodges and V. K. La Mer, *J. Am. Chem. Soc.* **70**, 722 (1948).

130. R. E. Föll, H. E. A. Kramer, and U. E. Steiner, *J. Phys. Chem.* **94**, 2476 (1990).

131. P.-A. Muller, C. Högeman, X. Allonas, P. Jacques, and E. Vauthey, *Chem. Phys. Lett.* **326**, 321 (2000).

132. M. Yang, S. Lee, and K. J. Shin, *J. Chem. Phys.* **108**, 117, 8557, 9069 (1998).

133. A. I. Burshtein, I. V. Gopich, and P. A. Frantsuzov, *Chem. Phys. Lett.* **289**, 60 (1998).

134. A. A. Kipriyanov, O. A. Igoshin, and A. B. Doktorov, *Physica A* **268**, 567 (1999); O. A. Igoshin, A. A. Kipriyanov, and A. B. Doktorov, *Chem. Phys.* **244**, 371 (1999).

135. I. V. Gopich and N. Agmon, *J. Chem. Phys.* **110**, 10433 (1999).

136. I. V. Gopich and A. I. Burshtein, *J. Chem. Phys.* **109** 2833 (1998).

137. I. V. Gopich, A. A. Kipriyanov, and A. B. Doktorov, *Chem. Phys. Rep.* **14**, 1443 (1995).

138. K. Allinger and A. Blumen, *J. Chem. Phys.* **72**, 4608 (1980).

139. S. Lee and M. Karplus, *J. Chem. Phys.* **86**, 1883 (1987).

140. S. Lee and M. Karplus, *J. Chem. Phys.* **86**, 1904 (1987).

141. S. Lee, M. Yang, K. J. Shin, K. J. Choo, and D. Lee, *Chem. Phys.* **156**, 339 (1991).

142. J. Sung, K. J. Shin, and D. Lee, *J. Chem. Phys.* **101**, 7241 (1994).

143. T. Bandyopadhyay, *J. Chem. Phys.* **102**, 9557 (1995).

144. T. Bandyopadhyay, *J. Chem. Phys.* **102**, 5049 (1997).

145. W. Naumann and A. Molski, *J. Chem. Phys.* **103**, 3474 (1995).

146. W. Naumann, *J. Chem. Phys.* **114**, 4791 (2001).

147. A. I. Burshtein, *J. Chem. Phys.* **103**, 7927 (1995).

148. H.-J. Wolff, D. Bürßner, and U. Steiner, *Pure and Appl. Chem.* **67 (1)**, 167 (1995).

149. E. B. Krissinel, A. I. Burshtein, N. N. Lukzen, and U. E. Steiner, *Mol. Phys.* **96**, 1083 (1999).

150. A. I. Burshtein, E. Krissinel, and U. E. Steiner, *Phys. Chem. Chem. Phys.* **3**, 198 (2001).

151. A. A. Neufeld, A. I. Burshtein, G. Angulo, and G. Grampp, *J. Chem. Phys.* **116**, 2472 (2002).

152. A. I. Burshtein and A. A. Neufeld, *J. Phys. Chem. B* **105**, 12364 (2001).

153. A. A. Zharikov and N. V. Shokhirev, *Chem. Phys. Lett.* **186**, 253 (1991).

154. N. N. Korst, *Theor. Math. Phys.* **6**, 196 (1971).

155. M. Tachiya, *J. Chem. Phys.* **71** 1276 (1979).

156. B. Sipp and R. Voltz *J. Chem. Phys.* **79**, 434 (1983).

157. M. Abramowitz and I. A. Stegun, eds., *Handbook of Mathematical Functions*, (Dover, New York, 1965).

158. A. I. Burshtein and N. V. Shokhirev, *J. Phys. Chem. A* **101**, 25 (1997).

159. A. Yoshimori, K. Watanabe, and T. Kakitani, *Chem. Phys.* **201**, 35 (1995).

160. E. Pines and D. Huppert, *J. Chem. Phys.* **84**, 3576 (1986).

161. E. Pines, D. Huppert, and N. Agmon, *J. Chem. Phys.* **88**, 5620 (1988).

162. N. Agmon, E. Pines, and D. Huppert, *J. Chem. Phys.* **88**, 5631 (1988).

163. D. Huppert, E. Pines, and N. Agmon, *J. Opt. Soc. Am. B* **7**, 1545 (1990).

164. E. Pines and G. Fleming, *J. Phys. Chem.* **95**, 10448 (1991).

165. E. Pines and G. Fleming, *Chem. Phys.* **183**, 393 (1994).

166. A. I. Burshtein and B. I. Yakobson, *Int. J. Chem. Kinet.* **12**, 261 (1980).

167. A. I. Burshtein, A. B. Doktorov, and A. V. Morozov, *Chem. Phys.* **104**, 1 (1986).

168. I. V. Gopich, K. M. Solntsev, and N. Agmon, *J. Chem. Phys.* **110**, 2164 (1999).

169. N. Agmon, *J. Chem. Phys.* **110**, 2175 (1999).

170. N. Agmon and A. Szabo, *J. Chem. Phys.* **92**, 5270 (1990).

171. H. Kim and K. J. Shin, *PRL* **82**, 1578 (1999).

172. N. Agmon and I. V. Gopich, *Chem. Phys. Lett.* **302**, 399 (1999).

173. J. Sung, K. J. Shin, and S. Lee, *J. Chem. Phys.* **109**, 9101 (1998).

174. A. Molski and J. Keizer, *J. Chem. Phys.* **96**, 1391 (1992).

175. W. Naumann, *J. Chem. Phys.* **111**, 2414 (1999).

176. A. Molski and W. Naumann, *J. Chem. Phys.* **103**, 10050 (1995).

177. A. Szabo, *J. Chem. Phys.* **95**, 2481 (1991).

178. A. I. Burshtein and P. A. Frantsuzov *J. Chem. Phys.* **107**, 2872, (1997).

179. A. I. Burshtein and P. A. Frantsuzov, *Chem. Phys.* **212**, 137 (1996).

180. I. V. Gopich and A. Szabo, *J. Chem. Phys. 117*, 507 (2002).

181. I. V.Gopich, Private communication.

182. E. B. Krissinel and N. V. Shokhirev, Differential approximation of spin-controlled and anisotropic diffusional kinetics (Russian), in Siberian Academy '*Mathematical Methods in Chemistry,*' Preprint 30 (1989); Diffusion-controlled reactions 22, in *DCR User's Manual* 11-20-1990.

183. D. Beckert, M. Plüschau, and K.P. Dinse, *J. Phys. Chem.* **96**, 3193 (1992).

184. M. Plüschau, G. Kroll, K. P. Dinse, and D. Beckert, *J. Phys. Chem.* **96**, 8820 (1992).

185. A. I. Burshtein and P .A. Frantsuzov, *Chem. Phys. Lett.* **263**, 513 (1996).

186. G. Grampp and G. Hetz, *Ber. Bunsenges. Phys. Chem.* **96**, 198 (1992).

187. A. I. Burshtein and A. Yu. Sivachenko, *Chem. Phys.* **235**, 257 (1998).

188. A. I. Burshtein, *J. Lumin.* **93**, 229 (2001).

189. A. I. Burshtein, A. A. Neufeld, and K. L. Ivanov, *J. Chem. Phys.* **115**, 2652 (2001).

190. M. Bixon and R. Zwanzig, *J. Chem. Phys.* **75**, 2354 (1981).

191. R. D. Mussell and D. Nocera, *J. Am. Chem. Soc.* **110**, 2764 (1988); *Inorg. Chem.* **29**, 3711 (1990).

192. P. Szrebowaty and A. Kapturkiewicz, *Chem. Phys. Lett.* **328**, 160 (2000).

193. A. Kapturkiewicz and P. Szrebowaty, *J. Chem. Soc. Dalton Trans.* 3219 (2002).

194. A. I. Burshtein, A. A. Neufeld, and K. L. Ivanov, *J. Chem. Phys.* **115**, 10464 (2001).

195. N. N. Lukzen, E. B. Krissinel, O. A. Igoshin, and A. I. Burshtein *J. Phys. Chem. A* **105**, 19 (2001).

196. A. Matsuyama, K. Maeda, and H. Murai, *J. Phys. Chem. A* **103**, 4137 (1999).

197. M. R. Eftink and C. A. Ghiron, *J. Phys. Chem.* **80**, 486 (1976).

198. S. G. Fedorenko and A. I. Burshtein, *J. Chem. Phys.* **97**, 8223 (1992).

199. P. A. Frantsuzov and A. I. Burshtein, *J. Chem. Phys.* **109**, 5957 (1998).

200. O. A. Igoshin and A. I. Burshtein, *J. Chem. Phys.* **112**, 10930 (2000).

201. W. Naumann and A. Szabo, *J. Chem. Phys.*, **107**, 402 (1997).

202. A. I. Burshtein and O. A. Igoshin, *J. Chem. Phys.* **111**, 2200 (1999).

203. P. A. Frantsuzov, O. A. Igoshin, and E. B. Krissinel, *Chem. Phys. Lett.* **317**, 481 (2000).

204. D. L. Dexter, *J. Chem. Phys.* **21**, 836 (1953).

205. M. Tomkievich and M. Cocivera, *Chem. Phys. Lett.* **8**, 595 (1971).

206. W. Bube, R. Haberkorn, and M. E. Michel-Beyerle, *J. Am. Chem. Soc.* **100**, 5993 (1978).

207. L. Sterna, D. Ronis, S. Wolfe, and A. Pines, *J. Chem. Phys.* **73**, 5493 (1980).

208. H. Hayashi and S. Nagakura, *Bull. Chem. Soc. Japan* **57**, 322 (1984).

209. A. A. Zharikov and N. V. Shokhirev, *Z. Phys. Chem.* **99**, 2643 (1992).

210. D. Bürßner, H-J. Wolff, and U. Steiner, *Z. Phys. Chem.* **182**, 297 (1993); *Angewandte Chemie Int. Ed. Engl.* **33**, 1772 (1994).

211. A. Abragam, *The Principles of Nuclear Magnetism* (Clarendon Press, Oxford, 1961).

212. M. Ottolenghi, *Charge Transf. Complexes* **6**, 153 (1973).

213. K. Kikuchi, M. Hoshi, T. Niwa, Y. Takahashi, and T. Miyashi, *J. Phys. Chem.* **95**, 38 (1991).

214. T. Niwa, K. Kikuchi, N. Matsusita, M. Hayashi, T. Katagiri, Y. Takahashi, and T. Miyashi *J. Phys. Chem.* **97**, 11960 (1993).

215. H. A. Montejano, J. J. Cosa, H. A. Garrera, and C. M. Previtali, *J. Photochem. Photobiol. A: Chem.* **86**, 115 (1995).

216. G. P. Zanini, H. A. Montejano, and C. M. Previtali, *J. Photochem. Photobiol. A: Chem.* **132**, 161 (2000).

217. K. Schulten, H. Staerk, A. Weller, H.-J. Werner, and B. Nickel, *Z. Phys. Chem.* **NF101**, 371 (1976).

218. H. -J. Werner, H. Staerk, and A. Weller, *J. Chem. Phys.* **68**, 2419 (1978).

219. A. Weller, H. Staerk, and R. Treichel, *Faraday Discuss. Chem. Soc.* **78**, 217 (1984).

220. V. S. Gladkikh, A. I. Burshtein, G. Angulo, and G. Grampp, *PCCP* **5**, 2581 (2003).

221. R. G. Mints and A. A. Pukhov, *Chem. Phys.* **87**, 467 (1984).

222. Z. Schulten and K. Schulten, *J. Chem. Phys.* **66**, 4616 (1977).

223. G. Angulo, V. S. Gladkikh, and A. I. Burshtein (in preparation).

224. The computer program QYield, developed by E. B. Krissinel, is available from his Website: *http://www.fh.huji.ac.il/ krissinel*.

225. K. Kikuchi, *J. Photochem. Photobiol. A: Chem.* **65**, 149 (1992).

226. W. Adam, D. Reinhardt, and C. R. Saha-Möller, *Analyst* **121**, 1527 (1996).

227. W. Adam, I. Bronstein, R. F. Vasil'ev, and A. V. Trofimov, *Russ. Chem. Bull.* **49**, 659 (2000).

228. W. Adam, I. Bronstein, A. V. Trofimov, and R. F. Vasil'ev, *J. Am. Chem. Soc.* **121**, 958 (1999).

229. W. Adam and A. Trofimov, *J. Org. Chem.* **65**, 6474 (2000).

230. W. Adam, M. Matsumoto, and A. V. Trofimov, *J. Am. Chem. Soc.* **122**, 8631 (2000).

231. A. I. Burshtein, *Chem. Phys.* **289**, 251 (2003).

232. H. Schomburg, H. Staerk, and A. Weller, *Chem. Phys. Lett.* **21**, 433 (1973).

233. V. A. Belyakov, R. F. Vasil'ev, and G. F. Fedorova, *Bull. Acad. Sci. USSR, Div. Phys. Sci.* **42**, 621 (1978).

234. P. Jacques and X. Allonas, *J. Photochem. Photobiol. A: Chem.* **78**, 1 (1994).

235. T. N. Inada, K. Kikuchi, Y. Takahasaki, H. Ikeda, and T. Miyashi, *J. Photochem. Photobiol. A: Chem.* **137**, 93 (2000).

236. A. A. Kipriyanov, I. V. Gopich, and A. B. Doktorov, *Physica* A **205**, 585 (1994); *Chem. Phys.* **187**, 251 (1994).

237. A. Molski and J. Keizer, *J. Phys. Chem.* **97**, 8707 (1993).

238. A. Molski, *Chem. Phys.* **182**, 203 (1994); *J. Phys. Chem.* **99**, 2353 (1995).

239. L. Monchick, J. L. Magee, and A. H. Samuel, *J. Chem. Phys.* **26**, 935 (1957).

240. T. R. Waite, *Phys. Rev.* **107**, 463 (1957).

241. W. Naumann, N. V. Shokhirev, and A. Szabo, *Phys. Rev. Lett.* **79**, 3074 (1997).

242. J. Sung, K. J. Shin, and S. Lee, *J. Chem. Phys.* **107**, 9418 (1997).

243. J. Vogelsang and M. Hauser, *J. Phys. Chem.* **94**, 7488 (1990); *Ber. Bunsenges. Phys. Chem.* **94**, 1326 (1990).

244. J. Vogelsang, *J. Chem. Soc. Faraday Trans.* **89**, 15 (1993).

245. M. N. Berberan-Santos and J. M. G. Martinho, *Chem. Phys. Lett.* **178** 1 (1991).

246. J. Sung and S. Lee, *J. Chem. Phys.* **111**, 796 (1999).

247. J. Sung and S. Lee, *J. Chem. Phys.* **111**, 10159 (1999).

248. B. Stevens and C. J. Biver III, *Chem. Phys. Lett.* **226**, 268 (1994).

249. A. I. Burshtein, I. V. Khudyakov, and B. I. Yakobson, *Prog. React. Kinet.* **13**, 221 (1984).

250. B. Stevens and D. N. McKeithan, *J. Photochem. Photobiol. A: Chem.* **40**, 1 (1987).

251. B. Stevens, C. J. Biver III, and D. N. McKeithan, *Chem. Phys. Lett.* **187**, 590 (1991).

252. I. V. Gopich and A. B. Doktorov, *J. Chem. Phys.* **105**, 2320 (1996).

253. I. V. Gopich and N. Agmon, *Phys. Rev. Lett.* **84**, 2730 (2000).

254. N. Agmon and I. V. Gopich, *J. Chem. Phys.* **112**, 2863 (2000).

255. I. V. Gopich, A. A. Ovchinnikov, and A. Szabo, *Phys. Rev. Lett.* **86**, 922 (2001).

256. B. U. Felderhof and R. B. Jones *J. Chem. Phys.* **103**, 10201 (1995).

257. I. V. Gopich and A. I. Burshtein *J. Chem. Phys.* **113**, 2932 (2000).

258. A. B. Doktorov and A. A. Kipriyanov, *Physica A* **319**, 253 (2003).

259. A. Szabo and R. Zwanzig, *J. Stat. Phys.* **65**, 1057 (1991).

260. W. Naumann, *J. Chem. Phys.* **101**, 10953 (1994).

261. A. V. Popov and N. J. Agmon, *Chem. Phys.* **115**, 8921 (2001) .

262. B. Nickel, H. E. Wilhelm, and A. A. Ruth, *Chem. Phys.* **188**, 267 (1994).

263. N. A. Efremov, S. G. Kulikov, E. I. Personov, and Yu. V. Romanovskii, *Chem. Phys.* **128**, 9 (1988); *Sov. Phys. Sol. State* **34**, 239 (1992).

264. A. I. Burshtein and P. A. Frantsuzov, *J. Lumin.* **78**, 33 (1998).

265. B. Nickel, H. E. Wilhelm, and C. P. Jaensch, *Opt. Spectrosc.* **83**, 541 (1997).

266. D. S. Northrop and O. Simpson, *Proc. Roy. Soc.* **244**, 377 (1958).

267. A. I. Burshtein, *Sov. Phys. Sol. State* **5**, 922 (1963).

268. A. I. Burshtein, *Kinetica i Kataliz* **5**, 414 (1964).

269. M. Pope and C. E. Swenberg, *Electronic Processes in Organic Crystals*, (Oxford Univ. Press, 1982), pp. 159, 479.

270. S. G. Ballard and D. C. Mauzerall, *J. Chem. Phys.* **72**, 933 (1980).

271. D. Mauzerall and S. G. Ballard, *Annu. Rev. Phys. Chem.* **33**, 377 (1982).

272. F. Elisei, G. Favaro, and H. Görner, *J. Photochem. Photobiol. A: Chem.* **59**, 243 (1991).

273. M. von Raumer, P. Suppan, and P. Jacques, *J. Photochem. Photobiol. A: Chem.* **105**, 21 (1997).

274. S. F. Kilin, M. S. Mikhelashvilli, and I. M. Rozman, *Izv. Akad. Nauk (Physika)* **42**, 414 (1978).

275. E. B. Krissinel, O. A. Igoshin, and A. I. Burshtein, *Chem. Phys.* **247**, 261 (1999).

276. S. I. Choi and S. A. Rice, *J. Chem. Phys.* **38**, 366 (1963).

277. V. A. Benderskii, V. Kh. Brikenshtein, A. I. Burshtein, A. G. Lavrushko, A. G. Prikhozhenko, and P. G. Filipov, *Phys. Stat. Sol.* **95**, 47 (1979).

278. A. I. Burshtein, *Sov. Phys. JETP* **57**, 1165 (1983).

279. K. Zachariasse, *Chemiluminescence from Radical Ion Recombination*, Ph.D. thesis, Free Univ., Amsterdam, 1972.

280. A. Weller and K. Zachariasse, in *Chemiluminescence and Bioluminescence*, M. J. Cormier, D. M. Hercules, and J. Lee, eds. (Plenum, New York, 1973), p. 169.

281. A. B. Doktorov, K. L. Ivanov, N. N. Lukzaen, and V. A. Morozov, *J. Chem. Phys.* **117**, 7995 (2002).

282. K. L. Ivanov, N. N. Lukzen, A. A. Kipriyanov, A. B. Doktorov *Phys. Chem. Chem. Phys.* **6**, 1706 (2004).

283. K. L. Ivanov, N. N. Lukzen, A. B. Doktorov *Phys. Chem. Chem. Phys.* **6**, 1719 (2004).

284. K. L. Ivanov, N. N. Lukzen, V. A. Morozov, and A. B. Doktorov, *J. Chem. Phys.* **117**, 9413 (2002).

NONLINEAR DYNAMIC SUSCEPTIBILITIES AND FIELD-INDUCED BIREFRINGENCE IN MAGNETIC PARTICLE ASSEMBLIES

YURIY L. RAIKHER AND VICTOR I. STEPANOV

Institute of Continuous Media Mechanics
Ural Branch of RAS, Perm, Russia

CONTENTS

Advances in Chemical Physics, Volume 129, edited by Stuart A. Rice
ISBN 0-471-44527-4   Copyright © 2004 John Wiley & Sons, Inc.

I. INTRODUCTION

The goal of this chapter is to present a comprehensive review of the work done since the 1980s on the low-frequency magnetodynamics and magnetoorientational behavior (field-induced birefringence) in the assemblies of single-domain (superparamagnetic) particles. The approach that we use—the Fokker–Planck equation, which is also known as the *rotary diffusion equation* or the *micromagnetic kinetic equation*—allows us to consider from a unified viewpoint the dynamical magnetization (its linear as well as nonlinear regimes) of such particles and their dilute assemblies as well as the evolution of the orientational order parameter in such assemblies.

Of course, the latter scope of problems refers to a specific kind of nanodisperse system, where the assemblies of the magnetic particles are orientable, that is, where, the particles possess the mechanical degrees of freedom. Such magnetic-field-sensitive ultradispersions are now widely known as *magnetic fluids* or *ferrocolloids* and represent an important class of composite media of multipurpose destination [1–5]. One of the most inspiring fields of application for magnetic fluids is medicine. In particular, the effect of an external alternating field on the ferrocolloid drops injected in a live organism is now under study as a prospective method of creating a local hyperthermia, see Ref. 6, for example, where such a method of tumor treatment is described. Moreover, it turns out that field-induced optical effects are a convenient method for diagnostics and probing of the magnetic fluids in medical applications [7,8]. Another fascinating example of orientable magnetic dispersions are *ferrolacquers*, that is, liquid mixtures containing single-domain particles that are used for manufacturing magnetic tapes and disks. Before such a lacquer will find its way on the given surface, where it will dry, there are a lot of possibilities for the particles in it to be oriented [9]. The same applies for the particles introduced in paints and inks for magnetic-field-controlled deposition [10,11].

All the new applications notwithstanding, the traditional situation, where the nanoparticles are considered as fully trapped in a solid matrix, still attracts the most attention. The "customers" for the knowledge of the behavior of fine magnetic particles in wide temperature ranges and under magnetic fields of arbitrary frequency and amplitude are the numerous community of researches and engineers who work on developing and improvement of materials and methods of magnetic recording. Being historically the first practical method of recording and data storage (RAM on magnetic rings and data storage on magnetic tapes), this method still remains the most convenient and less expensive. The progress achieved since the early 1950s years is impressive: nowadays hard disks of PCs, which are 2–3 in. in diameter, store dozens of gigabytes of information. Notably, the magnetic records are long-living and do not need any energy to their maintenance. It is not surprising that research in this field has high priority, each real advance step brings to its authors fame and wealth and to all the human community—enhanced ability of processing information. Right now we are witnessing a new breakthrough caused by wide commercial production of supersensitive reading heads, which for their work employ the giant magnetoresistance effect in multilayers. This generation of so-called spinotropic devices is considered to be very promising for magnetic recording.

According to the existing estimates [12], we are approaching the epoch when the areal density of records in commercially available devices will reach the value 1 Tbit (terabit) $= 10^{12}$ bits per square inch. But the researchers are already eager to look over this barrier to farther future. What it there? The answer has been known for a long time, however. There looms the unshaken limit that nature had set to the human desire to miniaturize the magnetic recording devices. This fundamental law of magnetic recording is expressed by the well-known single-line formula that for the first time had been derived in 1949 by Louis Néel [13]. The result obtained led Néel to reach certain conclusions on the universal nonhysteretic (*superparamagnetic*) behavior acquired by any ferrite or ferromagnet had it been transformed to a state of nanosize particles. The simple formula has the form

$$\tau = \tau_0 \exp\left(\frac{E_a}{kT}\right) \tag{4.1}$$

and by its physical meaning it is the estimate of the time lapse that the magnetic moment $\mathbf{\mu}$ of a single-domain particle that possesses the easy-axis anisotropy and, accordingly, the energy E_a, spends in a state with definite orientation at a given temperature T. For longer times, $t > \tau$, due to thermal fluctuations drastically increases the probability that μ will turn upside down at least once. In this way, the memory of the initial state will be lost. The prefactor time τ_0 in Eq. (4.1) has the meaning of a microscopic time and is the parameter that characterizes a particular magnetic material; its typical values lie in the range $\tau_0 \sim 10^{-9}$–10^{-10} s.

Therefore, superparamagnetism [14], that is, interference of the thermo-fluctuational effects in the magnetization processes, is a fundamental obstacle toward enhancing the density of magnetic recording. Imagine that we use a PC with a hard disk, where the size of the magnetic grains is constantly diminishing. At first, we will see a positive effect; with smaller particles the resolution of record will grow, and the PC will be able to write and read a growing amount of data. Then, as a result of the particle size diminution, the time τ will become comparable to the mean time interval between writing and reading, and certain effects, analogous to the symptoms found in a mild case of Alzheimer's disease, such as fits of forgetfulness and occasional memory blackouts, will emerge. Further "grinding" of the particles in the working layer will lead to the terminal stage—after writing down huge lumps of data, the PC will forget them immediately. When reading is attempted, only monotonous (white) noise will emerge, analogous to complete amnesia.

Let us take a slightly deeper look at the problem. Consider an ideal situation where each bit of information is associated with the magnetic state of a single particle. The magnetic anisotropy energy E_a that fixes the direction of the magnetic moment of a particle (stores the record) may be of various origins. One can distinguish the crystallographic, magnetostatic [i.e., induced by the nonsphericity (anisometricity) of the sample shape], striction (induced by tension), stoichiometry (imposed by the anomalies in the atomic environment of spins), and other types of anisotropy [15]. However, in the overpowering majority of cases one encounters one of two simple variants:

$$E_{av} = KV \quad \text{or} \quad E_{as} = K_S S \qquad (4.2)$$

where V is the particle volume while S is its lateral area. Accordingly, the quantities K and K_S are called, respectively, the *volume* and *surface densities* of magnetic anisotropy. A sufficiently high value of the volume magnetic anisotropy is $K \sim 1.5 \times 10^6$ erg/cm^3. Using this formula, we find that at room temperature ($kT \sim 4 \times 10^{-14}$ erg), the ratio $E_{av}/kT \sim 40$ (stability of the record during 10 years) is ensured with the particles of volume $V \gtrsim 10^{-18}$ cm^3. Assuming the particle to be spherical, for its linear size (diameter) one finds $d \sim 12$ nm.

Now let us estimate the density of particle distribution over the surface. The area that occupies one particle at a plane is $S_1 \sim d^2 \sim 1.5 \times 10^{-12}$ cm^2.

Consider the grains arranged in a square lattice with the minimal center-to-center interparticle distance equal $2d$ in order to reduce the stray fields of the neighbors. With such a pattern the number density of the particles is $\sim 10^{12}$ per square inch, that is about 1 Tbit/in.^{2}, about the same value one finds if a surface anisotropy with the density $K_S \sim 1$ erg/cm^2 is assumed for the particles. Note that this simple estimate coincides by an order of magnitude with that derived in a more sophisticated way [12]. However, one should consider the 1 Tbit level as an ideal rather that a real assertion, since it is based on the assumption that the dimensions of the writing and reading heads are of the order of d. As yet, with allowance for a finite size of a head and the fact that one bit is written at the cluster of several grains, the highest densities reached are 40 Gbit/in.^{2} for the longitudinal recording [16] and up to 60 Gbit/in.^{2} for the transversal one [17].

We remark, however, that the above-discussed transition to the regime of spontaneous remagnetization of particles is an undesirable effect only if considered from the perspective of magnetic recording technology. To treat the problem more widely, from the position of general physics, it becomes clear that to investigate both the spontaneous motion of the magnetic moment (free orientational diffusion) and its forced diffusion motion would be no less interesting than the magnetodynamic studies restricted effectively to the $T = 0$ domain. The main merit of the physics of small particles is that the theory being developed allows one to go study all the important size ranges and to get an idea of the role of thermofluctuational effects. In this way, one obtains a description of the phenomena from its dynamies driven end (large particles and/or low temperatures) to its essentially diffusional limit (small particles and/or high temperatures). In particular, with respect to the magnetic recording, this indicates the limit beyond which the data cannot feasibly be condensed further.

The fluctuational domain, where noise effects are very pronounced, has been found very interesting as such and very rich in novel effects [18]. In the systems, where the ability for oscillations is suppressed by strong noisy motion (overdamped oscillator), the famous *stochastic resonance* has been discovered [19,20], as have the perhaps less widely known, but equally fascinating, noise- and force-induced resonances [21–24]. The fact that single-domain magnetic particles in the superparamagnetic regime $(E_a/kT \sim 1)$ are extremely convenient (one may say, ideal) objects for observation and investigation of such phenomena had been mentioned many times [25–31]. We would like to add that theoretical models equivalent to that of a single-domain particle were introduced to describe the functional properties of magnetoresistive heads [32–34].

The mathematical basis for our work is formed by the technique of solving the orientational rotary diffusion (Fokker–Planck) equation by reducing this partial derivative equation to an infinite set of ordinary differential equations that describe the time evolution of the statistical moments of the distribution function. The latter renders either the probability density $W(\mathbf{\mu}, t)$ for the angle coordinates of the magnetic moment $\mathbf{\mu}$ or $W(\mathbf{n}, t)$ that is the distribution for the components of the orientational tensor of the particle geometry axis $\mathbf{n}$. In a number of cases the joint distribution function $W(\mathbf{\mu}, \mathbf{n}, t)$ is considered. The equations obtained are finally solved numerically by sweep methods. The algorithm described above has proved to be very efficient for this type of problem since the emerging matrices are rather sparse. A completely different (at first glance) method has been described in the book [35]. It uses the formalism of the continued fractions and provides expressions for solutions of the Fokker–Planck equation, which when written symbolically, appear to be exact [36]. However, any conversion of a continued fraction into some particular result necessitates a numerical procedure. In fact, both methods (sweep and continued fraction) are identical in terms of content, meaning, and ability.

II. GENERAL FRAMEWORK: THE MICROMAGNETIC FOKKER–PLANCK EQUATION

A. Reference Timescales

1. Magnetodynamic Equation

The phenomenological equation that determines the motion of the magnetic moment of a ferromagnetic sample is known generically as the *Landau–Lifshitz–Gilbert equation* and has two basic modifications. The first form

$$\frac{d\mathbf{e}}{dt} = -\gamma(\mathbf{e} \times \mathbf{H}_{\text{eff}}) - \alpha\gamma(\mathbf{e} \times (\mathbf{e} \times \mathbf{H}_{\text{eff}})) \tag{4.3}$$

was proposed in Ref. 37 and in historical aspect, it should be called the *Landau–Lifshitz equation* proper. When addressed to a single-domain particle, the notations in Eq. (4.3) mean: $\mathbf{e} \equiv \mathbf{\mu}/\mu$ is the unit vector of the magnetic moment, γ is the gyromagnetic ratio for electrons, and α is the nondimensional constant (the constant of spin–lattice relaxation) that determines the rate of damping of the Larmor precession of the vector $\mathbf{\mu}$. The magnitude of the latter is set to be $\mu = I_s V_m$, where I_s is the saturation magnetization of the ferromagnetic (or ferrite) substance and V_m is the volume of that part of the particle that possesses bulk ferromagnetic properties. Thus we allow for the fact that a particle of a total volume V may bear a "dead" shell on its surface [38,39] or that the particle surface layer may be effectively demagnetized because of its specific magnetic structure; see Refs. 40–44 for particular examples. The effective magnetic field $\mathbf{H}_{\text{eff}}$ in Eq. (4.3) sums up all the internal and external fields acting on the vector $\mu\mathbf{e}$.

The second form of the magnetodynamic equation was introduced by Gilbert [45]:

$$\frac{de}{dt} = -\gamma^*(e \times H_{\text{eff}}) + \alpha_0 \left(e \times \frac{de}{dt}\right) \qquad (4.4)$$

As one sees from comparison, the Landau–Lifshitz and Gilbert equations do differ in the form of the damping term. Specifically, the Gilbert equation embodies explicitly the well-known linear hypothesis of irreversible thermodynamics, which establishes that dissipation is proportional to the rate of change of the corresponding dynamic variable. The pioneering role of Eq. (4.4) is that Gilbert was the first to show how this principle works for a gyrotropic medium; note the presence of the time derivative of e in both parts of the equation. However, soon afterward it was proved [46] that Eqs. (4.3) and (4.4) are in fact identical and transform into each other on the following renormalization of the parameters:

$$\gamma^* = \gamma(1 + \alpha_0^2), \qquad \alpha = \alpha_0 \qquad (4.5)$$

Therefore, on writing

$$\frac{de}{dt} = -\frac{\gamma^*}{1 + \alpha^2}(e \times H_{\text{eff}}) - \frac{\alpha\gamma^*}{1 + \alpha^2}(e \times (e \times H_{\text{eff}})) \qquad (4.6)$$

we present the Gilbert equation in the Landau–Lifshitz form; owing to coincidence of α and α_0 further on, we omit the subscript at the damping parameter. The above mentioned equivalence between both equations notwithstanding, it had been established that for the magnetic materials with enhanced level of dissipation the Gilbert form is more appropriate than the other one. The main point of this advantage is that the resonance frequency (field) determined in terms of the Gilbert material parameters weakly depends on α while the same parameter obtained with the aid of the Landau–Lifshitz equation reduces considerably with α. To prove this, let us take the Gilbert equation for a simple isotropic ferromagnet under the combination of a constant (magnetizing) field and a weak probing field $\propto \exp(-i\omega t)$ and derive from it the linear dynamic susceptibility $\chi(\omega)$. Setting $H_{\text{eff}} = H_0$ and associating (as in a standard experiment) the resonance field with the value of H_0, which delivers maximum to the out-of-phase (imaginary) component of χ at a fixed value $\omega = \omega_0$, we find

$$H_0^{(\text{res})} = \frac{\omega_0}{\gamma^*}\sqrt{2\sqrt{1 + \alpha^2} - (1 + \alpha^2)} \simeq \frac{\omega_0}{\gamma^*}\left(1 - \tfrac{1}{8}\alpha^4\right) \qquad (4.7)$$

where the last simplification corresponds to the frequently used limit $\alpha \ll 1$. With regard to this, one may consider the dependence of H_{res} on α as a weak one. Meanwhile, in the Landau–Lifshitz form one has

$$H_0^{(res)} = \frac{\omega_0}{\gamma} \sqrt{\frac{2}{(1+\alpha^2)^{3/2}} - \frac{1}{1+\alpha^2}} \simeq \frac{\omega_0}{\gamma(1+\alpha^2)} \qquad (4.8)$$

where the last result refers to the $\alpha \ll 1$ limit. Comparing (4.7) and (4.8), one sees that the Landau–Lifshitz resonance field contains an additional factor $\propto 1/(1+\alpha^2)$, and thus is more sensitive to α than is the Gilbert one. Of course, at $\alpha \to 0$ both approaches coincide. The behaviors of the Gilbert and Landau–Lifshitz resonance fields at arbitrary α are shown in Figure 4.1. Turning of H_0^{res} into zero at $\alpha = \sqrt{3}$ and its nonexistence at greater α values means that in this range the function $\chi''(H_0)$ lacks a maximum and, as its argument grows, diminishes monotonically.

In the framework of the standard linear response theory, the relaxation times of precession are determined in the $\alpha \ll 1$ limit according to the generic formula $\tau = (\alpha\gamma H)^{-1}$. Setting the field equal to its resonance value, we get

$$\tau = \begin{cases} 1/\alpha\omega_0 & \text{Gilbert case} \\ (1+\alpha^2)/\alpha\omega_0 & \text{Landau–Lifshitz case.} \end{cases} \qquad (4.9)$$

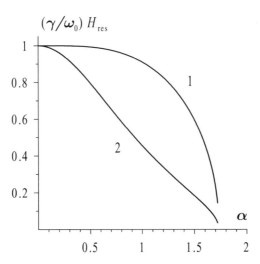

Figure 4.1. Comparison of the resonance fields as functions of the damping parameter in the Gilbert (1) and Landau–Lifshitz (2) representations.

For the case when the external magnetizing field is absent, the reference relaxation time of precession is defined on the assumption that the particle possesses uniaxial magnetic anisotropy with the energy density K. Then the Larmor precession in the intrinsic anisotropy field is considered, and the generic formula for the relaxation time emerges as $\tau = I_s/2\alpha\gamma K$. Specifying it for the two representations under discussion, one gets

$$
\tau_0 = \begin{cases} I_s(1+\alpha^2)/2\alpha\gamma^* K & \text{Gilbert case} \\ I_s/2\alpha\gamma^* K & \text{Landau–Lifshitz case,} \end{cases} \tag{4.10}
$$

which expressions, as they should, coincide at $\alpha \to 0$. Therefore, τ_0 provides a basic reference timescale for high-frequency magnetic processes in a single-domain particle in the absence of external fields and thermal fluctuations. To estimate τ_0 by the order of magnitude, we take $I_s \sim 400\,\mathrm{G}$ and $K \sim 10^6\,\mathrm{erg/cm}^3$, which is typical for a nanodisperse ferrite; set $\alpha = 0.1$ (moderate quality factor of the precession); and recall that $\gamma \simeq 1.7 \times 10^7\,\mathrm{rad/Oe\cdot s}$. This yields $\tau_0 \sim 10^{-10}\,\mathrm{s}$.

2. Rotary Mobility Coefficient for the Magnetic Moment

For our calculations we choose a spherical coordinate framework with the polar axis directed along the vector $\boldsymbol{H}_{\text{eff}}$. Then the basic vectors of the problem are expressed as

$$
\boldsymbol{H}_{\text{eff}} = (0,0,H_{\text{eff}}), \qquad \boldsymbol{e} = (\sin\vartheta\cos\varphi, \sin\vartheta\sin\varphi, \cos\vartheta) \tag{4.11}
$$

As Eqs. (4.3) and (4.4) show, if deviated from its equilibrium position $\boldsymbol{e} \parallel \boldsymbol{H}_{\text{eff}}$ or $\vartheta = 0$, the magnetic moment $\boldsymbol{e}$ begins to precess around the direction of $\boldsymbol{H}_{\text{eff}}$. If there is no dissipation ($\alpha = 0$), this precession is perpetual. To see that, it suffices to multiply Eq. (4.3) scalarly by $\boldsymbol{H}_{\text{eff}}$. At a nonzero damping the same multiplication yields

$$
\left(\boldsymbol{H}_{\text{eff}} \cdot \frac{d\boldsymbol{e}}{dt} \right) = -\alpha\gamma(\boldsymbol{e} \cdot \boldsymbol{H}_{\text{eff}})^2 + \alpha\gamma H_{\text{eff}}^2 \tag{4.12}
$$

whence after dividing by H_{eff}, one gets

$$
\frac{de_z}{dt} = -\alpha\gamma H_{\text{eff}}(\cos^2\vartheta - 1) \quad \text{or} \quad \frac{d\vartheta}{dt} = -\alpha\gamma H_{\text{eff}}\sin\vartheta \tag{4.13}
$$

The quantity $d\vartheta/dt$ is the component of the angular velocity $\boldsymbol{\Omega}$ of the magnetic moment in the direction perpendicular to the Oz axis, specifically, to the field $\boldsymbol{H}_{\text{eff}}$. In Eq. (4.13) it is the dissipative torque that strives to bring the magnetic moment back to its equilibrium position $\vartheta = 0$. With allowance for the constancy of the length μ of the magnetic moment, we present Eq. (4.13) in vector form as

$$\mathbf{\Omega}_{\perp} = \frac{\alpha\gamma}{\mu}(\mu e \times \mathbf{H}_{\text{eff}})_{\perp} \qquad (4.14)$$

Note that this relation is an exact one. As it indicates, the coefficient

$$b^{(LL)} = \frac{\alpha\gamma}{\mu} \qquad (4.15)$$

has the meaning of the rotary mobility of the magnetic moment with respect to its relaxational (i.e., toward its equilibrium position) motion. Indeed, in the Landau–Lifshitz equation this coefficient establishes proportionality between the torque exerted on the vector μ and the angular velocity of the latter normal to the precession axis. Using the correspondence relations (4.5), one finds that the mobility coefficient rendered by the Gilbert equation (4.4) has the form

$$b^{(G)} = \frac{\alpha\gamma^{*}}{\mu(1 + \alpha^2)} \qquad (4.16)$$

Remarkably, in the forms of the magnetodynamic equation presented above, the mobility coefficients depend in a different way on the precession damping parameter. While the one found in the Landau–Lifshitz equation is directly proportional to the parameter α [see formula (4.15)], the respective coefficient from the Gilbert equation [see Eq. (4.16)], turns out to be a nonmonotonic function: $b^{(G)}(\alpha)$ passes through a maximum at $\alpha \sim 1$. As Eqs. (4.3) and (4.4) show, in this range of the damping parameter the angular velocities $d\varphi/dt$ of the precessional (azimuthal) and $d\vartheta/dt$ of the relaxational (meridional) motions become comparable. This is a crossover between two regimes, each of which is rather clear. Indeed, at low dissipation ($\alpha \ll 1$) the magnetic moment before settling in the equilibrium position makes many turns around the direction $\mathbf{H}_{\text{eff}}$. In the opposite case $\alpha \gtrsim 1$ the motion of the vector e does not at all resemble precession; it hardly has time to accomplish a single turn. In qualitative terms, one may say that at small α the nonequilibrity $\vartheta \neq 0$ "lives" because of the gyromagnetic properties of the magnetic moment; one may treat gyromagnetism as a kind of rotary inertia. Consequently, in this limit the growth of α accelerates relaxation. In the opposite case, at large α, according to formula (4.16), gyromagnetism is irrelevant, and vector e relaxes exclusively by way of the meridional "crawl" to the direction of the field. In this situation the parameter α plays the same role as the viscosity in an ordinary fluid; its growth stretches the nonequilibrity lifetime, so it is quite natural that the mobility is inversely proportional to α.

3. Rotary Diffusion Coefficient for the Magnetic Moment

In the presence of thermal fluctuations, that is, at a finite temperature T, the magnetic moment, quiescent or moving, experiences perturbations with the reference energy $\sim kT$ per each degree of freedom. To take the thermofluctuational factor into account, one has to insert a random torque to the magnetodynamic equation. This can be done by adding formally a random magnetic field H_f to the regular one. Then the random torque will take the form $\mu(e \times H_f)$. At the next step, assuming that the noise is white, the intensity of the correlation function $\langle H_f^2(t) \rangle$ is evaluated. This approach was first used by Brown [47] and later was advanced in the works of Raikher and Shliomis (see their review article [48]) and Coffey and Kalmykov (see their book [36]). However, as we are interested in brevity and simplicity, we shall use another method based on the equation of motion for the distribution function of the magnetic moment. Under the action of fluctuations the magnetic moment, regardless of whether it is entrained in some regular motion, is necessarily subjected to a Brownian process and undergoes random thermal walks. Given that, the angular coordinates of e are no longer dynamical variables; for instance, the notion of an angular trajectory of e becomes meaningless. Instead, the description is done in terms of the distribution function $W(e, t)$ of the coordinates of the vector e. Namely, $W(e, t)de$ equals the probability of finding the magnetic moment vector in the element de of the solid angle; it is convenient to define this function on the surface of a unit sphere. The relation describing the evolution of $W(e, t)$ (the Fokker–Planck equation) emerges as the continuity equation for the probability flux at this surface:

$$\frac{\partial W}{\partial t} = -\text{Div } Q \tag{4.17}$$

where the operator Div involves only the angular variables.

On constructing the probability flux vector Q, we start with the relation

$$H_{\text{eff}} = -\frac{1}{\mu} \frac{\partial U}{\partial e} \tag{4.18}$$

where $U(e)$ is the part of the magnetic energy of the particle that depends solely on the orientation of the magnetic moment. Using these notations, formula (4.14) becomes

$$\Omega = -\frac{\alpha\gamma}{\mu}(e \times \nabla)U \tag{4.19}$$

where the gradient is taken only with respect to the angular variables. Evidently, the angular velocity Ω pertains to the regular (drift) motion of e so that the full probability flux is the sum

$$Q = \Omega W + Q_{\text{diff}} \tag{4.20}$$

Here the second term is the diffusion flux. We write the latter in the commonly used weak irreversibility approximation, namely, as the Fick law

$$Q_{\text{diff}} = -D(e \times \nabla)W \tag{4.21}$$

with the diffusion coefficient D, which is yet to be evaluated. In Eq. (4.21), while writing the spatial derivative, we have accounted for the fact that the function W is defined at the surface of a unit sphere and its radial (parallel to e) gradients do not exist. Thence the gradient operator reduces to

$$\hat{J} = (e \times \nabla) \tag{4.22}$$

which coincides, as one can easily find (e.q., see Ref. 49, Sec. 26), with the variational derivative of the infinitesimal rotation operator or (adding a numeric factor) with the angular momentum operator in quantum mechanics.

Substituting the explicit expressions for Ω and Q_{diff} in Eq. (4.20), we transform it to

$$Q = -\frac{\alpha\gamma}{\mu} W \hat{J} U - D \hat{J} W = -W \hat{J} \left[\frac{\alpha\gamma}{\mu} U + D \ln W \right] \tag{4.23}$$

In equilibrium the Boltzmann distribution

$$W_0 \propto \exp\left[-\frac{U}{T} \right] \tag{4.24}$$

holds, and the probability flux vanishes. (Note that from now on we set the Boltzmann constant k equal to unity.) Substituting Eq. (4.24) in Eq. (4.23), we obtain

$$-W_0 \hat{J} U \left[\frac{\alpha\gamma}{\mu} - \frac{D}{T} \right] = 0$$

which is evidently satisfied at

$$\left[\frac{\alpha\gamma}{\mu} - \frac{D}{T} \right] = 0$$

From here the rotary diffusion coefficient is expressed as

$$D^{(LL)} = \frac{\alpha\gamma T}{\mu} = \frac{\alpha\gamma T}{I_s V_m} \qquad (4.25)$$

Obviously, using the Einstein relation , Eq. (4.25) might have been written down right away as soon as the rotary mobility coefficient had been found. This is equally valid, of course, for both the Landau–Lifshitz and Gilbert representations of the magnetodynamic equation. Using formula (4.16) one finds

$$D^{(G)} = \frac{\alpha\gamma^* T}{I_s V_m (1 + \alpha^2)} \qquad (4.26)$$

Identifying the operation Div as a scalar multiplication of the operator $\hat{\boldsymbol{J}}$ by the flux vector, one rewrites the continuity equation (4.17) as

$$\frac{\partial W}{\partial t} = -\hat{\boldsymbol{J}}\boldsymbol{Q}$$

and then with the aid of Eq. (4.23) obtains the Fokker–Planck kinetic equation (FPE) in the form

$$2\tau_D \frac{\partial W}{\partial t} = \hat{\boldsymbol{J}} W \hat{\boldsymbol{J}} \left[\frac{U}{T} + \ln W \right] \qquad (4.27)$$

here the rotary diffusion (Debye) time is defined as

$$\tau_D^{(LL)} = (2D)^{-1} = \frac{I_s V_m}{2\alpha\gamma T} \qquad (4.28)$$

This relaxation time—which, to be specific, we have written in the Landau–Lifshitz representation—has the anticipated behavior: the smaller the precession damping constant (the higher the quality factor of the oscillations), the slower does the particle magnetic moment approach its equilibrium position. For ferromagnet or ferrite nanoparticles the typical values of the material parameters are Is $I_s \lesssim 10^3$ G, $V_m \sim 10^{-18}$ cm^3, and $\alpha \sim 0.1$. Substituting them in formula (4.28) at $\gamma \approx 2 \times 10^8$ rad/Oe $\cdot$ s and room temperature, one obtains $\tau_D \sim 10^{-9}$ s.

Note the close resemblance between the reference time of the internal rotary diffusion of the magnetic moment and the Debye time

$$\tau_B = \frac{3\eta V}{T} \qquad (4.29)$$

of the "external" rotary diffusion of a spherical particle of the volume V in an isotropic liquid with viscosity η. As one can see from comparison of formulas (4.28) and (4.29), it is possible to introduce the coefficient of effective magnetic viscosity of a ferromagnet [48]:

$$\eta_m^{(LL)} = \frac{I_s}{6\alpha\gamma} \tag{4.30}$$

In the Landau–Lifshitz representation the orientational diffusion time (4.28) behaves in a simple way: it is the greater the weaker is the Larmor precession damping time. However, as was noted at the end of Section II.A., such a behavior appears quite reasonable at small damping, but is difficult to understand at strong damping when the motion is aperiodic. In this respect, the interpretation that the magnetic viscosity acquires in the Gilbert representation seems rather intelligible—using the diffusion coefficient from Eq. (4.26), one finds for the orientation relaxation time the expression

$$\tau_D^{(G)} = (2D)^{-1} = \frac{I_s V_m (1 + \alpha^2)}{2\alpha\gamma^* T} \tag{4.31}$$

Accordingly, from comparison of Eqs. (4.31) and (4.29) for the effective magnetic viscosity, we get

$$\eta_m^{(G)} = \frac{I_s(1 + \alpha^2)}{6\alpha\gamma} \propto \begin{cases} 1/\alpha & \text{for } \alpha \ll 1 \\ \alpha & \text{for } \alpha \gg 1 \end{cases} \tag{4.32}$$

which means different behavior in the weak and strong precession damping limits.

From FPE (4.27) it follows that in the absence of internal and external fields $(U = 0)$ the parameter τ_D exhaustively determines the response time of the system to a weak external excitation. For example, it is only τ_D that determines the dispersion of the magnetic susceptibility in a magnetically isotropic particle.

4. Relaxation of the Magnetic Moment in an Anisotropic Particle

Thermofluctuational behavior of the magnetic moment in a magnetically anisotropic particle acquires new features in comparison with the isotropic case and, accordingly, extends the set of relaxation times. Let us consider a single-domain particle with the easy-axis anisotropy defined by

$$U_a = -KV_m(\boldsymbol{e} \cdot \boldsymbol{n})^2 = -KV_m \cos^2 \vartheta \tag{4.33}$$

where $\boldsymbol{n}$ is the unit vector of the easy axis, $K > 0$ is the volume energy density of magnetic anisotropy, and ϑ is the angle between the direction of the particle magnetic moment and the easy axis. As formula (4.33) indicates, in the internal configurational space of a particle at $\vartheta = \pi/2$ there exists the energy barrier of the height KV_m, which separates the two minima, $\vartheta = 0$ and $\vartheta = \pi$ corresponding to the orientations $e \parallel n$ and $e \parallel -n$. At finite temperatures, that is, in the presence of fluctuations, the characteristic parameter for the system is the nondimensional ratio

$$\sigma = \frac{KV_m}{T} \qquad (4.34)$$

of the magnetic energy to the thermal one. Evidently, at $\sigma \ll 1$ the motion of the particle magnetic moment greatly resembles the case of free diffusion $(U = 0)$; thence in this limit the relaxation time $\tau(\sigma)$ is close to τ_D.

Essentially different situation is encountered in the case of a high potential barrier or, equivalently, at low temperatures; both these conditions are expressed by the relation $\sigma \gg 1$. Applying the Boltzmann law (4.24) to the two-well potential (4.33), we arrive at the conclusion that the orientational probability is almost totally localized in exponentially small vicinities of the directions $\vartheta = 0$ and $\vartheta = \pi$. It is also obvious that in a system with the energy function (4.33) at full equilibrium, the populations of both wells are equal.

Kinetics of establishing of orientational equilibrium of a magnetic moment of a single-domain particle in the presence of thermal fluctuations is described by FPE (4.27). We express it in spherical coordinates at the surface of a unit sphere. Assuming that all the functions depend only on the meridional angle ϑ, we obtain, [47]

$$\frac{\partial W}{\partial t} = -\mathrm{Div}\, \boldsymbol{Q} = -\frac{1}{\sin \vartheta} \frac{\partial}{\partial \vartheta} [\sin \vartheta \cdot Q_\vartheta]$$
$$= \frac{1}{2\tau_D} \frac{1}{\sin \vartheta} \frac{\partial}{\partial \vartheta} \left[\sin \vartheta \left(\frac{W}{T} \frac{dU_a}{d\vartheta} + \frac{dW}{d\vartheta} \right) \right) \qquad (4.35)$$

The expression for the probability flux vector that we use here

$$\boldsymbol{Q} = Q_\vartheta \boldsymbol{i}_\vartheta, \qquad Q_\vartheta = -\frac{1}{2\tau_D} \left(\frac{W}{T} \frac{dU_a}{d\vartheta} + \frac{dW}{d\vartheta} \right) \qquad (4.36)$$

(in the situation under consideration it has only the ϑ component), as it should have been expected, coincides with that introduced above in definition (4.23).

Full solution of FPE (4.35) is a rather complicated task. There exists a numerous literature devoted to the analysis of the problem for various forms of the potential function, see, for example, the book [36] or the review article by Coffey et al. [50]. Here we will restrict ourselves by a qualitative discussion of the situation and following mostly Ref. 47, will show how it is possible to estimate the relaxation time of the orientational populations (i.e., that of the observed magnetic moment) at $\sigma \gg 1$. In the kinetic theory such a problem is known as the *Kramers problem* [50,51]. The main point of this approximation (as well as that of its multidimensional extension, the Langer method [52,53]) is that the full nonstationary problem is replaced by an effective steady one. One assumes that inside (in the depths of) the potential wells the Boltzmann orientational equilibrium is achieved in a short time and is never affected by the interwell diffusion process by which the system responds to a small perturbation of the well populations against their equilibrium values. Meanwhile the model assumes that between the wells there exists a stationary flux of the particles (or their representing points) as it is required by the balance between the diffusion and regular terms in FPE. One may say that the Kramers approximation implies that by some nonphysical way the particles fleeing the well in the kinetic process return to it in order to again and again repeat their motion toward equilibrium. (We note that the conservation of the number of representing points is the direct consequence of the probability conservation law.) Evidently, the assumptions adopted are justified for the cases where the energy wells are sufficiently deep and well localized; that is, the potential barriers are high in comparison with the thermal energy. Indeed, the opposite case—intense thermal diffusion—entirely contradicts the approximations we rely on here.

Therefore, we assume that the barrier is high and the probability flux Q couples two compact spots in the orientational space that are localized at the poles of the unit sphere. Accordingly, on the total flux the requirement of nondivergency is imposed; that is, it is assumed that in the whole coordinate interval except for the vicinities $[0, \vartheta_1]$ and $[\vartheta_2, \pi]$ of the poles, the quantity $B \equiv \sin \vartheta \cdot Q_\vartheta$ is constant. Applying this condition to Q_ϑ defined by relation (4.37), one comes out with the equation that couples the gradients of the energy and of the distribution function:

$$\frac{dW}{d\vartheta} + \frac{W}{T}\frac{dU_a}{d\vartheta} = -\frac{2\tau_D B}{\sin \vartheta} \tag{4.37}$$

As might have been expected, the flux of particles between the wells is nonzero, vanishing identically only in the case of true equilibrium, which corresponds to the system where the global Boltzmann distribution is established ($B = 0$).

Let us denote the populations of the potential wells (energy minima) as n_1 (that around $\vartheta = 0$) and n_2 (that around $\vartheta = \pi$), for the distribution function $W(\vartheta)$, the natural normalizing condition is used

$$\int W(\vartheta)d\Gamma = 2\pi \int_0^\pi W(\vartheta) \, \sin \, \vartheta d\vartheta = 1 \tag{4.38}$$

so that $n_1 + n_2 = 1$. Now we multiply Eq. (4.35) by $\sin \vartheta$ and integrate the result over all ϑ, over the interval $[0, \pi]$. Then the left-hand part of the equation turns into zero as a derivative of a constant. On the same grounds—where B is constant—will turn into zero the right-hand part as well. In the high-barrier approximation the obtained probability conservation law means the balance between the populations of the polar regions:

$$\frac{dn_1}{dt} + \frac{dn_2}{dt} = 0 \qquad (4.39)$$

On the other hand, in line with the hypothesis we exploit and in accordance with Eq. (4.39), the probability flux is

$$\frac{dn_1}{dt} = -\frac{dn_2}{dt} = -2\pi B \qquad (4.40)$$

Indeed, integration of the right-hand side of Eq. (4.35) on transforming the volume integral into a surface one reduces to a sum of products of the quantity B and the unit outer normal vectors: on entering the region (close to $\vartheta = 0$) and on exiting it (close to $\vartheta = \pi$). Moreover, the exact position of the boundary over which the flux is calculated doesnot matter much.

We transform Eq. (4.37) as follows. We multiply it by $\exp(U_a/T)$ and integrate from ϑ_1 to ϑ_2, taking the latter as the respective borders of the populated polar regions. Thus we find

$$W(\vartheta_2)e^{U_a(\vartheta_2)/T} - W(\vartheta_1)e^{U_a(\vartheta_1)/T} = -2\tau_D B \int_{\vartheta_1}^{\vartheta_2} \frac{e^{U_a/T}d\vartheta}{\sin \vartheta} \qquad (4.41)$$

On the right-hand side of this relation we make the following adjustments:

1. Replace the potential by its expansion in series in the vicinity of the summit of the potential curve:

$$\frac{U_a(\vartheta)}{T} = \frac{U_a(\pi/2)}{T} + \frac{1}{2T}\frac{d^2 U_a}{d\vartheta^2}(\pi/2 - \vartheta)^2 = -\sigma\left(\frac{\pi}{2} - \vartheta\right)^2 \qquad (4.42)$$

2. Replace $\sin \vartheta$ by its value at the summit, that is, by unity.
3. Assuming drastic diminution of the integrand under deviation of the integration variable from $\pi/2$, we proceed to infinite integration limits $(-\infty, \infty)$. After that the integral evaluates easily, and Eq. (4.41) assumes the form

$$W(\vartheta_2)e^{U_a(\vartheta_2)/T} - W(\vartheta_1)e^{U)_a(\vartheta_1)/T} = 2\tau_D B\sqrt{\frac{\pi}{\sigma}} \qquad (4.43)$$

Substituting here the explicit form of the energy (4.42), we present the velocity of the population change as

$$\frac{\tau_D}{\sqrt{\pi\sigma}}\frac{dn_1}{dt} = W(\vartheta_2)e^{U_a(\vartheta_2)/T} - W(\vartheta_1)e^{U_a(\vartheta_1)/T} \qquad (4.44)$$

To close this last equation, we use the above-introduced hypothesis that in each well there exist an independent local Boltzmann distribution. To be specific, let us take the region around the pole $\vartheta = 0$ and present the distribution function at the interval $[0, \vartheta_1]$ in the form

$$W(\vartheta)|_{\vartheta \le \vartheta_1} = W(0)e^{-U_a/T} = W(0)e^{\sigma}e^{-\sigma\vartheta^2} \qquad (4.45)$$

where $W(0)$ is a constant. In the last of equalities (4.45), utilizing the proximity to the minimum, we have replaced the exact potential by its expansion in a series and limited ourselves by the quadratic accuracy. The distribution function integrated over the interval $[0, \vartheta_1]$ must equal the population n_1:

$$n_1 = 2\pi \int_0^{\vartheta_1} W(\vartheta)|_{\vartheta \le \vartheta_1} \sin\vartheta d\vartheta = 2\pi W(0)e^{\sigma} \int_0^{\infty} e^{-\sigma\vartheta^2}\vartheta \, d\vartheta = \frac{\pi}{\sigma}W(0)e^{\sigma} \qquad (4.46)$$

In this way we have evaluated the coefficient $W(0)$ in formula (4.45). Repeating the same considerations, for the potential minimum located near $\vartheta = \pi$, one finds finally for the "localized" distribution functions:

$$W(\vartheta)_{\vartheta \le \vartheta_1} = \frac{\sigma}{\pi}n_1 e^{-\sigma\vartheta^2} \qquad (\vartheta_1 \ll 1)$$
$$W(\vartheta)|_{\vartheta \ge \vartheta_2} = \frac{\sigma}{\pi}n_2 e^{-\sigma(n-\vartheta)^2} \qquad (\pi - \vartheta_2 \ll 1) \qquad (4.47)$$

We use formulas (4.47) for transformation of the right-hand part of Eq. (4.44):

$$W(\vartheta_2)e^{U_a(\vartheta_2)/T} - W(\vartheta_1)e^{U_a(\vartheta_1)/T}$$
$$= \frac{\sigma}{\pi}n_2 e^{-\sigma\vartheta_2^2}e^{-\sigma}e^{\sigma\vartheta_2^2} - \frac{\sigma}{\pi}n_1 e^{-\sigma\vartheta_1^2}e^{-\sigma}e^{\sigma\vartheta_1^2} = \frac{\sigma}{\pi}e^{-\sigma}(n_2 - n_1) \qquad (4.48)$$

As a result, we arrive at an elementary kinetic equation that describes the evolution of populations in a two-level system:

$$\frac{e^{\sigma}}{\sigma}\sqrt{\frac{\pi}{\sigma}}\tau_D\frac{d}{dt}n_1 = -\frac{e^{\sigma}}{\sigma}\sqrt{\frac{\pi}{\sigma}}\tau_D\frac{d}{dt}n_2 = (n_2 - n_1)$$

Closing it, that is, expressing it in terms of the nonequilibrium population $(n_1 - n_2)$, we finally obtain

$$\frac{d}{dt}(n_1 - n_2) = -\frac{2\sigma}{\tau_D e^\sigma}\sqrt{\frac{\sigma}{\pi}}(n_1 - n_2) \qquad (4.49)$$

From here it immediately follows that the expression for the characteristic time of magnetic relaxation is

$$\tau(\sigma) = \frac{e^\sigma}{2\sigma}\sqrt{\frac{\pi}{\sigma}}\tau_D \qquad (4.50)$$

Evidently, because of its exponential dependence on the parameter $\sigma = KV_m/T$, the quantity $\tau(\sigma)$ may grow virtually unboundedly. This is the specific property of the solution of any Kramers problem: that the probability of transition between the potential minima (wells) depends exponentially on the barrier height related to the thermofluctuational energy. In Section III.B we revisit the problem of calculation of the overbarrier relaxation time and using a mathematically correct asymptotic expansion procedure, show how the preexponential factor in the overbarrier relaxation time may be evaluated with the accuracy far higher than that in Eq. (4.50).

In conclusion, let us note the conditions under which the Kramers approximation may be applied. First, this is a sufficient height of the barrier $(\sigma \gg 1)$ and, in gyromagnetic problems, a sufficiently high value of the precession damping parameter. Recalling in this connection the preceding discussion of the expressions for the orientation diffusion time in the Landau–Lifshitz and Gilbert representations, we settle on the conclusion that the Gilbert representation seems more adequate. This, in turn, means that τ_D should be expressed with the aid of Eq. (4.31). In fact, the problem of the intrinsic relaxation time in a magnetic nanoparticle is much wider. For example, one may justifiably ask what would happen if the magnetic damping were sufficiently low, and the Larmor precession were a pronounced and a longliving process. A well-written account on the theory of thermally activated phenomena including the low- and high-damping limits and the crossover behavior bridging them is given in the review article by Coffey et al. [50].

B.   Set of the Moment Equations

Equations for macroscopic characteristic of magnetic particles can be obtained by averaging of the corresponding microscopic quantities with a distribution function satisfying Eq. (4.27). For example, for the dimensionless magnetization $\langle e \rangle$ of an assembly of magnetically isotropic particles, where

$$U = -\mu(e \cdot H)$$

one obtains from Eq. (4.27)

$$2\tau_D \frac{\partial}{\partial t}\langle e_i \rangle = \xi(\delta_{ik} - \langle e_i e_k \rangle)h_k - 2\langle e_i \rangle \tag{4.51}$$

where $\xi = \mu H/T$ and $\boldsymbol{h} = \boldsymbol{H}/H$ is the unit vector of the external field. Equation (4.51) describes the evolution of the first moment of the distribution function. However, it contains the second moment, and thus is coupled with higher terms. This chain linking turns out to be infinite: the equation for the second moment includes the third one, and so on. Thus one may find that, as usually happens in kinetic problems, the Fokker–Planck equation (partial derivative equation) for the distribution function is equivalent to an infinite set of ordinary differential equations for the moments of this function.

Another useful and important example is a single-domain magnetic particle with a uniaxial anisotropy described by the expression

$$U = -KV_m(\boldsymbol{e} \cdot \boldsymbol{n})^2 - \mu(\boldsymbol{e} \cdot \boldsymbol{H}) \tag{4.52}$$

which will be in use all throughout this chapter. With the energy function (4.52), the equation for the dimensionless magnetization following from Eq. (4.27) is

$$2\tau_D \frac{\partial}{\partial t}\langle e_i \rangle = 2\sigma(n_i n_k\langle e_k \rangle - n_k n_l\langle e_i e_k e_l \rangle) + \xi(\delta_{ik} - \langle e_i e_k \rangle)h_k - 2\langle e_i \rangle \tag{4.53}$$

where the right-hand side incorporates both second and third statistical moments.

To write down Eq. (4.53) as well as (4.51), we used, after Ref. 54, the distribution function moments presented as Cartesian tensors. However, when solving the orientational problem, it is more natural to use the set of spherical functions. Choosing spherical coordinates for the unit vectors $\boldsymbol{e}$, $\boldsymbol{n}$, and $\boldsymbol{h}$ as (θ, φ), $(0,0)$, $(\psi,0)$, respectively, that is, taking $\boldsymbol{n}$ as the polar axis of the framework, one gets

$$(\boldsymbol{e} \cdot \boldsymbol{h}) = \cos\psi\cos\theta + \sin\psi\sin\theta\cos\varphi \tag{4.54}$$

so that Eq. (4.52) may be rewritten as

$$U = -KV_m\cos^2\theta - \mu H(\cos\psi\cos\theta + \sin\psi\sin\theta\cos\varphi) \tag{4.55}$$

Then the nonstationary solution of the kinetic equation (4.27) with the energy function (4.55) is sought in the form of the spherical harmonics expansion

$$W_e(\theta, \varphi, t) = \sum_{l=0}^{\infty}\sum_{m=-l}^{l} b_{l,m}(t)\frac{2l+1}{4\pi}\frac{(l-|m|)!}{(l+|m|)!}X_l^{m*}(\boldsymbol{e},\boldsymbol{n}) \tag{4.56}$$

In this equation the symbols $X_l^m(a,b)$ denote the so-called nonnormalized spherical harmonics defined as

$$X_l^m(a,b) = P_l^m(\cos\alpha)e^{im\beta} \tag{4.57}$$

where P_l^m are the associated Legendre polynomials. The angles α and β are the coordinates of the unit vector a in the spherical coordinate framework with the polar axis set along the unit vector b. Functions (4.57) are connected to the conventional (normalized) spherical harmonics by the relationship

$$Y_l^m = \sqrt{\frac{2l+1}{4\pi}\frac{(l-|m|)!}{(l+|m|)!}}X_l^m \tag{4.58}$$

The time dependence in Eq. (4.56) is determined by the quantities $b_{l,m}(t)$, which appear to be the moments of the distribution function:

$$b_{l,m}(t) = \langle P_l^m(\cos\theta)e^{im\varphi}\rangle \tag{4.59}$$

One finds these complex coefficients by solving numerically the infinite set of differential recurrence relations obtained by substitution of expansion (4.56) into Eq. (4.27):

$$
\begin{aligned}
2\tau_D\frac{d}{dt}b_{l,m} &+ l(l+1)b_{l,m} - 2\sigma\left[\frac{(l+1)(l+m-1)(l+m)}{(2l-1)(2l+1)}b_{l-2,m}\right.\\
&+ \frac{l(l+1)-3m^2}{(2l-1)(2l+3)}b_{l,m} - \left.\frac{l(l-m+2)(l-m+1)}{(2l+1)(2l+3)}b_{l+2,m}\right]\\
&- \frac{\xi\cos\psi}{2l+1}[(l+1)(l+m)b_{l-1,m} - l(l-m+1)b_{l+1,m}]\\
&- \frac{\xi\sin\psi}{2(2l+1)}[(l+1)(l+m-1)(l+m)b_{l-1,m-1}\\
&+ l(l-m+2)(l-m+1)b_{l+1,m-1} - (l+1)b_{l-1,m+1} - lb_{l+1,m+1}] = 0
\end{aligned}
\tag{4.60}
$$

Equations (4.60) are valid for $m \geq 1$, and one does not need to consider the negative values of m because of the symmetry of Eq. (4.56) with respect to the replacement $m \rightarrow -m$ that yields $b_{l,-m} = b_{l,m}$. However, the case $m = 0$ is somewhat special, and the corresponding equation that closes the set (4.60) is to be derived separately.

On doing so, it is expressed in the following form:

$$2\tau_D \frac{d}{dt}b_{l,0} + l(l+1)b_{l,0} - 2\sigma \left[\frac{(l-1)l(l+1)}{(2l-1)(2l+1)}b_{l-2,0} \right.$$

$$+ \frac{l(l+1)}{(2l-1)(2l+3)}b_{l,0} - \left. \frac{l(l+1)(l+2)}{(2l+1)(2l+3)}b_{l+2,0} \right]$$

$$- \frac{\xi\cos\varphi}{2l+1}l(l+1)(b_{l-1,0} - b_{l+1,0}) + \frac{\xi\sin\varphi}{2l+1}[(l+1)b_{l-1,1} + lb_{l+1,1}] = 0$$

$$(4.61)$$

The kinetic equation (4.27) with the energy function (4.52), first studied by Brown [47], has since been investigated extensively [48,54–59]. However, due to mathematical difficulties, the case of arbitrary orientation of the external and anisotropy fields (i.e., vectors h and n) has been addressed only relatively recently. The numerical solution of the relaxation problem for h and n crossed under an arbitrary angle for the first time was given in Ref. 60.

C. Solution Methods

In general case Eqs. (4.60) and (4.61) present infinite sets of the five-term (pentadiagonal) recurrence relations with respect to the index l. In certain special cases ($\xi = 0$ or $\sigma = 0$), they reduce to three-term (tridiagonal) recurrence relations. In this section the sweep procedure for solving such relations is described. This method, also known as the *Thomas algorithm*, is widely used for recurrence relations entailed by the finite-difference approximation in the solution of differential equations (e.g., see Ref. 61). In our case, however, the recurrence relation follows from the exact expansion (4.60) of the distribution function in the basis of orthogonal spherical functions and free of any seal of proximity, inherent to finite-difference method. Moreover, in our case, as explained below, the sweep method provides the numerical representation of the exact solution of the recurrence relations.

1. Solution of Three-Term Scalar Recurrence Relations

Consider the three-term inhomogeneous recurrence relation

$$A_l b_{l-1} + B_l b_l + C_l b_{l+1} = f_l, \qquad l = 1, 2, 3 \ldots \qquad (4.62)$$

where b_0 is a given quantity (initial condition). The sweep method is realized by Eq. (4.62) as a two-term recurrence equation by means of the sweep coefficients, which relate to each other the neighboring terms as

$$b_l = \alpha_l b_{l-1} + \gamma_l, \quad b_{l+1} = \alpha_{l+1} b_l + \gamma_{l+1} \qquad (4.63)$$

Substituting Eq. (4.63) into (4.62), one obtains

$$A_l b_{l-1} + (B_l + C_l \alpha_{l+1}) b_l + C_l \gamma_{l+1} = f_l \tag{4.64}$$

Comparison of Eqs. (4.64) and (4.63) yields the two-term backward recurrence for the sweep coefficients:

$$\alpha_l = -\frac{A_l}{B_l + C_l \alpha_{l+1}}, \qquad \gamma_l = \frac{f_l - C_l \gamma_{l+1}}{B_l + C_l \alpha_{l+1}} \tag{4.65}$$

For practical calculations one has to choose a sufficiently large enough $l = L$ and set α_{L+1} and γ_{L+1} to zero. Then, on implementing the backward recurrence relations (4.65) until $l = 1$, one evaluates the full set of the sweep coefficients. By an upward iteration of (4.63), one can find all the terms b_l with the indices $l = 1, \ldots L$. Due to the stability of both iteration processes, the initial error, caused by truncation of the infinite recurrences (4.65), dissolves in the course of iterations and finally becomes less than computer zero. The desired accuracy is provided by varying the cutoff index number L; normally, the higher σ or ξ in Eqs. (4.60)–(4.61), the higher L is to be used. In this sense one may call the solutions obtained *numerically* exact.

2. Solution of Five-Term Scalar Recurrence Relations

Now we consider a five-term inhomogeneous recurrence relation

$$A_l b_{l-2} + B_l b_{l-1} + C_l b_l + D_l b_{l+1} + E_l b_{l+2} = f_l, \quad l = 1, 2, 3 \ldots, \tag{4.66}$$

where b_0 is a given quantity (initial condition). In the literature on the solution of the Fokker–Planck Equation (see, for example, Refs. [35,36]) the five-term *scalar* recurrence relation is usually reduced to a three-term *vector* one, by pairing the neighboring even and odd terms into a two-component columnar matrices (vectors). Here we show how to solve Eq. (4.66) by the sweep method, which is more feasible for realization. For this purpose we introduce three sets of sweep coefficients relating the neighboring terms:

$$b_l = \alpha_l b_{l-1} + \beta_l b_{l-2} + \gamma_l, \quad b_{l+1} = \alpha_{l+1} b_l + \beta_{l+1} b_{l-1} + \gamma_{l+1}$$
$$b_{l+2} = (\alpha_{l+2} \alpha_{l+1} + \beta_{l+2}) b_l + \alpha_{l+2} \beta_{l+1} b_{l-1} + \gamma_{l+2} + \alpha_{l+2} \gamma_{l+1} \tag{4.67}$$

Substituting relations (4.67) into Eq. (4.66), one obtains

$$A_l b_{l-2} + (B_l + D_l \beta_{l+1} + E_l \alpha_{l+2} \beta_{l+1}) b_{l-1} + [C_l + D_l \alpha_{l+1} + E_l (\alpha_{l+2} \alpha_{l+1} + \beta_{l+2})] b_l$$
$$= f_l - (D_l + E_l \alpha_{l+2}) \gamma_{l+1} - E_l \gamma_{l+2} \tag{4.68}$$

On comparing Eqs. (4.68) and (4.67), one finds the three-term backward recurrences for sweep coefficients in the form

$$\alpha_l = -\frac{B_l + D_l\beta_{l+1} + E_l\alpha_{l+2}\beta_{l+1}}{C_l + D_l\alpha_{l+1} + E_l(\alpha_{l+2}\alpha_{l+1} + \beta_{l+2})}$$

$$\beta_l = -\frac{A_l}{C_l + D_l\alpha_{l+1} + E_l(\alpha_{l+2}\alpha_{l+1} + \beta_{l+2})} \tag{4.69}$$

$$\gamma_l = \frac{f_l - (D_l + E_l\alpha_{l+2})\gamma_{l+1} - E_l\gamma_{l+2}}{C_l + D_l\alpha_{l+1} + E_l(\alpha_{l+2}\alpha_{l+1} + \beta_{l+2})}$$

As in the case described above, when performing practical calculations of sweep coefficients, one has to choose a large enough $l = L$ and set all the sweep coefficients of the higher orders to zero. Then implementation of the backward recurrence relations (4.69) until $l = 1$ will render the full sets of the sweep coefficients. With the aid of the upward iteration of (4.67), one can find then all the terms b_l within the interval $l = 1, \ldots, L$.

3. Solution of Vector Recurrence Relations

Above we mentioned a method used to solve a pentadiagonal scalar recurrence relation by reducing it to tridiagonal vector recurrence relations. After that the problem may be solved as it is described in Section II.C.1 with the only exception that all the coefficients of Eq. (4.62) must be treated as matrices. The matrix sweep coefficients, calculated according to the rules of the matrix algebra, now take the form

$$\boldsymbol{\alpha}_l = -(\boldsymbol{B}_l + \boldsymbol{C}_l\boldsymbol{\alpha}_{l+1})^{-1}\boldsymbol{A}_l, \quad \boldsymbol{\gamma}_l = -(\boldsymbol{B}_l + \boldsymbol{C}_l\boldsymbol{\alpha}_{l+1})^{-1}(\boldsymbol{f}_l - \boldsymbol{C}_l\boldsymbol{\gamma}_{l+1}) \tag{4.70}$$

where the superscript -1 denotes inversion of the matrix. By the upward iteration of

$$\boldsymbol{b}_{l+1} = \boldsymbol{\alpha}_{l+1}\boldsymbol{b}_l + \boldsymbol{\gamma}_{l+1} \tag{4.71}$$

one can find all the vectors $\boldsymbol{b}_l$.

Equations (4.60) and (4.61) are also able to yield the vector recurrence relations for the case of a skew bias field, that is, when vectors $\boldsymbol{h}$ and $\boldsymbol{n}$ are not parallel. In this case one should ascribe to each b_l as many as $2l + 1$ components, corresponding to different values of the azimuthal index m. Another problem, involving vector recurrence relations, is a steady-state nonlinear oscillations of b_l in a high-AC field. To study the harmonic content of the nonlinear response, one has to expand all the moments $b(t)_l$ in the Fourier series. Then the Fourier coefficients may be treated as components of a columnar vector. This situation will be considered more than once throughout the chapter.

We conclude this section with formulas delivering the solution of the pentadiagonal recurrence relations (4.66) and (4.67), where coefficients must be treated as matrices and the variables b_l as column vectors. Then, for the matrix sweep coefficients, one obtains

$$\alpha_l = -\Delta^{-1}(B_l + D_l\beta_{l+1} + E_l\alpha_{l+2}\beta_{l+1}), \quad \beta_l = -\Delta^{-1}A_l$$

$$\gamma_l = \Delta^{-1}(f_l - (D_l + E_l\alpha_{l+2})\gamma_{l+1} - E_l\gamma_{l+2})$$

(4.72)

where

$$\Delta = C_l + D_l\alpha_{l+1} + E_l(\alpha_{l+2}\alpha_{l+1} + \beta_{l+2})$$

III.   LOW-FREQUENCY NONLINEAR SUSCEPTIBILITIES OF SUPERPARAMAGNETIC PARTICLES IN SOLID MATRICES

A.   Linear and Cubic Dynamic Susceptibilities: Numerical Solutions

1.   Magnetic Granulometry from the Dynamic Susceptibility Data

Since the very first studies of fine-particle systems [14], the development of the micromagnetic science was inspired mainly by the necessity to predict the magnetic properties and response of a ferromagnetic particulate media. Beyond argument, in this objective the fundamental and applicational aspects are tied up very closely, if not inseparably.

The problem of prime interest while performing experiments on or manufacturing fine-particle magnetic systems is to characterize the magnetic content of the sample with as few measurements as possible. Magnetic granulometry by means of a quasistatic magnetization curve is very well known and widely used [14,62]. The dynamic approach, where linear and nonlinear susceptibilities are simultaneously taken into account, is newer, most probably having been inspired by its use in the spin-glass science [63]. To justify the method, one should process a good deal of experimental data with the aid of an appropriate theory. Such a work has been attempted in Refs. 64,65 with a precipitating Cu-Co alloy as a test object. The authors had no difficulties in fitting the linear susceptibility measurements with the aid of the super-paramagnetic blocking model, assuming that (1) the particles are single-domain and their magnetization does not depend on temperature, (2) the magnetic anisotropy is uniaxial and has one and the same value for all the particles, and (3) the magnetic dipole–dipole interaction is negligible. It was fitting the nonlinear (cubic) susceptibility data where a problem arose, since there was no theory for it thus far, consistent with the aforementioned assumptions. To fill the gap, Bitoh and colleagues [64,65] employed the formulas originally derived for an *isotropic* superparamagnet. They were adjusted by replacing the pertinent relaxation time with a one exponential in the *magnetic anisotropy* constant K. However, the resulting agreement turned out to be poor. From that the authors [64,65] concluded that some of the basic assumptions (viz., 1–3) are wrong. From our viewpoint, in the first place this reproach should be addressed not to the classic superparamagnetic theory proper but to a rather "intuitive" manner of its usage.

The incentive and the main goal of this section are to consistently extend the conventional theory on the case of a nonlinear response and by that to confirm its validity. While doing that we propose practical schemes (both exact and approximate) to handle linear and cubic dynamic responses in the framework of classical superparamagnetism. Applying our results to the reported data on the nonlinear susceptibility of Cu-Co precipitates, we demonstrate that a fairly good agreement may be achieved easily.

2. Static Susceptibilities

As a starting point we take an isolated single-domain particle of a ferro- or ferrimagnetic material rigidly trapped in the bulk of a solid nonmagnetic matrix. *Single-domainness* means a spatial uniformity of the spin alignment over the grain, and enables us to describe it by the net magnetic moment $\mathbf{\mu} = \mu\mathbf{e}$, whose direction is given by a unit vector $\mathbf{e}$. As defined in Section II.A, the magnetic moment magnitude is $\mu = I_s V_m$. Besides that, we assume the particle to possess a uniaxial magnetic anisotropy with an energy density K and a direction defined by a unit vector $\mathbf{n}$.

If the external magnetic field $\mathbf{H}$ is not so high as to affect the atomic magnetic structure, its only effect on a single-domain grain is the magnetic moment rotation. Then the corresponding orientation-dependent part of the particle energy may be written as

$$U = -KV_m(\mathbf{e} \cdot \mathbf{n})^2 - \mu(\mathbf{e} \cdot \mathbf{H}) \tag{4.73}$$

The equilibrium distribution function of the particle magnetic moment or (if we neglect interactions) of an assembly of magnetic moments is determined by the Gibbs law

$$W_{eq}(\mathbf{e}) = Z^{-1} \exp[\sigma(\mathbf{e} \cdot \mathbf{n})^2 + \xi(\mathbf{e} \cdot \mathbf{h})]$$
$$Z(\sigma, \xi) = \int \exp[\sigma(\mathbf{e} \cdot \mathbf{n})^2 + \xi(\mathbf{e} \cdot \mathbf{h})]d\mathbf{e} \tag{4.74}$$

where

$$\sigma = \frac{KV_m}{T}, \qquad \xi = \frac{\mu H}{T} \tag{4.75}$$

and h is a unit vector along the external magnetic field.

In what follows we shall consider a situation where the external magnetic field is weak enough and does not change the basic state of the system too strongly. Assuming the value of ξ to be small, one gets, by expanding Eq. (4.74)

$$W_{eq}(e) = W_0 \frac{1 + \xi(e \cdot h) + \frac{1}{2}\xi^2(e \cdot h)^2 + \frac{1}{6}\xi^3(e \cdot h)^3}{1 + \frac{1}{2}\xi^2 \langle (e \cdot h)^2 \rangle_0} \tag{4.76}$$

where the distribution and partition functions

$$W_0(y) = Z_0^{-1} \exp(\sigma y^2), \quad Z_0(\sigma) = 2\pi \int_{-1}^{1} \exp(\sigma y^2), \quad y = (e \cdot n) \tag{4.77}$$

describe the unperturbed state $H = 0$. Accordingly, the angular brackets with a subscript 0 in Eq. (4.76) designate averaging with respect to the distribution (4.77).

Due to evenness of W_0, all the odd moments of the equilibrium distribution (4.77) vanish. In particular, it means zero net magnetization. The latter may appear only as a response to an applied field. Assuming that the interparticle interaction may be neglected, one finds that $M/c\mu$, the reduced magnetization in the direction of the field, equals the mean cosine $\langle (e \cdot h) \rangle$. With the distribution (4.76), it reads

$$\frac{M}{c\mu} = \langle (e \cdot h) \rangle = \langle (e \cdot h)^2 \rangle_0 \xi + \left[\frac{1}{6} \langle (e \cdot h)^4 \rangle_0 - \frac{1}{2} \langle (e \cdot h)^2 \rangle_0^2 \right] \xi^2 \tag{4.78}$$

where c is the particle number concentration.

To evaluate the averages like those in Eq. (4.78), it is very convenient to pass from cosines $\langle (en)^k \rangle$ to the set of corresponding Legendre polynomials for which a spherical harmonics expansion (addition theorem)

$$P_l(e \cdot h) = \frac{4\pi}{2l+1} \sum_{m=-l}^{l} Y_{lm}^*(n \cdot h) Y_{lm}(e \cdot n) \tag{4.79}$$

takes place.

Taking into account a uniaxial symmetry of the distribution function W_0, on averaging of Eq. (4.79), one gets a generic formula

$$\langle P_l(e \cdot h)\rangle_0 = S_l P_l(e \cdot h), \quad l \text{ even}; \quad \langle P_l(e \cdot h)\rangle_0 = 0, \quad l \text{ odd} \quad (4.80)$$

where

$$S_l(\sigma) = \langle P_l(e \cdot n)\rangle_0 = \int_{-1}^{1} P_l(y) W_0(y) dy \quad (4.81)$$

The order parameters grow from zero with σ and saturate at the unity value at $\sigma \to \infty$. The corresponding asymptotic relations read

$$S_l(\sigma) = \begin{cases} \dfrac{(l-1)!!}{(2l+1)!!} (2\sigma)^{l/2} + \cdots & \text{for } \sigma \ll 1 \\[3mm] 1 - \dfrac{l(l+1)}{4\sigma} + \cdots & \text{for } \sigma \gg 1 \end{cases} \quad (4.82)$$

We remark that the first term of this set, S_2, has the meaning of the internal orientational magnetic order parameter, and as such coincides with the normalized Edwards–Anderson parameter q, which is in use in the spin-glass theory. The main difference is that in a spin glass it is caused by the exchange interaction, while here—by the magnetic anisotropy energy, that is, either spin–orbit or spin dipole–dipole coupling. The particular representation of S_2 is as follows:

$$S_2 = \frac{3}{2}\left(\frac{R'}{R} - \frac{1}{3}\right), \quad R = \int_0^1 \exp(\sigma y^2) dy, \quad R' \equiv \frac{dR}{d\sigma} \quad (4.83)$$

With the notations introduced, the dimensionless magnetization (4.78) transforms into

$$\langle (e \cdot h)\rangle = \frac{1 + 2S_2 P_2(n) \cdot h)}{3} \xi$$
$$- \frac{7 + 70[S_2 P_2(n \cdot h)]^2 + 40 S_2 P_2(n \cdot h) - 12 S_4 P_4(n \cdot h)}{315} \xi^3 \quad (4.84)$$

Comparison of Eqs. (4.78) and (4.84) with the standard definition of the magnetic response

$$M = \chi^{(1)} H + \chi^{(3)} H^3 + \chi^{(5)} H^5 + \cdots \quad (4.85)$$

yields explicit expressions for the first two terms: the linear [$\chi^{(1)}$ or simply χ] and cubic [$\chi^{(3)}$] susceptibilities of the system in question.

From Eq. (4.84) it is apparent that in a solid system (immobilized grains) all the susceptibilities depend on the implemented distribution of anisotropy axes. Let us consider some important limiting cases.

Longitudinal alignment: $n \parallel h$. All the angular functions $P_l(n \cdot h)$ turn into unity. From Eq. (4.84) one obtains for the susceptibilities

$$\chi_\| = \frac{c\mu^2}{T}\frac{1 + 2S_2}{3}, \quad \chi_\|^{(3)} = -\frac{c\mu^4}{T^3}\frac{7 + 70S_2^2 + 40S_2 - 12S_4}{315} \tag{4.86}$$

Transversal alignment: $n \perp h$. The angular functions turn into $P_2(n \cdot h) = -\frac{1}{2}$ and $P_4(n \cdot h) = \frac{3}{8}$, which yields

$$\chi_\perp = \frac{c\mu^2}{T}\frac{1 - S_2}{3}, \quad \chi_\perp^{(3)} = -\frac{C\mu^4}{T^3}\frac{14 + 35S_2^2 - 40S_2 - 9S_4}{630} \tag{4.87}$$

Random orientation: angular averaging shows that $P_2(n \cdot h) = P_4(n \cdot h) = 0$ and $[P_2(n \cdot h)]^2 = \frac{1}{5}$, which results in

$$\widetilde{\chi} = \frac{c\mu^2}{3T}, \quad \widetilde{\chi}^{(3)} = -\frac{c\mu^4}{T^3}\frac{1 + 2S_2^2}{45} \tag{4.88}$$

Note that we use a tilde to mark the random orientation averages, saving the customary overline for the future to denote size averaging.

From Eqs. (4.86)–(4.88) one finds that in a random system the linear part of the susceptibility obeys the superposition rule $\widetilde{\chi} = [\chi_\| + 2\chi_\perp]/3$. Therefore the particle anisotropy drops out of the result. However, the cubic susceptibility appears to be rather sensitive to the anisotropy factor. Indeed, according to Eq. (4.88), for an assembly of magnetically rigid grains ($S_2 \to 1$), the susceptibility $\widetilde{\chi}^{(3)}$, it is 3 times greater than that of an isotropic ($S_2 = 0$) system.

Let us compare the formulas for a solid random system with those for an assembly of orientationally free grains, such as a magnetic suspension. In the latter case, the extension of the configurational space in the single-particle partition function, specifically, adding integration over n in Eq. (4.74), removes the effect of the internal anisotropy on the macroscopic magnetization. This leads, as had been noticed for the first time in Ref. 66, to an isotropic (Langevin) equilibrium magnetization curve that expands, yielding

$$\chi_0 = \frac{c\mu^2}{3T}, \quad \chi_0^{(3)} = -\frac{c\mu^4}{45T^3} \tag{4.89}$$

Compare this with Eq. (4.88). Coinciding in linear parts—a fact that has been mentioned in Ref. 14—the susceptibilities of a fluid and random solid assemblies differ in the cubic contributions unless one deals with magnetically isotropic particles for which $S_l = 0$. This important fact has been overlooked in Refs. 64 and 65, where the authors have taken Eq. (4.89) as a starting point to study a solid system. Right from the comparison of the static formulas (4.88) and (4.89) for $\chi^{(3)}$, underestimation of the predicted values, when Eqs. (4.89) are used, becomes apparent.

We remark that the results presented above, results first published in Ref. 67 in a very short time, had been rederived in Ref. 68.

3. Dynamic Susceptibilities: General Scheme

The rotary diffusion (Fokker–Planck) equation for the distribution function $W(e, t)$ of the unit vector of the particle magnetic moment was derived by Brown [47]. As shown in other studies [48,54], it may be reduced to a compact form

$$2\tau_D \frac{\partial W}{\partial t} = \hat{J} W \hat{J} \left[\frac{U}{T} + \ln W \right] \tag{4.90}$$

which we have already presented as Eq. (4.27). In this notations $\hat{J}$ is the operator of infinitesimal rotations with respect to the components of e while the relaxation time $\tau_D \propto T^{-1}$ of the internal rotary diffusion of the particle magnetic moment is given by Eqs. (4.28) and (4.31). Combining the reference damping time for the Larmor precession introduced by Eq. (4.10) and either of Eqs. (4.28) or (4.31), one finds a simple representation [48,54]

$$\tau_D = \sigma \tau_0 \tag{4.91}$$

where τ_0 is assumed to be temperature-independent. Below we make use of the relationship (4.91) when it becomes necessary to single out temperature dependencies.

In the case of isotropic magnetic particles, that is, $U = -\mu(e \cdot H)$, both linear and cubic dynamic susceptibilities may be obtained analytically. To show this, we first transform Eq. (4.90) into an infinite set of differential recurrence relations:

$$\frac{2\tau_D}{l(l+1)} \frac{d}{dt} \langle P_l \rangle + \langle P_l \rangle - \frac{\xi}{2l+1} (\langle P_{l-1} \rangle - \langle P_{l+1} \rangle) = 0 \tag{4.92}$$

In the absence of the external field ($\xi = 0$, zeroth-order solution) the magnetic moments are distributed at random, and

$$P_0 = 1, \quad \langle P_l \rangle = 0 \quad \text{for} \quad l > 0 \tag{4.93}$$

The probing field is taken to change harmonically as

$$H = \frac{1}{2}H_0(e^{i\omega t} + e^{-i\omega t}) \tag{4.94}$$

Transforming it in the dimensionless form $\xi = \mu H/T$ and then substituting in Eqs. (4.92), one arrives at a problem that is nonlinear with respect to the field amplitude. However, being in a probing regime implies that $\xi \ll 1$ and allows us to build up a perturbation approach, taking Eq. (4.93) as the initial step. On doing so, the result obtained in the first order in ξ is

$$M^{(1)}/c\mu = \langle P_1 \rangle^{(1)} = \frac{1}{6}\xi\left(\frac{e^{i\omega t}}{1 + i\omega\tau_D} + \frac{e^{-i\omega t}}{1 - i\omega\tau_D}\right)$$

The second-order correction, due to the parity conditions, does not contribute to magnetization, whereas the third order leads to

$$\frac{M^{(3)}}{c\mu} = \langle P_1 \rangle^{(3)} = -\frac{\xi^3}{360}\left[\frac{e^{3i\omega t}}{(1 + i\omega\tau_D)(1 + \frac{2}{3}i\omega\tau_D)(1 + 3i\omega\tau_D)} + \text{c.c.}\right]$$
$$-\frac{\xi^3}{360(1 + \omega^2\tau_D^2)}\left[\frac{(3 + \frac{1}{3}i\omega\tau_D)e^{i\omega t}}{(1 + i\omega\tau_D)(1 + \frac{2}{3}i\omega\tau_D)} + \text{c.c.}\right] \tag{4.95}$$

where c.c. stands for complex conjugates. From Eq. (4.95) it follows that with respect to frequency the term cubic in ξ incorporates two harmonics. One oscillates with the single frequency ω and hence yields just a small correction to the amplitude of the basic harmonic, where the major contribution is provided by the linear response term. The other oscillation entirely determines the response at 3ω. The corresponding complex susceptibility reads

$$\chi_{3\omega}^{(3)} = \frac{1}{4}\chi_0^{(3)}\frac{1}{(1 + i\omega\tau_D)(1 + \frac{2}{3}i\omega\tau_D)(1 + 3i\omega\tau_D)} \tag{4.96}$$

where $\chi_0^{(3)}$ is the static value given by Eq. (4.89).

Formulas like Eqs. (4.92)–(4.96) are well known in the theory of rotary molecular diffusion in dipolar fluids [69]. Here we recall their magnetic analogs in order to clarify the difference between our theory and the approach once proposed by Bitoh et al. [64,65]. The analysis of that attempt is instructive since it gives a good example of a case where intuitive considerations turn out to be misleading. The story is that the authors of Refs. 64 and 65, who were needed to extend the superparamagnetic blocking model to a nonlinear case, have done so in the following way. They simply replaced τ_D in Eq. (4.96) by the Néel asymptotic expression

$$\tau_N = \tau_0 \exp(\sigma) = \frac{\tau_D}{\sigma} \exp(\sigma) \tag{4.97}$$

compare it to Brown's result in (4.50), and used the resulting form as $\chi^{(3)}$ for a solid system of randomly oriented uniaxial particles.

We remark that even at the same intuitive level the counterarguments turn up easily: (1) Eq. (4.96) with τ_N from Eq. (4.97) yields an incorrect value for the static ($\omega \to 0$) susceptibility [cf. Eq. (4.88)] and (2) Eq. (4.96) ignores the fact that the magnetic anisotropy imparts the exponential evolution only in relaxation of $\langle P_1 \rangle$, leaving the quadrupole one ($\langle P_2 \rangle$) unchanged.

This does not mean, however, that the intuitive way is completely impossible. For example, the approximate form

$$\chi_{3\omega}^{(3)} = \frac{1}{4} \chi_0^{(3)} \frac{1 + 2S_2^2}{(1 + i\omega\tau_N)(1 + \frac{2}{3} i\omega\tau_D)(1 + 3i\omega\tau_N)} \tag{4.98}$$

is free of the aforementioned qualitative drawbacks and because of that, in principle, has much more grounds to be called a blocking model approximation for $\chi^{(3)}$ than does the model of Refs. 64 and 65. Below we show that Eq. (4.98) is indeed rather close to the best possible approximation (4.112).

Resuming the main line of our consideration, let us show how to consistently take into account the effect of the particle magnetic anisotropy by solving the Brown equation (4.90). Taking n as the polar axis of the coordinate framework, we recover the situation considered as an illustration in Section II.B. Namely, the dimensionless particle magnetization is expressed as Eq. (4.54) and the particle energy as Eq. (4.55). Then the nonstationary solution of the kinetic equation (4.90), which is equivalent to Eq. (4.27), is presented in the form of expansion (4.56) whose amplitudes satisfy Eqs. (4.60) and (4.61).

In terms of the approach adopted, the observed magnetization induced by an external field is found by averaging of Eq. (4.54) with the distribution function (4.56). This done, the magnetization takes the form

$$\frac{M}{c\mu} = \langle (e \cdot h) \rangle = b_{10} \cos \psi + b_{1,1} \sin \psi \tag{4.99}$$

The recurrence relations (4.60) and (4.61) are convenient for constructing a perturbative calculational scheme. Specifically, Eqs. (4.60) and (4.61) are rewritten in the matrix form as

$$2\tau_D \frac{\partial b_{l,m}^{(k)}}{\partial t} + \hat{\Lambda}_{l,l'}^m b_{l',m}^{(k)}$$
$$= \xi e^{\pm i\omega t} \left[\hat{V}_{l,l'}^{m,m} b_{l',m}^{(k-1)} \cos \psi + \left(\hat{V}_{l,l'}^{m,m+1} b_{l',m+1}^{(k-1)} + \hat{V}_{l,l'}^{m,m-1} b_{l',m-1}^{(k-1)} \right) \sin \psi \right] \tag{4.100}$$

where $\hat{\Lambda}^m$ is the relaxational matrix for the basic state, and $\hat{V}_{l,l'}^{m,m'}$ are the matrix elements of the perturbation operator; summation is implied over the repeating indices. Notation (4.100) explicitly separates the contributions with different m indices in each order of the perturbation procedure. That means that the solution of Eq. (4.100) may be constructed as a linear combination of particular solutions of the inhomogeneous equations

$$2\tau_D \frac{\partial a_{l,m}^{(k)}}{\partial t} + \hat{\Lambda}_{l,l'}^m a_{l',m}^{(k)} = \hat{V}_{l,l'}^{m,m'} a_{l',m'}^{(k-1)} e^{\pm i\omega t}, \quad m' = m, m+1, m-1 \qquad (4.101)$$

which are tridiagonal in index l. Equations (4.401) may be solved using the sweep method with any desired accuracy, as described in Section II.C.1.

To distinguish the solutions obtained in Eq. (4.101), we use the following convention. Let $a_{l,m_1 \ldots m_{j-1}}$ be the value obtained in the preceding iteration. At the present step it enters the right-hand side of Eq. (4.100) as a perturbation. Thence the solution of the current iteration will be designated as $a_{l,m_1 \ldots m_{j-1} m_j}$, that is, we simply add the current value of m alongside the second index of a. In this way it is easy to trace back the iteration sequence and recover a correct angular dependence in the final formulas for b_l. As one can see from Eq. (4.100), every time when the added number differs from its left neighbor, the corresponding contribution to b_l emerges bearing $\sin \psi$ as a factor, otherwise it will bear $\cos \psi$.

4. Dynamic Susceptibilities: Linear and Cubic Terms

To be able to obtain cubic susceptibilities, the described above sequence of calculations must be carried out down to the third order.

a. Zeroth-Order. Only even harmonics of the distribution function work, and they yield the static contributions

$$b_{l,0}^{(0)} = a_{l,0} = S_l(\sigma)$$

compare with Eq. (4.81), governed by the dimensionless parameter σ defined by Eq. (4.75).

b. First-Order. The linear contributions take the form

$$b_{l,0}^{(\omega)} = \frac{1}{2}\xi \cos \psi (a_{l,00}e^{i\omega t} + \text{c.c.}), \quad b_{l,1}^{(\omega)} = \frac{1}{2}\xi \sin \psi (a_{l,01}e^{i\omega t} + \text{c.c.})$$

Using Eq. (4.99), one gets the linear complex susceptibility as

$$\chi = 3\chi_0 (a_{1,00} \cos^2 \psi + a_{1,01} \sin^2 \psi) \qquad (4.102)$$

The frequency dependence of the linear susceptibility (4.102) is determined by a superposition of the Debye relaxation modes as

$$a_{1,0m} = a_{1,0m}^{(w=0)} \sum_{k=1}^{\infty} \frac{w_k^{(m)}}{1 + i\omega\tau_{km}}, \qquad \sum_{k=1}^{\infty} w_k^{(m)} = 1 \qquad (4.103)$$

The relaxation spectrum of the operator $\hat{\Lambda}$ is known to a sufficiently good detail. It consists of one interwell mode τ_{10} becoming exponential in σ at $\sigma \gg 1$ and of an infinite number of intrawell modes τ_{1m} with weak dependence on σ. As shown in Refs. 29 and 70, in superparamagnets with a rather good accuracy one may set

$$a_{1,00} = \frac{1 + 2S_2}{3} \frac{1}{1 + i\omega\tau_{10}}, \qquad a_{1,01} = \frac{1 - S_2}{3} \frac{1}{1 + i\omega\tau_{11}} \qquad (4.104)$$

In Eq. (4.104) the relaxation times are either evaluated numerically or taken in the form of the approximate expressions

$$\tau_{10} = \tau_D \frac{e^\sigma - 1}{2\sigma} \left[\frac{1}{1 + 1/\sigma} \sqrt{\frac{\sigma}{\pi}} + 2^{-\sigma-1} \right], \qquad \tau_{11} = 2\tau_D \frac{1 - S_2}{2 + S_2} \qquad (4.105)$$

These compact forms were proposed by Coffey et al. [71] and Shliomis and Stepanov [72], respectively. We emphasize that both formulas (4.105) are *interpolations*, not merely asymptotics. As such, they are valid globally, namely, for the entire range of the parameter σ. This circumstance considerably facilitates all the approximate calculations.

For an assembly of particles with randomly oriented anisotropy axes, Eq. (4.102) yields

$$\tilde{\chi} = 3\chi_0 A_\omega, \qquad A_\omega(\sigma, \omega\tau_0) = \frac{1}{3}(a_{1,00} + 2a_{1,01}) \qquad (4.106)$$

where A_ω may be called a *dispersion factor* for the linear susceptibility.

We emphasize that the description delivered by Eqs. (4.101)–(4.105) is much more general than the commonly known superparamagnetic blocking model. Indeed, to recover the latter, two limiting transitions are to be done. First, one sets $\omega\tau_D \ll 1$. This assumption is not that strong since this condition on frequency usually holds on up to the MHz range. Thence one may neglect the frequency dispersion of $a_{1,01}$ in Eq. (4.104). For a random assembly this gives

$$\tilde{\chi} = \frac{1}{3}\chi_0 \left[\frac{1 + 2S_2}{1 + i\omega\tau_{10}} + 2(1 - S_2) \right] \qquad (4.107)$$

The second transition is the low-temperature/massive-particle limit: $\sigma \gg 1$. With the aid of Eq. (4.82) and on plain replacement of τ_{10} by τ_N from Eq. (4.97), one obtains

$$\tilde{\chi} = \chi_0 \frac{1 + i\omega\tau_N/\sigma}{1 + i\omega\tau_N} \qquad (4.108)$$

which is entirely equivalent to the blocking model expression for the linear dynamic susceptibility first proposed in Ref. 73 [see formulas (6) and (7) there]. However, now all the terms are completely specified.

According to Eqs. (4.102)–(4.108), the blocking model is in fact just the low-frequency end of the low-temperature/massive-particle asymptotics of the superpara-magnetic theory. However, this approach is still widely used as if it were the only theoretical tool used to analyze the micromagnetic relaxation. We remark that the thus imparted incorrectness is easy to trace back if one considers a single particle or an assembly of identical particles. But the situation changes when one has to deal with experimental results and perform size averages, that is, simultaneously superpose calculations for a wide range of σ. In such a case, of which the situation considered in Refs. 64 and 65 is a direct example, the implanted mistakes are well hidden and may become quite harmful.

c. *Second-Order.* The corresponding coefficients of the expansion (4.56) are

$$b_{l,0}^{(2\omega)} = \frac{1}{4}\xi^2[(a_{l,000}\cos^2\psi + a_{l,010}\sin^2\psi)e^{2i\omega t} + \text{c.c.}]$$

$$b_{l,1}^{(2\omega)} = \frac{1}{4}\xi^2[(a_{l,011} + a_{l,001})\sin\psi\cos\psi e^{2i\omega t} + \text{c.c.}]$$

$$b_{l,2}^{(2\omega)} = \frac{1}{4}\xi^2[a_{l,012}\sin^2\psi e^{2i\omega t} + \text{c.c.}]$$

Although these quantities do not enter the expression for magnetization directly, one needs them to proceed to the next order.

d. *Third-Order.* With the accuracy of the order of ξ^3, one obtains

$$b_{l,0}^{(3\omega)} = \frac{1}{8}\xi^3\{[a_{l,0000}\cos^3\psi + (a_{l,0100} + a_{l,0110} + a_{l,0010})\sin^2\psi\cos\psi]e^{3i\omega t} + \text{c.c.}\}$$

$$b_{l,1}^{(3\omega)} = \frac{1}{8}\xi^3\{[(a_{l,0111} + a_{l,0011} + a_{l,0001})\sin\psi\cos^2\psi$$

$$+ (a_{l,0101} + a_{l,0121})\sin^3\psi]e^{3i\omega t} + \text{c.c.}\}$$

$$(4.109)$$

Using that, for the cubic contribution to the dynamic susceptibility at the triple frequency, we find

$$\chi_{3\omega}^{(3)} = \frac{c\mu^4}{4T^3} A_{3\omega}, \quad A_{3\omega}(\psi, \sigma, \omega\tau_0) = a_{1,0000} \cos^4 \psi + (a_{1,0100}$$

$$+ a_{1,0110} + a_{1,0010} + a_{1,0111} + a_{1,0011} + a_{1,0001}) \sin^2 \psi \cos^2 \psi$$

$$+ (a_{1,0101} + a_{1,0121}) \sin^4 \psi \qquad (4.110)$$

and for a randomly oriented assembly, one gets

$$\tilde{\chi}_{3\omega}^{(3)} = \frac{c\mu^4}{4T^3} \tilde{A}_{3\omega}, \quad \tilde{A}_{3\omega}(\sigma, \omega\tau_0) = \frac{1}{5} a_{1,0000} + \frac{2}{15}(a_{1,0100} + a_{1,0110}$$

$$+ a_{1,0010} + a_{1,0111} + a_{1,0011} + a_{1,0001}) + \frac{8}{15}(a_{1,0101} + a_{1,0121}) \quad (4.111)$$

The modes a_1 contributing to $A_{3\omega}$ differ in symmetry and, therefore, in the type of their dependence on the dimensionless parameters σ and $\omega\tau_0$. Only for an isotropic case $(\sigma \to 0)$ do all of them have the same high-frequency dispersion (4.96). As the interwell potential barrier grows $(\sigma \gtrsim 1)$, the dispersion of the modes whose subscripts have zeros in even positions rapidly shifts to lower frequencies. Those modes describe the interwell transition, and for them the frequency parameter tends to $\omega\tau_N$ with the exponential τ_N from Eq. (4.97). For the intrawell modes the frequency parameter tends to $\omega\tau_0$ and thus remains in the high-frequency range. As in the linear susceptibility case, the amplitudes of the transverse modes (the last digit in the subscript is 1) vanish with σ, whereas those of the longitudinal modes saturate at some finite values.

One may surmise that the low-frequency limit, introduced while discussing the linear relaxation, would also lead to a reliable simplification in the nonlinear case since the process is governed mainly by the relaxation time τ_{10}. As we were tempted by this idea, in Ref. 67 we have supposed that the approximate expression

$$\tilde{\chi}_{3\omega}^{(3)} = \frac{1}{4} \chi_0^{(3)} \frac{(1 + 2S_2^2)(1 - i\omega\tau_{10})}{(1 + i\omega\tau_{10})(1 + 3i\omega\tau_{10})} \qquad (4.112)$$

should fairly well describe the frequency dependence of $\chi^{(3)}$ in a randomly oriented assembly. This conclusion was confirmed by comparison with the results of particular numerical calculations. That is why we used formula (4.112) for estimations and approximate calculations. At the same time, absence of rigorous justification was quite a compromising circumstance. This situation

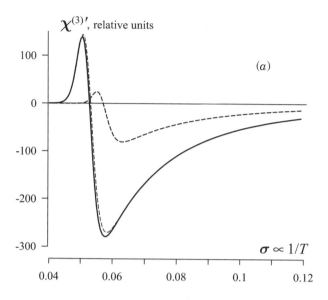

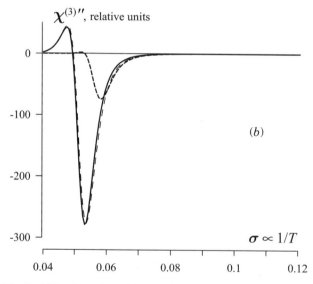

Figure 4.2. Real (*a*) and imaginary (*b*) parts of the cubic susceptibility for $\omega\tau_0 = 10^{-8}$. Solid lines show the numerical solution; thick dashed line corresponds to the theory by Refs. 64 and 65; thin dashed line close to the solid one represents approximation (4.112). Actual numbers on vertical axes render the corresponding χ values divided by coefficient C_3 introduced in Eq. (4.121).

lasted until 2002, when we finally worked out a proof to the fact that Eq. (4.112) is in fact the *exact* low-frequency limit of the cubic susceptibility. This proof is a part of the general consideration of the high-barrier limit ($\sigma \gg 1$) given in the following text (see also Section III.2.6).

The results of numerical calculations of the dynamic nonlinear suscepti-bilities for a random assembly of monodisperse grains along formulas (4.99)–(4.111) are presented in Figure. 4.2. We show them as the functions of the temperature parameter $\sigma \propto 1/T$, taking $\omega\tau_0 = 10^{-8}$ as the reference value. This choice seems reasonable since the low-frequency magnetic measurements are typically carried out at ω about $10^2 - 10^3$ Hz, and the customary reference value of τ_0 lies in the $10^{-10} - 10^{-9}$ s range. The curves obtained by the approximate formula (4.112) are plotted with thin dashed lines that visually almost coincide with the curves corresponding to the exact solution.

For comparison in the same figures we present the results of the theory proposed in Refs. 64 and 65. We remark that this approach suffers from two main drawbacks: (1) the authors neglect the dependence of the static susceptibilities on the internal order parameter S_2 introduced by our Eq. (4.81) and (2) they define the frequency dispersion of the cubic term χ_2 taking formula (4.96)—which is correct only for isotropic particles—and "manually" replacing there the Debye relaxation time τ_D by the Néel asymptotical one τ_N given by Eq. (4.97). From Figure. 4.2 the consequences of these modifications are apparent. Resembling the exact results in general, the cubic susceptibilities by the theory of Refs. 64 and 65 lead to considerable quantitative deviations. In particular, the relative heights of the extrema are much smaller that those rendered by the exact solution. Also, the nodes of the curves are considerably shifted along the temperature axis.

5. Susceptibilities of a Monodisperse Assembly

The set of formulas derived in the preceding section is self-sufficient for numerical evaluation of the linear and cubic magnetic responses as soon as the material parameters are known. Here we assume that the system is dilute enough and interparticle interaction may be neglected. In this framework, to construct the dependencies to be compared to the experimental data, one has, first, to calculate the responses for a subsystem of identical grains with the volume V_m, and then perform the averaging with an appropriate volume-distribution function $f(V_m)$.

Before doing that, however, it is useful to understand qualitatively the anticipated behavior of the susceptibilities. Let us consider the Debye-type expression Eq. (4.107) for the linear susceptibility. The drastic change of the exponential factor reduces all the limiting behavior to just two cases. The first is $\omega\tau_{10} \gg 1$, specifically, the system is rather "cold and stiff." Thus the response is close to zero. The second case is $\omega\tau_{10} \ll 1$. The system is "warm and soft," and readily responds to a signal of almost any frequency. Estimating on this basis the frequency dependence of the susceptibility (ω grows from zero at given finite T), we conclude on a stepwise-like change of the real part, $\chi'(\omega)$, and a peak at the imaginary component $\chi''(\omega)$. For both curves the characteristic points are determined by the same condition $\omega\tau_{10} \approx 1$.

The temperature behavior of the said quantities is as easy to foresee. One has just to note that the temperature growth at given frequency is roughly equivalent to reducing of ω in formula (4.108). The main difference is that the static susceptibility entering Eq. (4.108) imposes the Curie law $\chi_0 \propto 1/T$ [see Eq. (4.88)] on either curve in the high-temperature range. Therefore, both $\chi'(T)$ and $\chi''(T)$ are expected to have similar contours, be positive, and display a peak at a temperature corresponding to $\omega\tau_{10} \approx 1$.

Qualitative behavior of the cubic susceptibility $\chi_{3\omega}^{(3)}$ may be analyzed with the aid of Eq. (4.112). Separating there the real and imaginary parts, one obtains

$$\frac{\chi^{(3)'}}{|\tilde{\chi}_0^{(3)}|} = \frac{7\omega^2\tau_{10}^2 - 1}{4(1 + \omega^2\tau_{10}^2)(1 + 9\omega^2\tau_{10}^2)}, \quad \frac{\chi^{(3)''}}{|\tilde{\chi}_0^{(3)}|} = \frac{\omega\tau_{10}(3\omega^2\tau_{10}^2 - 5)}{4(1 + \omega^2\tau_{10}^2)(1 + 9\omega^2\tau_{10}^2)}$$

$$(4.113)$$

where $\tilde{\chi}_0^{(3)}$ stands for the static cubic susceptibility of a random assembly introduced by Eq. (4.88).

According to Eq. (4.113), the nonlinear susceptibilities may invert their signs with both frequency and temperature at some points in the $\omega\tau \sim 6$ range. This change is from negative to positive with the frequency growth, and to the contrary with the temperature increase. A simple calculation proves that on the temperature dependencies the node of $\chi^{(3)'}$ always resides at higher T than that of $\chi^{(3)''}$.

To completely clarify the effect of anisotropy on susceptibility of a random assembly, in Figure 4.3 and 4.4 we present 3D plots. Slicing them along the frequency axes, one may observe how do the susceptibility curves transform when the anisotropy ($\sigma \propto K$) increase. The isolines $\chi = $ const drawn in the basic planes help to understand the limiting behavior. In the low-anisotropy range (here it is $\sigma \lesssim 9$) the "exponentiality" of the relaxation time is but weakly pronounced. This yields in each plot the short parts of isolines parallel to the σ axes. As σ grows, there occurs a narrow crossover range. After it all the other relaxation modes become insignificant, and the isolines turn into linear dependencies $\sigma \propto -\ln \omega\tau_0$. For a given ω, this reflects the growth of the reference relaxation time with σ. Accordingly, in all the plots the parts of characteristic (abrupt) changes of $\chi(\omega)$ move to smaller frequencies approximately logarithmically. Note that since $\sigma \propto 1/T$, cross sections along the σ axes present the temperature dependencies at constant K and frequency.

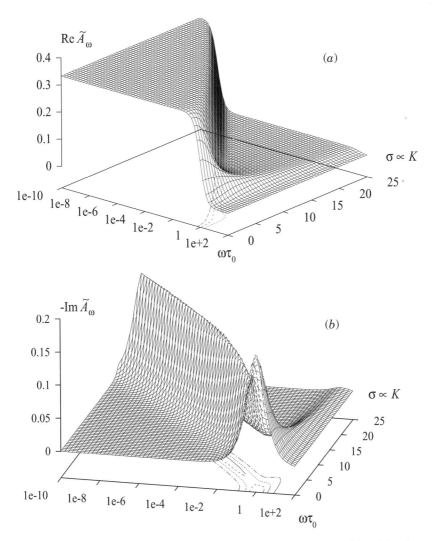

Figure 4.3. Monodisperse random assembly. Transformations of the real (a) and imaginary (b) parts of the reduced linear susceptibility with the anisotropy growth as rendered by Eq. (4.106) for $\omega\tau_0 = 10^{-8}$.

Some complicated details—inflections and inflations of the representing surfaces—occur in the nearest corner of the plots in Figures 4.3 and 4.4, namely, at $\omega\tau_0 > 1$ and $\sigma \sim 3$. This is the range that is beyond the scope of the low-frequency approximation, and much so the blocking model. Those results are available only by numerical evaluation since one has to take into account on

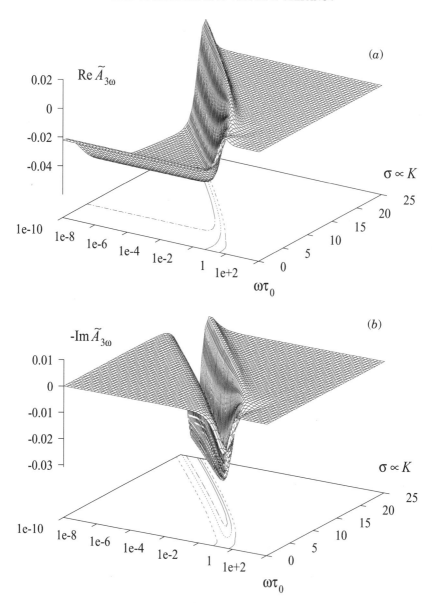

Figure 4.4. Monodisperse random assembly. Transformations of the real (a) and imaginary (b) parts of the reduced cubic susceptibility as rendered by Eq. (4.111) for $\omega\tau_0 = 10^{-8}$.

equal basis a wide set of relaxation modes, both transverse and longitudinal. Physically, this means that at those frequencies the intrawell modes yield a noticeable contribution.

The surfaces shown in Figures 4.3 and 4.4 are interesting as well from the viewpoint that they visualize the theoretical data arrays involved when one passes to experiment interpretation. Indeed, performing size (or volume) averagings of the susceptibilities means superposing of a number of curves, each corresponding to its own $\sigma = $ const cross section with appropriate weights given by the distribution function.

6. Polydisperse Systems: Trial Distribution Functions

In the majority of cases when one deals with nanosize superparamagnetic grains, polydispersity seems to be an inherent feature. The independent measurement of the size distribution function, such as by electron micrography, is a painstaking and rare opportunity. Besides, even when it is done, from the statistical viewpoint a set of available measurements ($10^3 - 10^4$ grains) for the particle number concentration even as small as $10^{10} - 10^{18}$, that is, 0.01% by volume at the particle size ~ 10 nm, is far from being statistically representative.

That is why the conventional approach is to choose a trial (model) size distribution function and determine its parameters on fitting theoretical curves to the experimental data. By obvious reasons, of the variety of the available distribution formulas, the two-parameter ones are most popular; and of the latter, two are known to work better than others:

1. The *lognormal distribution*, given by

$$f_l(V) = \frac{1}{\sqrt{2\pi} s V} \exp\left[-\frac{\ln^2(V/V_{0l})}{2s^2}\right] \tag{4.114}$$

which is the most famous. Its closeness to real histograms of disperse systems has been reported in thousands of papers. The most probable value v_p of the argument [the position of $f_l(V)$ maximum] and the nth moment are, respectively

$$V_{pl} = V_{0l}\exp(-s^2), \quad \overline{V_l^n} = V_{0l}^n \exp\frac{(ns)^2}{2} \tag{4.115}$$

2. The *gamma distribution*, expressed as

$$f_g(V) = \frac{1}{\Gamma(\beta+1)V_{0g}}\left(\frac{V}{V_{0g}}\right)^{\beta}\exp\left(-\frac{V}{V_{0g}}\right) \tag{4.116}$$

where $\Gamma(x)$ is the Euler gamma function. The most probable value V_{pg} and the nth moment are

$$V_{pg} = (\beta + 1)V_{0g}, \quad \overline{V_g^n} = (V_{0g})^n \frac{\Gamma(\beta + n + 1)}{\Gamma(\beta + 1)} \qquad (4.117)$$

The Γ distribution is the generalization of the Poisson one that is also in use [73]. A detailed demonstration of its usefulness for interpretation of the low-frequency magnetic spectra of magnetic fluids (dispersions of nanosize ferroparticles) was given in Ref. 74.

Note, that in given representations (4.114) and (4.116) of either of the distribution functions, the V_0 parameter is a reference value and does not have a direct meaning of the mean volume. The latter is formed as a particular combination—[see Eqs. (4.115) and (4.116) of corresponding V_0 and the distribution width.

When passing to the averaging formulas, one must treat the susceptibility expression like Eq. (4.108) as yielding a contribution of the jth fraction of the particles with a volume V_j and number concentration c_j. By definition, the latter is expressed

$$c_j = c\left(\frac{c_j}{c}\right) = cf(V)dV \qquad (4.118)$$

where summation over j is equivalent to integration over V. However, the meaning of c for a polydisperse system is ambiguous, and we replace it by introducing the dimensionless total particle volume fraction

$$\varphi = c\overline{V} \qquad (4.119)$$

We emphasize that for a polydisperse system the replacement described by Eqs. (4.118) and (4.119) is the only correct way to eliminate c. Plain substitution of φ instead of cV in formulas (4.89), as in Refs. 64, 65, and 73, would have been correct only for monodisperse systems.

Let us do that taking Eq. (4.110) as an example. With the aid of Eqs. (4.118) and (4.119), it transforms into

$$d\overline{\chi}_{3\omega}^{(3)} = \frac{\varphi I^4 V^4}{\overline{V}T^3} \tilde{A}_{3\omega}(\sigma, \omega\tau_0)f(V)dV \qquad (4.120)$$

It is convenient to scale the current volume as $y = V/V_0$, which does not change the integration limits. Then the current value of the anisotropy parameter becomes $\sigma = (KV_0/T)y = \sigma_0 y$, and the assembly-averaged cubic susceptibility is written

$$\overline{\chi}_{3\omega}^{(3)} = C_3 \sigma_0^3 \frac{V_0}{\overline{V}} \int_0^\infty y^4 f(y) \tilde{A}_{3\omega}(\sigma_0 y, \omega \tau_0) dy, \qquad C_3 = \frac{\varphi I^4}{K^3} \qquad (4.121)$$

Similarly, for the linear susceptibility, one gets

$$\overline{\chi} = C_1 \sigma_0 \frac{V_0}{\overline{V}} \int_0^\infty y^2 f(y) \tilde{A}_\omega(\sigma_0 y, \omega \tau_0) dy, \quad C_1 = \frac{\varphi I^2}{K} \qquad (4.122)$$

where now the temperature dependence is rendered by the argument $\sigma_0 \propto 1/T$. Evidently, formulas (4.121) and (4.122) show the general way of volume averaging for χ and $\chi_{3\omega}^{(3)}$ and hold for the exact dispersion factors A as well as for their approximations.

In Figure 4.5 we visualize the averaging procedure described by Eqs. (4.121) and (4.122). The dispersion factors are evaluated by the numerical procedure. The plots show the evolution of the susceptibility curves $\chi'(T)$ with the widening of the lognormal distribution. The front plane $s = 0$ corresponds to the monodisperse case $f(V) \propto \delta(V - V_0)$ considered in Figure 2. Besides the expected smearing down of sharp peaks, from the sequence of the level lines one observes that as the distribution width grows, the positions of the main extrema drift to the higher temperature range. Very qualitatively, this tendency may be understood from the relation $T \propto K\overline{V}$ where the mean value is determined by Eq. (4.115). We remark that for a great many of dispersed systems, the typical values of s fall into the interval 0.5 ± 0.2.

7. Polydisperse Systems: Comparison with Experiment

For practical comparison we take an ample set of experimental data reported in Ref. 64 (and confirmed in Ref. 65) on linear and cubic susceptibilities of Cu-Co precipitating alloys. The given data cover the frequency domain 38–840 Hz and the temperature range 10–190 K. Synthesis of the samples as well as the method of magnetic measurement are described. However, apart from the observation that Cu-Co alloys precipitate, yielding a dispersion of cobalt nanosize particles in a copper matrix, no particular structural information on the system is given.

When the superparamagnetic theory is applied for interpretation of any measured susceptibility line, it means that some model function $\chi(V, \omega, T)$ depending on several material parameters, is processed through averaging like such as Eq. (4.121) or (1.122). In our case the basic set of the material parameters comprises magnetization I, anisotropy energy density K, relaxation time τ_0, and the particle volume fraction φ. Obviously, for nanosize-dispersed systems the effective values of I, K, and τ_0 do not coincide with those for a bulk material. The size/volume averaging itself introduces two independent

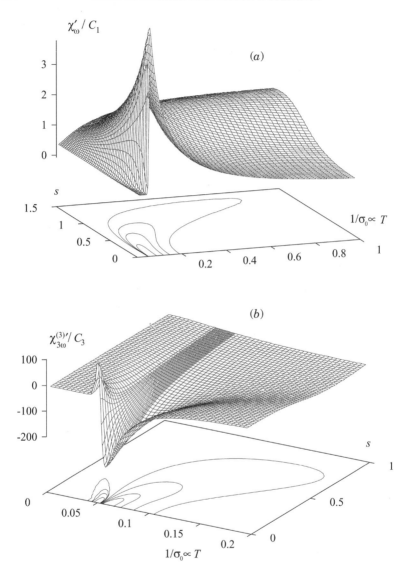

Figure 4.5. Modeling of volume-averaging effect with the lognormal distribution function on the real parts of the linear (*a*) and cubic (*b*) susceptibilities.

statistical parameters. Therefore, the fitting procedure turns into an optimization problem in a multidimensional parameter space with rather loose restrictions.

Consequently, one should not expect that the analysis of simply the susceptibility temperature–frequency behavior is self-sufficient for unambiguous determination of the magnetic material parameters. But it is reasonable to ask what is the real extent to which one could reduce the uncertainty while fitting the measured data.

First we remark that in the framework of the proposed theory the number of independent parameters determining the susceptibilities of a superparamagnetic assembly equals 5. Indeed, from Eqs. (4.121),(4.122) and (4.115),(4.117) it follows that the reference volume as such does not enter the results. It appears only in the combination KV_0. Denoting it as T_0, one may rewrite Eqs. (4.115) and (4.122) as

$$\overline{\chi}_{3\omega}^{(3)} = C_3 \left(\frac{T_0}{W}\right)^3 \cdot \mathscr{A}_3 \left(\frac{A_0}{T}, \omega\tau_0, s\right), \quad \overline{\chi} = C_1 \left(\frac{T_0}{T}\right) \cdot \mathscr{A}_1 \left(\frac{T_0}{T}, \omega\tau_0, s\right) \quad (4.123)$$

(for the Γ distribution, one would have β instead of s). In Eqs. (4.123) the quantities $\mathscr{A}_1$ and $\mathscr{A}_3$ stand for the integrals of the dispersion factors multiplied by the ratio $V_0/\overline{V}$, which depends solely on the distribution width s or β.

Thus the set of parameters yield by fitting comprises T_0, τ_0, s, and two amplitudes C_1 and C_3. From the definitions of the latter [see Eqs. (4.121) and (4.122)] one sees that they incorporate three material parameters of the system. That means that either I, K, or φ may be taken arbitrary. Note that if one would have dealt with just linear susceptibility measurements, then C_3 would not appear, and the uncertainty of the choice would have been enhanced, allowing two of the three parameters as arbitrary. In our case, we set the magnetization of the cobalt particles equal to its bulk value $I = 1460\,G$. On doing that, one gets explicit relationships

$$V_0 = \frac{T}{I}\sqrt{C_3/C_1}, \quad K = T_0 I \sqrt{C_1/C_3}, \quad \varphi = \frac{C_1}{I}\sqrt{C_1/C_3} \quad (4.124)$$

relating the customary physical parameters to the results of the fitting procedure.

We perform nonlinear fitting using the Levenberg–Marquardt method implemented in the MRQMIN routine [75]. From the experimental end, eight families of data are involved, namely, $\chi'(T)$ and $\chi'_{3\omega}(T)$ at four frequencies, taken from Ref. 64. From the theory end, we employ formulas (4.121)–(4.124) with the numerical dispersion factors. The results of fitting are presented in Figures 4.6 and 4.7 and Table I.

8. Discussion

The fitting process worked sufficiently stably, which means that the set of susceptibilities as a function of its magnetic and statistical parameters does not have too many local minima. So, the sets of values presented in the first two lines

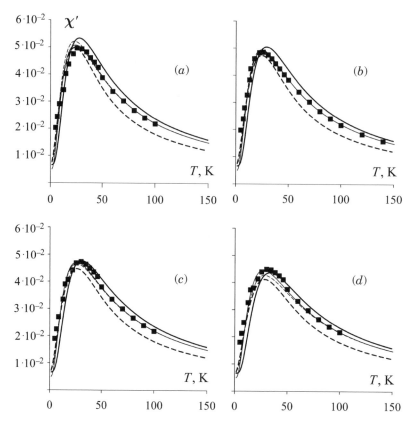

Figure 4.6. Linear susceptibility, real part. Comparison with the experiment [64] for the frequencies 30 Hz (a), 80 Hz (b), 220 Hz (c), 840 Hz (d). Volume averaging is performed with the magnetic and statistical parameters according to Table I. Entry 1 indicated by thick solid lines, entry 2 indicated by thick dashed lines, entry 3 indicated by thin solid lines, entry 4 indicated by thin dashed lines. Almost everywhere the two latter visually coincide. The vertical scale coincides with the one adopted in Refs. 64 and 65 for the experimental data.

of Table I are unique in a wide vicinity of the parameter space. The results given in Figures 4.6 and 4.7 have almost equal statistical residuals for the lognormal and gamma functions. However, from the simple *chi-by-eye* viewpoint, the lognormal distribution seems far more satisfactory. For the Γ distribution two flaws are apparent: noticeable deviations at high T and misplacements of the peak positions for both linear and cubic curves.

The main difference between the lognormal and Γ distributions is their reduction rate at the right end, that is, at $V \to \infty$. As one may see from Eqs. (4.121) and (4.122), the higher is the susceptibility order, the greater is the enhancement of the right-end effect, since under the integral $f(y)$ is multiplied

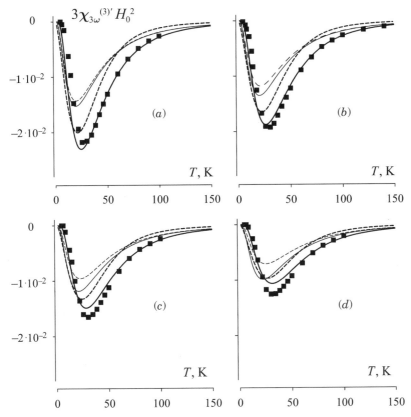

Figure 4.7. Cubic susceptibility, real part. Comparison with the experiment [64] for the frequencies 30 Hz (a), 80 Hz (b), 220 Hz (c), 840 Hz (d). The line drawing convention is the same as in Figure 6. The vertical scale (ordinate) coincides with the one adopted in Refs. 64 and 65 for the experimental data.

TABLE I
Magnetic and Statistical Parameters Obtained by Fitting[a]

	Fitting Attempt	Reference Relaxation Time (s)	Reference Volume (cm)^3	Standard Deviation	Volume Fraction (%)	Anisotropy Constant (erg/cm^3)
1	Lognormal, present fit	2.13×10^{-10}	3.39×10^{-20}	0.608	0.734	0.92×10^5
2	Gamma, present fit	5.32×10^{-11}	2.30×10^{-20}	$\beta = 0.585$	0.738	8.36×10^5
3	Lognormal, after Ref. 64	1.65×10^{-14}	5.6×10^{-27}	0.72	0.10	8.8×10^3
4	Lognormal, after Ref. 65	1.00×10^{-10}	5.6×10^{-20}	0.72	0.30	5.2×10^1

[a] The saturation magnetization is $I = 1460$ G for all the entries.

by yet higher powers of y. Apparently, one has no basis to expect that a histogram of a real system would obey any simple two-parameter function. But since χ and $\chi^{(3)}$ are determined by different moments of the distribution, their simultaneous consideration is an effective method to single out the appropriate and inappropriate ones. According to our evidence, for the Co-Cu systems the lognormal distribution is far better.

At the same time, we remark that if we estimate the difference between the two undertaken fittings in terms of statistical moments, the deviations that occur are rather moderate. Indeed, evaluating mean volume and mean-square volume from formulas using the numbers from Table I (entries 1 and 2), one gets for the lognormal and Γ cases $\overline{V_l} = 4.43$ nm, $\sqrt{\overline{V_l^2}} = 4.71$ nm, $\overline{V_g} = 4.11$ nm, and $\sqrt{\overline{V_g^2}} = 4.49$ nm, respectively. Thus one sees that visual comparison happens to be a sensitive test. The differences between Figures 4.6 and 4.7 are rather pronounced, whereas the deviations in mean and mean-square particle diameters do not exceed 8% and 6%, respectively, which is a fairly good accuracy.

It is essential to emphasize, however, that even a very good fitting may be completely misleading unless it matches physical considerations. By that we mean that presently there exists enough knowledge on the static and dynamic properties of magnetic nanosize systems to impose a reasonable frame on the orders of magnitude of the desired parameters. From this viewpoint, the numerical values obtained by fitting with both lognormal and Γ distributions are quite reasonable (entries 1 and 2 in Table I). On the contrary, attempting a quantitative explanation with the aid of a lognormal distribution, the authors of Ref. 64 have obtained some agreement between their theory and experiment setting $\tau_1 \sim 10^{-14}$ s (see the third entry of Table I). The contradiction is apparent—whatever is the it chi-by-eye agreement in the graphs, the set of parameters should be discarded. The matter is that so short a time is completely forbidden in superparamagnetic particles [76]. Probably, just those considerations caused the same authors to reinterpret their data in Ref. 65 (see the fourth entry of Table I) with $\tau_0 \sim 10^{-10}$. However, the result of fitting given in Ref. 65 (see thin lines in Figs. 4.6 and 4.7), although rather good for χ', suffers from a serious flaw with respect to $\chi_{3\omega}^{(3)'}$. Namely, the position of the theoretical temperature peak is considerably shifted from the experimental one to the low-temperature region. Supposedly, this discrepancy results from the joint effect of using a not completely consistent theory and a not too thorough fitting procedure.

In this connection we note a fact that we have run into during our own fitting work. As long as we tried the procedure involving just the linear susceptibility data, the best-fit routine, whatever was the initial set, "stubbornly" lead us deep into the unphysical region of far too small reference τ values. As soon as the data arrays were extended to include the cubic susceptibilities as well, the best-fit parameter set (entries 1 and 1 of Table I) immediately became reasonable.

To summarize this part of the chapter, we have constructed a consistent theory of linear and cubic dynamic susceptibilities of a noninteracting superparamagnetic system with uniaxial particle anisotropy. The scheme developed was specified for consideration of the assemblies with random axis distribution but may be easily extended for any other type of the orientational order imposed on the particle anisotropy axes. A proposed simple approximation is shown to be capable of successful replacement of the results of numerical calculations.

The theory was tested with the aid of an ample data array on low-frequency magnetic spectra of solid Co-Cu nanoparticle systems. In doing so, we combined it with the two most popular volume distribution functions. When the linear and cubic dynamic susceptibilities are taken into account simultaneously, the fitting procedure yields a unique set of magnetic and statistical parameters and enables us to conclude the best appropriate form of the model distribution function (histogram). For the case under study it is the lognormal distribution.

The achieved agreement yields one more ample argument in favor of the idea that magnetic measurements may work as a sensitive and reliable quantitative material science test for nanosize systems.

B. High-Barrier Asymptotics for Nonlinear Susceptibilities

In the preceding section we showed how the nonlinear superparamagnetic theory enables one to obtain valuable information on the material parameter of nanodisperse systems and especially on the granulometric content of nanopowders or solid dispersions. This was done by presenting a detailed numeric calculation of the linear and cubic susceptibilities for a solid system of uniaxial fine particles. With allowance for the polydispersity of real samples, the description provided a fairly good agreement with the dynamic magnetic measurements taken on Co-Cu nanocomposites. This approach was used successfully [77] for the linear and cubic susceptibilities of the samples of randomly oriented γ-Fe_2O_3 nanoparticles. In the present section we carry on the buildup of the nonlinear superparamagnetic relaxation theory by working out a set of compact and accurate analytical expressions that considerably facilitate calculations as well as experiment interpretation.

This section is based mostly on the results presented in Ref. 78 and is arranged in the following way. In Section III.B.1 we note mentioned the problem of superparamagnetic relaxation, which has been already tackled by means of the Kramers method, in the in Section II.A), and show how to obtain the analytical solution (in the form of asymptotic series) for the micromagnetic Fokker–Planck equation in the uniaxial case. In Section III.B.1 the perturbative expansions for the orientational distribution function are obtained, which are used in Sections III.B.2–III.B.5 to construct asymptotic expressions for the nonlinear dynamic susceptibilities. The explicit forms of those expansions are given, and their accuracy is proved by comparison with the results of numerical calculations. In Section III.B.6 the enveloping discussion is given.

1. General Scheme

The cornerstone of the superparamagnetic relaxation theory is the Arrhenius-like law (4.50) for the relaxation rate of a magnetic moment of a single-domain particle. This classical problem considers an immobile single-domain grain of vo!ume V_m. This particle possesses a uniaxial bulk magnetic anisotropy with the energy density K and the easy axis n. The temperature T is assumed to be much lower than the Curie point, and due to that the particle magnetization I_s, as a specific parameter, is practically constant so that the particle magnetic moment is expressed as $\mu = I_s V_m$ and its direction is denoted by the unit vector e. Thence, the orientation-dependent part of the particle energy (in the absence of external magnetic fields) is given by Eq. (4.33), according to which the function U_a has two equal minima separated by the potential barrier of the height KV_m and correspond to the configurations $e \parallel \pm n$ because for the magnetic moment e the directions n and $-n$ are equivalent. At zero temperature, the magnetic moment e, once located in a particular potential well, is confined there forever. At finite temperature, the probability of an overbarrier (interwell) transition becomes nonzero.

Brown [47] shaped up those semiqualitative considerations into a rigorous Sturm–Liouville eigenvalue problem by deriving the micromagnetic kinetic equation

$$2\tau_D \frac{\partial W}{\partial t} = \hat{J} W \hat{J} \left(\frac{U}{T} + \ln W \right) \tag{4.125}$$

for the orientational distribution function of the magnetic moment. For the case of a sufficiently high interparticle barrier, Brown obtained for the reference time of the particle remagnetization the exponential law (4.50).

Generally speaking, Eq. (4.125) is incomplete since a gyromagnetic term is absent there. This means that the consideration is limited by the frequency range $\omega\tau_0 \ll 1$, where τ_0 is the relaxation time of the Larmor precession of the particle magnetic moment in the internal anisotropy field $H_a \sim 2K/I_s$, where K may incorporate possible shape contribution. Comparing this condition with the other one, $\omega_L\tau_0 \lesssim 1$, confirms a low-to-moderate quality factor of the Larmor precession for real nanodisperse ferrites, one estimates the allowed frequency as $\omega \ll \omega_L$ that means, in fact, a fairly wide range. However, if a transverse field is applied $(e \perp n)$, the lowest eigenvalue and the corresponding eigenvector of the

FPE become sensitive to α, demonstrating crossovers between different damping-dependent regimes [50,79]. The consequences of this effect for the nonlinear susceptibility were first found in numerical simulations [80] and later confirmed by an analytically calculation in Ref. 81. The results given below are valid for the high-damping regime, that is, $\alpha \gtrsim 1$.

In the statistical description delivered by Eq. (4.125), the observed (macroscopic) magnetic moment per particle is given by the average

$$m(t) = \mu \langle e \rangle = \int e W(e,t)\,de \tag{4.126}$$

As with allowance for Eq. (4.33) the function W has a parametric dependency on the vector n, and the actual angular argument of W is $(e \cdot n)$.

As mentioned in Section II, the magnetodynamic equation underlying the Brown kinetic equation (4.125) can be either that by Landau and Lifshitz or that by Gilbert. To be specific, we adopt the former one, noting their equivalence established by formulas (4.5). Thence, the reference relaxation time in Eq. (4.125) is given by Eq. (4.24).

Assuming uniaxial symmetry of the time-dependent solution and separating the variables in Eq. (4.125) in the form

$$W(e,t) = \frac{1}{2\pi} \sum_{l=0}^{\infty} A_l \psi_l(e \cdot n) \exp\left[\frac{-\lambda_l t}{2\tau_D}\right] \tag{4.127}$$

where the amplitudes A_l depend on the initial perturbation, one arrives at the spectral problem

$$\hat{L}\psi_l = \lambda_l \psi_l, \quad \hat{L} \equiv \hat{J}\left[2\sigma(e \cdot n)(e \times n) - \hat{J}\right] \tag{4.128}$$

where the nonnegativity of the decrements λ_l can be proved easily. Expanding the eigenmodes ψ_l in the Legendre polynomial series

$$\psi_l = \frac{1}{2} \sum_{k=1}^{\infty} (2k+1) b_k^{(l)} P_k(\cos\theta), \quad k = 1, 3, 5, \ldots \tag{4.129}$$

where θ is the angle between e and n, one arrives at the homogeneous tridiagonal recurrence relation

$$\left[1 - \frac{\lambda_l}{k(k+1)}\right] b_k^{(l)} - 2\sigma\left[\frac{k-1}{(2k-1)(2k+1)} b_{k-2}^{(l)} + \frac{1}{(2k-1)(2k+3)} b_k^{(l)}\right.$$
$$\left. - \frac{k+2}{(2k+1)(2k+3)} b_{k+2}^{(l)}\right] = 0 \tag{4.130}$$

Note that Eqs. (4.127)–(4.130) describe only the longitudinal (with respect to the easy axis) relaxation of the magnetic moment. We remark that under condition $\omega \ll \omega_L$, that is, far from the ferromagnetic resonance range, the transversal components of $\boldsymbol{m} = \mu \langle \boldsymbol{e} \rangle$ are of minor importance.

2. Interwell Mode

Spectral equation (4.128) describes the temperature-induced (fluctuation) motions of the vector $\boldsymbol{e}$ in the orientational potential with a symmetric profile (4.33). With respect to the time dependence, the set of possible eigenmodes splits into two categories: interwell (overbarrier) transitions and intrawell wanderings. In the spectral problem (4.128) the interwell transitions of the magnetic moment are associated with the single eigenvalue λ_1. As rigorous analysis shows [82], it drastically differs from the others; whereas for $l \geq 1$ all the λ_l gradually *grow* with σ, the decrement λ_1 exponentially *falls down* proportionally to $\exp(-\sigma)$.

In the opposite limit, $\sigma \to 0$, all the decrements, including λ_1, tend to the sequence $\lambda_l = l(l+1)$ and thus become of the same order of magnitude. This regime corresponds to a vanishing anisotropy so that the difference between the inter- and intrawell motions disappear, and the magnetic moment diffuses almost freely over all the 4π radians with the reference time τ_D given by Eq. (4.91).

From Eqs. (4.126) and (4.127) one finds that the longitudinal component of the magnetic moment evolves according to

$$m(t) = \mu \sum_{l=1}^{\infty} A_l e^{-\lambda_l t / 2\tau_D} \int_{-1}^{1} x \psi_l dx \qquad (4.131)$$

where $x = \cos\theta = (\boldsymbol{e} \cdot \boldsymbol{n})$. For a symmetric potential like (4.33), the equilibrium value m_0 of the particle magnetic moment is zero.

With the abovementioned structure of the eigenvalue spectrum, the term with $l = 1$ in Eq. (4.131), which is proportional to $\exp(-e^{-\sigma}t/\tau_D)$, at $\sigma > 1$ is far more long-living than any other one. The dominating role of the decrement λ_1 had been proved by Brown, who derived [47] the following asymptotic expression for it:

$$\lambda_B = (4/\sqrt{\pi})\sigma^{3/2} e^{-\sigma} \qquad (\sigma \gg 1) \qquad (4.132)$$

A short time later, using a continued fraction method, Aharoni constructed [83] for λ_1 a fairly long power series in σ and also showed numerically that Brown's expression (4.132) resembles the exact one with the accuracy of several percent for $\sigma \gtrsim 3$. In the 1990s the eigenvalue λ_1 became a subject of extensive studies. Efficient numerical procedures were developed [84], and a number of extrapolation formulas with a good overall accuracy were proposed [71,85–87].

3. Asymptotic Solution of the Brown Equation

The study that we describe below was inspired by our work on fitting the dynamic susceptibilities measurements for real assemblies of fine particles. Those data typically describe polydisperse systems in the low-frequency bandwidth $\omega/2\pi = 1 - 10^3$ Hz. As $\tau_0 \sim 10^{-9}$ s or smaller, then, using formula (4.132) for estimations, one concludes that the frequency interval mentioned becomes a dispersion range for the interwell (superparamagnetic) mode at $\omega\tau_0 e^\sigma \gtrsim 1$, that is, $\sigma \gtrsim 10$. For temperatures up to 300 K, this condition holds for quite a number of nanomagnetic systems.

Application of the best-fit procedure to a set of experimental data implies numerous recalculations of the linear and nonlinear susceptibility curves $\chi^{(k)}$ of the assembly. Any such curve, due to a considerable polydispersity of the particles, is a superposition of a great number of partial curves $\chi^{(k)}(\sigma)$ spread over a wide size or, in the dimensionless form, σ range. For successful processing, one needs a fast and very accurate algorithm to evaluate $\chi^{(k)}(\sigma)$ everywhere including the domain $\sigma \gg 1$. The existing extrapolation formulas are unsuitable for that purpose because of their illcontrollable error accumulation. A plausible way out is to use solution of Eq. (4.128) in the form of an asymptotic series in σ^{-1}. In the course of the fitting procedure, this approximation can be easily matched in the intermediate σ range with the well-known expansions for the small σ end.

It is noteworthy that Brown himself resumed [88,89] the studies on λ_1 and modified the preexponential factor in Eq. (4.132), transforming it into an asymptotic series in σ^{-1}. On the basis of Eq. (4.128), he had constructed an integral recurrence procedure, and evaluated λ_1 down to terms $\propto 1/\sigma^{10}$. What we do below, is, in fact, carry on this line of analysis that had not been touched since then. Our method advances Brown's results in two aspects: (1) for λ_1 it is simpler, and (2) it provides not only the eigenvalue but the eigenfunction as well. Possessing only the latter, one is able to obtain theoretical expressions for the directly measurable quantities, that is, the susceptibilities $\chi^{(k)}$.

Taking Eq. (4.128) as the starting point, we note that its equilibrium solution

$$\psi_0 = Z_0^{-1}\exp(\sigma x^2), \quad Z_0 = 2\mathscr{R}(\sigma), \quad \mathscr{R}(\sigma) = \int_0^1 \exp(\sigma x^2)dx \qquad (4.133)$$

corresponds to $l = 0$ and $\lambda_0 = 0$ and note the asymptotic expansion for the partition integral $R(\sigma)$ found in Ref. 54:

$$\mathscr{R}(\sigma) = \frac{e^\sigma G}{2\sigma}, \quad G(\sigma) \equiv 1 + \frac{1}{2\sigma} + \frac{3}{4\sigma^2} + \frac{15}{8\sigma^3} + \cdots + \frac{(2n-1)!!}{2^n\sigma^n} + \cdots$$

$$(4.134)$$

The operator $\hat{L}$ in Eq. (4.128) is not self-conjugated and thus produces two sets of eigenfunctions, which obey the respective equations

$$\hat{L}\psi_k = \lambda_k\psi_k, \quad \hat{L}^+\varphi_j = \lambda_j\varphi_j \qquad (4.135)$$

here + denotes Hermitian conjugation. The eigenfunctions of these two families are orthonormalized and related to each other in a simple way:

$$\psi_k = \psi_0\varphi_k, \quad \int_{-1}^{1} dx\varphi_j\psi_k = \delta_{jk} \qquad (4.136)$$

Qualitatively, from Eq. (4.136) one may say that φ_k are the same eigenfunctions but "stripped" of the exponential equilibrium solution ψ_0. Substituting Eq. (4.136) in Eq. (4.128), one gets two useful relationships

$$-\hat{J}\psi_0\hat{J}\varphi_k = \lambda_k\psi_0\varphi_k, \quad \int \psi_0(\hat{J}\varphi_j)(\hat{J}\varphi_k)dx = \lambda_k\delta_{jk} \qquad (4.137)$$

where the second one follows from the first after multiplication by φ_j and integration by parts. Note that in the second formula action of each operator reaches no farther than the nearest closing parenthesis.

On rewriting the first formula in Eq. (4.137) in terms of a single orientational variable $x = (e \cdot n)$, the spectral problem takes the form

$$\frac{d}{dx}\left[\psi_0(1 - x^2)\frac{d\varphi_k}{dx}\right] = -\lambda_k\psi_0\varphi_k \qquad (4.138)$$

In the equilibrium state Eq. (4.138) reduces to

$$\frac{d}{dx}\left[\psi_0(1 - x^2)\frac{d\varphi_0}{dx}\right] = 0 \qquad (4.139)$$

whose normalized solution is $\varphi_0 = 1$. This solution, which is a truly equilibrium one, turns the inner part of the brackets, that is, the probability flux in the kinetic equation (4.125), into identical zero.

As mentioned in Section III.B.1, at $\sigma \gg 1$ the most long-living nonstationary solution of Eq. (4.138) is the eigenfunction with $l = 1$, whose eigenvalue is exponentially small; see Brown's estimation (4.132). We use this circumstance for approximate evaluation of φ_1 in the $\sigma \gg 1$ limit by neglecting the right-hand side of Eq. (4.138) for $l = 1$. On so doing, the equation obtained for the function φ_1 formally coincides with Equation (4.139) for φ_0. However, the essential difference is that now the content of the bracket is nonzero

$$\psi_0(1-x^2)\frac{d\varphi_1}{dx} = \frac{1}{2}C \qquad (4.140)$$

where $\frac{1}{2}C$ is the integration constant. Note also that, contrary to φ_0, the solution φ_1 is odd in x.

Using the explicit form of ψ_0 from Eq. (4.133) and integrating, one gets for $x > 0$

$$\varphi_1 = C\mathscr{R}\int_0^x \frac{e^{-\sigma x^2}}{1-x^2}dx = C\mathscr{R}\int_0^x e^{-\sigma x^2}(1+x^2+x^4+x^6+\cdots)dx \qquad (4.141)$$

The integrals in expansion (4.141) are akin. Denoting

$$F_n = \int_0^x x^{2n}e^{-\sigma x^2}dx$$

one can easily write for them the recurrence relation and "initial" condition as

$$F_n = -\frac{\partial}{\partial\sigma}F_{n-1}, \quad F_0 = \frac{\sqrt{\pi}}{2\sqrt{\sigma}}\mathrm{erf}(\sqrt{\sigma}x) \qquad (4.142)$$

respectively. Using the asymptotics of the error integral, with the exponential accuracy in σ, one finds

$$F_n = \left[\frac{(2n-1)!!}{2^n\sigma^n}\right]F_0, \quad F_0 \simeq \frac{\sqrt{\pi}}{2\sqrt{\sigma}} \qquad (4.143)$$

Comparing this with expression (4.134) for the function G, we get the representation

$$\varphi_1(x > 0) \simeq C\mathscr{R}F_0G \qquad (4.144)$$

Applying to Eq. (4.144) the normalizing condition (4.136), one evaluates the constant as $C = 1/\mathscr{R}F_0G$. Therefore, from Eqs. (4.142)–(4.144) the principal relaxational eigenmode determined with the $\exp(-\sigma)$ accuracy emerges as an odd step function

$$\varphi_1(x) \simeq \begin{cases} -1 & \text{for } x < 0 \\ 1 & \text{for } x > 0 \end{cases} \qquad (4.145)$$

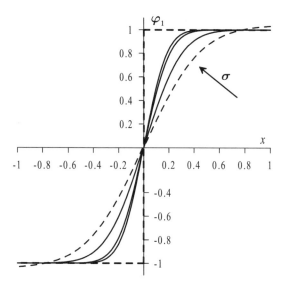

Figure 4.8. Eigenmode $\varphi_1(x)$ determined with the aid of the numerical solution of Eq. (4.130) for the dimensionless barrier height σ: 5 (dashed line), 10, 20, 25 (solid lines); the arrow shows the direction of σ growth. Thick dashes show the step-wise function that is the limiting contour for φ_1 at $\sigma \to \infty$.

In Figure 4.8 the limiting contour (4.145) is shown against the exact curves $\varphi_1(x)$ obtained by solving numerically Eq. (4.130) for several values of σ. We remark that in the statistical calculations carried out below the typical integrals are of two kinds. In the first, the integrand consists of the product of $\varphi_1 \psi_0$ and some nonexponential function. As $\psi_0 \propto \exp \sigma x^2$, the details of behavior of φ_1 in the vicinity of $x = 0$ are irrelevant because the approximate integral will differ only exponentially from the exact result. The integrals of the second type contain $d\varphi_1/dx$ in the integrand. For these integrals, a stepwise approximation [Eq. (4.145)] with its derivative equal identical zero everywhere except for $x = 0$ is an inadmissible choice. So, to keep the exponential accuracy in this case, one has to get back to Eq. (4.140).

The eigenvalue λ_1 corresponding to the approximate eigenfunction φ_1 from Eq. (4.145) is evaluated via formula (4.137), which can be rewritten as

$$\lambda_1 = \int_{-1}^{1} \psi_0 (\hat{J}\varphi_1)^2 dx = \frac{1}{\mathcal{R}} \int_0^1 e^{\sigma x^2} (1 - x^2) \left(\frac{d\varphi_1}{dx} \right)^2 dx \qquad (4.146)$$

Substituting the derivative from Eq. (4.140), one finds

$$\lambda_1 = C = \frac{2}{\mathcal{R}G} \sqrt{\frac{\sigma}{\pi}}$$

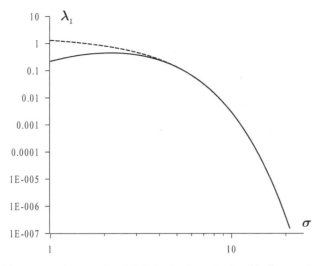

Figure 4.9 Asymptotic expression (4.147) for the eigenvalue λ_1 with allowance for terms up to σ^{-9} (solid line) compared to the exact numeric value (dashed line).

and using expression (4.134) for $\mathscr{R}$ finally arrives at

$$\lambda_1 = \frac{4\sigma^{3/2}e^{-\sigma}}{\pi^{1/2}G^2} = \frac{\lambda_B}{G^2} \qquad (4.147)$$

With G expanded in powers of σ^{-1} [see Eq. (4.136)], this formula reproduces the asymptotic expression derived by Brown [88]. At $G = 1$ it reduces to his initial result [47] corresponding to Eq. (4.134) here. Function $\lambda_1(\sigma)$ from Eq. (4.147) is shown in Figure 4.9 for comparison with the exact result obtained by a numerical solution. Indeed, at $\sigma \gtrsim 3$ the results virtually coincide.

According to expansion (4.127), each decrement λ_l defines the reference relaxation time

$$\tau_l = 2\tau_D/\lambda_l \qquad (4.148)$$

Then, from Eq. (4.147), we get

$$\tau_1 = \frac{2\tau_D}{\lambda_1} = \tau_B G^2, \qquad \tau_B \equiv \frac{2\tau_D}{\lambda_B} \qquad (4.149)$$

where τ_B denotes the asymptotic relaxation time obtained by Brown [47]. Substituting in Eq. (4.149) the explicit asymptotic series (4.134) for G, one gets

$$\tau_1 = \tau_D \frac{\sqrt{\pi}e^{\sigma}}{2\sigma^{3/2}}\left(1 + \frac{1}{\sigma} + \frac{7}{4\sigma^2} + \frac{9}{2\sigma^3} + \cdots\right) \qquad (4.150)$$

4. Asymptotic Integral Time

The decrements λ_l or, equivalently, relaxation times τ_l, which are characteristic of the eigenfunctions of the distribution function, are not observable if taken as separate quantities. However, in combination they are involved in a useful, directly measurable quantity, the so-called integral relaxation time. In terms of correlation functions this characteristic is defined as

$$\tau_{\text{int}} = \int_0^\infty \frac{\langle m(t)m(0)\rangle_0}{\langle m^2(0)\rangle_0} \, dt = \int_0^\infty \frac{\langle x(t)x(0)\rangle_0}{\langle x^2(0)\rangle_0} \, dt \qquad (4.151)$$

where the angular brackets stand for the statistical ensemble averaging over the equilibrium distribution (4.134). As follows from Eq. (4.151), the integral relaxation time equals the area under the normalized decay of magnetization.

The Green function of Eq. (4.125), that is, the probability density of a state (x, t), provided the initial state is $(x_0, 0)$, is expressed as

$$W(x, t; x_0, 0) = \sum_{l=0}^\infty \psi_l(x)\varphi_l(x_0)e^{-\lambda_l t} \qquad (4.152)$$

Similarly to Eq. (4.129), we expand the eigenfunctions in Legendre polynomials as

$$\psi_l = \frac{1}{2}\sum_{k=1}^\infty (2k+1)b_k^{(l)}P_k(x), \qquad \varphi_l = \sum_{k=1}^\infty a_k^{(l)}P_k(x) \qquad (4.153)$$

and introduce special notations for the first two functions:

$$\psi_0 = \frac{1}{2}\sum_{k=0}^\infty (2k+1)S_kP_k(x), \qquad \psi_1 = \frac{1}{2}\sum_{k=0}^\infty (2k+1)Q_kP_k(x) \qquad (4.154)$$

The procedures employed to evaluate the coefficients S_k and Q_k and the explicit asymptotic forms for Q_1 and S_2 are given in Section III.B.8.a; note representation (4.133) for the equilibrium function ψ_0.

Due to Eq. (4.136), the coefficients in formulas (4.153) are related to each other by $b_k^{(l)} = \langle P_k P_{k'}\rangle_0 a_{k'}^{(l)}$. In those terms one obtains for the correlator in Eq. (4.136):

$$\langle\langle x(t)x(0)\rangle\rangle_0 = \iint xx_0\psi_0 W(x, t; x_0, 0)dx\, dx_0 = \sum_{l=1}^\infty [b_1^{(l)}]^2 e^{-\lambda_l t/2\tau_D} \qquad (4.155)$$

where the averaging over the current coordinate x is performed with the function W from Eq. (4.152) whereas that over the initial conditions—with the equilibrium function ψ_0. Substituting expression (4.155) in Eq. (4.151), one gets the integral time in the form

$$\tau_{\text{int}} = \frac{\sum_{l=1}^{\infty} \tau_l [b_1^{(l)}]^2}{\sum_{l=1}^{\infty} [b_1^{(l)}]^2} = \frac{\sum_{l=1}^{\infty} \tau_l [b_1^{(l)}]^2}{\langle x^2 \rangle_0} \tag{4.156}$$

Unlike τ_1, which in principle cannot be evaluated analytically at arbitrary σ [90] for τ_{int} an exact solution is possible for arbitrary values of the anisotropy parameter. Two ways were proposed to obtain quadrature formulas for τ_{int}. One method [91] implies a direct integration of the Fokker–Planck equation. Another method [58] involves solving three-term recurrence relations for the statistical moments of W. The emerging solution for τ_{int} can be expressed in a finite form in terms of hypergeometric (Kummer's) functions. Equivalence of both approaches was proved in Ref. 92.

In the present study, as mentioned, we are dealing in the high-barrier approximation. In this limiting case λ_1 is exponentially small, so that the term with $l = 1$ in the numerator in Eq. (4.156) is far greater than the others. With allowance for Eq. (4.154) it can be written as

$$\tau_{\text{int}} = \tau_1 \frac{[b_1^{(1)}]^2}{\langle x^2 \rangle_0} = \frac{\tau_1 Q_1^2}{\langle x^2 \rangle_0} \tag{4.157}$$

The equilibrium moment calculated by definition is written as

$$\langle x^2 \rangle_0 = \frac{1}{2\sigma} (e^{\sigma} - 1) = \frac{1}{G} - \frac{1}{2\sigma} \tag{4.158}$$

and for $\sigma \gg 1$, using formula (4.206) (below), we get

$$Q_1 \simeq \frac{1}{G} \tag{4.159}$$

Substitution of Eqs. (4.158) and (4.159) in (4.157) with allowance for relationships (4.134), (4.147) and (4.149) gives the asymptotic representation in the form

$$\tau_{\text{int}} = \tau_B \frac{2\sigma G}{(2\sigma - G)} = \tau_D \frac{\sqrt{\pi} e^{\sigma}}{2\sigma^{3/2}} \left(1 + \frac{1}{\sigma} + \frac{3}{2\sigma^2} + \frac{13}{4\sigma^3} + \cdots \right) \tag{4.160}$$

As is seen from formulas (4.150) and (4.160), written with accuracy up to σ^{-3}, the asymptotic expressions for the interwell and integral times deviate beginning with the term $\propto \sigma^{-2}$. This contradicts the only known (to us) asymptotic expansion of τ_{int}, given in Eq. (60) of Ref. 58 and repeated in Eq. (7.4.3.22) of the book [36]. The latter expression, written with the accuracy $\propto \sigma^{-2}$, instead of turning into Eq. (4.160), coincides with Brown's expression (4.150) for τ_1. Meanwhile, as follows from formula (4.157), such a coincidence is impossible and therefore Eqs. (60) of Ref. 58 and (7.4.3.22) of Ref. 36 are misleading. The necessity to rectify this issue made us begin to demonstrate our approach with the case of the integral relaxation time. Further on we consistently apply our procedure for description of the nonlinear (third- and fifth-harmonic) dynamic susceptibilities of a solid superparamagnetic dispersion.

5. Perturbative Expansions for the Distribution Function

a. Static Probing Field. To find the nonlinear susceptibilities, one has to take into account the changes that the probing field induces in the basic state of the system. In the limit $\sigma \gg 1$, which we discuss, the relaxation time τ_1 of the interwell mode ψ_1 is far greater than all the other relaxation times τ_k. This means that with respect to the intrawell modes, the distribution function is in equilibrium. So it suffices to determine the effect of the probing field $\boldsymbol{H} = H\boldsymbol{h}$ just on ψ_0 and ψ_1. Assuming the energy function in the form

$$U + U_H = -KV_m(\boldsymbol{e} \cdot \boldsymbol{n})^2 - I_s V_m H(\boldsymbol{e} \cdot \boldsymbol{h}) \qquad (4.161)$$

comparing with Eq. (4.33), and separating variables in Eq. (4.125), one arrives at the eigenfunction problem

$$\hat{L}f_\beta = \xi \hat{V}f_\beta \qquad (4.162)$$

where $\xi = IV_mH/T$ and notation f_β refers to the distribution function modes that stem from ψ_0 or ψ_1 at $H \neq 0$, namely, $\beta = 0$ or 1. In Eq. (4.162) operator $\hat{L}$ is as defined by Eq. (4.128) while $\hat{V} = -\xi \hat{\boldsymbol{J}}(\boldsymbol{e} \times \boldsymbol{h})$ is the operator caused by the energy term U_H in Eq. (4.161). As above, for the non-self-conjugated spectral problem (4.162), we introduce the family of conjugated functions g_β and set $f_\beta = g_\beta \psi_0$.

Following our approach, in the low-temperature limit ($\sigma \gg 1$) we *set to zero* the eigenvalues corresponding to both f_0 and f_1; compare with Eqs. (4.139) and (4.140) for ψ_0 and ψ_1. Assuming the temperature-scaled magnetic field ξ to be small, we treat U_H as a perturbation Hamiltonian and expand the principal eigenfunctions as

$$f_0 = \sum_{n=0} \xi^n f_0^{(n)}, \qquad f_1 = \sum_{n=0} \xi^n f_1^{(n)} \qquad (4.163)$$

Then for the field-free $(H = 0)$ case one has $f_0^{(0)} = \psi_0$ and $f_1^{(0)} = \psi_1$. The same kind of expansion is assumed for g_β with $g_0^{(0)} = 1$ and $g_1^{(0)} = \varphi_1$. Note also that in order to retain the normalizing condition, we require that $f_\beta^{(n)}$ have zero averages.

Substituting expansion (4.163) in Eq. (4.162) and collecting the terms of the same order in ξ, we arrive at the recurrence relation

$$\hat{L}f_\beta^{(n)} = \hat{V}f_\beta^{(n-1)} \tag{4.164}$$

which for the particular cases $\beta = 0$ and 1 with the aid of the identity $e \times h = \hat{J}(e \cdot h)$ takes the forms

$$\hat{J}\psi_0\hat{J}g_0^{(n)} = \hat{J}\psi_0 g_0^{(n-1)}\hat{J}(e \cdot h)$$
$$\hat{J}\psi_0\hat{J}g_1^{(n)} = \hat{J}\psi_0 g_1^{(n-1)}\hat{J}(e \cdot h) \tag{4.165}$$

respectively. Set (4.165) solves easily for g_0 since $g_0^{(0)} = \varphi_0 = 1$. Starting with $n = 0$, one obtains sequentially

$$g_0^{(1)} = (e \cdot h)$$

$$g_0^{(2)} = \frac{1}{2}[(eh)^2 - \langle(e \cdot h)\rangle_0^2]$$

$$g_0^{(3)} = \frac{1}{6}(e \cdot h)^3 - \frac{1}{2}(e \cdot h)\langle(e \cdot h)^2\rangle_0$$

$$g_0^{(4)} = \frac{1}{24}[(e \cdot h)^4 - \langle(e \cdot h)^4\rangle_0] - \frac{1}{4}[(e \cdot h)^2\langle(e \cdot h)^2\rangle_0 - \langle(e \cdot h)^2\rangle_0^2]$$

$$g_0^{(5)} = \frac{1}{120}(e \cdot h)^5 - \frac{1}{12}(e \cdot h)^3\langle(e \cdot h)^2\rangle_0 - \frac{1}{24}(e \cdot h)[\langle(e \cdot h)^4\rangle_0$$
$$- 6\langle(e \cdot h)^2\rangle_0^2] \tag{4.166}$$

All the functions obtained are constructed in such a way that the corresponding $f_\beta^{(n)}$ satisfy the abovementioned zero average requirement. We remark also that there is no problem continuing the calculational procedure to any order.

Evaluation of g_1 is done in two steps. In the first one, we set $g_1^{(0)}$ equal to the antisymmetric stepwise function (4.145) and its derivative equal zero. After that from the second formula of Eqs. (4.165), we can express $g_1^{(k)}$ in closed form.

Taken up to the fourth order, these "zero-derivative" solutions are expressed as follows:

$$g_1^{(1)} = \varphi_1(e \cdot h) - \langle \varphi_1(e \cdot h) \rangle_0$$

$$g_1^{(2)} = \tfrac{1}{2}\varphi_1(e \cdot h)^2 - (e \cdot h)\langle \varphi_1(e \cdot h) \rangle_0$$

$$g_1^{(3)} = \tfrac{1}{6}[\varphi_1(e \cdot h)^3 - \langle \varphi_1(e \cdot h)^3 \rangle_0] - \tfrac{1}{2}\langle \varphi_1(e \cdot h) \rangle_0[(e \cdot h)^2 - \langle (e \cdot h)^2 \rangle_0]$$

$$g_1^{(4)} = \tfrac{1}{24}\varphi_1(e \cdot h)^4 - \tfrac{1}{6}\langle \varphi_1(e \cdot h) \rangle_0[(e \cdot h)^3 - 3(e \cdot h)\langle (e \cdot h)^2 \rangle_0]$$

$$- \tfrac{1}{6}(e \cdot h)\langle \varphi_1(e \cdot h)^3 \rangle_0$$

$$(4.167)$$

Note the alternating parity in e with the term order growth in both Eqs. (4.166) and (4.167).

It is instructive to compare the approximate expressions (4.167) with the numerical results obtained without simplification of $g_1^{(0)}$. To be specific, we consider the case when a probing field is applied along the particle easy axis n. Then Eqs. (4.165) become one-dimensional and the second of them is expressed as

$$\frac{dg_1^{(n)}}{dx} = g_1^{(n-1)} \tag{4.168}$$

Its "zero derivative" solutions up to the second order follow from the first two lines of Eqs. (4.167):

$$g_1^{(1)} = \varphi_1 x - \langle \varphi_1 x \rangle_0, \qquad g_1^{(2)} = \tfrac{1}{2}\varphi_1 x^2 - x\langle \varphi_1 x \rangle_0 \tag{4.169}$$

In Figures 4.10 and 4.11 these functions are compared to the numerical solutions of Eq. (4.168). For our calculation, the most important is the behavior of those functions near $x = \pm 1$ since these regions yield the main contribution when integrated with the weight function ψ_0. As one can see from the figures, the zero-derivative solution $g_1^{(1)}$ agrees well with the exact one, while $g_1^{(2)}$ deviates significantly. This discrepancy is due to the change of the barrier height that occurs in the second order with respect to the probing field amplitude, and manifests itself in all the even orders of the perturbation expansion. Correction of solution (4.169) constitutes the second step of our procedure. For that we

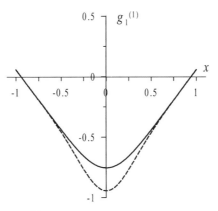

Figure 4.10. Function $g_1^{(1)}$ found numerically (solid line) and evaluated in the "zero derivative" approximation (dashed line).

integrate Eq. (4.168) two times by parts and substitute there the zero-derivative form of $g_1^{(1)}$ from Eq. (4.169):

$$g_1^{(2)} = \tfrac{1}{2}x^2\varphi_1 - x\langle x\varphi_1\rangle + \tfrac{1}{2}\int x^2 \tfrac{d\varphi_1}{dx}\,dx \qquad (4.170)$$

Thus one finds that the corrected $g_1^{(2)}$ differs from that of Eq. (4.169) by adding a stepwise [like that of Eq. (4.145)] term

$$g_1^{(2)} = \tfrac{1}{2}x^2\varphi_1 - x\langle x\varphi_1\rangle + D_2\varphi_1 \qquad (4.171)$$

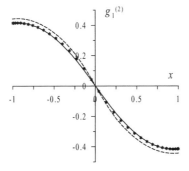

Figure 4.11. Function $g_1^{(2)}$ found numerically (solid line) and evaluated in the "zero derivative" approximation (dashed line). Asterisks show a corrected calculation with allowance for the coefficient D_2 [see Eq. (4.171)].

with the amplitude

$$D_2 = \frac{1}{2}\int_0^1 x^2 \frac{d\varphi_1}{dx}\,dx \qquad (4.172)$$

We remark that the results of evaluation of the integrals $I_{2k} = \int_0^1 x^{2k}(d\varphi_1/dx)dx$ can be tabulated as follows

k	0	1	2
I_{2k}	1	$1 - G^{-1}$	$1 - (1 + 1/2\sigma)G^{-1}$

$$(4.173)$$

so that Eq. (4.173) gives

$$D_2 = \frac{1}{2}I_{2k} = \frac{G-1}{2G} = \frac{1}{4\sigma} + \frac{1}{4\sigma^2} + \frac{5}{8\sigma^3} + \frac{37}{16\sigma^4} + \cdots \qquad (4.174)$$

Function $g_1^{(2)}$ corrected in such a way is shown in Figure 4.11 by asterisks. It is seen that the corrected dependence with a fairly good accuracy follows the numerically obtained curve.

In a similar way one can prove that the corrected function $g_1^{(4)}$ has the form

$$g_1^{(4)} = \frac{1}{24}\varphi_1 x^4 - \frac{1}{6}[\langle\varphi_1 x\rangle_0 x^3 - 3x\langle x^2\rangle_0] - \frac{1}{6}x\langle\varphi_1 x^3\rangle_0 + D_2 g_1^{(2)} + D_4\varphi_1$$

$$(4.175)$$

where the corrected function $g_1^{(2)}$ given by Eq. (4.171) is used and

$$D_4 = \frac{1}{24}I_4 - D_2^2 = -\frac{10\sigma G^2 - 22\sigma G + G + 12\sigma}{48\sigma G^2}$$

$$= -\frac{1}{32\sigma^2} - \frac{1}{16\sigma^3} - \frac{5}{32\sigma^4} - \frac{29}{64\sigma^5} + \cdots \qquad (4.176)$$

In the general case, when the direction of the probing field does not coincide with the particle anisotropy axis, the corrected functions $g_1^{(n)}$ still can be written as

$$g_1^{(2)} = \frac{1}{2}\varphi_1(e \cdot h)^2 - (e \cdot h)\langle\varphi_1(e \cdot h)\rangle_0 + D_2\varphi_1$$

$$g_1^{(3)} = \frac{1}{6}[\varphi_1(e \cdot h)^3 - \langle\varphi_1(e \cdot h)^3\rangle_0]$$

$$\quad - \frac{1}{2}\langle\varphi_1(e \cdot h)\rangle_0[(e \cdot h)^2 - \langle(e \cdot h)^2\rangle_0] + D_2 g_1^{(1)} \qquad (4.177)$$

$$g_1^{(4)} = \frac{1}{24}\varphi_1(e \cdot h)^4 - \frac{1}{6}\langle\varphi_1(e \cdot h)\rangle_0[(e \cdot h)^3 - 3(e \cdot h)\langle(e \cdot h)^2\rangle_0]$$

$$\quad - \frac{1}{6}(e \cdot h)\langle\varphi_1(e \cdot h)^3\rangle_0 + D_2 g_1^{(2)} + D_4\varphi_1$$

But since Eqs. (4.165) cannot be reduced to a form like Eq. (4.168), the correcting coefficients D_2 and D_4 cannot be presented in a closed form. In this case the corrected solutions taking into account the behavior of function φ_1 around zero are built up as power series near $x = 0$; such a procedure for the coefficients D_2 and D_4 is described in Section III.B.8.b.

b. Dynamic Probing Field. To obtain the dynamic susceptibilities, one has to find the distribution function W in the oscillating probing field $\xi \exp(i\omega t)$. For this situation the kinetic equation (4.125) takes the form

$$\left(2\tau_D \frac{\partial}{\partial t} + \hat{L}\right) W(t) = \xi \hat{V} e^{i\omega t} W(t) \tag{4.178}$$

where the operators $\hat{L}$ and $\hat{V}$ have been introduced in Eqs. (4.128) and (4.162), respectively. Assuming that the exciting field amplitude is not too high, we expand the steady-state oscillatory solution of Eq. (4.178) in a power series with respect to ξ:

$$W(t) = \sum_{n=0} \xi^n W^{(n)} e^{in\omega t} \tag{4.179}$$

Note that, mathematically, representation (4.179) is not complete. Indeed, in a general case the exact amplitude of the $n\omega$ mode must contain, along with the contribution $\sim \xi^n$, an infinite set of terms $\sim \xi^{n+2}, \xi^{n+4}$, and so on. However, in a weak-field limit $\xi < 1$ the terms with higher powers are of minor importance so that the main contribution to the magnetization response signal filtered at the frequency $n\omega$ is proportional to ξ^n.

Substituting Eq. (4.179) in (4.178), we arrive at the recurrence set

$$(2in\omega\tau_D + \hat{L})W^{(n)} = \hat{V}W^{(n-1)} \tag{4.180}$$

which we solve sequentially starting from $n = 1$. At the first step the function on the right-hand side corresponds to the equilibrium case ($\xi = 0$). Therefore, $W^{(0)} = \psi_0$, where the latter function is as defined by Eq. (4.133) and is frequency-independent. Combining Eq. (4.164) taken for $\beta = 0$ and $n = 1$ and Eq. (4.180), we eliminate the operator $\hat{V}$ and obtain

$$(2i\omega\tau_D + \hat{L})W^{(1)} = \hat{L}f_0^{(1)} \tag{4.181}$$

Now we expand the functions subjected to operator $\hat{L}$ with respect to the set $\{\psi_k\}$ of its eigenfunctions [see Eq. (4.128)]

$$W^{(1)} = \sum c_j^{(1)}(\omega)\psi_j \quad f_0^{(1)} = \sum (\varphi_j \,|\, f_0^{(1)})\psi_j \tag{4.182}$$

here $(\varphi \mid f)$ denotes functional scalar multiplication, that is, the integral of the product φf over all the orientations of e. Substitution of Eq. (4.182) in (1.181), then multiplying it from the left by ψ_k, followed by integration, render the expansion coefficient as

$$c_k^{(1)}(\omega) = (\varphi_k \mid f_0^{(1)})[1 + i\omega\tau_k]^{-1} \qquad (4.183)$$

where the reference relaxation times are as defined by Eq. (4.148).

In the low-frequency limit only $\omega\tau_1$ is set to be nonzero while all the higher modes are taken at equilibrium ($\omega\tau_k = 0$). Thence, when constructing $W^{(1)}$ via Eq. (4.182), by adding and subtracting a term with $c_1^{(1)}(0)$, one can present the first-order solution in the form

$$W^{(1)} = f_0^{(1)} - \frac{i\omega\tau_1}{1 + i\omega\tau_1}(\varphi_1 \mid f_0^{(1)})\psi_1 \qquad (4.184)$$

where $f_0^{(1)}$, as seen from Eq. (4.181), is the equilibrium solution for the same value of the field amplitude ξ. We remind the reader that the functions without upper index (superscript) belong to the fundamental set defined by Eqs. (4.128), whereas those with an upper index are evaluated in the framework of the perturbation scheme described in Section III.B.5.

In the next order in ξ the function $W^{(1)}$ is substituted in the right-hand side of Eq. (4.180) and by a procedure similar to that leading to Eqs. (1.181)–(1.183), the function $W^{(2)}$ is found. We carry on this cycle up to $k = 5$. The results are

$$W^{(2)} = f_0^{(2)} - \frac{i\omega\tau_1}{1 + i\omega\tau_1}(\varphi_1 \mid f_0^{(1)})f_1^{(1)} \qquad (4.185)$$

$$W^{(3)} = f_0^{(3)} + [(\varphi_1 \mid f_0^{(1)})(\varphi_1 \mid f_1^{(2)}) - (\varphi_1 \mid f_0^{(3)})]f_1^{(0)} - (\varphi_1 \mid f_0^{(1)})f_1^{(2)}$$

$$+ \frac{1}{1 + i\omega\tau_1}[(\varphi_1 \mid f_0^{(1)})f_1^{(2)} - \frac{3}{2}(\varphi_1 \mid f_0^{(1)})(\varphi_1 \mid f_1^{(2)})f_1^{(0)}]$$

$$+ \frac{1}{1 + 3i\omega\tau_1}[(\varphi_1 \mid f_0^{(3)}) + \frac{1}{2}(\varphi_1 \mid f_0^{(1)})(\varphi_1 \mid f_1^{(2)})]f_1^{(0)} \qquad (4.186)$$

$$W^{(4)} = f_0^{(4)} + [(\varphi_1 \mid f_0^{(1)})(\varphi_1 \mid f_1^{(2)}) - (\varphi_1 \mid f_0^{(3)})]f_1^{(1)} - (\varphi_1 \mid f_0^{(1)})f_1^{(3)}$$

$$+ \frac{1}{1 + i\omega\tau_1}[(\varphi_1 \mid f_0^{(1)})f_1^{(3)} - \frac{3}{2}(\varphi_1 \mid f_0^{(1)})(\varphi_1 \mid f_1^{(2)})f_1^{(1)}]$$

$$+ \frac{1}{1 + 3i\omega\tau_1}[(\varphi_1 \mid f_0^{(3)}) + \frac{1}{2}(\varphi_1 \mid f_0^{(1)})(\varphi_1 \mid f_1^{(2)})]f_1^{(1)} \qquad (4.187)$$

$$W^{(5)} = f_0^{(5)} - (\varphi_1 \mid f_0^{(5)})f_1^{(0)} + (\varphi_1 \mid f_0^{(1)})[(\varphi_1 \mid f_1^{(4)})f_1^{(0)} - f_1^{(4)}]$$

$$+ [(\varphi_1 \mid f_0^{(1)})(\varphi_1 \mid f_1^{(2)}) - (\varphi_1 \mid f_0^{(3)})][f_1^{(2)} - (\varphi_1 \mid f_1^{(2)})f_1^{(0)}]$$

$$+ \frac{1}{1 + i\omega\tau_1}(\varphi_1 \mid f_0^{(1)})\left\{ f_1^{(4)} - \frac{3}{2}(\varphi_1 \mid f_1^{(2)})f_1^{(2)} \right.$$

$$+ \left[\frac{15}{8}(\varphi_1 \mid f_1^{(2)})^2 - \frac{5}{4}(\varphi_1 \mid f_1^{(4)}) \right] f_1^{(0)} \right\}$$

$$+ \frac{1}{1 + 3i\omega\tau_1}\left[(\varphi_1 \mid f_0^{(3)}) + \frac{1}{2}(\varphi_1 \mid f_0^{(1)})(\varphi_1 \mid f_1^{(2)}) \right]$$

$$\times \left[f_1^{(2)} - \frac{5}{2}(\varphi_1 \mid f_1^{(2)})f_1^{(0)} \right]$$

$$+ \frac{1}{1 + 5i\omega\tau_1}\left[(\varphi_1 \mid f_0^{(5)}) + \frac{1}{4}(\varphi_1 \mid f_0^{(1)})(\varphi_1 \mid f_1^{(4)}) \right.$$

$$\left. + \frac{3}{8}(\varphi_1 \mid f_0^{(1)})(\varphi_1 \mid f_1^{(2)})^2 + \frac{3}{2}(\varphi_1 \mid f_0^{(3)})(\varphi_1 \mid f_1^{(2)}) \right] f_1^{(0)} \qquad (4.188)$$

We note an important feature of Eqs. (4.185)–(4.188)—they do not contain dispersion factors of even orders. This ensures that the frequency dependence of the full distribution function W incorporates only dispersion factors with odd multiples of the basic frequency. Qualitatively, this is the result of absence of the interwell mode for the statistical moments of even orders. Technically, it is due to vanishing of the products $(\varphi_1 \mid f_k^{(l)})$ entering Eqs. (4.184)–(4.188) if the sum $k + l$ is even. This rule follows immediately from combination of the oddity of φ_1 (see Section III.B.1) with the parity properties of the functions $f_k^{(l)}$ introduced in Section III.B.1.

For actual calculations one needs the values of the scalar products entering Eqs. (4.184)–(4.188). In Section III.B.8.c we obtain their representations in terms of the moments Q_k and S_k of the functions ψ_0 and ψ_1, respectively. The procedures of asymptotic expansion of Q_k and S_k are given in Section III.B.8.a.

6. Dynamic Susceptibilities

The set of magnetic susceptibilities of an assembly of non-interacting particles is defined by relation (4.85), which refers to the magnetization of the system in the direction of $\boldsymbol{H} = H\boldsymbol{h}$. Therefore, of all the components of the corresponding susceptibility tensors, we retain only the combinations that determine the projection on the probing field. With representation (4.179) for the distribution function, this magnetization component takes the form

$$M = cIV\langle(e \cdot h)\rangle$$
$$= c\sum_{n=1} H^n \frac{I_s^{n+1} V_m^{n+1}}{(T)^n} e^{in\omega t} \int (e \cdot h) W^{(n)} de \qquad (4.189)$$

The susceptibilities are then derived by direct comparison with Eq. (4.85)

$$\chi^{(n)} = \frac{cI_s^{n+1} V_m^{n+1}}{(T)^n} e^{in\omega t} \int (e \cdot h) W^{(n)} de \qquad (4.190)$$

thus relating the set of $\chi^{(n)}$ to the perturbation functions $W^{(n)}$ found in the preceding section. Therefore, evaluation of $\chi^{(n)}$ becomes a tedious but simple procedure. Remarkably, the final expressions come out rather compact.

a. Linear Susceptibility. The resulting expression can be presented in the form

$$\chi_\omega^{(1)} = \chi_0^{(1)} \left(B_0^{(1)} + \frac{B_1^{(1)}}{1 + i\omega\tau_1} \right), \quad \chi_0^{(1)} = \frac{cI_s^2 V_m^2}{3T} \qquad (4.191)$$

which follows from substituting Eq. (4.184) in (4.189). Each of the two frequency-independent coefficients $B^{(1)}$, which are the result of statistical averaging over the orientational variable e (see Section III.B.8.c), expands into a series of Legendre polynomials with respect to β, the angle between the direction h of the probing field and the particle easy axis n. This can be written as

$$\begin{cases} B_0^{(1)} = b_{00}^{(1)} + b_{02}^{(1)} P_2(\cos\beta) \\ B_1^{(1)} = b_{10}^{(1)} + b_{12}^{(1)} P_2(\cos\beta) \end{cases} \qquad (4.192)$$

where

$$\begin{pmatrix} b_{00}^{(1)} & b_{02}^{(1)} \\ b_{10}^{(1)} & b_{12}^{(1)} \end{pmatrix} = \begin{pmatrix} 1 - Q_1^2 & 2S_2 - 2Q_1^2 \\ Q_1^2 & 2Q_1^2 \end{pmatrix}$$

and definitions of functions Q_1 and S_2 and their explicit asymptotic representations are given in Section III.B.8.a. The asymptotic series for the coefficients $b_{\alpha\beta}^{(1)}$

derived on the basis of expansion (4.134) and Eq. (4.159) are

$$b_{02}^{(1)} = -\frac{1}{\sigma} + \frac{1}{4\sigma^3} + \frac{13}{8\sigma^4} + \frac{165}{16\sigma^5} + \frac{2273}{32\sigma^6} + \frac{34577}{64\sigma^7} + \frac{581133}{128\sigma^8} + \cdots$$

$$b_{10}^{(1)} = 1 - \frac{1}{\sigma} - \frac{3}{4\sigma^2} - \frac{2}{\sigma^3} - \frac{31}{4\sigma^4} - \frac{153}{4\sigma^5} - \frac{3629}{16\sigma^6} - \frac{1564}{\sigma^7} - \frac{785931}{64\sigma^8} + \cdots$$

$$(4.193)$$

The other components, namely, $b_{00}^{(1)}$ and $b_{12}^{(1)}$, may be constructed straightforwardly using their relations with the given ones [see Eqs. (4.192)]. For a random system, that is, for an assembly of noninteracting particles with a chaotic distribution of the anisotropy axes, the average of any Legendre polynomial is zero, so that $\tilde{B}_k^{(1)} = b_{k0}^{(1)}$, and the linear dynamic susceptibility reduces to

$$\tilde{\chi}_\omega^{(1)} = \chi_0^{(1)} \frac{1 + i\omega\tau_1 b_{00}^{(1)}}{1 + i\omega\tau_1} \tag{4.194}$$

which is the asymptotic representation of the full expression given by formula (39) of Ref. 67.

b. Cubic Susceptibility. As follows from definitions (4.85) and (4.189), the third-order susceptibility is defined through the response at the triple frequency that at weak H scales is H^3. Performing calculations along the same scheme as for $\chi^{(1)}$, one may arrange the result in the form of the sum of relaxators as

$$\chi_{3\omega}^{(3)} = -\frac{45}{4}\chi_0^{(3)}\left(B_0^{(3)} + \frac{B_1^{(3)}}{1 + i\omega\tau_1} + \frac{B_3^{(3)}}{1 + 3i\omega\tau_1}\right)$$

$$\chi_0^{(3)} = -\frac{cI_s^4 V_m^4}{45(T)^3} \tag{4.195}$$

where the coefficients expand as

$$B_k^{(3)} = b_{k0}^{(3)} + b_{k2}^{(3)}P_2(\cos\beta) + b_{k4}^{(3)}P_4(\cos\beta), \quad k = 0, 1, 3, \ldots \tag{4.196}$$

up to the fourth Legendre polynomial in $\cos\beta$. Note that in Eq. (4.195) the numerical coefficients are divided between static and dynamic susceptibilities in such a way that the definition of $\chi_0^{(3)}$ coincides with that established above by Eq. (4.189). The coefficients for higher orders are constructed in the same manner, only requiring that the factor at $\chi_0^{(n)}$ coincide with the corresponding number in the expansion of the Langevin function with respect to its argument; for instance, in the fifth order this yields 2/945.

The explicit expansions for the amplitudes $b_{\alpha\beta}^{(3)}$ are

$$b_{00}^{(3)} = -\frac{1}{30\sigma^3} - \frac{47}{240\sigma^4} - \frac{49}{40\sigma^5} - \frac{815}{96\sigma^6} - \frac{7837}{120\sigma^7} - \frac{355391}{640\sigma^8} + \cdots$$

$$b_{02}^{(3)} = -\frac{1}{42\sigma^3} - \frac{2}{21\sigma^4} - \frac{4}{7\sigma^5} - \frac{1385}{336\sigma^6} - \frac{11231}{336\sigma^7} - \frac{19083}{64\sigma^8} + \cdots$$

$$b_{04}^{(3)} = \frac{2}{35\sigma^3} + \frac{8}{35\sigma^4} + \frac{41}{35\sigma^5} + \frac{50}{7\sigma^6} + \frac{1756}{35\sigma^7} + \frac{63749}{160\sigma^8} + \cdots$$

$$b_{10}^{(3)} = -\frac{1}{15} + \frac{1}{6\sigma} + \frac{23}{240\sigma^2} + \frac{61}{192\sigma^3} + \frac{1357}{960\sigma^4} + \frac{235447}{30720\sigma^5} + \frac{11962691}{245760\sigma^6}$$
$$+ \frac{694849241}{1966080\sigma^7} + \frac{15133953221}{5242880\sigma^8} + \cdots$$

$$b_{12}^{(3)} = -\frac{13}{84} + \frac{65}{168\sigma} + \frac{25}{168\sigma^2} + \frac{863}{1344\sigma^3} + \frac{3931}{1344\sigma^4} + \frac{698911}{43008\sigma^5}$$
$$+ \frac{35309123}{344064\sigma^6} + \frac{2061480665}{2752512\sigma^7} + \frac{45071465669}{7340032\sigma^8} + \cdots$$

$$b_{14}^{(3)} = -\frac{1}{35} + \frac{1}{14\sigma} - \frac{2}{35\sigma^2} + \frac{1}{112\sigma^3} + \frac{73}{560\sigma^4} + \frac{17033}{17920\sigma^5} + \frac{1007549}{143360\sigma^6}$$
$$+ \frac{64390439}{1146880\sigma^7} + \frac{4493994417}{9175040\sigma^8} + \cdots$$

$$b_{30}^{(3)} = \frac{2}{15} - \frac{3}{10\sigma} - \frac{1}{16\sigma^2} - \frac{337}{960\sigma^3} - \frac{449}{320\sigma^4} - \frac{85309}{10240\sigma^5} - \frac{2563751}{49152\sigma^6}$$
$$- \frac{245269747}{655360\sigma^7} - \frac{47628510799}{15728640\sigma^8} + \cdots$$

$$b_{32}^{(3)} = \frac{29}{84} - \frac{43}{56\sigma} - \frac{11}{56\sigma^2} - \frac{1279}{1344\sigma^3} - \frac{1881}{448\sigma^4} - \frac{320765}{14336\sigma^5} - \frac{48133699}{344064\sigma^6}$$
$$- \frac{920146163}{91750\sigma^7} - \frac{178560431695}{22020096\sigma^8} + \cdots$$

$$b_{34}^{(3)} = \frac{11}{105} - \frac{47}{210\sigma} - \frac{2}{21\sigma^2} - \frac{559}{1680\sigma^3} - \frac{2419}{1680\sigma^4} - \frac{409499}{53760\sigma^5}$$
$$- \frac{4080395}{86016\sigma^6} - \frac{1166954357}{3440640\sigma^7} - \frac{75334335763}{27525120\sigma^8} + \cdots$$

$$(4.197)$$

For a random system, the averages of Legendre polynomials drop out and $\tilde{B}_k^{(3)} = b_{k0}^{(3)}$. With respect to formalism constructed in Section III.A, these expressions yield the asymptotic representations for formulas (4.110) and (4.111) there.

c. Fifth-Order Susceptibility. The susceptibility of the fifth order writes in an expectable way as a sum of three relaxators:

$$\chi_{5\omega}^{(5)} = \frac{945}{8}\chi_0^{(5)}\left(B_0^{(5)} + \frac{B_1^{(5)}}{1 + i\omega\tau} + \frac{B_3^{(5)}}{1 + 3i\omega\tau} + \frac{B_5^{(5)}}{1 + 5i\omega\tau}\right), \qquad \chi_0^{(5)} = \frac{2cI^6V_m^6}{945T^5}$$

$$(4.198)$$

with the coefficients

$$B_k^{(5)} = b_{k0}^{(5)} + b_{k2}^{(5)}P_2(\cos\beta) + b_{k4}^{(5)}P_4(\cos\beta) + b_{k6}^{(5)}P_6(\cos\beta), \quad k = 0, 1, 3, 5$$

$$(4.199)$$

The explicit forms for the asymptotic series are

$$b_{00}^{(5)} = \frac{1}{80\sigma^5} + \frac{367}{2240\sigma^6} + \frac{123}{70\sigma^7} + \frac{41233}{2240\sigma^8} + \cdots \qquad (4.200)$$

$$b_{10}^{(5)} = \frac{1}{96} - \frac{19}{420\sigma} + \frac{1}{120\sigma^2} - \frac{65}{1792\sigma^3} - \frac{79}{336\sigma^4} - \frac{85913}{57344\sigma^5} - \frac{72636131}{6881280\sigma^6}$$
$$- \frac{4543038053}{55050240\sigma^7} - \frac{14938598691}{20971520\sigma^8} + \cdots$$

$$b_{30}^{(5)} = -\frac{47}{560} + \frac{11}{35\sigma} - \frac{29}{280\sigma^2} + \frac{437}{1920\sigma^3} + \frac{5473}{4480\sigma^4} + \frac{1046209}{143360\sigma^5}$$
$$+ \frac{169435283}{3440640\sigma^6} + \frac{684614895}{1835008\sigma^7} + \frac{230861266333}{73400320\sigma^8} + \cdots$$

$$b_{50}^{(5)} = \frac{311}{3360} - \frac{137}{420\sigma} + \frac{13}{105\sigma^2} - \frac{5911}{26880\sigma^3} - \frac{2141}{1920\sigma^4} - \frac{1874309}{286720\sigma^5}$$
$$- \frac{299470403}{6881280\sigma^6} - \frac{17964831133}{55050240\sigma^7} - \frac{400677748549}{146800640\sigma^8} + \cdots$$

$$b_{02}^{(5)} = \frac{1}{112\sigma^5} + \frac{3}{28\sigma^6} + \frac{507}{448\sigma^7} + \frac{5377}{448\sigma^8} + \cdots$$

$$b_{12}^{(5)} = \frac{13}{504} - \frac{19}{168\sigma} + \frac{23}{672\sigma^2} - \frac{737}{8064\sigma^3} - \frac{2959}{5376\sigma^4} - \frac{99733}{28672\sigma^5}$$
$$- \frac{50499149}{2064384\sigma^6} - \frac{350973527}{1835008\sigma^7} - \frac{72765921299}{44040192\sigma^8} + \cdots$$

$$b_{32}^{(5)} = -\frac{5}{21} + \frac{149}{168\sigma} - \frac{193}{672\sigma^2} + \frac{5245}{8064\sigma^3} + \frac{18677}{5376\sigma^4} + \frac{1785635}{86016\sigma^5}$$
$$+ \frac{289305193}{2064384\sigma^6} + \frac{5846947361}{5505024\sigma^7} + \frac{394448762615}{44040192\sigma^8} + \cdots$$

$$b_{52}^{(5)} = \frac{139}{504} - \frac{27}{28\sigma} + \frac{109}{336\sigma^2} - \frac{1343}{2016\sigma^3} - \frac{9203}{2688\sigma^4} - \frac{431321}{21504\sigma^5} - \frac{9839105}{73728\sigma^6}$$
$$- \frac{196654913}{196608\sigma^7} - \frac{30690812563}{3670016\sigma^8} + \cdots$$

$$b_{04}^{(5)} = -\frac{3}{140\sigma^5} - \frac{1563}{6160\sigma^6} - \frac{7767}{3080\sigma^7} - \frac{613353}{24640\sigma^8} + \cdots$$

$$b_{14}^{(5)} = \frac{15}{2464} - \frac{183}{6160\sigma} + \frac{713}{24640\sigma^2} - \frac{433}{19712\sigma^3} - \frac{409}{4928\sigma^4} - \frac{319665}{630784\sigma^5}$$
$$- \frac{8222083}{2293760\sigma^6} - \frac{5744848239}{201850880\sigma^7} - \frac{403943151013}{1614807040\sigma^8} + \cdots$$

$$b_{34}^{(5)} = -\frac{29}{280} + \frac{293}{770\sigma} - \frac{47}{385\sigma^2} + \frac{7081}{24640\sigma^3} + \frac{74647}{49280\sigma^4} + \frac{7137293}{788480\sigma^5}$$
$$+ \frac{385804437}{6307840\sigma^6} + \frac{4682760003}{10092544\sigma^7} + \frac{1580817298041}{403701760\sigma^8} + \cdots$$

$$b_{54}^{(5)} = \frac{1713}{12320} - \frac{2929}{6160\sigma} + \frac{2551}{24640\sigma^2} - \frac{34863}{98560\sigma^3} - \frac{92061}{49280\sigma^4}$$
$$- \frac{34432191}{3153920\sigma^5} - \frac{23756287}{327680\sigma^6} - \frac{15647080587}{28835840\sigma^7} - \frac{7317549380671}{1614807040\sigma^8} + \cdots$$

$$b_{06}^{(5)} = -\frac{1}{616\sigma^6} - \frac{3}{77\sigma^7} - \frac{1467}{2464\sigma^8} + \cdots$$

$$b_{16}^{(5)} = -\frac{1}{1584} + \frac{1}{1848\sigma} + \frac{7}{1056\sigma^2} + \frac{337}{88704\sigma^3} + \frac{53}{1848\sigma^4} + \frac{51433}{315392\sigma^5}$$
$$+ \frac{2188103}{2064384\sigma^6} + \frac{471762913}{60555264\sigma^7} + \frac{4428495037}{69206016\sigma^8} + \cdots$$

$$b_{36}^{(5)} = -\frac{1}{84} + \frac{10}{231\sigma} - \frac{17}{924\sigma^2} + \frac{103}{3168\sigma^3} + \frac{2489}{14784\sigma^4} + \frac{236615}{236544\sigma^5}$$
$$+ \frac{38344237}{5677056\sigma^6} + \frac{776232845}{15138816\sigma^7} + \frac{52467158027}{121110528\sigma^8} + \cdots$$

$$b_{56}^{(5)} = \frac{1207}{55440} - \frac{661}{9240\sigma} + \frac{17}{12320\sigma^2} - \frac{5525}{88704\sigma^3} - \frac{1169}{3520\sigma^4} - \frac{9116467}{4730880\sigma^5}$$
$$- \frac{131486063}{10321920\sigma^6} - \frac{1918435847}{20185088\sigma^7} - \frac{639291980689}{807403520\sigma^8} + \cdots$$

and for a random system, as for the lower orders, $\tilde{B}_k^{(5)} = b_{k0}^{(5)}$.

7. Calculational Advantages and Accuracy of the Asymptotic Approach

The formulas derived above, despite their cumbersome look, are very practical. Indeed, they present the nonlinear initial susceptibilities of a superparamagnetic particulate medium as analytical expressions of arbitrary accuracy. Another remarkable feature of the formulas of Section III.B.6 is that with respect to the frequency behavior they give the exact structure of the susceptibilities and demonstrate that those dependencies are quite simple. This makes our formulas a handy tool for analytical studies. Yet they are more convenient for numerical work because with their use the difficult and time-consuming procedure of solving the differential equations is replaced by a plain summation of certain power series. For example, if to employ Eqs. (4.194)–(4.200), a computer code that fits simultaneously experimental data on linear and a reasonable set of nonlinear susceptibilities (say, the 3th and the 5th) taking into account the particle polydispersity of any kind (easy-axes directions, activation volume, anisotropy constants) becomes a very fast procedure.

Some graphic examples justifying these statements are presented in Figures 4.12 and 4.13, where the components of two nonlinear complex susceptibilities are plotted as the functions of the parameter σ. For a given sample, σ in a natural way serves as a dimensionless inverse temperature. In those figures, the solid lines correspond to the above-proposed asymptotic formulas where we retain the terms, including σ^{-3}. The circles show the results of numerical solutions obtained by the method described in Ref. 67. Note that even at σ ∼ 5 the accuracy is still rather high.

It is worth noting that in the literature there exists a model, that may be justly called the *predecessor* of the model presented above. In Ref. 93 the authors had calculated the initial susceptibilities up to the seventh order in $1/\sigma$, having replaced a superparamagnetic assembly by a two-level macrospin system. The interrelation between the present work and Ref. 93 closely resembles the situation with evaluation of the rate of a superparamagnetic process. In 1949 Néel [13] and then, 10 years later, Brown [94] had evaluated the super-paramagnetic time in the framework of a two-level model. In such a framework, one allows for the magnetic moment flips but totally neglects, its possible diffusion over energetically less favorable directions; the result is simple; in the terms we use, it is described by Eq. (4.97). In 1963 Brown [47] went beyond this restriction and took into account the possibility for the magnetic moment to wander over all 4π radians. This lead, as discussed in Section II.A, to a spectrum of relaxation times stemming from the eigenvalue problem for the Fokker–Planck equation (4.97).

In the present case, the obtained V/T dependencies of the nonlinear susceptibilities and those from Ref. 93 are qualitatively the same. Their most typical feature is the double-peak shape. Quantitatively, however, the

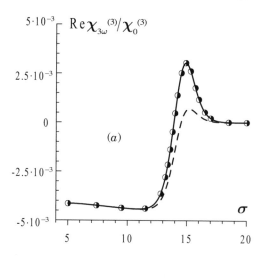

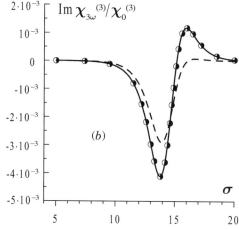

Figure 4.12. Real (*a*) and imaginary (*b*) components of the cubic susceptibility of a superparamagnetic assembly with coherently aligned easy axes; the direction of the probing field is tilted with respect to the alignment axis at $\cos \beta = 0.5$; the dimensionless frequency is $\omega \tau_0 = 10^{-6}$. Solid lines show the proposed asymptotic formulas taken with the accuracy σ^{-3}; circles present the result of numerically exact evaluation; dashed lines correspond to the "zero derivative" approximation (4.167). The discrepancy of the curves is mentioned in the text following Eq. (4.220).

corresponding lines differ and do not reduce to one another in any case. Indeed, as long as the temperature is finite (however low), the configurational space for the unit vector e of the magnetic moment is the full (4π-radian) solid angle; its reduction to just two directions along a bidirectional axis could not be done

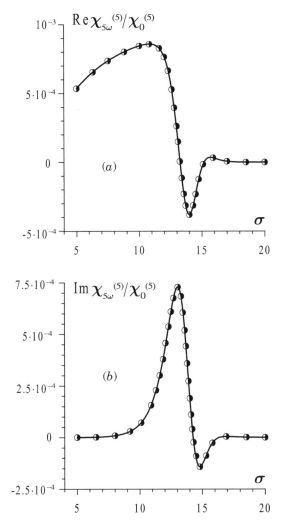

Figure 4.13. Real (a) and imaginary (b) components of the fifth-order susceptibility of a random superparamagnetic assembly; the dimensionless frequency is $\omega\tau_0 = 10^{-6}$. Solid lines show the proposed asymptotic formulas with the accuracy σ^{-3}; circles present the result of a numerical evaluation.

otherwise than "by hand." This is exactly what the two-level Ising-like model does; it forcibly imposes a quantum property (discrete spin projections) on a macrospin assembly. From the calculational viewpoint, another essential demerit of the results of Ref. 93 is that the coefficients in the susceptibility formulas are not given in an analytical form. The authors propose evaluating them by solving an infinite set of recurrence equations. Hence, the procedure [93] does not provide any gain with respect to the former ones, neither in analytical considerations nor in constructing fitting codes.

In the presented framework the results by Klik and Yao (including the analytical formulas for them missing in Ref. 93) can be obtained immediately if to take the function φ_1 in a stepwise form (4.145) and not to allow for the corrections caused by the finiteness of its derivative at $x = 0$. In our terms this means to stop at set (4.167), namely, zero-derivative solution, and not to go further. The emerging error is, however, uncontrollable and not at all small. As an illustration, in Figure 4.12 we show the result obtained with this model (dashed lines) for the cubic susceptibility $\chi_{3\omega}^{(3)}$ in a textured system where the particle common axis $\boldsymbol{n}$ is tilted under the angle $\beta = \pi/3$ to the probing field. One can see that deviations are substantial.

Finally, we remark that the formulas of Section III.B.6 do justify the intuitive expression first proposed in Ref. 67 and repeated here in Section III.A as Eq. (4.112). When "invented," its reliability was confirmed only by comparison with the results of some particular numerical calculations. In a short time after, Eq. (4.112) was used with success for fitting experimental data in Ref. 77. However, until relatively recently its good practical "work" had not been supported by a clear theoretical derivation. Now, using the theory presented in Sections III.B.1–III.B.6, one finds that this expression follows from Eqs. (4.195)–(4.197) if we expand the coefficients $b_{i0}^{(3)}$ up to the zeroth order with respect to σ^{-1}. This procedure rederives Eq. (4.112) in its original form, thus proving it to yield the exact cubic susceptibility of a random assembly at low temperatures. The cause of its applicability at high temperatures (where it is by definition approximate) is the exponential dependence of τ_1 on σ. Indeed, in the frequency range $\omega\tau_0 \ll 1$, where we work, the condition $\sigma \lesssim 1$ means $\tau_1 \to \tau_0$, and all the dispersion factors in Eq. (4.112) drop out, thus transforming it into a correct static susceptibility $\chi_0^{(3)}$ from Eq. (4.89).

Applying a procedure similar to Eqs. (4.198)–(4.200), we get the fifth-order susceptibility in the form

$$\chi_{5\omega}^{(5)} = \frac{1}{16}\chi_{(0)}^{(5)}\frac{(1 + 6S_2^2 + 2S_2^3)(1 - \frac{21}{8}\omega\tau_1 - \frac{3}{4}\omega^2\tau_1^2)}{(1 + i\omega\tau_1)(1 + 3i\omega\tau_1)(1 + 5i\omega\tau_1)} \tag{4.201}$$

As this equation is in a certain way the analog of the already tested Eq. (4.112), it has a good chance (although not yet tested) to be a good approximation for $\chi_{5\omega}^{(5)}$ in the whole temperature interval. As we have already ascertained in Section III.A.4, the best interpolation expression for the relaxation time τ_{10} to be used in the susceptibility formulas is Eq. (4.105) first introduced in Refs. 71 and 87.

In summary, in this section (Section III.B) we give a consistent procedure yielding the integral relaxation time and initial nonlinear susceptibilities for an assembly of noninteracting superparamagnetic particles in the low-to-moderate temperature range. The results are presented as asymptotic series with respect to the dimensionless parameter σ that is the uniaxial anisotropy barrier height scaled with temperature. High-order expansion terms are easily accessible which enable one to achieve any desirable extent of accuracy. This is proved by comparison of the proposed approximation with the results of numerical solutions. The susceptibilities incorporate angular dependencies that allow one to consider the particle assemblies with any extent of orientational texture— from perfectly aligned to random. The proposed formulas stand much closer to reality than those for a two-level system and facilitate considerably both analytical and numerical calculations in the theory of superparamagnetic relaxation in single-domain particles.

8.   Mathematical Details of Calculations

a. Evaluation of the Expansion Coefficients for the Eigenfunctions ψ_0 *and* ψ_1. Both functions ψ_0 and ψ_1 are uniaxially symmetrical about the anisotropy axis n and can be expanded in the Legendre polynomial series, see Eq. (4.154):

$$\psi_0 = \frac{1}{2}\sum_{k=0}(2k+1)S_k P_k(x), \quad k = 0, 2, 4, \ldots$$
$$\psi_1 = \frac{1}{2}\sum_{k=1}(2k+1)Q_k P_k(x), \quad k = 1, 3, 5, \ldots \tag{4.202}$$

where in accordance with the parity properties of the eigenfunctions nonzero terms are

$$S_0 = 1, \quad S_k = (P_k(x) \mid \psi_0), \quad k = 2, 4, \ldots$$
$$Q_k = (P_k(x) \mid \psi_1), \quad k = 1, 3, 5, \ldots \tag{4.203}$$

Taking into account that $\psi_1 = \psi_0\varphi_1$, where ψ_0 in a finite form is given by Eq. (4.133), one arrives at the general formula

$$\mathscr{F}_k = \frac{1}{\mathscr{R}}\int_0^1 P_k(x)e^{\sigma x^2}dx \tag{4.204}$$

where $\mathscr{F}$ is S_k for even and is Q_k for odd values of the index, and the function $\mathscr{R}(\sigma)$ is defined by Eq. (4.133). In particular

$$Q_1 = \frac{1}{\mathscr{R}}\int_0^1 xe^{\sigma x^2}dx = \frac{e^\sigma - 1}{2\sigma\mathscr{R}} \tag{4.205}$$

Using asymptotic expansion (4.134) for $\mathscr{R}$, one gets

$$Q_1 = 1/G = 1 - \frac{1}{2\sigma} - \frac{1}{2\sigma^2} - \frac{5}{4\sigma^3} - \frac{37}{8\sigma^4} - \frac{353}{16\sigma^5} - \frac{4881}{32\sigma^6}$$
$$- \frac{55205}{64\sigma^7} - \frac{854197}{128\sigma^8} + \cdots \qquad (4.206)$$

Knowing Q_1, one can derive all the other moments Q_k with the aid of the three-term recurrence relation obtained from Eq. (4.136) by setting $b_k = Q_k$ and $\lambda = 0$ there. The same relation can be used to find the equilibrium order parameters S_k. This is a head-to-tail procedure, where $S_0 = 1$ and S_2 is determined by the integral

$$S_2 = \frac{1}{2\mathscr{R}} \int_0^1 (3x^2 - 1) e^{\sigma x^2} dx \qquad (4.207)$$

Taking the latter by parts, one obtains

$$S_2 = \frac{3(e^\sigma - \mathscr{R})}{4\sigma\mathscr{R}}$$

On comparison with Eq. (4.305), we find

$$S_2 = \frac{6Q_1 - 3(3 - 2\sigma)}{4\sigma}$$

that upon substituting asymptotic series (4.206), transforms into

$$S_2 = 1 - \frac{3}{2\sigma} - \frac{3}{4\sigma^2} - \frac{15}{8\sigma^3} - \frac{111}{16\sigma^4} - \frac{1059}{32\sigma^5} - \frac{12243}{64\sigma^6} - \frac{165615}{128\sigma^7} - \frac{2562591}{256\sigma^8} + \cdots$$
$$(4.208)$$

b. Evaluation of Correcting Coefficients D_n in the General Case. Let us present the solution of Eq. (4.164) in the form

$$f_1^{(n)} = \psi_0 g_1^{(n)} + u^{(n)} \qquad (4.209)$$

where the functions $g_1^{(n)}$ are rendered by formulas (4.167) and are not corrected with respect to the derivative $d\varphi_1/dx$. Substituting Eq. (4.209) in (4.164) and taking into account Eqs. (4.167), we get a recurrent sequence of equations for the corrections $u^{(n)}$:

$$\hat{L}u^{(n)} = \hat{V}u^{(n-1)} + \hat{J}\psi_0 \frac{(e \cdot h)^n}{n!} \hat{J}\varphi_1 \qquad (4.210)$$

With allowance for the fact that function $\varphi_1^{(0)}$ depends only on x, Eq. (4.210) is rewritten as

$$\hat{L}u^{(n)} = \hat{V}u^{(n-1)} + \frac{d}{dx}\left[\psi_0(1-x^2)\frac{(\boldsymbol{e}\cdot\boldsymbol{h})^n}{n!}\frac{d\varphi_1}{dx}\right]$$

Finally, making use of the relation

$$\frac{d\varphi_1}{dx} = \frac{\lambda_1}{2\psi_0(1-x^2)} \tag{4.211}$$

which follows from Eq. (4.140), we get

$$\hat{L}u^{(n)} = \hat{V}u^{(n-1)} + \frac{\lambda_1}{2}\frac{d}{dx}\left[\frac{(\boldsymbol{e}\cdot\boldsymbol{h})^n}{n!}\right] \tag{4.212}$$

In particular, at $n = 1$ Eq. (4.212) takes the form

$$\hat{L}u^{(n)} = \frac{\lambda_1}{2}\frac{d}{dx}(\boldsymbol{e}\cdot\boldsymbol{h}) \tag{4.213}$$

Equations (4.212) are solved sequentially beginning from Eq. (4.213) by expanding in a power series with respect to x. The right-hand sides of Eqs. (4.212) and (4.213) are proportional to an exponentially small parameter λ_1. For this reason alone, we did not take into account the corrections of the order $u^{(n)}$ when deriving Eqs. (4.167). However, the quantities

$$D_n = (\varphi_1 \mid u^{(n)}) \quad n = 2, 4, \ldots$$

have finite values. To show that, let us multiply Eq. (4.212) by φ_1 and integrate. This yields

$$(\varphi_1 \mid \hat{L}u^{(n)}) = (\varphi_1 \mid \hat{V}u^{(n-1)}) + \frac{\lambda_1}{2}\left(\varphi_1 \mid \frac{d}{dx}\left[\frac{(\boldsymbol{e}\cdot\boldsymbol{h})^n}{n!}\right]\right) \tag{4.214}$$

In the left part we make use of the fact that φ_1 is the left eigenfunction of the operator $\hat{L}$; in the right part the integrals are taken by parts and yield

$$\lambda_1 D_n = 2\int_0^1 (1-x^2)u^{(n-1)}\left[\frac{d\varphi_1}{dx}\right]\frac{d}{dx}(\boldsymbol{e}\cdot\boldsymbol{h})dx$$
$$- \lambda_1 \int_0^1 \frac{d\varphi_1}{dx}\frac{(\boldsymbol{e}\cdot\boldsymbol{h})^n}{n!}dx \tag{4.215}$$

Replacing the derivative $d\varphi_1/dx$ in the first term of the right-hand side with the aid of Eq. (4.211), we arrive at the representation of the coefficient D_n as

$$D_n = \int_0^1 \frac{u^{(n-1)}}{\psi_0} \frac{d}{dx} (e \cdot h) dx - \int_0^1 \frac{d\varphi_1}{dx} \frac{(e \cdot h)^n}{n!} dx \qquad (4.216)$$

Since $\psi_0 \propto \exp(\sigma x^2)$, the first integral in Eq. (4.216) can be presented as an asymptotic series if the power expansion of the function $u^{(n-1)}$ in the vicinity of $x = 0$ is known. A closed form for the second integral can be found with the aid of the table given in Eq. (4.172), (see Section III.B.1).

As an example, we calculate the coefficient D_2. As from the addition theorem

$$(e \cdot h) = \cos\theta\cos\beta + \sin\theta\sin\beta\cos\varphi$$

we seek the solution of Eq. (4.213) with the sum

$$u^{(1)} = \cos\beta \sum_k C_k^{(0)} x^k + \sin\beta e^{i\varphi}(1-x^2)^{\frac{1}{2}} \sum_k C_k^{(1)} x^k \qquad (4.217)$$

Here the upper index of the C coefficients corresponds to the azimuthal number m of the spherical harmonic $e^{im\varphi}$. Operator $\hat{L}$ now includes the azimuthal coordinate and takes the form

$$-\hat{L} = (1-x^2)\frac{d}{dx^2} - [2\sigma x(1-x^2) + 2x]\frac{d}{dx}$$
$$+ \left[2\sigma(3x^2 - 1) - \frac{m^2}{1-x^2}\right]$$

Substitution of expansion (4.217) in Eq. (4.213) leads to the set of equations

$$2\sigma(k + m + 1)C_{k-2}^{(m)} - [k(k+1+2m+2\sigma) + m(m+1) + 2\sigma]C_k^{(m)}$$
$$+ (k+1)(k+2)C_{k+2}^{(m)} = N_k^{(m)} \qquad (4.218)$$

where $m = 0, 1$ and the numbers on the right-hand side are

$$N_k^{(0)} = \begin{cases} -1 & \text{for } k = 0, \\ 0 & \text{for } k \neq 0, \end{cases} \qquad N_k^{(1)} = \begin{cases} 1 & \text{for } k \text{ odd} \\ 0 & \text{for } k \text{ even} \end{cases}$$

In reality, one retains in expansion (4.217) only a finite number of terms so that Eqs. (4.218) could be easily solved analytically by any computer algebra solver.

In terms of expansion (4.217) expression (4.216) at $n = 2$ is written as

$$D_2 = \cos^2 \beta \sum_{k=0} C_{2k}^{(0)} \frac{(2k-1)!!}{2^k \sigma^k G}$$
$$- \frac{1}{2} \sin^2 \beta \sum_{k=1} C_{2k-1}^{(1)} \frac{(2k-1)!!}{2^k \sigma^k G} - \frac{1}{6} - \frac{2G-3}{6G} P_2(\cos \beta) \qquad (4.219)$$

Since the coefficients C found from Eq. (4.218) are functions of σ, one has to perform asymptotic expansion in Eq. (4.219). This gives finally

$$D_2 = \frac{1}{4\sigma} + \frac{1}{4\sigma^2} + \frac{5}{8\sigma^3} + \frac{37}{16\sigma^4} + \cdots$$
$$- \sin^2 \beta \left(\frac{1}{4} + \frac{1}{8\sigma} + \frac{1}{16\sigma^2} + \frac{7}{64\sigma^3} + \frac{19}{64\sigma^4} + \cdots \right) \qquad (4.220)$$

As it should be, at $\beta = 0$ this formula reduces to Eq. (4.174), which was obtained for a one-dimensional case. We remark, however, that in a tilted situation ($\beta \neq 0$) the coefficient D_2 acquires a contribution independent on σ that assumes the leading role. This effect is clearly due to admixing of transverse modes to the set of eigenfunctions of the system, and it is just it that causes so a significant discrepancy between the zero-derivative approximation and the correct asymptotic expansion for $\chi^{(3)}$ curves in Figure 4.12. Evaluation of the coefficient D_4 is done according to the same scheme and requires taking into account a number of the perturbation terms that makes it rather cumbersome.

c. Evaluation of Integrals. Before proceeding to the integrals (scalar products) in Eqs. (4.184)–(4.188) and (4.189), let us consider the "primitive" ones:

$$X_n = ((e \cdot h)^n \mid \psi_0), \quad Y_n = ((e \cdot h)^n \mid \psi_1)$$

The functions ψ_0 and ψ_1 are originally defined in terms of the polar angle that is $\theta = \arccos(e \cdot h)$. Thus, before performing integration, one needs to transform both integrands to the same coordinate set. Doing this with the aid of the addition theorem for Legendre polynomials, one finds

$$X_2 = \frac{1}{3} [2S_2 P_2(\cos \beta) + 1]$$
$$X_4 = \frac{1}{35} [8S_4 P_4(\cos \beta) + 20S_2 P_2(\cos \beta) + 7] \qquad (4.221)$$
$$X_6 = \frac{1}{231} [16S_6 P_6(\cos \beta) + 72S_4 P_4(\cos \beta) + 110S_2 P_2(\cos \beta) + 33]$$

and

$$Y_1 = Q_1 \cos \beta, \quad Y_3 = \frac{1}{5} [2 Q_3 P_3 (\cos \beta) + 3 Q_1 \cos \beta]$$

$$Y_5 = \frac{1}{63} [8 Q_5 P_5 (\cos \beta) + 28 Q_3 P_3 (\cos \beta) + 27 Q_1 \cos \beta] \tag{4.222}$$

where $\cos \beta = (\boldsymbol{n} \cdot \boldsymbol{h})$ and the parameters S_k and Q_k are the expansion coefficients introduced by formulas (4.154).

Now, using the expressions for functions $f_0^{(n)}$ and $f_1^{(n)}$ derived in Section III.B.1, one sees that the relevant integrals of Eqs. (4.185)–(4.189) are expressed in terms of X_k and Y_k as

$$((\boldsymbol{e} \cdot \boldsymbol{h}) \,|\, f_0^{(1)}) = X_2$$

$$(\varphi_1 \,|\, f_0^{(1)}) = Y_1$$

$$((\boldsymbol{e} \cdot \boldsymbol{h}) \,|\, f_0^{(3)}) = \frac{1}{6} X_4 - \frac{1}{2} X_2^2$$

$$(\varphi_1 \,|\, f_0^{(3)}) = \frac{1}{6} Y_3 - \frac{1}{2} X_2 Y_1$$

$$((\boldsymbol{e} \cdot \boldsymbol{h}) \,|\, f_0^{(5)}) = \frac{1}{120} X_6 - \frac{1}{8} X_4 X_2 + \frac{1}{4} X_2^3$$

$$(\varphi_1 \,|\, f_0^{(5)}) = \frac{1}{120} Y_5 - \frac{1}{12} Y_3 X_2 + \frac{1}{4} X_2^2 Y_1 - \frac{1}{24} X_4 Y_1 \tag{4.223}$$

$$((\boldsymbol{e} \cdot \boldsymbol{h}) \,|\, f_1^{(2)}) = \frac{1}{2} Y_3 - Y_1 X_2 + D_2 Y_1$$

$$(\varphi_1 \,|\, f_1^{(2)}) = \frac{1}{2} X_2 - Y_1^2 + D_2$$

$$((\boldsymbol{e} \cdot \boldsymbol{h}) \,|\, f_1^{(4)}) = \frac{1}{24} Y_5 - \frac{1}{6} Y_3 X_2 - \frac{1}{6} X_4 Y_1 + \frac{1}{2} X_2^2 Y_1 + D_4 Y_1$$
$$+ D_2 ((\boldsymbol{e} \cdot \boldsymbol{h}) \,|\, f_1^{(2)})$$

$$(\varphi_1 \,|\, f_1^{(4)}) = \frac{1}{24} X_4 - \frac{1}{3} Y_3 Y_1 + \frac{1}{2} X_2 Y_1^2 + D_4 + D_2 (\varphi_1 \,|\, f_1^{(2)})$$

IV. MAGNETIC STOCHASTIC RESONANCE AND RELATED NONLINEAR PHENOMENA

A. Stochastic Resonance: Linear Response Theory

1. Resonance Where Nothing Resounds

The phenomenon called in modern thesaurus *the stochastic resonance* (SR) by now has shaped up in a general concept appealing to a great many of researchers in diverse fields. Nowadays the number of papers on stochastic resonance counts in thousands, including several very good review articles [18,20,95].

Stochastic resonance is a kinetic effect universally inherent to bi- or multistable dynamic systems exposed to either white or color noise. Its main manifestation is the appearance of a maximum on the noise intensity dependencies of the signal-to-noise ratio in a system subject to a weak driving force. Essentially, this behavior is due to the presence of an exponential Kramers time $\tau \propto \exp(\Delta U/\mathcal{D})$ of the system switching between energy minima; here ΔU is the effective height of the energy barrier separating the potential wells and $\mathcal{D}$ is the noise intensity.

In several papers [96–103] devoted to SR studies, which have already become classical, the main model under investigation is an overdamped anharmonic oscillator with the potential

$$U(q) = -aq^2 + bq^4$$

where both coefficients, a and b, are positive. Although possible for experimental realization [98], this model does not at all exhaust the variety of systems displaying the SR behavior. Relevant particular examples from physics and biophysics are numerous.

The idea of the magnetic stochastic resonance has emerged in a natural way first as a theoretical issue [25–27] and shortly afterward was supported by some experimental evidence [28,31]. A consistent theoretical treatment of magnetic SR in a superparamagnetic particle in the framework of the linear response theory was developed in Refs. 29 and 30.

A single-domain particle of a ferromagnet or ferrite with a uniaxial anisotropy provides a perfectly understandable and simple in realization example of a bistable object whose properties are sensitive to thermal noise. As had been shown experimentally in the mid-1950s [14,104] and since then many times verified (more recently, in precise experiments on isolated nanograins [105,106]), with the particles of a size $\sim 10\,\mathrm{nm}$ in the temperature range about 100 K, one may easily proceed from the conventional ferromagnetism (where the particles are tiny permanent magnets) to *superparamagnetism*: the intense fluctuational motion of a magnetic moment inside the particle. The parameter responsible for the change of regimes is the ratio of the magnetic anisotropy barrier to the thermal energy; its definition $\sigma = KV_m/T$ was already introduced by Eqs. (4.34) and (4.75).

Note also that, when a contribution due to a uniaxial magnetic anisotropy is included in the particle energy [see Eq. (4.73)], the emerging parameter h, on changing, in a natural way guides the system through the variety of oscillation situations: from the absence of any symmetric potential (an isotropic particle) to a twin pair of infinitely deep wells (a magnetically hard particle). In still other words, one can easily move between the Heisenberg (continuous) and Ising (two-level) models for the orientational behavior of a classical spin.

The meaning of the generic SR terms as applied to a single-domain superpara-magnetic particle (or an assembly of those) is as follows. The dynamic variable is connected to the orientation of magnetic moment, the background noise is thermal and thus white, and the excitation is created by a weak external harmonic magnetic field $H_p(t) = H_{p0}\exp(i\Omega t)$. The signal, or the response, is the dynamic magnetization $M(t)$ in the direction of $\boldsymbol{H}_p$ and the signal-to-noise ratio is defined in terms of the spectral power density $Q(\omega)$ of $M(t)$ at some given frequency. We remark that from the experimental viewpoint all the quantities involved are feasible to measure by conventional technique. In below we present a theory of SR for an assembly of single-domain particles embedded in a solid nonmagnetic matrix. With some minor changes it is also valid for magnetic fluids, specifically, suspensions of anisotropic ferroparticles.

In a more formal way, a single-domain particle is a rotationally bistable system with the potentials

$$U(\vartheta) = -a\cos^2\vartheta \qquad \text{or} \qquad U(\vartheta) = -a\cos^2\vartheta - b\cos\vartheta$$

respectively, depending on whether one assumes that the external (bias) magnetic field is zero or not. We employ the superparamagnetic approach based on the Fokker–Planck rotary kinetic equation (4.90) to analyze the magnetic stochastic resonance. Dealing in terms of dynamic susceptibilities is a very convenient way since we can use the general linear-response theory specified for SR in [96,97,99]. The following consideration, unlike the former attempts on magnetic SR [26,29], retains a sufficiently large part of the relaxation spectrum, thus making our calculations virtually exact. In other words, all the significant contributions of intrawell modes are always taken into account. In this aspect our chapter is closely related to Refs. 107 and 108, where the same potential has been analyzed numerically in the case of color-noise-driven stochastic relaxation.

2. Magnetic Nanoparticle as a SR Object

Consider a uniformly magnetized (single-domain) particle with a magnetic moment μe. The particle has a uniaxial magnetic anisotropy (crystalline or other) with the energy density K; the direction of the anisotropy axis is denoted by a unit vector $\boldsymbol{n}$. Let the particle (or an assembly of identical particles) be fixed inside some nonmagnetic solid matrix. In the case of an assembly, we assume that the particle concentration is small enough to allow us to neglect their magnetic dipole–dipole interaction.

The orientation-dependent part of the particle energy in the absence of external magnetic fields already has the familiar form

$$U = -KV_m(e \cdot n)^2 \qquad (4.224)$$

[see Eq. (4.33), where the variable $(e \cdot n) = \cos \vartheta$ renders the normalized projection of the magnetic moment on the direction of the anisotropy axis. In the presence of a thermal bath, the orientational distribution function $W(e, t)$ of vector e obeys the Fokker–Planck equation (4.90), which we present here in the form

$$\frac{\partial}{\partial t} W + \hat{\Lambda} W = 0, \qquad \hat{\Lambda} = -\frac{1}{2\tau_D} \hat{J} \left[\frac{1}{T} (\hat{J} U) + \hat{J} \right] \qquad (4.225)$$

where the operator $\hat{J}$ in parenthesis acts only on U. The equilibrium solution of Eq. (4.225) has the Gibbs form (4.77), which we rewrite here as

$$W_0 = Z_0^{-1} \exp(\sigma \cos^2 \vartheta), \quad Z_0 = 4\pi \mathscr{R}(\sigma), \quad \mathscr{R}(\sigma) = \int_0^1 \exp(\sigma x^2) dx \quad (4.226)$$

thus introducing the reduced partition function $\mathscr{R}(\sigma)$.

The stochastic resonance is determined by the longitudinal (with respect to n) modes of the relaxational problem (4.90). Since $\hat{\Lambda}$ is not a self-adjoint operator, it produces, together with the spectrum of eigenvalues $\{\lambda_i\}$, two sets of eigenfunctions defined as

$$\hat{\Lambda}\psi_i = \lambda_i \psi_i, \qquad \hat{\Lambda}^+ \varphi_j = \lambda_j \varphi_j$$

where $+$ denotes Hermitian conjugation. These eigenfunctions are connected by relation

$$\psi_i = W_0 \varphi_i \qquad (4.227)$$

and orthonormalized. Their expansion in the Legendre polynomial series

$$\psi_i = \frac{1}{2} \sum_{l=0}^{\infty} (2l + 1) b_l^{(i)} P_l(\cos \vartheta), \qquad \varphi_j = \sum_{l=0}^{\infty} a_l^{(j)} P_l(\cos \vartheta) \qquad (4.228)$$

produces two sets of eigenvectors, $\{a_l\}^{(i)}$ and $\{b_l\}^{(j)}$, the components of which, as follows from Eqs. (4.227) and (4.228), are related to each other by

$$b_l^{(k)} = [\langle P_l P_{l'} \rangle_0 - \langle P_l \rangle_0 \langle P_{l'} \rangle_0] a_{l'}^{(k)}$$

where the subscript 0 denotes the averaging over the equilibrium distribution W_0 from Eq. (4.226).

The Green function of Eq. (4.225), namely, the probability density of a state (x, t), providing the initial state is $(x_0, 0)$, reads

$$W(x, t \mid x_0) = \sum_{k=0}^{\infty} \psi_k(x) \varphi_k(x_0) e^{-\lambda_k t} \tag{4.229}$$

hereafter we denote $x = \cos \vartheta$. The subject of our interest is the correlator $\mu^2 \langle\langle x(t) x(0) \rangle\rangle_0$, where the averaging over $x(t)$ is performed with the function W from Eq. (4.229), and that over the initial conditions—with the equilibrium function W_0. Using formulas (4.227)–(4.228), one gets

$$\mu^2 \langle\langle x(t) x(0) \rangle\rangle_0 = \mu^2 \int\int dx \, dx_0 x x_0 W(x, t \mid x_0) W_0 = \mu^2 \sum_{k=1} [b_1^{(k)}]^2 e^{-\lambda_k t} \tag{4.230}$$

the summation here is taken only over the odd values of k.

Transforming the correlator (4.230) by the Kubo formula, one arrives at the longitudinal dynamic susceptibility

$$\chi(\Omega) = \chi' - i\chi'' = \frac{\mu^2 B}{T} \sum_{k=1} \frac{w_k \lambda_k}{\lambda_k + i\Omega}$$

$$B = [\langle \cos^2 \vartheta \rangle_0 - \langle \cos \vartheta \rangle_0^2] \tag{4.231}$$

of a single-domain particle with respect to the external field $H(t) = H \exp(i\Omega t)$; here we have introduced a notation $w_k = B^{-1}[b_1^{(k)}]^2$ so as $\sum w_k = 1$.

According to the fluctuational dissipation theorem, the spectral density function in terms of χ reads [96,97] as follows:

$$Q(\omega) = \pi H^2 |\chi|^2 \delta(\omega - \Omega) + 4 \frac{T}{\omega} \chi''(\omega) \tag{4.232}$$

Setting $\omega = \Omega$ (the necessary condition of the onset of SR) and comparing the signal induced (δ-function) and the noise (proportional to T) contributions in expression (4.232), one gets with the aid of Eq. (4.231) the signal-to-noise ratio as

$$S = \frac{\pi}{4\tau_0} \left(\frac{I_s H}{K}\right)^2 R(\sigma, \Omega)$$

$$R(\sigma, \Omega) = \sigma^2 B \frac{\left[\sum \dfrac{w_k}{1 + (\Omega\tau_k)^2}\right]^2 + \left[\sum \dfrac{\Omega w_k \tau_k}{1 + (\Omega\tau_k)^2}\right]^2}{\sum \dfrac{w_k \tau_k / \tau_0}{1 + (\Omega\tau_k)^2}} \tag{4.233}$$

and the phase shift

$$\phi(\sigma, \Omega) = -\arctan\left(\frac{\chi''}{\chi'}\right) = -\arctan\frac{\sum\dfrac{\Omega w_k \tau_k}{1 + (\Omega \tau_k)^2}}{\sum\dfrac{w_k}{1 + (\Omega \tau_k)^2}} \tag{4.234}$$

where we have introduced the spectrum of relaxation times $\tau_k = 1/\lambda_k$, and made use of the relations $\mu = I_s V_m$ and $\tau_D = \sigma \tau_0$. As long as summations in Eqs. (4.233) and (4.234) are infinite, the corresponding expressions represent exact results for SR in a single-domain particle assembly.

3. Magnetic SR by the Linear Response Theory

The sets of relaxation times τ_k and weight coefficients w_k entering Eqs. (4.233) and (4.234) were evaluated numerically. Substitution of expansions (4.228) into the Fokker–Planck equation (4.225) yields a homogeneous tridiagonal recurrence relation

$$\left[1 - \frac{\lambda_i}{l(l+1)}\right] b_l^{(i)} - 2\sigma\left[\frac{l-1}{(2l-1)(2l+1)} b_{l-2}^{(i)}\right.$$
$$\left. + \frac{1}{(2l-1)(2l+3)} b_l^{(i)} - \frac{l+2}{(2l+1)(2l+3)} b_{l+2}^{(i)}\right] = 0 \tag{4.235}$$

and the one corresponding to the conjugate equation reads

$$[l(l+1) - \lambda_i] a_l^{(i)} + 2\sigma\left[\frac{(l-2)(l-1)l}{(2l-3)(2l-1)} a_{l-2}^{(i)}\right.$$
$$\left. - \frac{l(l+1)}{(2l-1)(2l+3)} a_l^{(i)} - \frac{(l+1)(l+2)(l+3)}{(2l+3)(2l+5)} a_{l+2}^{(i)}\right] = 0 \tag{4.236}$$

For each i, the eigenvalue λ_i and the sets of corresponding eigenvectors $\{a_l\}^{(i)}$ and $\{b_l\}^{(j)}$ have been found with the aid of the continued fraction algorithm [35].

Our calculations show that to achieve good accuracy with Eqs. (4.233) and (4.234) in a wide range of temperature and frequency variations, it is necessary to retain at least five (odd $k = 1, 3, \ldots, 9$) lower modes of the spectrum. We remark that the first three relaxational modes have once been evaluated both numerically [109] and analytically [82] in studies of dielectric relaxation in nematic liquid crystals, where the forms of the potential and of the basic equation coincide with those given by our Eqs. (4.224) and (4.225), respectively.

The dependencies of the signal-to-noise ratio S on the temperature parameter $\sigma^{-1} = T/KV_m$ obtained are shown in Figures 4.14 and 4.15. Note that within the framework of the linear response theory, our approach removes all the

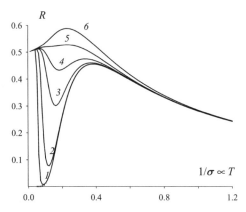

Figure 4.14.   Signal-to-noise ratio as a function of a dimensionless temperature. For $\Omega\tau_0 = -$ see dotted curve at the bottom of the minimum of curve 1. Solid curves: $\Omega\tau_0 = 0.01$ (1), 0.1 (2), 0.5 (3), 1 (4), 2 (5), 10 (6).

restrictions on the frequency range, that is, the SR characteristics are evaluated correctly for any Ω.

As might have been expected, the SR, that is, the main maximum of $R(\sigma)$ at $\sigma \sim 1$, is most pronounced at $\Omega \to 0$ and zero bias field—see curve 1 of Figure 4.14. This case is relatively simple, since in the $\Omega = 0$ limit Eq. (4.233) reduces to

$$R(\sigma) = \frac{\sigma^2 B\tau_0}{\tau_{\mathrm{eff}}}, \qquad \tau_{\mathrm{eff}} = \sum_1^\infty w_k\tau_k \qquad (4.237)$$

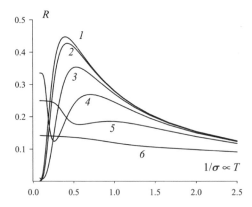

Figure 4.15.   Signal-to-noise ratio at non-zero values of the applied constant field $\xi = \varepsilon\sigma$ for $\varepsilon = 0.1$ (1), 0.2 (2), 0.5 (3), 1 (4), 2 (5), 5 (6). All the results correspond to the limit $\Omega\tau_0 = 0$.

The last expression justifies the treatment of magnetic SR in the $\Omega = 0$ limit as a process characterized by just one effective relaxation time τ_{eff}. As has been shown in Refs. 29 and 58, the time τ_{eff} is very close to τ_1 in a wide range of σ.

However, the case of Ω being exactly zero is to some extent an exception. At any finite Ω, one has to remember that the time τ_1 is exponential in σ; that is, it grows infinitely with cooling the system. Therefore, at low temperatures the interwell transition is completely "frozen," and the situation is governed by intrawell relaxation. The latter is sensitive to the details of the potential near the bottom of the well, and for the system in question is determined by the infinite eigenvalue spectrum λ_k of the kinetic equation (4.225) for $k \geq 3$.

The low-temperature limit of the function $R(\sigma)$ may be derived with the aid of asymptotic expressions for χ given in Ref. 82

$$\chi = \frac{\mu^2}{T} \left[\left(1 - \frac{1}{\sigma} - \frac{3}{4\sigma^2} \right) \frac{1}{1 + i\Omega\tau_1} \right.$$

$$\left. + \frac{1}{8\sigma^2} \left(\frac{1}{1 + i\Omega\tau_3} + \frac{1}{1 + i\Omega\tau_5} \right) + O(\sigma^{-3}) \right] \tag{4.238}$$

where $\tau_1 = \tau_0 \sqrt{\pi/4\sigma} e^{\sigma}$ and $\tau_3 = \tau_5 = \frac{1}{2}\tau_0(1 + 5/2\sigma + \cdots)$. The first term of this expansion describes the interwell relaxation process responsible for SR proper—the right maximum in Figure 4.14. But at low temperatures, $\sigma \gg 1$, and for this reason τ_1 is exponentially large. Hence, one has to set $\Omega\tau_1 \gg 1$ for any finite frequency, and the first term in Eq. (4.236) drops out. The second and following terms there render the contribution of intrawell relaxation whose rate remains finite at $T \to 0$. Substituting the truncated Eq. (4.238) into Eqs. (4.232) and (4.233), instead of zero predicted by formula (4.237), one obtains $R(\sigma \to \infty) = \frac{1}{2}$. Yielding the correct limiting value, Eq. (4.238) however, is not accurate enough for a correct description of $R(\sigma)$ at large σ, that is, low temperatures. Indeed, it is easy to show that it gives $(dR/d\sigma^{-1})_0$ negative whereas the exact numerical treatment proves that the initial temperature slope of the $R(1/\sigma)$ curve is positive (see Fig. 4.14). This low-temperature increase of the function $R(1/\sigma)$ causes the additional maximum at the signal-to-noise ratio (SNR) in superparamagnetic systems.

With minor complications our method provides a direct way to study the case of nonequal potential wells, the importance of which for the case of magnetic SR was first mentioned in Ref. 28. Physically this means merely that some bias constant field H_0 is applied to the particle parallel to its anisotropy axis n. Then the energy function (4.224) is replaced by

$$U = -\mu H_0(e \cdot n) - K V_m (e \cdot n)^2 \tag{4.239}$$

where now the first term $\propto \cos \vartheta$ breaks the bidirectional symmetry of the potential. However, the two-minima pattern for the potential $U(\vartheta)$ exists as long as the bias field is smaller than $H_a = 2K/I_s$—the maximum coercive force of a single-domain particle. In the range $H_0 < H_a$, though, the wells' equilibrium populations and transition rates are different, and the magnetic moment still undergoes interwell motions.

For this case the Fokker–Planck equation (4.225) redefined with regard to Eq. (4.238) after a procedure similar to that yielding Eqs. (4.235) and (4.236) turns into the pentadiagonal recurrence relationship

$$\left[1 - \frac{\lambda_i}{l(l+1)}\right]b_l^{(i)} - \frac{\xi}{2l+1}\left[b_{l-1}^{(i)} - b_{l+1}^{(i)}\right]$$

$$- 2\sigma\left[\frac{l-1}{(2l-1)(2l+1)}b_{l-2}^{(i)} + \frac{1}{(2l-1)(2l+3)}b_l^{(i)}\right.$$

$$\left. - \frac{l+2}{(2l+1)(2l+3)}b_{l+2}^{(i)}\right] = 0 \tag{4.240}$$

The conjugated one reads

$$[l(l+1) - \lambda_i]a_l^{(i)} + \xi\left[\frac{(l-1)l}{2l-1}a_{l-1}^{(i)} - \frac{(l+1)(l+2)}{2l+3}a_{l+1}^{(i)}\right]$$

$$+ 2\sigma\left[\frac{(l-2)(l-1)l}{(2l-3)(2l-1)}a_{l-2}^{(i)} - \frac{l(l+1)}{(2l-1)(2l+3)}a_l^{(i)}\right.$$

$$\left. - \frac{(l+1)(l+2)(l+3)}{(2l+3)(2l+5)}a_{l+2}^{(i)}\right] = 0 \tag{4.241}$$

These recurrence relations are solved using the matrix-continued fraction method [35,107,108]. Figure 4.15 shows how the decay of SNR takes place as a result of gradual shoaling of one of the minima under growth of the bias field H_0. To characterize the latter, in Eqs. (4.240) and (4.241) we have defined the dimensionless parameter $\xi = \mu H_0/T$ with respect to the bias field. Under that choice, the ratio

$$\varepsilon = \frac{\xi}{\sigma} \tag{4.242}$$

does not depend on temperature and renders the strength of the external field in the units of the internal (i.e., anisotropy) one.

As the field strength grows, the position of the SNR maximum in Figure 4.15 moves rightward, to higher temperatures. Normally, one would have expected a shift to the opposite side because of the increase of the net relaxation rate. To explain the "reversed" shift, one has to note that the presence of the field cardinally changes the temperature behavior of the coefficient B—the static susceptibility of the system. At $H_0 = 0$, when $\langle \cos \vartheta \rangle_0 = 0$, with the temperature growth it *diminishes* from 1 to $\frac{1}{3}$, rendering the initial susceptibility. However, under nonzero field the effect of saturation of the longitudinal magnetization, yielding $B \to 0$ at $T \to 0$, becomes essential, and B *grows* from zero at $T \to 0$ to $\frac{1}{3}$ at $T \to \infty$. Therefore, at $H_0 \neq 0$ the combination $\sigma^2 B$ [see Eq. (4.233)] acquires a temperature maximum of its own. It is this specific (static) temperature-dependent factor that reverses the direction of the SR maxima shift.

4. Phase Shifts in the SR Situation

The approach developed above is very convenient for obtaining a comprehensive account of the frequency and temperature behavior of the phase shift in the system in question. The problem has special interest since those dependencies for bistable systems have been put under discussion with contradicting conclusions [101,102]. Some particular numerical simulations on the same subject reported in Ref. 103, although interesting, are insufficient to draw out a final clarification.

Before proceeding to discussion of the details of the phase behavior, we would like to emphasize, after the authors of Ref. 110, that neither lack nor presence of the phase maximum may be taken as a "signature" of the SR proper. Actually, the maxima of R and $|\phi|$ have essentially different origins. Whereas the first is caused by the thermofluctuational nature of the interwell hopping, the second depends just of the very existence of the intrawell degrees of freedom. A large difference in the positions of these maxima (compare Figs. 4.14 and 4.16) and the existence of the $|\phi|$ peak in the range $\Omega\tau_0 \gg 1$ where SR is clearly absent (compare curves 5 and 6 in Fig. 4.14 with curves 2 and 3 in Fig. 4.17), are particular manifestations of this fact. However, the fact that there is no such an effect as *a stochastic resonance of the phase shift* notwithstanding, the behavior of $|\phi|$ under SR conditions is definitely worth consideration.

The main point of the argument set out in Refs. 101 and 102 is whether $|\phi|$ increases or decreases in the low-temperature limit at high frequencies (i.e., $\Omega\tau_0 \geq 1$). Note that the reversed characteristic intrawell time τ_0^{-1}, which remains finite at $T \to 0$, is the only natural frequency scaling parameter here, since the interwell hopping rate τ_1^{-1}, which is exponential in T, may not be used for this purpose.

The temperature–frequency behavior of the phase shift in magnetic SR evaluated by rigorous numerical procedure is presented in Figure 4.16 and with special emphasis on high frequencies,—in Figure 4.17. The asymptotic

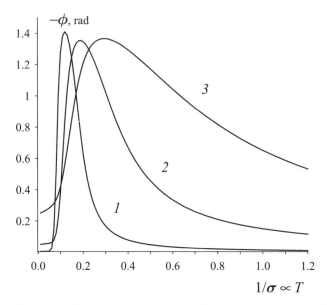

Figure 4.16. Phase shift as a function of a dimensionless temperature for $\Omega\tau_0 = 0.01$ (1), 0.1 (2), 0.5 (3).

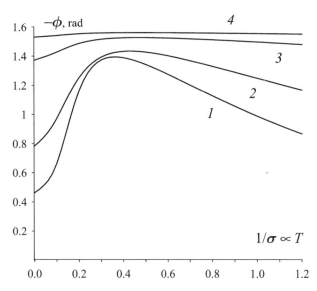

Figure 4.17. Phase shift as a function of dimensionless temperature at high frequencies; $\Omega\tau_0 = 1$ (1), 2 (2), 10 (3), 50 (4).

representation (4.239) helps to understand the origin of the limit $\phi(0) = \phi(T \to 0)$ and the form of the curves at low temperatures, that is, at $\Omega \tau_1 \to \infty$ and $\sigma \gg 1$. Under these conditions, Eq. (4.239) yields for the phase shift

$$-\phi\Big|_{1/\sigma=0} = \arctan\left[\frac{1}{2}\Omega\tau_0\left(1 + \frac{5}{2\sigma}\right)\right] \tag{4.243}$$

which is in a full agreement with the numerical data presented in Figures 4.16 and 4.17.

Differentiating Eq. (4.241) with respect to $1/\sigma$, that is, dimensionless temperature, one gets for the initial slope of the $|\phi(T)|$ curve

$$\frac{d|\phi|}{d\sigma^{-1}}\Big|_{1/\sigma=0} = \frac{5\Omega\tau_0}{4(1 + \frac{1}{4}\Omega^2\tau_0^2)} \tag{4.244}$$

This means that at any finite frequency the absolute value of the phase always *increases* with temperature. This result is consistent with the simulations of Ref. 102. Apparently, the effect is entirely due to intrawell motions and plays the main role in the temperature range where the interwell process may be neglected. But as soon as the latter is activated, it immediately becomes the dominating one. Together with creating conditions for SR, the interwell process tends function $|\phi(T)|$ down to zero. Given that, Eq. (4.239) is a direct proof of the existence of a maximum at the temperature dependence however large Ω may be.

Overview of the curves in Figures 4.16 and 4.17 suggests that at $\Omega\tau_0 \gg 1$ the position $(1/\sigma)_m$ of this maximum rather. weakly, if ever, depends on Ω. Numerical investigation confirms this conclusion. As is shown in Figure 4.18, the value of $(1/\sigma)_m$ first grows rapidly, then passes through a maximum, and finally descends very slowly to the limit $(1/\sigma)_m = 0$ at $\Omega \to \infty$.

Dealing in terms of intra- and interwell transitions, it is easy to understand also the conclusion of Ref. 101 prescribing a monotonous decrease and no maximum of $|\phi(T)|$. Although wrong for our case, it is valid for a system of a special type—the one that completely lacks any intrawell degrees of freedom. For such a model, instead of formula (4.235) or (4.241), the phase is exhaustively described by a relationship $\phi = -\arctan \Omega\tau_1$. Then for any Ω, however small, at $T \to 0$ it tends to $-\pi/2$ because of the rapid growth of the response time. Any heating causes reduction of τ_1 and hence diminution of $|\phi|$. Comparison of the two cases proves that with respect to the phase behavior, the systems with or without intrawell processes, although both capable of SR, are qualitatively different.

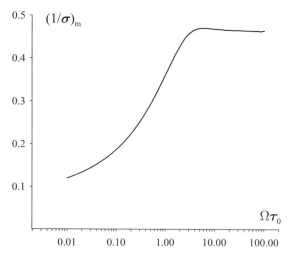

Figure 4.18. Position of the maximum of the function $|\phi(T)|$ against the dimensionless frequency.

The main results of this section may be summarized in the following way:

- Single-domain particle assemblies offer an easily realizable possibility to study all the scope of effects pertinent to SR.
- The kinetic (rotary diffusion) equation for the particle magnetic moment may be solved with high precision, thus taking into account contributions from the intrawell motions that are essential for a correct description of SR, especially at low temperatures.
- The kinetic model developed here does not use any kind of adiabatic assumption, and thus is capable of rendering frequency dependencies of the SR characteristics.
- By application of the external constant field of arbitrary strength, the symmetric double-well potential may be gradually transformed into a single-well (dipolar) one; on growth of the bias field, SR maximum noticeably shifts to higher temperatures.
- In the low-temperature limit the phase of magnetic oscillations $|\phi|$ increases with temperature as a result of intrawell relaxation processes.

Some estimations on the magnitude of the SR effect are given in Refs. 111 and 112. We would like also mention an extension of the framework presented for the case of the excited field nonparallel to the particle anisotropy axis. As shown in Ref. 113, despite the absence of interwell transitions for the components of the particle magnetic moment perpendicular to the anisotropy axis, a small admixture of the transverse susceptibility to the longitudinal one results in considerable increase of the peak at the temperature dependence of the function R.

B. Magnetic Stochastic Resonance in the Presence of a Bias Field

1. SR with Nonsymmetrical Wells

In the preceding section we have begun to study the effect of a bias constant field on the magnetic SR in single-domain particles. As once remarked in Ref. 28, this effect may be very important for realization of magnetic SR. Indeed, due to exponential dependence of the interwell hopping rate on the ratio $\Delta U/T$, at a plausible temperature the interwell frequency may well become too low, that is, the transitions may be effectively frozen. A promising way to at least partially "melt" the interwell process without changing the particle temperature is to impose a bias field along the anisotropy axis.

The bias field effect has two main aspects of interest: (1) it significantly affects the usual SR behavior determined by the linear response $M(t) \propto \exp(i\Omega t)$ and (2) the break in symmetry adds even terms into the frequency content of $M(t)$, which otherwise consists only of odd harmonics. Thus one gets an opportunity to examine a nonlinear response in its most simple (quadratic) form, $M^{(2)}(t) \propto H_{p0}^2 \exp(2i\Omega t)$. The latter is a subject of particular interest as the nonlinear susceptibilities make a much more sensitive tool to investigate static and dynamic properties of magnetic assemblies than the linear ones. Unlike the studies of Section III, where the odd-order susceptibilities inherent to the bias-field-free case are studied, here the even-order susceptibility of the lowest order (the quadratic one) is investigated for the same as in the preceding system of identical single-domain anisotropic ferromagnetic particles dispersed in a solid nonmagnetic matrix.

This is done by a numerical solution of the micromagnetic Fokker–Planck equation that grants a complete allowance for both magnetodynamic and thermal noise effects. Besides its direct range of application, the problem in question makes a case that contributes to understanding of the generic SR behavior of a bistable system subject to a constant force. In this aspect the results presented should be considered in line with a general theoretical treatment of nonlinear SR given in Ref. 114 and particular predictions on the nonlinear behavior of optical bistable systems [21] and a Josephson junction shortcircuited by a superconducting loop [115].

This section is arranged in the following way. The first subparts discuss the static susceptibility, and the details of the quadratic expansion of the kinetic equation, respectively. Thus, the relevant material parameters and all the necessary mathematical schemes are introduced and explained. In Section IV.B.4, the framework obtained is used to derive, calculate, and analyze the set of the dynamic susceptibilities, linear and quadratic, which are the basic quantities to deal with SR. Also, recalling the concept of the effective relaxation time, we propose a new simple expression for the quadratic susceptibility, and give its justifications. In Section IV.B.5, in terms of the dynamic susceptibility, the properties of the linear magnetic SR in a bias field are discussed. In Section IV.B.6, we formulate the definition for a nonlinear SR and present the dependencies of the magnetic quadratic SR on the temperature and the bias field strength. By comparison with the numerical results, we show that for the most part of the range of interest, the approximate expression of Section IV.B.4 works very well. When used, it yields the signal-to-noise ratio in a compact form, revealing all its essential temperature, bias field, and frequency dependencies.

2. Static Quadratic Susceptibility

In the presence of a probing field H_p and a constant bias field H, the orientation-dependent part of the energy function of a single-domain ferromagnetic grain is

$$U = -KV_m(e \cdot n)^2 - \mu[(e \cdot n) + (eH_p)] \qquad (4.245)$$

where e and n are, as already established, the unit vectors of the particle magnetic moment and anisotropy axis, respectively. The other parameters have their usual meanings explained in the preceding sections.

Here we focus on the longitudinal situation and assume that the imposed fields are collinearly directed along the anisotropy axis n. Then the set of the angular variables reduces to the polar angle ϑ of e with respect to n. Setting $\cos \vartheta = (e \cdot n) = x$, at $H_p = \text{const}$ for the equilibrium distribution function of the particle magnetic moment, one gets

$$W(e) = W(x) = Z^{-1} \exp[\sigma x^2 + (\xi + \xi_p)x] \qquad (4.246)$$

where the values of the internal (anisotropy) and external magnetic energies are scaled with the thermal energy:

$$\sigma = \frac{KV_m}{T}, \qquad \xi = \frac{\mu H}{T}, \qquad \xi_p = \frac{\mu H_p}{T} \qquad (4.247)$$

The partition integral in Eq. (4.246) is

$$Z(\sigma, \xi, \xi_p) = 2\pi \int_{-1}^{1} \exp[\sigma x^2 + (\xi + \xi_p)x]dx \qquad (4.248)$$

To avoid any further confusion in the future, we remark that under the definitions we use, all the parameters (viz., σ, ξ, and ξ_p) are essentially nonnegative.

Expanding expression (4.246) to the second order in ξ_p, one obtains

$$W(e) = W_0 \frac{1 + \xi_p x + \frac{1}{2}\xi_p x^2}{1 + \xi_p \langle x \rangle_0 + \frac{1}{2}\xi_p^2 \langle x^2 \rangle_0} \qquad (4.249)$$

here the angular brackets labeled with 0 denote the averaging over the basic state rendered by Eqs. (4.226).

For an assembly of identical noninteracting particles, that is our model, the longitudinal magnetization assumes the form $M = c\mu\langle x \rangle$, i.e., a single-particle contribution times particle number concentration. Here the unlabeled angular brackets denote the averaging with the distribution function W from Eq. (4.246). When the perturbing (probing) field is weak, one may present the magnetization as

$$M - M_0(H) = \chi H_p + \chi^{(2)} H_p^2, \quad M_0(H) = c\mu\langle x \rangle_0 \qquad (4.250)$$

thus introducing, along with the customary susceptibility χ, the quadratic one $\chi^{(2)}$. Note that owing to the magnetization saturation, $\chi^{(2)}$ must be negative.

On the other hand, using expansion (4.249), one obtains the representation

$$\langle x \rangle = \langle x \rangle_0 + \xi_p [\langle x^2 \rangle_0 - \langle x \rangle_0^2] + \frac{1}{2}\xi_p^2 [\langle x^3 \rangle_0 - 3\langle x^2 \rangle_0 \langle x \rangle_0 + 2\langle x \rangle_0^3] \qquad (4.251)$$

From Eqs. (4.249)–(4.251) the explicit relations follow:

$$\chi_0 = \frac{\phi I_s^2}{K} \sigma A, \quad A(\xi, \sigma) = \langle x^2 \rangle_0 - \langle x \rangle_0^2 \qquad (4.252)$$

$$\chi_0^{(2)} = \frac{\phi I_s^3}{K^2} \sigma^2 B \qquad (4.253)$$

$$B(\xi, \sigma) = \frac{1}{2}\left[\langle x^3 \rangle_0 - 3\langle x^2 \rangle_0 \langle x \rangle_0 + 2\langle x \rangle_0^3\right]$$

Hereafter, instead of the number concentration c, the dimensionless particle volume fraction $\phi = cV$ is used. Note that by definition

$$A = \frac{\partial \langle x \rangle_0}{\partial \xi}, \quad B = \frac{\frac{1}{2}\partial^2 \langle x \rangle_0}{\partial^2 \xi} \qquad (4.254)$$

In the limiting cases, simple expressions for the coefficients A and B are available. For high temperatures $(\sigma, \xi, \ll 1)$, one finds

$$A = \left(\frac{1}{3} + \frac{4}{45}\sigma\right) - \left(\frac{1}{15} + \frac{544}{11025}\sigma\right)\xi^2, \quad B = -\left(\frac{1}{15} + \frac{544}{11025}\sigma\right)\xi \qquad (4.255)$$

The low-temperature asymptotics was obtained by Garanin in Ref. 116. Namely,

$$A = A_{\text{inter}} + A_{\text{intra}} = \frac{1}{\cosh^2 \xi} \frac{(2\sigma + \xi)}{(2\sigma - \xi)} + \frac{1}{(2\sigma + \xi)^2} \quad \text{for} \quad 2\sigma - \xi \gg 1$$

$$(4.256)$$

and the corresponding B may be found with the aid of definitions (4.254). The two terms of the righthand side of Eq. (4.256) have clear meanings. The first one is caused by redistribution of the particle magnetic moments between the potential minima located at $x = \pm 1$. It may be called the *interwell* contribution. The other is the *intrawell* one, and it accounts for the field-induced orientation inside the deeper well $x = 1$, where at $\xi \gg 1$ virtually all the magnetic moments dwell.

According to Eq. (4.251), the quadratic term is determined by the odd-rank moments of the equilibrium distribution function, and is absent if the latter is even in x. Thus, for the existence of $\chi^{(2)}$ the presence of a bias field in Eq. (4.245) is mandatory. Otherwise, the next-to-linear response term would be cubic in the probing field amplitude: see Section III.

In Figure 4.19 the results of numerical evaluation of static χ and $\chi^{(2)}$ are given for an assembly of uniaxial grains with their axes completely aligned. As the field strength grows, the maxima of the curves move to higher temperatures. A qualitative explanation for this shift is provided by Eq. (4.256). Note that the field strongly affects the temperature behavior of the coefficient A_{inter}. At $H = 0$ the interwell contribution is unity and entirely dominates A_{intra}. Then Eq. (4.252) gives $\chi \propto \sigma \propto 1/T$, specifically, the Curie law shown by a dashed line in Figure 4.19(*a*). For $H \neq 0$, the term A_{inter} in Eq. (4.256) acquires an exponential factor. Then, on the temperature decrease, an abrupt fall $A_{\text{inter}} \propto \exp(-2\xi)$ takes place. For χ this means that the Curie law gives up, yielding a characteristic maximum. Since $\xi = \mu H/T$, the higher is the field, the greater is the temperature of the maximum. With further cooling the coefficient A_{inter} becomes negligible, and the control on the behavior of χ is overtaken by A_{intra}. The occurring crossover—from the exponential to the power law in $1/T$— manifests itself as an inflection point at the leftmost parts of the curves in Figure 4.19(*a*). The same qualitative tendency holds for the quadratic static susceptibilities [see Fig. 4.19(*b*)].

The static responses discussed above are determined for the complete equilibrium. For this reason, they are the easiest to calculate but not at all easy to observe. Indeed, at low temperatures the time needed to achieve the interwell equilibrium becomes exponentially large [55].

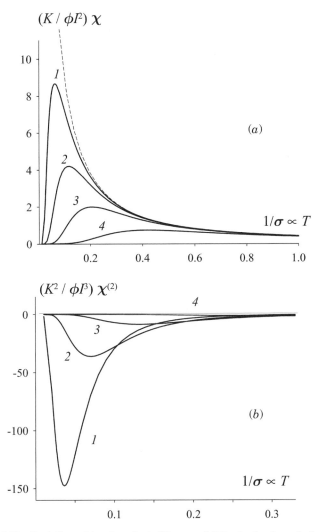

Figure 4.19.   Static linear (*a*) and quadratic (*b*) susceptibilities for the dimensionless bias field $\varepsilon = 0.05$ (1), 0.1 (2), 0.2 (3), 0.5 (4). Broken line shows the limiting behavior for $H = 0$. Note the threefold times difference between the abscissa scales of the (*a*) and (*b*) plots.

3.   Quadratic Expansion of the Time-Dependent Distribution Function

In a nonequilibrium situation, evolution of the distribution function $W(e, t)$ is governed by the rotary diffusion (Fokker–Planck) equation (4.90). The conventional representation for the diffusion time $\tau_D = \sigma\tau_0$ is given by Eq. (4.91), where τ_0 [see Eqs. (4.28) and (4.31)] accounts for the damping effect of the spin–lattice interactions on the orientational motion of e. Therefore, τ_0 is the material parameter of a particular system. For us, it yields the decay rate for the intrawell processes, those occurring near the bottom of the potential wells of $U(x)$. In the presence of the bias field, the intrawell relaxation time is modified as

$$\tau_{\text{intra}} = \tau_0 \left(1 + \frac{I_s H}{2K} \right)^{-1} \tag{4.257}$$

Thermal noise strongly affects the relaxation processes in a superparamagnetic system. At high temperatures, when the presence of the potential wells is insignificant ($\sigma \lesssim 1$), the magnetization relaxation time directly coincides with τ_D [see Eq. (4.264)]. In the low-temperature limit ($\sigma \gg 1$), the magnetization reversal takes the form of interwell transitions, and the its rate by the order of magnitude is given by Eq. (4.97):

$$\tau_N = \tau_0 \exp(\sigma) \tag{4.258}$$

where τ_0 plays the role of the preexponential factor. Note that the time τ_0 is of a dynamic, not diffusion, origin, and thus does not depend explicitly on temperature, which makes it a convenient noise-independent timescale. Then $\Omega \tau_0$ emerges as a natural criterion with regard to which any frequency Ω should be considered to be either high or low. For ferromagnetic particles the reference values of τ_0 are usually taken [117–119] as $10^{-10} - 10^{-9}$ s. Hence, all the frequencies up to the range of at least several MHz may be treated as being low.

Let us set the probing field H_p to vary harmonically

$$H_p = \tfrac{1}{2} H_{p0} \left(e^{i\Omega t} + e^{-i\Omega t} \right) \tag{4.259}$$

and seek the solution of the uniaxial Brown equation (4.90) in the form of the series

$$W(x, t) = \frac{1}{2\pi} \sum_{l=0}^{\infty} \frac{2l+1}{2} \langle P_l \rangle P_l(x) \tag{4.260}$$

where the function $\langle P_l \rangle$, with P_l as the Legendre polynomial, is in fact a conventional notation for the expansion coefficients. As has been mentioned, we restrict ourselves by the low-frequency range $\Omega \tau_0 \ll 1$. In this limit, the Larmor precession of the particle magnetic moment may be neglected, and Eq. (4.90) describes an overdamped angular oscillator in the bistable potential (4.245).

Substituting Eq. (4.260) into Eq. (4.90), and then integrating over e, one arrives at the pentadiagonal set of differential recurrence relations for the mean

Legendre polynomials.

$$2\tau_D \frac{d}{dt}\langle P_l\rangle + l(l+1)\langle P_l\rangle - 2\sigma\left[\frac{(l-1)l(l+1)}{(2l-1)(2l+1)}\langle P_{l-2}\rangle\right.$$

$$\left. + \frac{l(l+1)}{(2l-1)(2l+3)}\langle P_l\rangle - \frac{l(l+1)(l+2)}{(2l+1)(2l+3)}\langle P_{l+2}\rangle\right]$$

$$- (\xi + \xi_p)\frac{l(l+1)}{2l+1}[\langle P_{l-1}\rangle - \langle P_{l+1}\rangle] = 0 \qquad (4.261)$$

this set was derived in Refs. 120 and 121. Taking into account that the dimensionless amplitude ξ_{p0} [see scaling (4.247)] of the probing field is small, the response of the periodically driven system may be evaluated with the aid of a perturbation method. To be able to obtain the quadratic susceptibility, we have to perform the pertinent calculation up to the second order in ξ_{p0}. Dealing with the harmonic field (4.259), we introduce the time-independent complex amplitudes as

$$\langle P_l\rangle - \langle P_l\rangle_0 = \tfrac{1}{2}(b_l^{(1)}e^{i\Omega t} + \text{c.c.}) + \tfrac{1}{4}(b_l^{(2)}e^{2i\Omega t} + \text{c.c.}) \qquad (4.262)$$

where c.c. stands for complex conjugates. On substituting this expansion into Eq. (4.261), one gets the equation for the complex amplitudes

$$2ik\Omega\tau_D b_l^{(k)} + \sum_{l'}\Lambda_{l,l'}b_{l'}^{(k)} = \xi_{p0}\frac{l(l+1)}{2l+1}[b_{l-1}^{(k-1)} - b_{l+1}^{(k-1)}] \qquad (4.263)$$

valid for the perturbations $b_l^{(k)} \propto \xi_{0p}^k$ from Eq. (4.262). The operator $\hat{\Lambda}$ in Eq. (4.263) is the pentadiagonal relaxational matrix whose definition follows from Eq. (4.261). As usual, the right-hand side of Eq. (4.263) contains the result of the preceding iteration.

One remark concerning the zeroth order is worthwhile. There one deals with the equilibrium quantities $\langle P_l\rangle$, which can be found from the equation

$$\sum_{l'}\Lambda_{l,l'}\langle P_{l'}\rangle_0 = 0 \qquad (4.264)$$

Its nontrivial solution is provided by the identity $\langle P_0\rangle_0 = 1$. The corresponding term may be passed to the right-hand side of Eq. (4.264), thus rendering this equation nonhomogeneous. In all the higher perturbation orders, $b_0^{(k)}$ vanish identically.

The sets of the pentadiagonal recurrence relations (4.263) and (4.264) truncated at some large enough $l = N$, are solved with the aid of the generalized Thomas algorithm, as described in Section II.C.2.

4. Dynamic Susceptibilities

Numerical solution of Eqs. (4.263) and (4.264) yields a representation for the distribution function (4.260) accurate up to the terms of the order ξ_{p0}^2. Using it to evaluate the reduced magnetization $\langle x \rangle$ and taking into account expansion (4.250), one can present the magnetic response as a sum of the frequency-dependent contributions as

$$M = M_0 + \tfrac{1}{2}\chi(\Omega)H_{p0}\exp(i\Omega t) + \tfrac{1}{4}\chi^{(2)}(2\Omega)H_{p0}^2\exp(2i\Omega t) + \text{c.c.} \quad (4.265)$$

thus specifying the linear and quadratic complex susceptibilities:

$$\chi = \frac{\phi I_s^2 \sigma}{K}b_1^{(1)}, \qquad \chi^{(2)} = \frac{\phi I_s^3 \sigma^2}{K^2}b_1^{(2)} \quad (4.266)$$

Each of them is a function of the bias field ξ and describes the harmonic of magnetization at the pertinent frequency. Note that in Eq. (4.265), as in Eq. (4.262), we omit the stationary contribution to $\chi^{(2)}$.

To facilitate understanding, let us recall the main features of the linear longitudinal susceptibility of a superparamagnetic system. The general solution of the linear problem (4.263) can be formally presented as the spectral expansion

$$\chi = \chi_0 \sum_{k=1}^{\infty} \frac{w_k}{1 + 2i\Omega\tau_D/\lambda_k}, \qquad \sum_{k=1}^{\infty} w_k = 1 \quad (4.267)$$

where the sets of the eigenvalues $\{\lambda_k\}$ of the relaxational operator $\hat{\Lambda}$ are introduced together with the weights $\{w_k\}$ rendering the contributions of the eigenmodes to the linear susceptibility. In Eq. (4.267) λ_1 is the lowest eigenvalue, and it is the only one that yields the rate of the interwell relaxation. The corresponding relaxation time formula is

$$\tau_{\text{inter}} = \frac{2\tau_0}{\lambda_1} \quad (4.268)$$

On the other hand, one can try to approximate the magnetization damping process by a single effective relaxation time as

$$\chi = \chi_0(1 + i\Omega\tau_{\text{eff}})^{-1} \quad (4.269)$$

In the lowfrequency limit the exact and approximate susceptibilities must coincide. Matching expansions (4.267) and (4.269) at $\Omega\tau_D \ll 1$, one gets

$$\tau_{\text{eff}} = \sum_{k=1}^{\infty} w_k \tau_k, \qquad \tau_k = \frac{2\tau_D}{\lambda_k} \quad (4.270)$$

The last relationship is in fact the quantitative definition of τ_{eff}. However, in practice it is more feasible to find τ_{eff} from a solution of the recurrence relation [29,59]

$$\sum_{l'} \Lambda_{l,l'} F_{l'} = 2\tau_D(\langle P_l P_1 \rangle_0 - \langle P_l \rangle_0 \langle P_1 \rangle_0) \qquad (4.271)$$

as $\tau_{\text{eff}} = F_1/A$.

For high temperatures $(\sigma, \xi \ll 1)$, the effective relaxation time defined by Eq. (4.270) reduces to

$$\tau_{\text{eff}} = \tau_D \qquad (4.272)$$

In the low-temperature limit, Garanin [116] derived the expression

$$\tau_{\text{eff}} = A^{-1}(A_{\text{inter}}\,\tau_{\text{inter}} + A_{\text{intra}}\,\tau_{\text{intra}}) \qquad (4.273)$$

where the parameters A are as defined by Eq. (4.256), and the temperature-independent time τ_{intra} was introduced by Eq. (4.257). The corresponding asymptotic form of the interwell relaxation time (4.268) is provided by the formula

$$\lambda_1 = 2\pi^{-1/2}\sigma^{3/2}(1-h^2)\{(1+h)\exp[-\sigma(1+h)^2] + (1-h)\exp[-\sigma(1-h)^2]\} \qquad (4.274)$$

obtained by Aharoni [55]; here $h = I_s H/2K$.

In Figure 4.20 we present the plot of τ_{eff} obtained from the numerical solution of Eq. (4.271). The coordinates used for this schematic representation

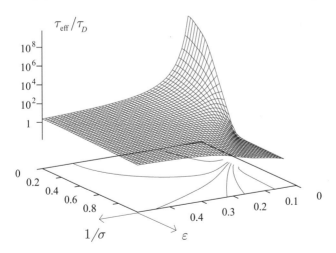

Figure 4.20. Integral relaxation time as a function of the dimensionless temperature $1/\sigma$ and bias field strength ε.

are the dimensionless temperature independent magnetic field

$$\varepsilon = \frac{\xi}{\sigma} = \frac{I_s H}{K}$$

already introduced by Eq. (4.242) and the dimensionless temperature $1/\sigma$. Note also the logarithmic scale of the vertical axis (ordinate). The plateau at $\sigma \lesssim 2$ shows that at enhanced temperatures the relaxation time does not depend on the details of the potential. At low temperatures, the effective time is strongly temperature-dependent. The form of the surface τ_{eff} in the interval $\sigma \gtrsim 5$ is very specific. At $\xi = 0$ and $1/\sigma \ll 1$, it reduces to τ_N given by Eq. (4.258). This yields the customary superparamagnetic blocking model with its exponential in $1/T$ behavior. A peculiar feature is the occurrence of a nonmonotonic dependence $\tau_{\text{eff}}(1/\sigma)$ when ε exceeds some finite value. This fact was discovered in Refs. 59 and 116, and evaluations gave $\varepsilon \approx 0.34$. Remarkable is that the similar inflection on the curves $2\tau_D/\lambda_1$ takes place [55], only at ε close to 2. This gives a strong direct argument in favor of using τ_{eff}, and not $2\tau_D/\lambda_1$, as the effective relaxation time for magnetization.

The temperature dependencies of χ' and χ'' found by the full numerical procedure [Eqs. (4.262)–(4.266)], and with a simple formula (4.269) but with inserted numerically exact τ_{eff}, are given and compared in Figure 4.21 for $\Omega\tau_0 = 10^{-4}$. We remark that the actual choice of Ω does not matter as long as we deal in the low-frequency range. However, the assumed value seems quite reasonable, since at $\tau_0 \sim 10^{-9}$ s it corresponds to the dimensional frequency $\sim 10^4$ Hz that is a convenient measurement range.

As one can see from Figure 4.21, the agreement is rather good. Besides other things, this implies that in the low-frequency domain one can propose a closed equation of the magnetization motion

$$\left(\tau_{\text{eff}}\frac{d}{dt} + 1\right)(M - M_0) = \chi_0 H_p \qquad (4.275)$$

where χ_0 is defined by Eq. (4.252).

From Figure 4.21 it follows that the presence of a bias field shifts the maxima of in and out-of-phase components of χ in opposite ways: the peaks of the real part move to the higher noise strength region, whereas those of the imaginary part display the reverse tendency. The behavior of the real part does not differ much from that of the static susceptibility, which is natural since the parameter $\Omega\tau_0$ is small; that is, we observe the dependence of the factor A on ξ described when discussing the static case. The imaginary part of the susceptibility possesses an additional factor $\sim \Omega\tau_{\text{eff}}$. For this reason, the χ'' behavior is affected mostly by the lowest barrier height, which decreases as the bias field grows.

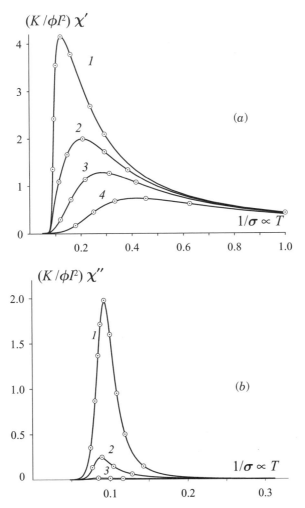

Figure 4.21. Real (a) and imaginary (b) parts of the linear dynamic susceptibility at $\Omega\tau_0 = 10^{-4}$ for the bias field $\varepsilon = 0.1$ (1), 0.2 (2), 0.3 (3), 0.5 (4). For figure (b) the curve with $\varepsilon = 0.5$ does not resolve. Circles show the result of the effective time approximation.

From the general viewpoint, the ability of a relationship such as Eq. (4.275) to describe the higher harmonics of magnetization is questionable. However, at least for semiquantitative results, one can obtain a useful approximation for $\chi^{(2)}$. For that purpose, let us rewrite Eq. (4.275) as

$$\left(\tau_{\text{eff}}\frac{d}{dt}+1\right)(M-M_0) = H_p\frac{\partial M}{\partial H} \tag{4.276}$$

which actually means replacement of M_0 by M in the right-hand side. Setting in accordance with Eq. (4.265)

$$M = M_0 + \delta M^{(1)} + \delta M^{(2)}, \quad \delta M^{(1)} = \chi_0(1 + i\Omega\tau_{\text{eff}})^{-1}H_p, \quad \delta M^{(2)} \sim H_{p0}^2$$

one gets from Eq. (4.276) the equation for the second-order correction in the form

$$\left(\tau_{\text{eff}}\frac{d}{dt} + 1\right)\delta M^{(2)} = H_p \frac{\partial}{\partial H} \delta M^{(1)}$$

This finally yields

$$\chi^{(2)} = \tfrac{1}{2}\chi_0^{(2)}[(1 + i\Omega\tau_{\text{eff}})(1 + 2i\Omega\tau_{\text{eff}})]^{-1} \tag{4.277}$$

where the static value $\chi_0^{(2)}$ is defined in Eq. (4.253).

The linear susceptibility plots of Figure 4.21 directly illustrate of these conclusions. What is less expected is that the effective time approach turns out to be very efficient for the nonlinear response. The justification is given in Figure 4.22, where we compare (similarly to Figs. 4.21) the quadratic susceptibilities evaluated by the numerically exact method and by means of Eq. (4.277). All the numerical evidence that we have (of which the presented graphical data are only a small portion) testifies to the effect that, except maybe for a rather narrow lowtemperature range, the effective time approximation rather closely follows the exact solution.

In the plots of quadratic susceptibilities presented in Figure 4.22 one can notice the same general features as for the linear χ. In the same way as for the linear terms, the maxima of $\chi^{(2)'}$ and $\chi^{(2)''}$ are comparable and display the same tendency in the peak shifts when the bias field is enhanced. Similar to the relation between Figures 4.19(a) and 4.19(b), the widths of the imaginary quadratic plots are smaller than those for the real parts.

Let us give some concluding remarks on the concept of the effective relaxation time:

1. The principle by which τ_{eff} is defined may be used to introduce similar effective relaxation parameters for perturbations of any higher symmetry as well. But the dipolar case, as δM is, seems to be the most natural.

2. Introduction of τ_{eff} does not simplify its evaluation. Indeed, according to Eq. (4.271) to find the effective time, one has to know the solution of the kinetic equation (4.90). Therefore, the real gain of this approach is that, as soon as $\tau_{\text{eff}}(\sigma, \xi)$ is found, all the dynamic response problems write very simply, making the obtained results compact and easy to analyze.

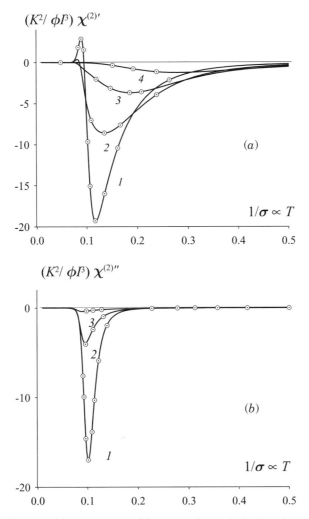

Figure 4.22. Real (a) and imaginary (b) parts of the quadratic dynamic susceptibility at $\Omega\tau_0 = 10^{-4}$ for the bias field $\varepsilon = 0.1$ (1), 0.2 (2), 0.3 (3), 0.5 (4). For figure (b) the curve with $\varepsilon = 0.5$ does not resolve. Circles show the result of the effective time approximation.

3. The virtual presence of all the spectral terms in the effective time makes this approach more adequate than the superparamagnetic blocking model. In the latter, the effective relaxation time of magnetization is identified just with the inverse of the decrement λ_1, which is the smallest at $\xi = 0$. Such a replacement reduces all the magnetic dynamics to the interwell transition, ignoring all the intrawell motions. For this reason, the blocking

model deviates significantly from the exact solution in the low-temperature limit ($\sigma \to \infty$) in the absence of the external field, and also at $\sigma \sim \xi$.

5. Magnetic SR in a Bias Field by the Linear Response Theory

Let us first analyze the effect of the bias field on the magnetic SR in the framework of the linear response theory formulated for SR [96,97,99]. The main idea of the linear response treatment is a direct use of the fluctuation–dissipation theorem, which expresses the thermal (fluctuational) power spectrum $Q_n(\omega)$ of the magnetic moment of the system, namely, *magnetic noise*, through the imaginary component of its linear dynamic susceptibility $\mathrm{Im}\chi = -\chi''$ to a weak probing ac field of an arbitrary frequency ω as

$$Q_n(\omega) = \frac{2V_t T}{\omega}\chi''(\omega) \tag{4.278}$$

Note that all throughout our consideration we use the usual definition of χ as the magnetic susceptibility of a unit volume of the disperse system. Therefore, to keep up with the meaning of the fluctuation–dissipation theorem, in formula (4.278) the total volume V_t of the system is introduced. In terms of the total number N of the magnetic particles $V_t = N/c = NV/\phi$.

The spectral power density of regular oscillations induced by a driving field of a frequency Ω, that is, the *response signal* in the same terms, is expressed as

$$Q_s(\omega) = \frac{3}{2}\pi V^2 H_{p0}^2 |\chi(\Omega)|^2 \delta(\omega - \Omega) \tag{4.279}$$

where H_{p0} stands for the ac field amplitude. For any real measurement, we get the spectral density in a certain finite frequency bandwidth Δ of the signal detection. This fact may be accounted for by replacing the delta function by the inverse of the bandwidth.

On setting $\omega = \Omega$ in Eqs. (4.278) and (4.279), the signal-to-noise ratio (SNR), which is the main issue investigated in the SR theory, may be presented as

$$S = \frac{Q_s}{S_n} = \frac{\pi N}{4\tau_0 \Delta}\left(\frac{I_s H_{p0}}{K}\right)^2 R(\sigma, \xi, \Omega) \tag{4.280}$$

where the dimensionless function R is introduced, as in Section IV.A.2 to take in all the temperature, frequency, and bias field strength dependencies. Note that in the front factor we use τ_0 instead of τ_D to entirely pass all the temperature dependence to R.

In the effective time approximation (4.270), from Eqs. (4.278)–(4.280), one has

$$R = \frac{\sigma^2 A \tau_0}{\tau_{\text{eff}}} \qquad (4.281)$$

revealing, in particular, the absence of the frequency dependence. For a zero bias field ($\xi = 0$) and $\sigma > 1$, one may set $\tau_{\text{eff}} \approx \tau_N$ [see Eq. (4.258)] and recover the most simple relation [25,26] for the linear magnetic SR:

$$R = \sigma^2 A \exp(-\sigma) \qquad (4.282)$$

The exact function $R(\sigma, \xi, \Omega)$ in a wide range of its arguments was investigated in Ref. 30 numerically. It had been done by solving the set comprising just Eqs. (4.263), that is, the linear framework. Comparison has shown that the superparamagnetic blocking model, that is, the assumption that the system is characterized by a single exponential interwell passage time $\sim \tau_N$, is valid for asymptotic considerations [25,26], but is insufficient for ascertaining certain essential details. In particular, Eq. (4.282) fails to describe SNR in the zero-temperature limit ($\sigma \to \infty$, $\Omega \tau_N \gg 1$), and as well cannot account for either the effect of a bias field or a finite probing frequency value [see Eq. (4.281)].

In Figures 4.23 we show (with the appropriate reduction of units) the exact 3D diagrams for Q_s, Q_n, and SNR as derived with the linear response approach from Eqs. (4.278), (4.279), and (4.280) respectively. Note that as the functions of the bias field, both Q terms are maximal at $H = 0$ and rapidly decrease with its growth. However, from their sharp peaks located at almost the same point, it is difficult to foresee the actual SNR behavior described by the function R shown in Figure 4.23(c). It turns out that its maximum is shifted considerably rightward (to higher temperatures) in comparison with those of the susceptibilities, and the rates of change along both axes are much smaller. Note also a massive low-temperature "shoulder" of any $R(T)$ curve caused by the intrawell motions [30]. As the bias field H grows, this low-temperature plateau widens, suppressing the maximum of $R(T)$ that means disappearance of SR as itself. This transformation is completed when the bias field exceeds the value $\varepsilon = 2$ above which the potential curve assumes a one-well shape.

The results for SNR evaluated with the effective relaxation time approxima-tion in the same linear response framework [i.e., by Eq. (4.281)], are shown in Figure 4.24. One observes a good agreement with respect to the main cusp of the function $R(T)$. However, in the low-temperature range the existing deviations from the exact solution [they do not resolve in the susceptibility graphs of Figure 4.21(a)] become noticeable; see the relative positions of the dots and curve 3 for $1/\sigma < 0.1$.

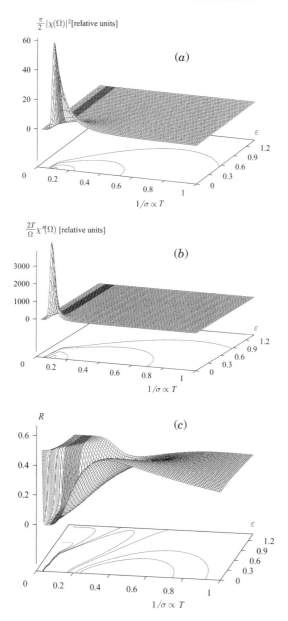

Figure 4.23. Signal (a) and noise (b) power densities, and SNR (c) for a superparamagnetic system in the linear response approximation. In the first two figures the vertical scales are chosen to retain only the susceptibility dependencies that really matter; namely, in (a) $(2T/\Omega)\chi'' = Q_n/V_m$ [see Eq. (4.278)]; in (b) $\frac{1}{2}\pi|\chi|^2 = Q_s/V_m^2 H_{p0}^2$ [see Eq. (4.279)]. In (c) the SNR is characterized by the function R [see Eq. (4.280)]. All the results are given for the low-frequency case $\Omega\tau_0 = 10^{-4}$.

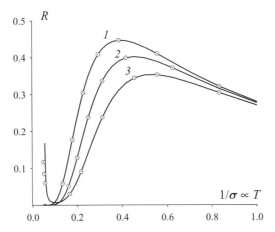

Figure 4.24. Comparison of the numerical solution (solid lines) and the effective time approximation (circles) with respect to SNR in the linear response approximation; $\varepsilon = 0.1$ (1), 0.3 (2), 0.5 (3). All the results are given for the low-frequency case $\Omega\tau_0 = 10^{-4}$.

Ending the discussion on linear SR in a bias field, we remark that it is practical to formulate explicitly the operational definition of the SR measurement that underlies the calculation presented. Namely, we suppose that, first, the equilibrium magnetic power spectrum (no driving AC field) of the system is measured at some frequency Ω, at which the driving field would be then applied. This quantity is assumed to be Q_n, the noise. Then the probing AC field is switched on, and the magnetic power density $Q(\Omega)$ is recorded anew. The change in the power density produced by the probing field is considered to be the signal. Accordingly, we write

$$Q(\Omega) = Q_n(\Omega) + Q_s(\Omega) \tag{4.283}$$

Since Q_n is known from the previous measurement, the signal-to-noise ratio is obtained as

$$S = \frac{Q - Q_n}{Q_n} \tag{4.284}$$

6.   Quadratic Magnetic SR

In Section IV.B.4 we have shown that the quadratic dynamic susceptibilities of a superparamagnetic system display temperature maxima that are sharper than those of the linear ones. If the maximum occurs as well at the temperature dependence of the signal-to-noise ratio, this should be called *the nonlinear stochastic resonance*. However, before discussing this phenomenon, one has to define what should be taken as the signal-to-noise ratio in a nonlinear case.

We shall do that with the same kind of the operational definition as in Section IV.B.5. Specifically, we assume that each signal-to-noise value emerges as a result of a three-step procedure: (1) one decides on the frequency Ω at which the test would be performed; then (2), at double this frequency, the noise spectral power density in the state with no driving field is evaluated, thus yielding

$$Q_n(2\Omega) = \frac{V_t T}{\Omega} \chi''(2\Omega) \tag{4.285}$$

compare this with Eq. (4.278); (3) The signal term may be derived directly from the definition

$$Q = \int_{-\infty}^{\infty} \langle \mu(t)\mu(0) \rangle e^{i\omega t} dt \tag{4.286}$$

Substituting there the magnetization expansion (4.265), one obtains

$$Q - Q_n = Q_s(\Omega) + Q_s(2\Omega) = \frac{1}{2} \pi V_t^2 [H_{p0}^2 |\chi(\Omega)|^2 \delta(\omega - \Omega)$$
$$+ H_{p0}^4 |\chi^{(2)}(2\Omega)|^2 \delta(\omega - 2\Omega)] \tag{4.287}$$

where now two terms are field-induced. Besides the one of the linear origin ($\propto H_{p0}^2$), there appears the fourth-order one stemming from the quadratic response, and rendering the component of Q at the double frequency. Using the recipe (4.284) as a model, we may define the quadratic SR as

$$S_2 = \frac{[Q(2\Omega) - Q_n(2\Omega)]}{Q_n(2\Omega)} \tag{4.288}$$

To extract the dimensional parameters that do not depend explicitly on temperature, we rearrange the quadratic signal-to-noise ratio to the form

$$S_2 = \frac{\pi N}{16\tau_0 \Delta} \left(\frac{I_s H_{p0}}{K} \right)^4 R_2(\sigma, \xi, \Omega) \tag{4.289}$$

Compare this with Eq. (4.280), which is in fact the definition for the function R_2. The latter we obtain through the numerical solution of the sets of equations (4.263) and (4.264) and its further substitution to Eqs. (4.266). Note that in the front factor, as in Eq. (4.280), we use τ_0 instead of τ_D to entirely single out the temperature dependence.

Using the effective time approximation described in Section IV.B.4, one can derive the explicit expression

$$S_2 = \frac{\pi N}{16\tau_0 \Delta} \left(\frac{I_s H_{p0}}{K} \right)^4 \frac{B^2 \sigma^4 \tau}{A \tau_{\text{eff}}} \frac{1}{1 + \Omega^2 \tau_{\text{eff}}^2} \tag{4.290}$$

Remarkably, even in a simplified approach, like Eq. (4.290), the quadratic SR turns out to be essentially frequency-dependent.

The inherent feature of the quadratic SR under study is that the bias field is the sole cause of even harmonics in the spectrum. Due to the symmetry considerations, they must vanish at $H \to 0$. This means that $S_2(\xi \to 0) = 0$. In Eq. (4.290) this limit is ensured by the proportionality of S_2 to B. According to the second of Eqs. (4.253), the coefficient B consists only of the odd equilibrium moments of the distribution (4.226). Since the function W_0 is even in x at $\xi = 0$, the odd moments vanish.

On the other hand, at sufficiently high H, the magnetization of the system saturates. This deprives the magnetic moment of any orientational freedom and eventually freezes it up. Thus, $H \to \infty$ must as well lead to a vanishing response at any harmonic. Under those circumstances, it is clear that the quadratic signal together with SNR, when plotted as the functions of the bias field strength, should pass a maximum. The existence of a maximum both at the temperature dependence of the signal-to-noise ratio R_2 (SR at constant H) and the occurrence of the above-described maximum at its bias field dependence, make it interesting to analyze the quadratic response in the whole field temperature coordinate plane. The numerical approach makes this possible, and in Figure 4.25 the results of such a consideration for $Q_s(2\Omega)$, $Q_n(2\Omega)$, and R_2 are shown. One sees that the 3D peaks of the spectral density components definitely do not coincide. For this reason, the position and height of the R_2 peak, that is, the optimized joint action of both dependencies, may be found but numerically. From Figure 4.25(c) it follows that the enhancement of R_2 occurs to be rather pronounced (see the contour lines at the baseplane of the figure). The dome-like shape of the maximum of R_2 supports the expectation that quadratic SR must be sharper than the linear one [compare Figures 4.23(c) and 4.25(c)].

Figure 4.26 presents a selection of cross-sections of the 3D plot of R_2 where we compare the exact (numerical) solution with the effective time approximation (4.290). In the low-temperature range, the deviations, although not resolved in the graph, are inevitable. But they are not the main issue of the present study. As to the quadratic SR proper, the agreement again is very good.

7. Summary

A consistent study of the linear and lowest nonlinear (quadratic) susceptibilities of a superparamagnetic system subjected to a constant (bias) field is presented. The particles forming the assembly are assumed to be uniaxial and identical. The method of study is mainly the numerical solution (which may be carried out with any given accuracy) of the Fokker–Planck equation for the orientational distribution function of the particle magnetic moment. Besides that, a simple heuristic expression for the quadratic response based on the effective relaxation

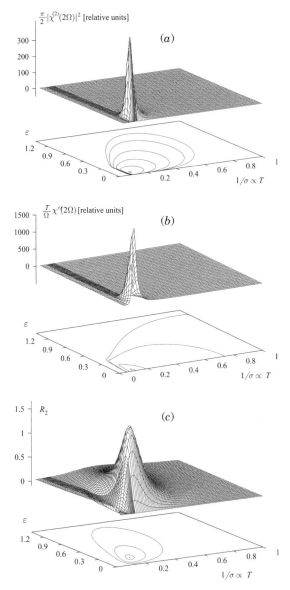

Figure 4.25. Signal (a) and noise (b) power densities, and SNR (c) for a superparamagnetic system at the doubled excitation frequency (i.e., quadratic response). In the first two figures the vertical scales are chosen to retain the susceptibility dependencies that really matter; specifically, in (a) $(T/\Omega)\chi''(2\Omega) = Q_n(2\Omega)/V_m$ [see Eq. (4.285)]; in (b) $\frac{1}{2}\pi|\chi^{(2)}(2\Omega)|^2 = Q_s(2\Omega)/V_m^2 H_{p0}^4$ [see Eq. (4.287)]. In (c) the SNR is characterized by the the function R_2 [see Eq. (4.289)]. The results are obtained numerically for the low-frequency case $\Omega\tau_0 = 10^{-4}$.

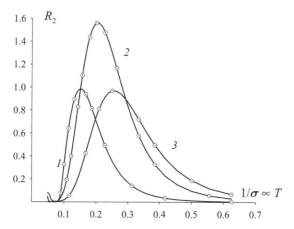

Figure 4.26. Comparison of the numerically exact solution (solid lines) and the effective time approximation (circles) with respect to the quadratic SNR; $\varepsilon = 0.1$ (1), 0.3 (2), 0.5 (3). Low-frequency case: $\Omega\tau_0 = 10^{-4}$.

time is proposed. The net result is a simple, highly accurate formula that is proved by comparison with the numerical solution.

We show that application of a constant force (bias field) results in shifting the position of the ordinary SR peak together with the anticipated reduction of its height and sharpness. For the quadratic SR the situation is more complicated. There, the joint action of the thermal noise and constant bias leads to formation of a mountain-like surface over the plane of those parameters. In other words, for each given value of the bias field there exists a unique value of the noise strength that maximizes SNR and vice versa. The discovered effect can be useful, for example, for evaluation of the parameters of bistable systems through susceptibility measurements. In addition, it has to be taken into account when designing any devices where the nonlinear SR is employed.

The studied quadratic SR of a superparamagnetic grain is caused by interaction of the periodic excitation with the thermal noise. However, there exists an analogy $T \to \gamma\hbar H_a$, where $H_a = 2K/I_s$ is the particle anisotropy field, which outlines a passage from thermal to quantum fluctuations. Thus, with certain caution, one may apply the results obtained to the nonlinear SR caused by the tunnel effect at $T = 0$.

C. Noise- and Force-Induced Resonances under a Bias Field

1. Nonlinear SR-Related Phenomena

Extensive studies of stochastic resonance have revealed a number of remarkable properties of noisy nonlinear oscillators. Among others, an unusual nonlinear effect dubbed *noise-induced resonance* (NIR) was discovered for both bistable [21] and monostable [22] potentials. Specifically, the term NIR refers to a noisy system driven by a periodic force and means that there occurs a strong selective suppression of higher harmonics under variation of the noise intensity. This phenomenon is under study in view of its prospects for development of new signal-processing schemes [122,123].

In the original paper [22], the theory of NIR was developed on the basis of perturbation formalism. The response spectrum was presented as a power series, and for the kth harmonic all the contributions of the powers higher than ξ^k were omitted; here ξ is the ratio of the intensity of excitation (μ times the amplitude of the exciting field) to the thermal energy. Therefore, the validity range of the theory [22] is essentially limited by the condition $\xi \ll 1$. In the literature [23,24,124,125] we have progressed beyond the restrictions of the perturbation method and analyzed the response spectra of a noisy system that is governed by an anharmonic potential and is exposed to both DC and AC fields of arbitrary amplitudes. We find that NIR turns out to be just a particular manifestation of a general nonlinear behavior of a thermalized oscillator. Namely, NIR always coexists with another effect that, to match the terms, we term *the force-induced resonance* (FIR). The latter also causes selective suppression of the higher harmonics, but for FIR the governing parameter is the intensity of excitation.

Customary stochastic resonance in superparamagnets in the framework of the linear response theory has been investigated [25,26,28,30,126,127] and is described in Section IV.A above. The response of noise-driven fine magnetic particles to a finite-amplitude excitation has also been discussed in the literature [67,120,128] and in Section IV.B above. A specific case of magnetodynamics in the presence of a high-DC field was analyzed in Ref. 129. The problem of magnetic NIR was first addressed in our papers [23,24]. Below we present the formulation and solution of the combined NIR/FIR problem for a solid superparamagnet (Section IV.C.2) and consider the manifestations of NIR in the case of a magnetic fluid excited by a strong ac magnetic field (Section IV.C.3).

2. Solid Systems: Generation of Harmonics of Magnetization

Consider a grain of a ferromagnet or ferrite that is small enough ($\lesssim 10$ nm) to be single-domain. Then, its magnetic state is characterized by a single magnetic dipole of the moment μ; far from the Curie point, the value of μ is constant. For a particle residing in a solid matrix, the motion of the dipole is exhausted by its internal rotations described by the unit vector e. Under given conditions, the orientation of e is influenced by the following factors: the crystallographic magnetic anisotropy, the external DC (bias) field H_0, and the AC (probing) field $H_p(t) = H_p \cos \omega t$. Both fields H_0 and H_p are directed along the particle easy axis; the excitation frequency ω is small in comparison with the Larmor one: $\omega \ll \gamma(H_0 + H_p)$, where γ is the gyromagnetic ratio.

Taking thermal fluctuations into account, the motion of the particle magnetic moment is described by the orientational distribution function $W(e, t)$ that obeys the Fokker–Planck equation (4.90). For the case considered here, the energy function is time-dependent:

$$\frac{U}{T} = -\sigma \cos^2 \vartheta - (\xi_0 + \xi \cos \omega t) \cos \vartheta \tag{4.291}$$

with the set of the dimensionless parameters defined in Eq. (4.247). At $\xi = 0$, Eq. (4.291) yields a potential that incorporates a symmetric and antisymmetric (with respect to $\cos \vartheta$) parts, and is bistable at $\xi < 2\sigma$ and monostable otherwise.

In a noisy system described by Eq. (4.90), observable are only the ensemble averages taken with the distribution function $W(e, t)$. The quantity of the prime physical interest is the magnetization $M(t) = I_s \phi \langle P_1(\cos \vartheta) \rangle$, where ϕ is the volume fraction of the particles and $P_l(\cos \vartheta)$ is a Legendre polynomial. As shown in Section IV.B.3, the kinetic equation (4.90) may be identically reduced to an infinite set of equations for the statistical moments $\langle P_l(\cos \vartheta) \rangle$, see Eq. (4.261).

To study its harmonic content, we expand the moments in the Fourier series

$$\langle P_l \rangle = \sum_{k=-\infty}^{\infty} b_{lk}(\omega) e^{ik\omega t} \tag{4.292}$$

On that, Eq. (4.261) transforms into a homogeneous pentadiagonal vector recurrence relation

$$\left[\frac{2i\omega\tau_D}{l(l+1)} + 1 \right] b_{lk} - 2\sigma \left[\frac{l-1}{(2l-1)(2l+1)} b_{l-2,k} + \frac{1}{(2l-1)(2l+3)} b_{lk} \right.$$

$$\left. - \frac{l-2}{(2l+1)(2l+3)} b_{l+2,k} \right] - \frac{\xi_0}{2l+1} (b_{l-1,k} - b_{l+1,k})$$

$$- \frac{\xi}{2(2l+1)} \left[b_{l-1,k-1} + b_{l-1,k+1} - b_{l+1,k-1} - b_{l+1,k+1} \right] = 0 \tag{4.293}$$

with the initial conditions $b_{00} = 1$, $b_{0k} = 0$ at $k \neq 0$. Solving it by the matrix sweep method (see Section II.C.3), the set of amplitudes $\{b_{lk}\}$ may be evaluated with any desired accuracy. According to Eq. (4.292), the subset $\{b_{1k}\}$ determines the magnetization, which we rewrite as

$$m(t) = \frac{M(t)}{I_s \phi} = \sum_{k \geq 0} m_k(\rho, \xi_0, \varepsilon, \omega\tau) \exp[i(k\omega t - \psi_k)] \tag{4.294}$$

By definition, the amplitudes m_k are real and nonnegative. The two additional dimensionless parameters on which the harmonics of magnetization depend are

$$\rho = \frac{H_p}{H_0}, \qquad \varepsilon = \frac{\xi_0}{\sigma} \qquad (4.295)$$

here the definition of the parameter ε is added for completeness; it merely repeats Eq. (4.242). Note that under the assumptions adopted, neither ρ nor ε depend explicitly on temperature.

Solving the matrix set (4.293) at various frequencies, one finds that the harmonic suppression effect is most pronounced under the adiabatic condition $\omega\tau \to 0$. Henceforth, we focus on it. In this limit, suppression of a kth harmonic means its complete vanishing: the value of $m_k(\xi_0, \rho)$ turns into zero. In Figure 4.27 the magnetization amplitudes m_{3-6} for the case $\sigma = 0$ (or $\varepsilon = \infty$), that is, for magnetically isotropic particles, are presented. Under a constant bias field H_0, it is convenient to treat the coordinate ξ_0 as the dimensionless inverse temperature and ρ as the temperature-independent excitation strength. With the logarithmic scale in the z axis, the lines of $m_k = 0$ are reflected as the grooves of infinite depth. The amplitudes $m_{1,2}$ are not shown; similar to the oscillators studied in Refs. 21 and 22, they turn to zero but trivially, along the ξ_0 and ρ axes. It should be noted that if one needs only the adiabatic solutions, there exists a much shorter way. Namely, it suffices to apply the equilibrium formalism with a quasi-Gibbs distribution $W \propto \exp[-U(t)/T]$, where the

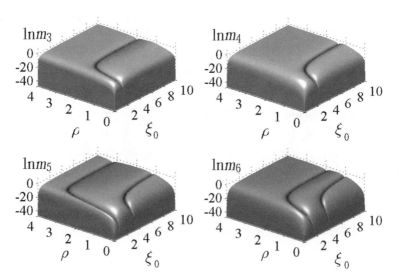

Figure 4.27. Response amplitudes of third to sixth harmonics of the superparamagnet magnetization for $\sigma = 0$ in the adiabatic limit; note the logarithmic scale of the vertical axis.

energy function is given by Eq. (4.291). We used both ways to cross-check the calculations.

In the graphs of Figure 4.27 the parameter domain corresponding to that investigated in Refs. 21 and 22 appears as a narrow band ($\rho \ll 1$) abutting on the ξ_0 axis. Our results, obtained by a nonperturbative method, provide a much wider overview and reveal at least three novel features of the response spectrum: (1) higher harmonic suppression exists in the range $\rho \gtrsim 1$ as well; (2) when a respective kth groove branches off the ξ_0 axis, the harmonic suppression line grows fast with ξ_0 only at the beginning—further on it bends and strives to turn parallel to this axis; and (3) the number of zero-level lines increases with the number of the harmonic. Namely, m_3 and m_4 have one zero-line per each, m_5 and m_6 have two lines, and so on, so that the pair m_{2k-1}, m_{2k} has as many as $k - 1$ zero-level lines per each. (We note that occurrence of two NIR points for the fifth harmonic was mentioned in Ref. 21, because it has emerged from the numeric simulations, but there it was left without discussion.)

In Figure 4.28 the projections $\rho_{kj}(\xi_0)$ of the zero-level lines onto the (ξ_0, ρ) plane are shown. Here we reflect two cases with respect to anisotropy. Solid

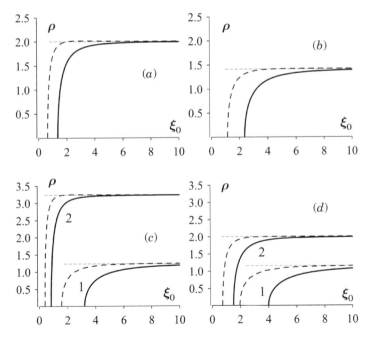

Figure 4.28. Zero-level lines for the magnetization amplitudes m_k at $k = 3$ (a), 4 (b), 5 (c), and 6 (d) for the cases $\varepsilon = \infty$ (solid lines; cf. Fig. 4.27) and $\varepsilon = 0$ (thick dashed lines); thin dashed straight lines show the asymptotic values rendered by Eq. (4.297). Numbers near the pairs of curves (one solid and one dashed) in (c) and (d) refer to the allowed indices j at a given k.

lines correspond to magnetoisotropic particles ($\varepsilon = \infty$) so that they visualize the bottoms of the grooves in Figure 4.27. Thick dashed lines describe the case of magnetically hard particles ($\varepsilon = 0$). The NIR effect—a resonance-like suppression of harmonics at certain values of the noise strength [18,21,22]—occurs when any of those lines is crossed while traversing the diagram in the direction parallel to the ξ_0 axis. Indeed, in Figure 4.28, any trajectory described by the equation $\rho = \text{const}$ means variation of the particle temperature (noise intensity). We remark that the greater ξ_0, the smaller is the angle between this trajectory and the suppression lines $\rho_{kj}(\xi_0)$. So, in the nonlinear region NIR "stabilizes" with respect to temperature and saturates with respect to the excitation strength ρ. Let us prove that. In the adiabatic limit ($\omega\tau \to 0$) the dependencies $\rho_{kj}(\xi_0, \varepsilon)$ follow from the condition that the kth Fourier coefficient of the expansion of the quasiequilibrium magnetization $m[\xi(t)]$ with respect to $\cos k\omega t$ turns to zero. Taking averages with the function $W \propto \exp[-U(t)/T]$, for the limiting cases of anisotropy one finds

$$m[\xi(t)] = \begin{cases} \cotan h[\xi_0(1 + \rho\cos\omega t)] - [\xi_0(1 + \rho\cos\omega t)]^{-1} & \text{for } \varepsilon \to \infty \\ \tanh[\xi_0(1 + \rho\cos\omega t)] & \text{for } \varepsilon = 0 \end{cases}$$

(4.296)

where the first line representats the Langevin function and the second equals the Brillouin one for the spin $\frac{1}{2}$ clearly exposing the case $\varepsilon = 0$ as a two-level system. At $\xi_0 \to \infty$ both expressions of Eq. (4.296) reduce to a step function. Asymptotic evaluation then yields

$$\rho_{kj}(\infty) = \sec\frac{j\pi}{k}, \quad 1 \leq j < \frac{k}{2}$$

(4.297)

where $k \geq 3$ and index j enumerates the zero-level branches at a given k. Therefore, according to Eq. (4.297), for the kth harmonic NIR ceases to exist as soon as ρ exceeds the maximum value

$$\rho_k^{\max} = \begin{cases} \text{cosec}(\pi/2k) & \text{for } k \text{ even} \\ \text{cosec}(\pi/k) & \text{for } k \text{ odd} \end{cases}$$

(4.298)

The asymptotes (4.297) are shown in Figure 4.28 by straight thin dashed lines.

Let us evaluate the positions where the lines $\rho_{kj}(\xi_0)$ branch off the ξ_0 axis, that is, where NIR takes place at an infinitesimal excitation. In Ref. 22 it was shown that this regime can be described in terms of cumulants Q_k of the equilibrium distribution function. Carrying out this adiabatic treatment, one finds that the branchoff points in the kth order are the zeros of

$$Q_{k+1}(\xi_0, \varepsilon) = \frac{\partial^k m}{\partial \xi_0^k} = \frac{2^{k-1} T^k}{I_s \phi \mu^k} \chi_k^{(a)}(\xi_0, \varepsilon)$$

(4.299)

TABLE II
Branchoff Points ξ_{0kj} for the Zero Lines of the Magnetization
Harmonics m_k

k	$j = 1$		$j = 2$	
	$\varepsilon = \infty$	$\varepsilon = 0$	$\varepsilon = \infty$	$\varepsilon = 0$
3	1.3722	0.6585	—	—
4	2.3607	1.1462	—	—
5	3.2119	1.5722	0.8499	0.4213
6	3.9937	1.9680	1.5208	0.7567

Here the last equality relates the cumulant to the adiabatic magnetic susceptibility $\chi_k^{(a)}$ at the frequency $k\omega$. Thus, the desired equation reduces to $\chi_k^{(a)}(\xi_0, \varepsilon) = 0$. In a real system the adiabatic susceptibility $\chi_k^{(a)}$ is a directly measurable quantity. Therefore, comparing the measured dimensional bias fields in the branching points with the theoretical numbers ξ_{0kj} one can, for example, evaluate the particle volume. Table II lists the calculated values of the branchoff points for the limiting cases $\varepsilon \to \infty$ and $\varepsilon = 0$, that is, for both families of lines shown in Figure 4.28. For intermediate values of ε, one finds that for every k and j the contours $\rho_{kj}(\xi_0, \varepsilon)$ fill in uniformly the area between the respective limiting curves.

The discovered overall behavior of the zero-level lines $\rho_{kj}(\xi_0)$ readily suggests that there is another, non-NIR, way to selective suppression of the higher harmonics. Let us take the AC field intensity ρ as the governing parameter. In terms of Figure 4.28, this means that we consider the trajectories, which are parallel to the ρ axis. Clearly, any straight line $\xi_0 = $ const that passes to the right of the first branch point inevitably crosses the zero-level curve at least once. Therefore, the harmonic suppression may equally be caused by varying not the temperature (noise) but the exciting field strength (force). To match the terms, we term this effect the *force-induced resonance* (FIR). From the evidence presented in Figures 27 and 28, it is apparent that NIR and FIR do not exist separately but are particular manifestations of a unified harmonic-suppression effect.

3. Fluid Systems: Generation of Harmonics of Orientation

Unlike solid systems, in magnetic fluids, due to the mechanical degrees of freedom of the particles, the response to an AC field is twofold. Provided the particles are anisometric (e.g., uniaxial), an external field induces not only the magnetic moment alignment (magnetization) but also the orientational ordering of the particle geometry axes. Given that, the magnetic fluid becomes a medium with an oscillating optical anisotropy, and the harmonic content of the transmitted light can be analyzed.

Let us introduce a unit vector $\mathbf{v}$ of the particle axis. In general, the order parameter is given by the second-rank tensor $p_{ik} \propto \frac{3}{2}\langle v_i v_k \rangle - \frac{1}{2}\delta_{ik}$, but in the uniaxial case it reduces to a scalar parameter $p = \langle P_2 \rangle$. Note that in a magnetic suspension p is a quantity independent of m; in a standard polarization setup the measured intensity of light is proportional to p^2.

For simplicity we assume that the particles are magnetically hard.[1] Then, the already developed formalism [Eqs. (4.90) and (4.293)] applies in full under two conditions: e is identified with $\mathbf{v}$ and the "internal" relaxation time τ_D is replaced by the external rotational diffusion (Debye) time τ_B. In the modified equations, the orientation order parameter is given by $\langle P_2 \rangle$. Similarly to (4.294), we set

$$p(t) = \sum_{k \geq 0} p_k(\rho, \xi_0, \omega\tau_D) \exp[i(k\omega t - \beta_k)] \qquad (4.300)$$

In Figure 4.29 the NIR/FIR maps for the order parameter are shown at $\omega\tau_B = 0$ for $k = 2 \cdots 6$. Again, one can see that the effect is present only for non-zero-bias fields. Qualitatively, the configurations of the zero lines look the same as in Figure 4.27. However, the correspondence rules between the order of a harmonic and the number of zero lines differ from those of Figure 4.28. Namely, p_2 and p_3 have one zero line each, p_4 and p_5 have two lines, and so forth. The asymptotic (low-temperature) analysis yields the zero lines of orientation (shown in Figure 4.30 by dashed lines) in the form

$$\rho_{kj}(\infty) = \begin{cases} \operatorname{cosec}[\pi(j/k)] & (1 \leq j < k/2) \quad \text{odd } k \\ \operatorname{cosec}[\pi(2j-1)/2k] & (1 \leq j \leq k/2) \quad \text{even } k \end{cases} \qquad (4.301)$$

but the convergence to the limiting shapes is much weaker as illustrated by the upper-right graph of Figure 4.30.

4. Summary

Resonance-like suppression of the higher harmonics in the response spectrum of a superparamagnetic particle is investigated. Using a nonperturbative approach, we analyze the steady processes that take place under arbitrary values of the DC (bias) and AC (excitation) strengths. The results show that the suppression effect is equally achieved on varying either noise (temperature) or force (external field intensity) or by a combination of both. In the fundamental aspect, we surmise

---

[1]The case of finite magnetic rigidity of the particles is addressed in Section V; the theory becomes much more complicated and, hence, cumbersome in form. Meanwhile, the case of magnetically hard particles suspended in a fluid is formally very close to the case of isotropic magnetic particles in a solid matrix. That is why we present it here.

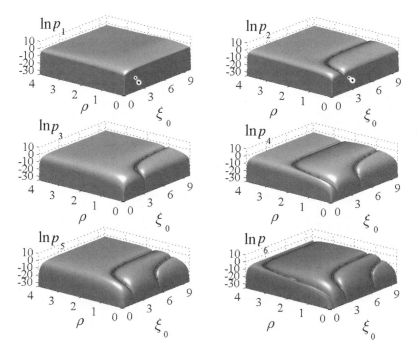

Figure 4.29. Response amplitudes of second to sixth harmonics of the magnetic fluid orientation under the rigid magnetic moment assumption ($\varepsilon = 0$). Note the logarithmic scale of the vertical axis.

that the found coexistence of the noise- and force-induced resonances is a generic feature of noisy nonlinear oscillators.

The suppression effect is most pronounced in the adiabatic (low-frequency limit). A typical zero-level curve $\rho(\xi_0)$ (see Fig. 4.28) may be, although roughly but reasonably, divided into three characteristic parts: the steep ascend with the noise strength (NIR branch), the bend (NIR–FIR crossover), and the noise-independent saturation (FIR branch). To evaluate the parameters of the suppression resonance, namely, the positions of the branchoff points ξ_{0kj} and the saturation values $\rho_{kj}(\infty)$ of the zero-level curves for particular harmonics, we have obtained simple but rather accurate approximate expressions.

In our model with the aid of parameter ε we continuously pass from a zero bistable potential (magnetically isotropic particle) to a pair of symmetric wells of infinite depth (highly anisotropic particle). For the magnetic case, as for those of Refs. 21 and 22, a crucial circumstance enabling the harmonic suppression is that an antisymmetric contribution (bias) should be present in the potential. On the other hand, the presence of a symmetric contribution turns out to be an

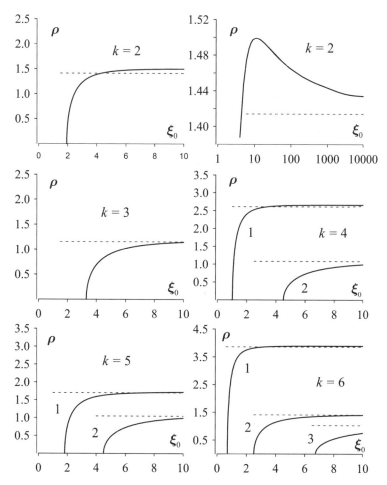

Figure 4.30. Solid: zero-level lines for the orientation amplitudes p_k in the adiabatic limit for k from 2 to 6; dashed: asymptotic values from Eq. (4.301). In multibranch graphs the numbers indicate index j; the upper right graph shows the case $k = 2$ in the extended ξ_0 scale.

optional feature, and, although producing some shifting of the reference points, has no qualitative effect on the harmonic suppression, however small or large the anisotropy parameter ε is.

Let us estimate the material parameter range that one needs to attain in order to observe the phenomena considered above. According to Figures 4.27 and 4.28, and Eqs. (4.297) and (4.298), the characteristic values for both NIR and FIR are ξ_0, $\rho \sim 1$. As a particular reference object we take a dispersion of gamma ferric oxide particles with mean size $a \sim 10\,\text{nm}$ and magnetization

$I_s = 400$ G. According to its definition, the parameter ξ_0 at room temperature becomes of the order of unity for $H_0 \sim 6T/\pi I_s a^3 \sim 200$ Oe. Recalling that $\rho = H_p/H_0$, one finds that of the same order of magnitude is the reference AC field amplitude. The frequency constraint $\omega\tau \ll 1$ for a solid dispersion of the abovementioned nanoparticles gives $\omega \ll 105 \text{ s}^{-1}$. Obviously, in the given frequency range the magnetic fields of estimated amplitudes (and several times greater) are feasible without serious problems.

In conclusion, a resonance-like selective suppression of both magnetization and orientation in magnetic fluids is predicted. The main features of this unusual field-induced behavior are as follows: (1) for the conventional magnetic fluids the effects discussed lie in a perfectly achievable magnetic field range and (2) considering the results presented with allowance for the polydispersity of magnetic fluids, one concludes that the use of the NIR/FIR effect offers an interesting way to get beyond the common practice of biparametric model size distributions, whose characteristics are extracted from the behavior of the static magnetization curve. For example, measuring simultaneously several harmonics of different orders and expressions, each of which is a function of the particle size histogram $f(V_m)$, makes it possible to determine this histogram as a function controlled by more than only two parameters.

V. NONLINEAR SUSCEPTIBILITIES OF MAGNETIC SUSPENSIONS: INTERPLAY OF MECHANICAL AND MAGNETIC DEGREES OF FREEDOM

A. The "Egg" Model of a Dipolar Particle in a Fluid

In this section we study the arbitrary relationship between the external and internal orientational relaxation reference times.

1. Torque Balance Equations

In this section we proceed to a study of dynamic magnetization in ferrofluids, namely, suspensions of single-domain particles in liquid matrices. Therefore, the particles "are given" an additional, mechanical set of the degrees of freedom. With the thus extended configurational space, the magnetic moment is involved simultaneously in two relaxation processes. One is related to the rotation of the particles (together with their magnetic moments) in a liquid with viscosity coefficient η. This is characterized by the rotational diffusion relaxation time (Brownian relaxation time)

$$\tau_B = \frac{3\eta V}{T} \qquad (4.302)$$

which we have already mentioned [see Eq. (4.29) and surrounding text]. Another relaxation process is the motion of the magnetic moment $\boldsymbol{\mu}$ inside the particle. If the liquid loses its fluidity as a result of freezing or polymerization, we have $\eta = \tau_B = \infty$, and only the internal motion of the vector $\boldsymbol{\mu}$ remains. This type of magnetodynamics in the presence of thermal noise is the subject of studies in Sections III and IV. They all are based on the energy expression (4.73), with the equilibrium distribution (4.77) and the kinetic Fokker–Planck equation (4.90). As shown, in a solid matrix, the reference time of the orientational diffusion of the vector $\boldsymbol{\mu}$ is $\tau_D = I_s V_m / 2\alpha\gamma T$, which may be presented as

$$\tau_D = \frac{3\eta_m V_m}{T}, \qquad \eta_m = \frac{I_s}{6\alpha\gamma} \tag{4.303}$$

Therefore, for the internal (Néel) relaxation the parameter, η_m plays the same role as the fluid viscosity η in the mechanism of the external (Brownian) diffusion. Note that the density of the anisotropy energy K is not included in τ_D. This means that τ_D can be considered as the internal relaxation time of the magnetic moment only for magnetically isotropic particles (where $K = \sigma = 0$). The sum of the rotations—thus allowing for both the diffusion of the magnetic moment with respect to the particle and for the diffusion of the particle body relative to the liquid matrix—determines the angle ϑ of spontaneous rotation of the vector $\boldsymbol{\mu}$ at the time moment t:

$$\langle \vartheta^2 \rangle = 2t(\tau_B^{-1} + \tau_D^{-1}) \tag{4.304}$$

For a typical magnetic fluid, estimation yields $\eta_m / \eta \leq 10^{-2}$, namely, $\tau_B \gg \tau_D$. Thus at the dimensionless anisotropy barrier height parameter $\sigma \ll 1$, the external diffusion is ineffective; the particle is "frozen" into the liquid, moving (or remaining stationary) in unison with it. In this limit the relaxation of the magnetic moment is *par excellence* due to its internal diffusion, see Eq. (5.3). The situation changes at larger values of σ. As explained in detail in Section (II), the internal relaxation regime changes drastically; the frequency of the thermally excited spontaneous flipping of the magnetic moment over the potential barrier with height KV_m separating the states with $\boldsymbol{e} = \boldsymbol{n}$ and $\boldsymbol{e} = -\boldsymbol{n}$ becomes proportional to $\exp(KV/T)$. In result, the time τ_D "is dressed" with a function depending exponentially on σ, transforming into $\tau(\sigma)$ [see Eq. (4.50)]. The latter asymptotic formula is fulfilled accurately whenever $\sigma > 2$. Rapid growth of τ with increase of the particle volume ($\sigma \propto V_m$) ensures that at $\sigma \geq 10$ one gets $\tau \gg \tau_B$. From expression (4.304), where τ_D must be replaced by τ, one can see that for large σ values the situation changes for the reciprocal one: the magnetic moment is "frozen" into the particle body, and only the Brownian mechanism of the rotary diffusion is operative. The pertinent details of this case were discussed in Ref. 130. There the FPE was obtained for the orientational distribution function $W(\boldsymbol{n}, t)$ of a magnetic fluid modeled by an assembly of noninteracting rigid $(\boldsymbol{e} = \boldsymbol{n})$ magnetic dipoles, and the equation of motion of the macroscopic magnetization was derived. Another limiting case has been considered in Ref. 54. The matrix was assumed to be solid (i.e., the particles are deprived of mechanical mobility, $\tau_B = \infty$) so that only the Néel relaxation works. The FPE for $W(\boldsymbol{e}, t)$ was obtained and the influence of the particle sizes on the shape and width of the magnetic resonance line in the solid colloidal solution was studied. In actual polydisperse magnetic fluids there are always particles with $\tau_B > \tau(\sigma)$ and with $\tau_B < \tau(\sigma)$. For this reason, the contributions of both diffusion processes to the magnetic relaxation are comparable. In order to facilitate study of this general case of the joint (internal and external) rotary diffusion, the FPE for the distribution function $W(\boldsymbol{e}, \boldsymbol{n}, t)$ of orientations of magnetic moments of particles and easy magnetization axes is derived below. Starting the derivation of the general Fokker–Planck equation, let us initially make two remarks in connection with the Landau–Lifshitz–Gilbert equation, which underlies all the kinetic considerations. To be specific, we take it in the form (4.3).

Remark 1. In the notations adopted above and with the energy function (4.73), this equation may be written as

$$\frac{d\boldsymbol{e}}{dt} = (\boldsymbol{\omega}_e \times \boldsymbol{e}), \qquad \boldsymbol{\omega}_e = \gamma \boldsymbol{H}_{\text{eff}} - \frac{\alpha \gamma}{\mu} \hat{\boldsymbol{J}}_e U \qquad (4.305)$$

where $\boldsymbol{\omega}_e$ is the angular velocity of the rotation of vector $\boldsymbol{e}$, and $\hat{\boldsymbol{J}}_e = (\boldsymbol{e} \times d/d\boldsymbol{e})$ is the infinitesimal rotation operator with respect to $\boldsymbol{e}$. As one can see from (4.305), $\boldsymbol{\omega}_e$ is made up of two parts: the velocity of the free precession of the magnetic moment around the direction of the effective field with Larmor frequency $\boldsymbol{\omega}_L = \gamma \boldsymbol{H}_{\text{eff}}$, and the relaxational (i.e., proportional to the damping constant α) component $\boldsymbol{\omega}_R$. The latter ensures cessation of the precession in the DC external field $\boldsymbol{H}$ after the time interval $(\alpha \omega_L)^{-1}$. Thence, the equilibrium orientation $\boldsymbol{e} \parallel \boldsymbol{H}$ of the magnetic moment is established. We note that the second of the expressions (4.305) may be equivalently presented in the form

$$6\eta_m V_m \boldsymbol{\omega}_R + \hat{\boldsymbol{J}}_e U = 0 \qquad (4.306)$$

Remark 2. For a mechanically unmovable particle, the Landau–Lifshitz equation (4.3) is written in the coordinate framework $Ox'y'z'$, which is quiescent with respect to the crystal axes of the sample. As soon as one considers a ferroparticle suspended in a liquid matrix, then the rotational degrees of freedom of $Ox'y'z'$ with respect to the laboratory reference system $Oxyz$ must be allowed for. Then, instead of using (4.305) and (4.306), the following equations must be used [131,132]

$$\frac{d}{dt}e = (\omega_e \times e), \quad \omega_e = \omega_L + \omega_R \tag{4.307}$$

$$6\eta_m V_m(\omega_R - \omega_n) + \hat{J}_e U = 0 \tag{4.308}$$

where ω_n is the angular velocity of rotation of the particle. Equation (4.308) may be considered as the balance equation between the friction torque (first term) and the magnetic torque (second term) exerted on a spherical permanent magnet with the volume V_m. This magnet rotates with the angular velocity ω_R in the model liquid, which has the viscosity η_m. Meanwhile, this "liquid" itself rotates with the angular velocity ω_n equal to that of the real particle. Let us now write (in the noninertial approximation) the equation of motion of the colloid particle. Outside, that is, in the liquid with viscosity η, the particle experiences the friction torque $6\eta V(\omega_n - \Omega)$, where $\Omega = \frac{1}{2}\text{curl } v$, namely, the local angular velocity rotation of the liquid. Inside, by Newton's third law, the particle experiences the braking torque $6\eta_m V_m(\omega_n - \Omega_R)$, which is equal in magnitude and opposite in sign to the torque acting on the "confined" particle magnetic moment [see Eq. (4.308)]. In result, the equation of motion of the particle is expressed as

$$6\eta V(\omega_n - \Omega) + 6\eta_m V_m(\omega_n - \omega_R) + \hat{J}_n U = 0 \tag{4.309}$$

where $\hat{J}_n = (n \times \partial/\partial n)$ is the infinitesimal rotation operator of the particle and $-\hat{J}_n U$ is the magnetic torque acting on the particle.

2. Kinetic Equation for Joint Rotary Diffusion

A simple heuristic model that helps one understand the physical mechanism underlying the set of balance equations (4.308) and (4.309) emerges if one likens a superparamagnetic particle to an egg [131,132]. In this context an "egg" of volume V is considered as floating inside a liquid with the viscosity η (Fig. 4.31). The magnetic moment of the particle is associated with the "yolk" of volume $V_m < V$. This magnetically rigid yolk (one may say, a permanent magnet) is surrounded by the "white" of effective viscosity η_m, thus modeling the intrinsic (magnetic) viscosity. Thence, the torque $-\hat{J}_e U$ created by the applied field [see Eq. (4.108)] acts directly on the yolk. However, owing to the presence of a viscous white, it is also transmitted to the "eggshell," which is the particle body itself. This transmitted torque is expressed in Eq. (4.109) by the second term.

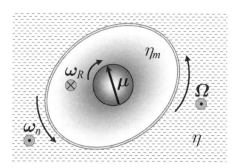

Figure 4.31. Schematic representation of a superparamagnetic particle in the "egg" model.

Eliminating it from (4.108) and (4.109), one arrives at the balance condition

$$6\eta V(\boldsymbol{\omega}_n - \boldsymbol{\Omega}) + (\hat{\boldsymbol{J}}_e + \hat{\boldsymbol{J}}_n)U = 0 \tag{4.310}$$

A specific feature of the operator $(\hat{\boldsymbol{J}}_e + \hat{\boldsymbol{J}}_n)$ is that when it acts at an arbitrary function of the product $(\boldsymbol{e} \cdot \boldsymbol{n})$, the result is zero. Therefore, in the absence of an external field, when $U = -KV_m(\boldsymbol{e} \cdot \boldsymbol{n})^2$, one finds from (4.310) that $\boldsymbol{\omega}_n = \boldsymbol{\Omega}$, that is, the particle rotates in unison with the surrounding liquid. In this case neither the torque acting on the yolk $(-\hat{\boldsymbol{J}}_e U)$ nor the one exerted on the eggshell $(-\hat{\boldsymbol{J}}_n U)$ equals zero. However, their sum vanishes identically. This is quite natural since the equilibrium of a system as a whole cannot be disturbed by the internal forces (or torques).

Let us proceed to the derivation of the pertinent Fokker–Planck equation. The probability density $W(\boldsymbol{e}, \boldsymbol{n}, t)$ of various orientations of the magnetic moment and the easy magnetization axis of a ferroparticle must satisfy the conservation law

$$\frac{\partial W}{\partial t} + \hat{\boldsymbol{J}}_e(\boldsymbol{\omega}_e W) + \hat{\boldsymbol{J}}_n(\boldsymbol{\omega}_n W) = 0 \tag{4.311}$$

Here $\boldsymbol{\omega}_e$ and $\boldsymbol{\omega}_n$ are delivered by the corresponding Langevin equations of the theory of the rotational Brownian motion. In order to obtain these equations, one must include in the dynamic equations (4.308) and (4.310) the random thermal torques. We do that in the following way:

$$6\eta_m V_m(\boldsymbol{\omega}_R - \boldsymbol{\omega}_n) + \hat{\boldsymbol{J}}_e(U + T \ln W) = 0$$
$$6\eta V(\boldsymbol{\omega}_n - \boldsymbol{\Omega}) + (\hat{\boldsymbol{J}}_e + \hat{\boldsymbol{J}}_n)(U + T \ln W) = 0 \tag{4.312}$$

Therefore, two types of random torque are accounted for:

1. The torque $-T\hat{J}_n \ln W$, which is exerted on the particle body on the part of the surrounding liquid and causes the rotational Brownian motion of the particle.

2. The torque $-T\hat{J}_e \ln W$, which acts on the magnetic moment and is the one responsible for the stochastic reorientation of the vector $\boldsymbol{\mu} = \mu\boldsymbol{e}$ inside the particle. In thermodynamic equilibrium, when $\boldsymbol{\omega}_R = \boldsymbol{\omega}_n = \boldsymbol{\Omega}$, that is, the yolk, the eggshell, and the liquid surrounding the egg rotate as a whole, and Eq. (4.312) reduces to the Gibbs distribution $W \propto \exp(-U/T)$. We interpret this fact as a heuristic justification for the above-used representation of the stochastic torques. On substituting $\boldsymbol{\omega}_e = \boldsymbol{\omega}_R + \boldsymbol{\omega}_L$ and $\boldsymbol{\omega}_n$ from (4.312) into (4.311), one gets the Fokker–Planck equation

$$\frac{\partial}{\partial t}W + (\hat{J}_e + \hat{J}_n)\Omega W + \hat{J}_e\omega_L W$$

$$= \left\{\frac{1}{2\tau_B}(\hat{J}_e + \hat{J}_n)W(\hat{J}_e + \hat{J}_n) + \frac{1}{2\tau_D}\hat{J}_e W\hat{J}_e\right\}\left(\frac{U}{T} + \ln W\right) \qquad (4.313)$$

In the absence of an external magnetic field, the solution of Eq. (4.313) factorizes as

$$W(\boldsymbol{e},\boldsymbol{n},t) = W_e(\boldsymbol{e}\cdot\boldsymbol{n},t)W_n(\boldsymbol{n},t) \qquad (4.314)$$

where W_e is the orientational distribution of vector $\boldsymbol{e}$ with respect to the anisotropy axis $\boldsymbol{n}$ and W_n is the orientational distribution of vector $\boldsymbol{n}$ with respect to the laboratory coordinate framework. For representation (4.314) the equations that govern the distribution functions do separate from one another and become self-closed. The equation for the internal degrees of freedom is

$$2\tau_D\frac{\partial W_e}{\partial t} = \hat{J}_e W_e \hat{J}_e \left[-\sigma(\boldsymbol{e}\cdot\boldsymbol{n})^2 + \ln W_e\right] \qquad (4.315)$$

where the parameter $\sigma = E_a/T$ is the height of the potential barrier of magnetic anisotropy scaled with temperature; note that this definition of σ is consistent with Eqs. (4.34) and (4.75). For the distribution function of the external rotations, one obtains

$$2\tau_B\frac{\partial W_n}{\partial t} = \hat{J}_n^2 W_n \qquad (4.316)$$

which is a standard rotary diffusion equation. Both Eqs. (4.315) and (4.316) are well known in the theory of the rotary Brownian motion. The eigenfunctions of Eq. (4.316) are the "external" spherical harmonics, specifically, the functions of the angles that vector n makes with the axis of the laboratory coordinate framework that is defined by the unit vector z:

$$W_n(\vartheta, \phi, t) = \sum X_l^m(n, z) \exp\left[\frac{-l(l+1)t}{\tau_B}\right] \qquad (4.317)$$

For the solution of Eq. (4.315) the basis of "internal" spherical harmonics

$$W_e(\theta, \varphi, t) = \sum_{l=0}^{\infty} \sum_{m=-l}^{l} b_{l,m}(t) \frac{2l+1}{4\pi} \frac{(l-|m|)!}{(l+|m|)!} X_l^{m*}(e, n) \qquad (4.318)$$

was proposed in Ref. 47. In Eqs. (4.317) and (4.318) symbols $X_l^m(a, b)$ denote the nonnormalized spherical harmonics defined in Section II.B by Eq. (4.57). Those functions are connected to the conventional, normalized, spherical harmonics by Eq. (4.58).

As soon as a magnetic field is switched on, it couples the internal and external degrees of freedom of the particle, and the variables in the kinetic equation (4.313) become inseparable. Therefore, the solution should be constructed in the functional space formed by direct products of the "internal" and "external" harmonics. A suitable representation for this case is

$$W(e, n, t) = \sum_{l=1}^{\infty} \sum_{l'=1}^{\infty} \sum_{m=-l'}^{l'} Q_{l,l'}^{m,m'}(t) \frac{(2l+1)(l-|m|)!}{4\pi(l+|m|)!}$$
$$\times X_l^{m*}(e, n) \frac{(2l'+1)(l'-|m'|)!}{4\pi(l'+|m'|)!} X_{l'}^{m'*}(n, h) \qquad (4.319)$$

where the time dependence is now determined by the four-index coefficients

$$Q_{l,l'}^{m,m'}(t) = \langle X_l^m(e, n) X_{l'}^{m'}(n, h)\rangle \qquad (4.320)$$

Note that the vector h, which is defined in the laboratory framework, in a natural way replaces the formerly arbitrary unit vector z in the subset of the "external" harmonics.

The general matrix equation of the problem that determines the amplitudes $Q_{ll'}^{mm'}$ is obtained by substitution of the spherical harmonic expansion (4.319) into the Fokker–Planck equation (4.313). After that the result is multiplied from the left by $X_l^m(e, n)X_{l'}^{m'}(n, h)$ and, finally, integrated with respect to both e and n. The emerging equation is

$$
\frac{d}{dt}Q_{l,l'}^{m,m'} + \frac{1}{2\sigma\tau_0}l(l+1)Q_{l,l'}^{m,m'} + \frac{1}{2\tau_B}l'(l'+1)Q_{l,l'}^{m,m'}
$$

$$
- \frac{1}{\tau_0}\left[\frac{(l+1)(l+m-1)(l+m)}{(2l-1)(2l+1)}Q_{l-2,l'}^{m,m'} + \frac{l(l+1)-3m^2}{(2l-1)(2l+3)}Q_{l,l'}^{m,m'}\right.
$$

$$
\left. - \frac{l(l-m+2)(l-m+1)}{(2l+1)(2l+3)}Q_{l+2,l'}^{m,m'}\right]
$$

$$
- \frac{\xi}{2\sigma\tau_0(2l+1)(2l'+1)}\left\{(l+1)(l+m)[(l'-m'+1)Q_{l-1,l'+1}^{m,m'}\right.
$$

$$
+ (l'+m')Q_{l-1,l'-1}^{m,m'}]
$$

$$
- l(l-m+1)[(l'-m'+1)Q_{l+1,l'+1}^{m,m'} + (l'+m')Q_{l+1,l'-1}^{m,m'}]
$$

$$
+ \frac{1}{2}(l+1)(l+m-1)(l+m)[(l'+m'-1)(l'+m')Q_{l-1,l'-1}^{m-1,m'-1}
$$

$$
- (l'-m'+1)(l'-m'+2)Q_{l-1,l'+1}^{m-1,m'-1}]
$$

$$
+ \frac{1}{2}l(l-m+2)(l-m+1)[(l'+m'-1)(l'+m')Q_{l+1,l'-1}^{m-1,m'-1}
$$

$$
- (l'-m'+1)(l'-m'+2)Q_{l+1,l'+1}^{m-1,m'-1}]
$$

$$
- \frac{1}{2}(l+1)(Q_{l-1,l'+1}^{m+1,m'+1} - Q_{l-1,l'-1}^{m+1,m'+1}) - \frac{1}{2}l(Q_{l+1,l'+1}^{m+1,m'+1} - Q_{l+1,l'-1}^{m+1,m'+1})\right\}
$$

$$
- \frac{\xi}{2\tau_B(2l+1)(2l'+1)}\left\{(l'+1)(l'+m')[(l-m+1)Q_{l+1,l'-1}^{m,m'}\right.
$$

$$
+ (l+m)Q_{l-1,l'-1}^{m,m'}]
$$

$$
- l'(l'-m'+1)[(l-m+1)Q_{l+1,l'+1}^{m,m'} + (l+m)Q_{l-1,l'+1}^{m,m'}]
$$

$$
+ \frac{1}{2}(l'+1)(l'+m'-1)(l'+m')[(l+m-1)(l+m)Q_{l-1,l'-1}^{m-1,m'-1}
$$

$$
- (l-m+1)(l-m+2)Q_{l+1,l'-1}^{m-1,m'-1}]
$$

$$
+ \frac{1}{2}l'(l'-m'+2)(l'-m'+1)[(l+m-1)(l+m)Q_{l-1,l'+1}^{m-1,m'-1}
$$

$$
- (l-m+1)(l-m+2)Q_{l+1,l'+1}^{m-1,m'-1}]
$$

$$
- \frac{1}{2}(l'+1)(Q_{l+1,l'-1}^{m+1,m'+1} - Q_{l-1,l'-1}^{m+1,m'+1}) - \frac{1}{2}l'(Q_{l+1,l'+1}^{m+1,m'+1} - Q_{l-1,l'+1}^{m+1,m'+1})\right\} = 0
$$

$$
(4.321)
$$

B. Linear and Cubic Dynamic Susceptibilities of a Dipolar Suspension: Effect of Mechanical Orientation

Linear dynamic susceptibility is one of the most relevant characteristics of any system providing fundamental information on it. For example, granulometry analysis based on the dynamic magnetic susceptibility curve is a customary tool used to investigate the particle content of magnetic suspensions (ferrofluids). With allowance for immobilization of the matrix (arrested mechanical degrees of freedom of the particles), it applies to solidified ferrofluids or, in general, to any solid superparamagnetic particle assemblies. However, extracting all the data from a single integral curve, this method preserves a good deal of uncertainty. As first mentioned in Section III.A, a plausible way (well known in the spin glass science) to make the information more specific implies that the linear and as many as possible nonlinear dynamic susceptibilities are measured and then fitted simultaneously. In magnetic nanoparticle dispersions for accomplishing such a procedure, a consistent nonlinear extension of the Néel–Brown super-paramagnetic theory [13,47] is needed. In other works [67,78,124,133,134] we have developed some appropriate models for superparamagnetic systems in solid matrices; this approach is described in Section III. Here we outline some relevant features of the nonlinear superparamagnetic theory for the case when magnetic nanoparticles are suspended in a liquid matrix.

Consider an isolated nanoparticle embedded in a nonmagnetic matrix of any kind. The particle is single-domain so that its magnetization I_m is uniform over its volume V_m. As in the previous sections, we describe the particle magnetic moment μ with the aid of the unit vector $e = \mu/\mu$ assuming that $\mu = I_s V_m$. Implicitly, this means that the system rests at the temperature well below the Curie point. The particle magnetic anisotropy is uniaxial with the energy density K and the easy axis n. Then the orientation-dependent part of the particle energy is given by Eq. (4.73), while the equilibrium distribution function depending on e and n is written as

$$W_0 = Z^{-1} \exp[\xi(e \cdot h) + \sigma(e \cdot n)^2]$$

$$Z = \int\int \exp[\xi(e \cdot h) + \sigma(e \cdot n)^2] de\, dn \tag{4.322}$$

where the dimensionless parameters σ and ξ are as defined by Eqs. (4.75), and h is the unit vector of the external magnetic field. Note that although we use the same function U as that defined by Eq. (4.73), the equilibrium distribution function W_0 is different from that given by Eq. (4.74). Indeed, for a particle in a solid matrix the configuration space is built on rotations of a single vector e, whereas in a liquid matrix the configuration space is the $e \otimes n$ manifold. In the partition function Z of Eq. (4.322) the presence of mechanical degrees of freedom is indicated by integration over n. For a trapped particle (solid matrix) in Z only integration over e is retained while n remains a parameter. Averaging over the latter, when necessary, is applied to get the free energy $F \propto -\ln Z$ of a solid assembly, such as a random system, where the distribution function is $f(n) = 1/4\pi$.

In the framework of statistical thermodynamics, the initial (equilibrium) susceptibilities are obtained as the even derivatives (2nd, 4th, ...) of the free energy F with respect to the external field at $H \to 0$. Expanding F in the power series with respect to ξ and using the addition theorem for spherical harmonics

$$P_l(e \cdot h) = \frac{4\pi}{2l+1} \sum_{m=-l}^{m=l} Y_l^{-m}(n,h) Y_l^m(e,n) \tag{4.323}$$

where P_l is a Legendre polynomial, one obtains the following [67,68]:

$$\langle e \cdot n \rangle = \frac{1 + 2S_2 \langle P_2(n \cdot h) \rangle}{3} \xi$$
$$- \frac{7 + 70S_2^2 \langle P_2(n \cdot h) \rangle^2 + 40S_2 \langle P_2(n \cdot h) \rangle - 12S_4 \langle P_4(e \cdot n) \rangle}{315} \xi^3 \tag{4.324}$$

In Eq. (4.323) notation of the type $Y(a,b)$ means that a spherical harmonic is built on the components of a unit vector a in the coordinate system whose polar axis points along the unit vector b. The functions S_2 and S_4 in Eq. (4.324) are the equilibrium parameters of the magnetic order of the particle defined, in general, by Eqs. (4.80)–(4.83).

From Eq. (4.324), taking into account that for a solid random system the equilibrium statistical moments of the particle anisotropy axes are

$$\langle P_2(n \cdot h) \rangle_0 = \langle P_4(n \cdot h) \rangle_0 = 0, \qquad \langle P_2^2(n \cdot h) \rangle_0 = \frac{1}{5} \tag{4.325}$$

one gets for the static susceptibility of such an assembly

$$\chi_{solid}^{(3)} = -\frac{c\mu^4}{45T^3}(1 + 2S_2^2) \tag{4.326}$$

where c is the particle number concentration. For a magnetic fluid, as mentioned, averaging with respect to the particle easy axes is done simultaneously with that over magnetic moments. Thus the result is determined by the third-order term of expansion of the Langevin function:

$$\chi_{fluid}^{(3)} = -\frac{c\mu^4}{45T^3} \tag{4.327}$$

So we see that in the limit $\sigma \gg 1$, which corresponds to magnetically rigid particles, the cubic susceptibility of a solid random assembly is thrice that of a magnetic fluid containing the same amount of identical particles. In this connection we recall a well-known fact [14,67,68] that for the cases compared, the respective static linear susceptibilities χ identically coincide, that is, are completely indistinguishable whatever σ. This confirms surmise that in nanomagnetic systems the cubic susceptibility is much more selective than the linear one.

The difference in configurational freedom between trapped and suspended nanoparticles manifests itself yet more pronouncedly in non-equilibrium (dynamic) susceptibilities of their assemblies. Let us first consider a mechanically fixed particle. Hereafter we restrict ourselves by the low-frequency range assuming that ω is sufficiently small in comparison with the Larmor frequency $\Omega_L = 2K\gamma/I_s$ of the precession of the particle magnetic moment in the internal anisotropy field; here γ is the gyromagnetic ratio. In this limit the motion of e occurs to be of a purely relaxational type. Its two principal modes are dubbed *intrawell* and *interwell* with regard to the two-well orientational potential imposed by the magnetic anisotropy; see the first term in Eq. (4.73). Macroscopically, in terms of the Landau–Lifshitz equation (LLE), the response time of the intrawell modes is of the order of the precession damping time $\tau_0 = I_s/2\alpha\gamma K$, where is the spin–lattice relaxation parameter of LLE. Normally, α is believed to be in the 0.01–0.1 range and weakly dependent on temperature.

In contrast, the interwell time $\tau(\sigma)$ is very temperature-sensitive and might change virtually unboundedly in a sufficiently narrow temperature interval. [Note that under constant particle volume the parameter σ according to its definition (4.75) may be treated as the inverse temperature.] Qualitatively, the behavior of $\tau(\sigma)$ is as follows. At low potential barriers ($\sigma \ll 1$) the interwell time is close to

$$\tau_D = \sigma\tau_0 = \frac{\mu}{2\alpha\gamma T} \tag{4.328}$$

which is the reference time of the free orientational diffusion of e inside the magnetically isotropic particle. At $\sigma > 1$ the interwell time $\tau(\sigma)$ grows drastically so that for $\sigma \gg 1$ it is very close to the Néel asymptotic expression (4.97). Rigorous description of the motion of e at arbitrary σ is achieved in terms of a time-dependent orientational distribution function $W(t)$ by the relevant Fokker–Planck equation [47]:

$$\frac{\partial W}{\partial t} = \frac{1}{2\tau_D}\hat{\boldsymbol{J}}_e W \hat{\boldsymbol{J}}_e \left(\frac{U}{T} + \ln W\right) \tag{4.329}$$

here $\hat{\boldsymbol{J}}_e = (e \times \partial/\partial e)$ is the operator of infinitesimal rotations of e.

In Section IV.B a procedure of numerical solution for Eq. (4.329) is described and enables us to obtain the linear and cubic dynamic susceptibilities for a solid system of uniaxial fine particles. Then, with allowance for the polydispersity of real samples, the model is applied for interpreting the magnetodynamic measurements done on Co-Cu composites [64], and a fairly good agreement is demonstrated. In our work we have proposed for the low-frequency cubic susceptibility of a randomly oriented particle assembly an interpolation (appropriate in the whole temperature range) formula

$$\chi_{3\omega}^{(3)} = \frac{1}{4} \chi_{0\,\text{solid}}^{(3)} \frac{(1 + 2S_2^2)(1 - i\omega\tau_{10})}{(1 + i\omega\tau_{10})(1 + 3i\omega\tau_{10})} \tag{4.330}$$

where τ_{10} is the inverse of that same eigenvalue of the Fokker–Planck operator (4.329) that corresponds to the interwell transition mode. A numerical frontfactor in Eq. (4.330) is not incidental. We keep it for the following reason. Actually, the cubic contribution to the response signal (magnetization) appears initially as a term proportional to the cube of the probing field, which is $\propto \cos^3 \omega t$. On splitting it into a sum of single- and triple-frequency terms, due to trigonometry, the coefficient $\frac{1}{4}$ emerges in front of the 3ω-contribution that we are interested in. Using numerical calculations as a verification tool, we have shown that formula (4.330) indeed works very well in the low-frequency range ($\omega\tau_0 < 1$) if it is used with the effective relaxation time

$$\tilde{\tau}_{10} = \tau_D \frac{e^\sigma - 1}{2\sigma} \left[\frac{1}{1 + 1/\sigma} \sqrt{\frac{\sigma}{\pi}} + 2^{-\sigma-1} \right] \tag{4.331}$$

proposed in Ref. 71; this expression for τ_{10} appeared earlier, in Eq. (4.105). On further investigation, while constructing a high-barrier ($\sigma \gg 1$) approximation for the set of odd-order dynamic susceptibilities of a solid system, we found [78] (see also Section III.B, above), that regarding the frequency dependence, Eq. (4.330) is in fact the exact form of $\chi_{3\omega}^{(3)}$. In Section III.B above we also gave the asymptotically exact expression for the relaxation time that should replace τ_{10} in formula Eq. (4.330) at $\sigma \gg 1$.

In a fluid system, the rotational freedom of the particles affects the susceptibilities in two ways: (1) the applied field (either AC or DC) deforms the orientational distribution function of the easy axes, which can never happen in a solid system with its fixed distribution of the particle axes; and (2) if out of equilibrium, in a magnetic fluid the orientational diffusion of the particle axes works as an additional channel of magnetic relaxation; that is, besides intrinsic processes, the magnetic moment can achieve equilibrium by rotating together with the particle in the suspended viscous liquid. Expressing the reference

(Brownian) time of the latter relaxation process by Eq. (4.29), one obtains [131,132] the extended Fokker–Planck equation in the form

$$\frac{\partial}{\partial t} W = \left[\frac{1}{2\tau_B} (\hat{\boldsymbol{J}}_e + \hat{\boldsymbol{J}}_n) W (\hat{\boldsymbol{J}}_e + \hat{\boldsymbol{J}}_n) + \frac{1}{2\tau_D} \hat{\boldsymbol{J}}_e W \hat{\boldsymbol{J}}_e \right] \left(\frac{U}{T} + \ln W \right) \tag{4.332}$$

Compare this with Eq. (4.329), where $\hat{\boldsymbol{J}}_n = (\boldsymbol{n} \times \partial/\partial \boldsymbol{n})$ is the operator of infinitesimal rotations over $\boldsymbol{n}$. We mention that, in our attempts to obtain low-frequency susceptibilities, we have omitted in Eq. (4.332) the term responsible for gyromagnetic effects.

Equations (4.329) for a solid assembly and (4.332) for a magnetic suspension are solved by expanding W with respect to the appropriate sets of functions. Convenient as such are the spherical harmonics defined by Eq. (4.318). In this context, the "internal" spherical harmonics used for solving Eq. (4.329) are written $X_l^m(\boldsymbol{e}, \boldsymbol{n})$. In the case of a magnetic fluid on this basis, a set of "external" harmonics is added, which are built on the angles of $\boldsymbol{e}$ with $\boldsymbol{h}$ as the polar axis. Application of a field couples [see the kinetic equation (4.332)] the internal and external degrees of freedom of the particle so that the dynamic variables become inseparable. With regard to this fact, the solution of equation (4.332) is constructed in the functional space that is a direct product of the "internal" and "external" harmonics:

$$W(\boldsymbol{e}, \boldsymbol{n}, t) = \sum_{l=0}^{\infty} \sum_{l'=0}^{\infty} \sum_{m=-l}^{l} \sum_{m'=-l'}^{l'} Q_{l,l'}^{m,m'}(t) \frac{(2l+1)(l-|m|)!}{4\pi(l+|m|)!}$$
$$\times X_l^{m*}(\boldsymbol{e}, \boldsymbol{n}) \frac{(2l'+1)(l'-|m'|)!}{4\pi(l'+|m'|)!} X_{l'}^{m'*}(\boldsymbol{n}, \boldsymbol{h}) \tag{4.333}$$

which time dependence is determined by the four-index coefficients

$$Q_{l,l'}^{m,m'}(t) = \langle X_l^m(\boldsymbol{e}, \boldsymbol{n}) X_{l'}^{m'}(\boldsymbol{n}, \boldsymbol{h}) \rangle \tag{4.334}$$

Hereafter, taking into account the axial symmetry of the problem that implies $m' = -m$ [see Eq. (4.333)], we will use only one upper index in the notation of the coefficient (4.334): $Q_{l,l'}^m$. Further on, however, when expanding Q with respect to the probing field amplitude ξ, we will have to indicate the order of the approximation by an additional upper index. To avoid confusion, this index will be kept inside curled braces [e.g. $Q_{l,l'}^{m\{2\}}$]. In a reduced form, namely, at $l' = m = m' = 0$, the parameters Q render customary uniaxial orientational order parameters of the particle axes in a suspension (magnetic fluid). In this case their meaning is illustrated by

$$Q_{0,l}^0 = \langle P_l(\cos \vartheta) \rangle \tag{4.335}$$

where ϑ is the angle between the particle major geometry axis and the applied field; note that the average is taken with a nonequilibrium distribution function.

In terms of the parameters (4.334), the dimensionless magnetization is expressed as

$$\langle e \cdot h \rangle = Q_{1,1}^0 + Q_{1,1}^1 \qquad (4.336)$$

The equation for $Q_{l,l'}^m$ is derived in Section V.C together with the scheme of its solution down to the third order in ξ. From the solutions obtained, the terms yielding the linear and cubic low-frequency responses to the probing magnetic field $H(t) = H \cos \omega t$ are extracted. In terms of linear and cubic susceptibilities those quantities evaluated numerically are compared in Figures 4.31 and 4.32; primes and double primes there denote, as usual, the in-phase and out-of-phase components of the dynamic susceptibilities.

One sees that at weak anisotropy the shapes of the respective graphs (a and c, b and d) in both figures nearly coincide. This is explained by the fact that the

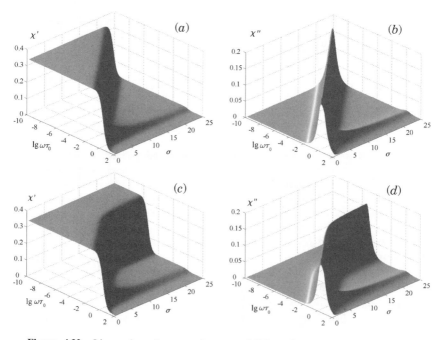

Figure 4.32. Linear dynamic magnetic susceptibilities of a randomly oriented superparamagnetic assembly (a, b) and of a magnetic fluid of the same particles (c, d); for all the graphs the ratio $\tau_D/\tau_B = 10^{-4}$. Vertical axes for χ are scaled in the units of $c\mu^2/T$ so that at $\omega \to 0$ both χ' tend to 1/3.

time τ_{10} [or, speaking rigorously, the exact relaxation time $\tau(\sigma)$, which τ_{10} approximates] in the region mentioned is much less than τ_B. Indeed, at $\sigma \ll 1$ one has $\tau_{10} \simeq \tau_D = \mu/2\alpha T$ which for ferrite particles of the size about 10 nm is normally several orders of magnitude smaller than τ_B given by Eq. (4.29). As the relaxation process always goes along the faster channel, this means that regardless of the aggregate state of the matrix, the orientational relaxation of the particle magnetic moment is always attained by means of the internal relaxation process.

It is at high anisotropy barriers ($\sigma \gg 1$) where the behavior of the susceptibilities becomes qualitatively different. For a solid system, τ_{10} is the only relaxation time and thus it largely determines the dynamics of any susceptibility at any ω. As a result, at $\sigma \gg 1$ the dispersion condition $\omega\tau_{10} \sim 1$ with exponential accuracy transforms into $\omega\tau_N = \omega\tau_0 \exp(\sigma) \sim 1$. This means that at given $\sigma \gg 1$ any process whose frequency ω is greater than a virtually infinitesimal value $(1/\tau_0)\exp(-\sigma)$ should be treated in the adiabatic (quasiequilibrium) limit $\omega \rightarrow 0$. In the coordinate plane $(\lg\omega\tau_0, \sigma)$ the condition $\omega\tau_{10} \sim 1$ holds along a diagonal straight line going to infinity.

In a magnetic suspension, the exponential freezing of interwell transitions that accompanies the enhancement of the anisotropy barrier σ does not indeed block the motion of the magnetic moment. Even having been tightly frozen into the particle, the magnetic moment retains its "external" degrees of freedom, that is, the ability to rotate together with the particle body. This mechanical motion has its own reference time which by order of magnitude equals the Brownian time (4.29). The latter may be effectively written as $\tau_B = 3\eta\sigma/K$. This dependence is far slower than the exponential behavior of the "magnetic" time τ_{10}; see its approximation (4.331), for example. Note also that the model under discussion does not take into account the possible freezing of the matrix, so the latter is assumed to stay fluid at any temperature. Therefore we conclude that as the anisotropy of the particles augments, in a magnetic fluid there occurs a crossover in the relaxation rate and at enhanced temperatures the latter instead of $1/\tau(\sigma)$ changes to $1/\tau_B$. The crossover condition takes the form of the transcedental equation

$$\frac{3\eta\sigma}{K} = \tau_0 \exp(\sigma) \tag{4.337}$$

whose single root may be found easily for any given set of material parameters of a superparamagnetic system.

In the images of Figures 4.32 and 4.33, where in any panel the transition from a solid to a fluid system indicates a vertical step downward ($a \rightarrow c$ or $b \rightarrow d$), the crossover in the relaxation process from τ_{10} to τ_B manifests itself as the change of the direction of the main "fold" (crest or canyon). Instead of a

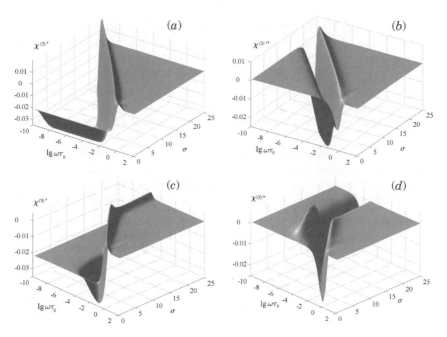

Figure 4.33. Cubic dynamic magnetic susceptibilities of a randomly oriented super-paramagnetic assembly (a,b) and of a magnetic fluid of the same particles (c,d); for all the graphs the ratio $\tau_D/\tau_B = 10^{-4}$. Vertical axes for $\chi^{(3)}$ are scaled in the units of $c\mu^4/4(T)^3$ so that at $\omega \to 0$ both $\chi^{(3)'}$ tend to $-1/45$.

straightforward unbounded propagation of the fold to the low-frequency region in the case of a solid system [Figs. 4.32(a),(b) and 4.33(a),(b)], in the case of a magnetic fluid the fold makes a turn and becomes seemingly parallel to the σ axis. As explained above, after the turning point (crossover), which is quite distinguishable in all the graphs, the condition determining the trajectory of the fold becomes $\omega\tau_B \sim 1$. As τ_B is not completely temperature independent [see Eq. (4.29)], the visual parallelism of the fold to the coordinate axis is merely an illusion due to the logarithmic scale of the ω axis. Indeed, in such a plot a simple power law $\tau_B \propto 1/T$ is rather slow and the tilt of its representing line against the σ axis does not resolve clearly.

Note that the graphs presented correspond to the ratio of the internal τ_D and external τ_B Brownian diffusion times—they are given by Eqs. (4.28) and (4.29), respectively—which amounts to a rather small value. This choice is deliberate. Indeed, using the representations of the times we have

$$\frac{\tau_D}{\tau_B} = \frac{I_s}{6\alpha\gamma\eta} \qquad (4.338)$$

which ratio incorporates only the material parameters of the substances forming a magnetic suspension. Setting the reference values $I_s \simeq 300\,\text{G}$ (a ferrite), $\alpha \sim 0.01$, $\eta \sim 10^{-2} - 10\,\text{P}$, one gets the values which spread from 10^{-3} (water) to 10^{-6} (glycerol) but always are quite small with respect to unity.

The behavior of all the graphs to the "right" of the fold is universal. For solid systems along the σ axis with the growth of the latter argument, the condition $\omega\tau_{10} \gg 1$, which may be rewritten as $\omega\tau_0 \gg \exp(-\sigma)$, holds for the increasingly widening interval of frequencies, thus causing the susceptibilities to vanish. For fluid systems the same zeroing of the susceptibilities takes place after the crossover as soon as the condition $\omega\tau_B \gg 1$ (equivalently written as $\omega\tau_0 \gg \tau_D/\sigma\tau_B$) is attained. Accordingly, in Figures 4.32 and 4.33 the zero-altitude "floor" (the flat area that in each graph lies to the right of the main fold) expands without bound to the low-frequency region as σ grows.

Besides the above mentioned universal features are inherent to all the susceptibilities, the plots have a complete pairwise resemblance $a \leftrightarrow c$, $b \leftrightarrow d$ in the $\omega\tau_0 \gg \tau_D/\sigma\tau_B$ region at any σ, (see Figs. 4.32 and 4.33). The explanation of this coincidence is simple. For $\tau_{10} \ll \tau_B$ the mechanical diffusion is "frozen," and there is no difference in the rates of magnetic relaxation between solid and fluid systems as long as there is no constant field imposed.

The following simple, qualitative expression further clarifies the resemblance between the linear susceptibilities: a

$$\chi(\omega) = \frac{c\mu^2}{3T}\frac{1}{1+i\omega\tilde{\tau}}, \qquad \tilde{\tau} = \frac{\tau_{10}\tau_B}{\tau_{10}+\tau_B} \tag{4.339}$$

Apparently, Eq. (4.339) yields a crossover at $\tau_{10} = \tau_B$.

For the cubic susceptibility in the limit $\tau_{10} \gg \tau_B$ ("frozen" internal motion of the magnetic moment), a simple analytical expression is valid:

$$\chi_{3\omega}^{(3)} = \frac{1}{4}\chi_{\text{fluid}}^{(3)}\frac{1}{(1+i\omega\tau_B)(1+\frac{2}{3}\omega\tau_B)(1+3i\omega\tau_B)} \tag{4.340}$$

This formula in an equivalent form (with τ_D instead of τ_B because a magnetic moment diffusion inside an isotropic particle was considered) had been obtained in Section III.A.3 as Eq. (4.96). Besides that, similar formulas, with τ_B indeed, are well known in the theory of rotary Brownian diffusion in dipolar fluids [69].

As follows from Figure 4.33, the $\sigma - \omega$ (one may say, temperature–frequency) behavior of the cubic susceptibilities of a magnetic fluid at low σ is more complicated than that of a solid dispersion. To clarify the origin of this, we recall expression (4.324) for the dimensional magnetization of a magnetic fluid. In a typical situation the response time of the magnetic degrees of freedom of the particles is much shorter than that of their mechanical rotation [see Eq. (4.338) and the comment following it]. Then Eq. (4.324) delivers a correct expression for the *instantaneous* magnetization at the time intervals $t > \tau_D$. Thence, considering Eq. (4.324), one sees that, in a suspension, unlike the case of a solid system, there exists a specific way to increase its magnetic response. It is the enhancement of the orientational order parameter $\langle P_2(\mathbf{n} \cdot \mathbf{h}) \rangle$ of the particle axes, [see Eq. (4.354) where, in terms of spherical components, it is denoted as $Q_{0,2}^{0,2}$]. In the presence of an oscillating magnetic field, this factor is modulated proportionally to $\xi^2(t)$ and with necessity contributes to the dispersion of the dynamic susceptibilities beginning with the cubic term $\chi_{3\omega}$. The corresponding frequency dispersion range is close to $\omega\tau_B \sim 1$. A more accurate formula follows from a derivation done in Ref. 70. It yields

$$\langle P_2(\mathbf{n} \cdot \mathbf{h}) \rangle = \frac{\xi^2}{15} \frac{S_2 \, e^{2i\omega t}}{1 + \frac{2}{3} i\omega\tau_B} \tag{4.341}$$

Substitution of Eq. (4.341) in (4.324) with allowance for relations (4.325), which are obviously valid (in the zeroth order in ξ) for a fluid system with a random orientation in the initial state, gives

$$\chi_{\text{fluid}}^{(3)}(\omega) = -\frac{c\mu^4}{45k^3T^3}\left(1 + 2S_2^2 - \frac{2S_2^2}{1 + \frac{2}{3}i\omega\tau_B}\right) = \chi_{\text{solid}}^{(3)} + \frac{c\mu^4}{45k^3T^3}\frac{2S_2^2}{1 + \frac{2}{3}i\omega\tau_B} \tag{4.342}$$

The last formula explicitly proves that, as already surmised, in comparison with a solid system, a fluid one has its own specific range of frequency dispersion around $1/\tau_B$.

C. Perturbative Equations for the Harmonic Amplitudes in the Case of Joint Diffusion

The general matrix equation of the problem that determines the amplitudes $Q_{l,l'}^m(t)$ is obtained by substitution of the spherical harmonic expansion (4.334) in the Fokker–Planck equation (4.332). After that, the equation is multiplied from the left by $X_l^m(\mathbf{e}, \mathbf{n})X_{l'}^{m'}(\mathbf{n}, \mathbf{h})$ and, finally, integrated with respect to both $\mathbf{e}$ and $\mathbf{n}$. The result can be written in a compact form as

$$2\tau_D\frac{d}{dt}Q_{l,l'}^m + \hat{R}Q_{l,l'}^m = \hat{V}Q_{l,l'}^m \tag{4.343}$$

Here we have introduced the relaxation matrix operator

$$
\hat{R}Q_{l,l'}^m \equiv l(l+1)Q_{l,l'}^m + \frac{\tau_D}{\tau_B}l'(l'+1)Q_{l,l'}^m
$$
$$
- 2\sigma\left[\frac{(l+1)(l+m-1)(l+m)}{(2l-1)(2l+1)}Q_{l-2,l'}^m + \frac{l(l+1)-3m^2}{(2l-1)(2l+3)}Q_{l,l'}^m\right.
$$
$$
\left. - \frac{l(l-m+1)(l-m+2)}{(2l-1)(2l+3)}Q_{l+2,l'}^m\right] \tag{4.344}
$$

and the perturbation matrix operator proportional to ξ:

$$
\hat{V}Q_{l,l'}^m \equiv \frac{\xi}{(2l+1)(2l'+1)}\left\{(l+1)(l+m)[(l'-m+1)Q_{l-1,l'+1}^m\right.
$$
$$
+ (l'+m)Q_{l-1,l'-1}^m] - l(l-m+1)[(l'-m+1)Q_{l+1,l'+1}^m
$$
$$
+ (l'+m)Q_{l+1,l'-1}^m]
$$
$$
+ \frac{1}{2}(l+1)(l+m-1)(l+m)[(l'+m-1)(l'+m)Q_{l-1,l'-1}^{m-1}
$$
$$
- (l'-m+1)(l'-m+2)Q_{l-1,l'+1}^{m-1}]
$$
$$
+ \frac{1}{2}l(l-m+2)(l-m+1)[(l'+m-1)(l'+m)Q_{l+1,l'-1}^{m-1}
$$
$$
- (l'-m+1)(l'-m+2)Q_{l+1,l'+1}^{m-1}]
$$
$$
- \frac{1}{2}(l+1)(Q_{l-1,l'+1}^{m+1} - Q_{l-1,l'-1}^{m+1}) - \frac{1}{2}l(Q_{l+1,l'+1}^{m+1} - Q_{l+1,l'-1}^{m+1})\right\}
$$
$$
- \frac{\xi}{(2l+1)(2l'+1)}\frac{\tau_D}{\tau_B}\left\{(l'+1)(l'+m)[(l-m+1)Q_{l+1,l'-1}^m\right.
$$
$$
+ (l+m)Q_{l-1,l'-1}^m]
$$
$$
- l'(l'-m+1)[(l-m+1)Q_{l+1,l'+1}^m + (l+m)Q_{l-1,l'+1}^m]
$$
$$
+ \frac{1}{2}(l'+1)(l'+m-1)(l'+m)[(l+m-1)(l+m)Q_{l-1,l'-1}^{m-1}
$$
$$
- (l-m+1)(l-m+2)Q_{l+1,l'-1}^{m-1}]
$$
$$
+ \frac{1}{2}l'(l'-m+2)(l'-m+1)[(l+m-1)(l+m)Q_{l-1,l'+1}^{m-1}
$$
$$
- (l-m+1)(l-m+2)Q_{l+1,l'+1}^{m-1}]
$$
$$
- \frac{1}{2}(l'+1)(Q_{l+1,l'-1}^{m+1} - Q_{l-1,l'-1}^{m+1}) - \frac{1}{2}l'(Q_{l+1,l'+1}^{m+1} - Q_{l-1,l'+1}^{m+1})\right\} = 0 \tag{4.345}
$$

These equations are valid for $m \geq 1$. Similar to what is presented in Section II.A, one does not need to consider the negative values of m because of the symmetry of Eq. (4.343) with respect to a replacement $m \Rightarrow -m$ that yields $Q_{l,l'}^{-m} = Q_{l,l'}^{m}$. The case $m = 0$ is, however, special, and the corresponding equation making the set (4.343) complete is to be derived and written separately:

$$
\hat{V} Q_{l,l'}^0 \equiv \frac{\xi}{(2l+1)(2l'+1)} \{ l(l+1)[(l'+1)(Q_{l-1,l'+1}^0 - Q_{l+1,l'+1}^0)
$$
$$
+ l(Q_{l-1,l'-1}^0 - Q_{l+1,l'-1}^0)] - (l+1)(Q_{l-1,l'+1}^1 - Q_{l-1,l'-1}^1)
$$
$$
- l(Q_{l+1,l'+1}^1 - Q_{l+1,l'-1}^1) \}
$$
$$
+ \frac{\xi}{(2l+1)(2l'+1)} \frac{\tau_D}{\tau_B} \{ l'(l'+1)[(l+1)Q_{l+1,l'-1}^0 - Q_{l+1,l'+1}^0)
$$
$$
+ l(Q_{l-1,l'-1}^0 - Q_{l-1,l'+1}^0)] - (l'+1)(Q_{l+1,l'-1}^1 - Q_{l-1,l'-1}^1)
$$
$$
- l'(Q_{l+1,l'+1}^1 - Q_{l-1,l'+1}^1) \} \tag{4.346}
$$

To be able to obtain cubic susceptibilities, the perturbation theory sequence of calculations with Eq. (4.343) must be carried out down to the third order in ξ. As the relaxation matrix $\hat{R}$ is tridiagonal, this can be done using the sweep method described in described in Section II.C.1.

1. Zeroth Order

The internal and external degrees of freedom are not coupled. The only nonzero components are

$$
Q_{l,0}^{(0)\{0\}} = S_l(\sigma) \tag{4.347}
$$

where the latter functions are the internal order parameters defined by Eq. (4.324) and depending solely on the dimensionless parameter σ. The perturbation order hereafter is indicated by the superscript in braces; see the comment following Eq. (4.334).

2. First Order

Here the components $Q_{l,1}^0$ and $Q_{l,1}^1$ are relevant, they are evaluated from the equations

$$
2\tau_D \frac{d}{dt} Q_{l,1}^{0\{1\}} + \hat{R} Q_{l,1}^{0\{1\}} = \frac{\xi}{3(2l+1)} \left\{ l(l+1)(S_{l-1} - S_{l+1}) \right.
$$
$$
\left. + 2\frac{\tau_D}{\tau_B}[(l+1)S_{l+1} + lS_{l-1}] \right\} \tag{4.348}
$$

and

$$2\tau_D \frac{d}{dt} Q_{l,1}^{1\{1\}} + \hat{R} Q_{l,1}^{1\{1\}} = \xi \frac{l(l+1)}{3(2l+1)} \left[(l+1)S_{l-1} + l S_{l+1} + 2 \frac{\tau_D}{\tau_B} (S_{l-1} - S_{l+1}) \right]$$

(4.349)

After solving this set of equations and then using Eq. (4.336), the linear susceptibility of a particle in the direction of the field is found.

3. Second Order

In this approximation one has to solve one equation with $l' = 0$ that is

$$2\tau_D \frac{d}{dt} Q_{l,0}^{0\{2\}} + \hat{R} Q_{l,0}^{0\{2\}} = \frac{\xi}{(2l+1)} \left[l(l+1) \left(Q_{l-1,1}^{0\{1\}} - Q_{l+1,1}^{0\{1\}} \right) \right.$$
$$\left. - (l+1) Q_{l-1,1}^{1\{1\}} - l Q_{l+1,1}^{1\{1\}} \right]$$

(4.350)

and three equations with $l' = 2$:

$$2\tau_D \frac{d}{dt} Q_{l,2}^{0\{2\}} + \hat{R} Q_{l,2}^{0\{2\}} = \frac{\xi}{5(2l+1)} [2l(l+1)(Q_{l-1,1}^{0\{1\}} - Q_{l+1,1}^{0\{1\}})$$
$$+ (l+1) Q_{l-1,1}^{1\{1\}} + l Q_{l+1,1}^{1\{1\}}] + \frac{\xi}{5(2l+1)} \frac{\tau_D}{\tau_B} \left\{ 6[(l+1) Q_{l+1,1}^{0\{1\}} \right.$$
$$\left. + l Q_{l-1,1}^{0\{1\}}] - 3(Q_{l+1,1}^{1\{1\}} - Q_{l-1,1}^{1\{1\}}) \right\}$$

(4.351)

$$2\tau_D \frac{d}{dt} Q_{l,2}^{1\{2\}} + \hat{R} Q_{l,2}^{1\{2\}} = \frac{3\xi}{5(2l+1)} \left\{ (l+1)^2 (Q_{l-1,1}^{1\{1\}} + l Q_{l-1,1}^{0\{1\}}) \right.$$
$$\left. - l^2 [Q_{l+1,1}^{1\{1\}} - (l+1) Q_{l+1,1}^{0\{1\}}] \right\} + \frac{9\xi}{5(2l+1)} \frac{\tau_D}{\tau_B} \left\{ (l+1)(Q_{l-1,1}^{1\{1\}} \right.$$
$$\left. + l Q_{l-1,1}^{0\{1\}}) + l[Q_{l+1,1}^{1\{1\}} - (l+1) Q_{l+1,1}^{0\{1\}}] \right\}$$

(4.352)

$$2\tau_D \frac{d}{dt} Q_{l,2}^{2\{2\}} + \hat{R} Q_{l,2}^{2\{2\}} = \frac{6\xi}{5(2l+1)} [(l+1)^2(l+2) Q_{l-1,1}^{1\{1\}} + (l-1)l^2 Q_{l+1,1}^{1\{1\}}]$$
$$+ \frac{18\xi}{5(2l+1)} \frac{\tau_D}{\tau_B} [(l+1)(l+2) Q_{l-1,1}^{1\{1\}} - (l-1)l Q_{l+1,1}^{1\{1\}}]$$

(4.353)

As a particular case, Eq. (4.351) at $l = 0$ renders the equation for the scalar order parameter of the particles along the field direction:

$$\frac{d}{dt} Q_{0,2}^{0\{2\}} + \frac{3}{\tau_B} Q_{0,2}^{0\{2\}} = \frac{\xi}{10\tau_B} (6 Q_{1,1}^{0\{1\}} - 3 Q_{1,1}^{1\{1\}})$$

(4.354)

In the main text for the same quantity we use the notation $\langle P_2(\boldsymbol{n} \cdot \boldsymbol{h}) \rangle$, which allows to directly associate it with the particle orientation.

4.   Third Order

As Eq. (4.336) shows, to evaluate the dimensionless magnetization, it suffices to solve the equations for $Q_{l,1}^0$ and $Q_{l,1}^1$:

$$2\tau_D \frac{d}{dt} Q_{l,1}^{0\{3\}} + \hat{R} Q_{l,1}^{0\{3\}}$$

$$= \frac{\xi}{3(2l+1)} \left\{ l(l+1) \left[2Q_{l-1,2}^{0\{2\}} + Q_{l-1,0}^{0\{2\}} - 2Q_{l+1,2}^{0\{2\}} - Q_{l+1,0}^{0\{2\}} \right] \right.$$

$$- (l+1)Q_{l-1,2}^{1\{2\}} - lQ_{l+1,2}^{1\{2\}} \right\} + \frac{\xi}{3(2l+1)} \frac{\tau_D}{\tau_B} \left\{ 2[(l+1)(Q_{l+1,0}^{0\{2\}} - Q_{l+1,2}^{0\{2\}}) \right.$$

$$\left. + l(Q_{l-1,0}^{0\{2\}} - Q_{l-1,2}^{0\{2\}})] - Q_{l+1,2}^{1\{2\}} - Q_{l-1,2}^{1\{2\}} \right\} \tag{4.355}$$

$$2\tau_D \frac{d}{dt} Q_{l,1}^{1\{3\}} + \hat{R} Q_{l,1}^{1\{3\}}$$

$$= \frac{\xi}{3(2l+1)} \left\{ (l+1)^2 Q_{l-1,2}^{1\{2\}} - l^2 Q_{l+1,2}^{1\{2\}} + l(l+1)^2 (Q_{l-1,0}^{0\{2\}} - Q_{l-1,2}^{0\{2\}}) \right.$$

$$+ l^2(l+1)(Q_{l+1,0}^{0\{2\}} - Q_{l+1,2}^{0\{2\}}) - \frac{1}{2}[(l+1)Q_{l-1,2}^{2\{2\}} + lQ_{l+1,2}^{2\{2\}}] \right\}$$

$$+ \frac{\xi}{3(2l+1)} \frac{\tau_D}{\tau_B} \left[lQ_{l+1,2}^{1\{2\}} + (l+1)Q_{l-1,2}^{1\{2\}} - l(l+1)(2Q_{l-1,0}^{0\{2\}}) \right.$$

$$\left. - 2Q_{l+1,2}^{0\{2\}} + Q_{l-1,2}^{0\{2\}} - Q_{l+1,2}^{0\{2\}}) + \frac{1}{2} (Q_{l+1,2}^{2\{2\}} - Q_{l-1,2}^{2\{2\}}) \right] \tag{4.356}$$

VI.   FIELD-INDUCED BIREFRINGENCE IN
MAGNETIC SUSPENSIONS

A.   Dynamic Field-Induced Birefringence in the Framework of the
"Egg" Model: Interplay between Mechanical and
Magnetic Degrees of Freedom of the Particles

1.   Coupled Rotational Degrees of Freedom of the Particles

Magnetic-field-induced birefringence is a convenient experimental tool [135] used to study magnetic fluids. The basic theoretical concept describing it is quite simple. It is assumed that the ferrite grains constituting the solid phase of magnetic fluid have a slightly anisotropic shape. The external field orients the particle magnetic moments, and thereby the particles themselves, thus establishing the orientational ordering. This imparts a uniaxial optical anisotropy to a system that (due to the Brownian rotary motion of the particles) is macroscopically isotropic in the field-free state. Since the magnetic moments of the particles are huge in comparison with atomic or molecular ones, the resulting birefringence is several orders greater than a usual Cotton–Mouton effect in liquids and may be easily distinguished from that.

For a dilute assembly of identical noninteracting single-domain particles with uniaxial magnetic anisotropy, the exhaustive description of the time-independent situation is delivered by the equilibrium distribution [see Eqs. (4.322)]. In a dynamic situation the induced orientational order (and, consequently, birefringence) of a magnetic fluid evolves with the response times of the particles. In the past, several models of the effect have been developed [136–138] but they were restricted by just three isolated limiting cases, with nothing in between. On the basis of the theory presented in Section V, we develop a model for the dynamic field-induced birefringence of a magnetic fluid. This model is valid for a wide frequency range and unites all the formerly known asymptotics.

In a magnetic fluid the phenomenon of dynamic birefringence, as compared with conventional molecular optics, is somewhat unique. To understand this, we recall the considerations of Section V. In single-domain particles of a colloidal (nanoscopic) size, due to superparamagnetism, the magnetic moments undergo internal rotary diffusion and are not rigidly aligned with the anisotropy and thus, geometric, axes of the particles. Hence, the magnetooptical response of a magnetic fluid emerges as a combination of two relaxational processes: (1) the orientational ordering in an assembly of magnetic moments $\boldsymbol{\mu}$ (i.e., magnetization of the suspension) and (2) the orientational ordering of the particle axes $\boldsymbol{n}$ striving to minimize their magnetic anisotropy energy by mechanical rotation of $\boldsymbol{n}$ towards $\boldsymbol{\mu}$, which is the ultimate cause of birefringence.

These processes have different timescales. For magnetic moments it is the time of internal superparamagnetic diffusion τ_D [see Eq. (4.28)], and for the axes alignment it is the time of mechanical rotary diffusion τ_B of a particle in a carrier liquid [see Eq. (4.29)]. As once noticed in Ref. 48 (see also Section II.A above), both parameters may be presented in a similar form

$$\tau_B = \frac{3\eta V}{T} \qquad \tau_D = \frac{3\eta_m V_m}{T} \qquad (4.357)$$

where η is the viscosity of the carrier liquid and η_m is the quantity that one may call an "effective" magnetic viscosity.

The Fokker–Planck equation governing the evolution of the orientational distribution function $W(\boldsymbol{e}, \boldsymbol{n}, t)$ for arbitrary τ_B and τ_D, that is, when the particle is mechanically free in the matrix and its configurational space, is a direct product of two two-dimensional vector spaces $e \otimes n$, has been derived in Refs. 72,132,139, and 140. Here it is presented as Eq. (4.309) in Section V.A.2.

In Section V.B we have introduced the orientation order parameter of the particle axes

$$\langle P_2(n \cdot h) \rangle = Q_{0,2}^0 \qquad (4.358)$$

[see Eq. (4.335)] and the comment preceding it. The equation of motion for $Q_{0,2}^0$ is derived in Section V.C and is written there as Eq. (4.354). Under the assumptions adopted, the orientation order parameter (4.358) is directly proportional to the birefringence of the magnetic fluid.

When probing dynamic birefringence of a such a system with a weak alternating field, which is usually the case [135,137], the leading term determining the induced orientation order parameter is proportional to ξ^2, where ξ is defined with respect to the probing field amplitude. Then the right-hand side of Eq. (4.354) may by evaluated to not higher than the first order in ξ. From equation (4.354) for a linearly polarized field $H(t) = H \exp(i\omega t)$, one obtains

$$Q_{0,2}^{0\{2\}} = \frac{\xi}{10} \frac{2Q_{1,1}^{0\{1\}}(t) - Q_{1,1}^{1\{1\}}(t)}{1 + \frac{2}{3}i\omega\tau_B} e^{i\omega t} \qquad (4.359)$$

which is a rigorous relation. However, it is too formal for a clear understanding of the solution obtained. A convenient approximate form of Eq. (4.359) is

$$Q_{0,2}^{0\{2\}} = \frac{\xi^2}{15} \frac{\chi_\parallel(\omega) - \chi_\perp(\omega)}{1 + \frac{2}{3}i\omega\tau_B} e^{2i\omega t} \qquad (4.360)$$

presented in terms of the reduced magnetic susceptibilities

$$\chi_\parallel(\omega) = \frac{1 + 2S_2}{3} \frac{1}{1 + i\omega\tau_\parallel}, \qquad \chi_\perp(\omega) = \frac{1 - S_2}{3} \frac{1}{1 + i\omega\tau_\perp} \qquad (4.361)$$

introduced in Refs. 72 and 140. Here $S_2 = \langle P_2(e \cdot n) \rangle_0$ (note the argument) is the internal order parameter of index 2 characterizing the internal magnetic order of the particle. In other words, this function shows to what extent the magnetic moment is "frozen" inside the particle; it is defined by Eq. (4.83). Notation (4.360) emphasizes that it is the response of the magnetic moments of the particles that is the very source of birefringence in a magnetic fluid.

The effective relaxation times in Eqs. (4.361)

$$\tau_\parallel^{-1} = \tau_B^{-1} + \tau_{10}^{-1}, \qquad \tau_\perp^{-1} = \tau_B^{-1} + \tau_{11}^{-1} \qquad (4.362)$$

are defined from the literature [72,131,132,139] and incorporate contributions from diffusions with regard to both external (mechanical) and internal (magnetic) rotary degrees of freedom of a particle. Although an exact analytical solution for τ_{10}^{-1} has been found [58], evaluation of τ_{11}^{-1} needs numerical work, anyway. However, a rather good accuracy is available as well with simple approximate expressions. For example, one may take τ_{10} and τ_{11} from Eq. (4.105) after Ref. 71 and Refs. 72 and 140, respectively.

2. Dynamic Birefringence

Establishing the description of oscillations in the amplitude–phase terms as

$$Q_{0,2}^0(t) = A_0 + A_{2\omega}\cos(2\omega t - \psi)$$

from Eqs. (4.360) and (4.361), one gets for the amplitudes

$$A_0 = \frac{\xi^2}{90}\left[\frac{1 + 2S_2}{1 + \omega^2\tau_\parallel^2} - \frac{1 - S_2}{1 + \omega^2\tau_\perp^2}\right]$$

$$A_{2\omega} = \frac{\xi^2}{90}\frac{1}{1 + \frac{4}{9}\omega^2\tau_B^2}$$

$$\left\{\left[\frac{(1 + 2S_2)(1 - \frac{2}{3}\omega^2\tau_\parallel\tau_B)}{1 + \omega^2\tau_\parallel^2} - \frac{(1 - S_2)(1 - \frac{2}{3}\omega^2\tau_\perp\tau_B)}{1 + \omega^2\tau_\perp^2}\right]\cos\psi \right.$$

$$\left. + \left[\frac{(1 + 2S_2)(\omega\tau_\parallel + \frac{2}{3}\omega\tau_B)}{1 + \omega^2\tau_\parallel^2} - \frac{(1 - S_2)(\omega\tau_\perp + \frac{2}{3}\omega\tau_B)}{1 + \omega^2\tau_\perp^2}\right]\sin\psi\right\}$$

$$(4.363)$$

and for the phase shift

$$\psi = \arctan$$

$$\times\left[\frac{(1 + 2S_2)(1 + \omega^2\tau_\perp^2)(\omega\tau_\parallel + \frac{2}{3}\omega\tau_B) - (1 - S_2)(1 + \omega^2\tau_\parallel^2)(\omega\tau_\perp + \frac{2}{3}\omega\tau_B)}{(1 + 2S_2)(1 + \omega^2\tau_\perp^2)(1 - \frac{2}{3}\omega^2\tau_\parallel\tau_B) - (1 - S_2)(1 + \omega^2\tau_\parallel^2)(1 - \frac{2}{3}\omega^2\tau_\perp\tau_B)}\right]$$

$$(4.364)$$

Equations (4.362)–(4.364) constitute the generalized model; formulas (4.105), which are approximate, are optional. Since for a weak-field (probing) regime both A_0 and $A_{2\omega}$ scale as ξ^2 (or H^2), for a monodisperse suspension the minimal set of relevant parameters may be chosen as $\omega\tau_B$, σ and the ratio $\epsilon = \tau_D/\tau_B$. Figures 4.34 and 4.35 show examples of frequency dependencies $A_0(\omega\tau_B)$ and $A_{2\omega}(\omega\tau_B)$; the corresponding phase curves are plotted in Figure 4.36.

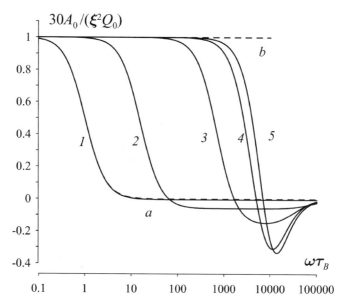

Figure 4.34. Time-independent part of the orientation parameter as a function of the field frequency for different values of the internal magnetic anisotropy of the particles. The ratio $\epsilon = \tau_D/\tau_B = 10^{-4}$; curves correspond to $\sigma = 100$ (1), $\sigma = 10$ (2), $\sigma = 5$ (3), $\sigma = 2$ (4), $\sigma = 0.1$ (5). Thin lines a and b resemble the limiting behavior predicted by the *rigid* and *soft* dipole models, respectively. Note that at this graph the lines a and 1 visually coincide.

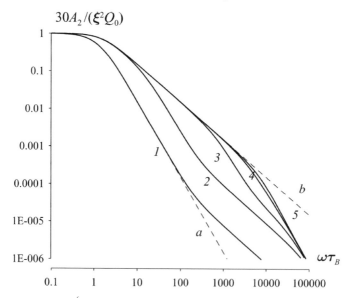

Figure 4.35. The second-harmonic amplitude of the orientation parameter for different values of the internal magnetic anisotropy of the particles. The set of material parameters and meanings of the thin curves are the same as those for Figure 4.34.

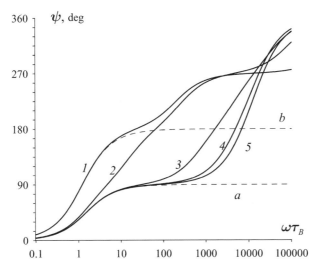

Figure 4.36. Birefringence phase shift for different values of the internal magnetic anisotropy of the particles. The set of material parameters and meanings of the thin lines are the same as those for Figure 4.34.

We remark that since in any experiment the phase shift between the square of the field is proportional to $\cos^2 \omega t$ and thus the second harmonic of birefringence is resolved with the accuracy of arbitrary addition or subtraction of $180°$, the definition of ψ is not unique. In principle, the curves in Figure 4.36 may be dissected at the points $\psi(\omega\tau_B) = 180°$ and replotted as if growing anew from zero at each cut.

The validity range of the developed approach is rather wide, since it allows for arbitrary σ and ϵ. Let us show how all the already known cases may be recovered on a unified basis.

a. Rigid-Dipole Model. There have been attempts [136] to describe the low-frequency ($\omega\tau_B \lesssim 1$) dynamic magnetic birefringence in suspensions of highly anisotropic particles. In this limit one sets $\sigma \to \infty$, which yields $S_2 = 1$ and $\tau_{10} \sim \exp(\sigma) \to \infty$. Then from Eqs. (4.362) and (4.105) follows $\tau_{\parallel} = \tau_B$ and $\tau_{\perp} = 0$, and on substitution to Eqs. (4.363) and (4.364), one gets

$$A_0 = \frac{\xi^2}{30}\frac{1}{1+\omega^2\tau_B^2}, \qquad A_{2\omega} = \frac{\xi^2}{30}\frac{1}{\left[(1+\frac{4}{9}\omega^2\tau_B^2)(1+\omega^2\tau_B^2)\right]^{1/2}}$$

$$\tan\psi = \frac{\frac{5}{3}\omega\tau_B}{1-\frac{2}{3}\omega^2\tau_B^2}$$

$$(4.365)$$

in accordance with the results of Skibin et al. [136] obtained directly from the Fokker–Planck equation, which took into account only the external diffusion. Here the maximum value of the phase is $\psi(\omega\tau_B \to \infty) = 180°$.

b. *Soft-Dipole model.* This approximation, once introduced in Ref. 137, describes the systems of ferroparticles with an intense internal orientational diffusion. Its essential assumptions are $\sigma \ll 1$ and $\epsilon \ll 1$. In this limit $S_2 = 2\sigma/15$, and $\tau_\parallel = \tau_\perp = \tau_D$. In the low-frequency range the terms $\sim \omega\tau_D$ may be neglected, which yields

$$A_0 = \frac{\sigma\xi^2}{225}, \qquad A_{2\omega} = \frac{\sigma\xi^2}{225}\frac{1}{\left(1 + \frac{4}{9}\omega^2\tau_B^2\right)^{1/2}}, \qquad \tan\psi = \tfrac{2}{3}\omega\tau_B$$

These formulas coincide with those derived in Ref. 137 and explain the intermediate asymptotics $A_0 = \text{const}(\omega)$ confirmed there experimentally. The limiting value of the phase is less than half that for rigid dipoles: $\psi(\omega\tau_B \to \infty) = 90°$.

c. *Rapidly Oscillating Dipole Model.* This case was addressed in Ref. 138 in connection with the peculiar behavior of the field-induced birefringence encountered experimentally [140,141] in suspensions of magnetically rigid particles subjected to a relatively high-frequency field. For the anisotropy parameter one sets $\sigma \gg 1$, which leads to $S_2 = 1 - 3/2\sigma$. For the relaxation times, expressions (4.362) and (4.105) then give $\tau_\parallel = \tau_B$, $\tau_\perp = \tau_D/\sigma$.

The frequency of the external field in other experiments [138,140,141] was chosen within the interval $\omega\tau_B \gg 1 \gg \omega\tau_D/\sigma$, which, given $\sigma \gg 1$ and $\epsilon \ll 1$, is several decades wide for viscous fluids such as glycerol. For the weak-field probing, a negative stationary and negligible oscillating birefringence had been predicted and confirmed. With the aid of our Eqs. (4.362) and (4.363), this is immediately recovered in the form

$$A_0 = -\frac{\xi^2}{60\sigma}, \qquad A_{2\omega} \sim \frac{A_0}{\omega\tau_B} \ll A_0$$

3. Summary

The proposed approach unites the results of previous studies and is valid for a wide range of material parameters affecting the dynamic magnetooptical response of a magnetic fluid. The essential feature of the new model is that it is sensitive to the internal magnetic relaxation of single-domain particles. That points out a way to test those processes with the aid of standard birefringence technique. Such measurements seem to be rather important since the problem of evaluation of τ_D (the basic quantity in the dynamic micromagnetism) is yet far from its ultimate solution in both theory and experiment—see discussions in Refs. 57 and 142.

B. Higher Harmonics Generation in Field-Induced Birefringence by a Suspension of Rigid Dipoles: Zero Bias Field

1. Introduction

In this section we get back to the case of the suspension of the particles whose dipolar (magnetic or electric) moments are "frozen" into their bodies. This case conceptually coincides with the situation of a magnetic fluid consisting of hard magnetic dipoles studied on the subject of higher harmonics generation in Section IV.C.3. The only reason why that optical problem is presented a bit "in advance" and not here is that it completely fits the formalism of Section IV.C. To postpone the solution of that problem to this section would have meant a hardly justifiable repetition of the pertinent schematization from Section IV.C. Meanwhile, generation of higher harmonics in a zero bias field has no magnetic analogs in this chapter, and it is logical to present it in a section devoted to orientational optical effects. We begin this section with a very brief introduction that describes the place of the model, which we develop below, relative to the other already existing models mentioned in the preceding section.

Electro- and magnetooptical phenomena in colloids and suspensions are widely used for structure and kinetics analysis of those media as well as practical applications in optoelectronics [143,144]. The basic theoretical model used to study optical anisotropy of the disperse systems is the noninteracting Brownian particle ensemble. In the frame of this general approximation, several special cases according to the actual type of particle polarization response to the applied field may be distinguished: (1) particles with permanent dipole moments, (2) linearly polarizable particles, (3) nonlinearly polarizable particles, and (4) particles with hysteretic dipole moment reorientation.

Models 1 and 2 are well known in molecular and colloid electromagneto-optics [145], while models 3 and 4 are pertinent to magnetooptics of ferrosus-pensions (magnetic fluids) and are specified for description of models 3—superparamagnetic particles [137] and 4—magnetically hard particles in a sufficiently high-frequency magnetic field [138]. Approaches 1–4 separately or in various combinations provide reasonable theoretical interpretation for quite a number of observations.

The goal of the following discussion is to study the low-frequency dynamic birefringence induced by an alternating field in a suspension of acicular particles with dipole moments, that is, the system of type 1 according to the classification given above. The results presented are obtained avoiding any small-parameter expansions and thus are valid for arbitrary values of the external field amplitude. As the assumptions adopted are not sensitive to the exact nature of the particle dipoles, the dependencies found equally account for either electrical or magnetic dipole suspensions, that is, for description of the amplitude–frequency characteristics of the Kerr or Cotton–Mouton effects [146].

We consider a system of acicular Brownian particles with the dipole moments μ suspended in a nonpolar liquid carrier with the viscosity η. If the interparticle interactions are negligible, then the main characteristic of such a statistical ensemble is the orientational distribution function $W(e,t)$ of the particle dipole moment. Provided the dipole is frozen into the particle body (rigid permanent dipole), the function $W(e,t)$ determines as well the angular distribution of the particle axes. Further on, for the sake of simplicity we assume that the suspended grains have a prolate spheroidal shape with vector n pointing along the major geometric axis. Generalization, if necessary, for any other case of relative alignment does not cause any essential difficulties.

In the absence of external fields the suspension under consideration is macroscopically isotropic ($W = $ const). The applied field h (we denote it in the same way as above but imply the electric field and dipoles as well as the magnetic ones), orienting, statically or dynamically, the particles, thus induces a uniaxial anisotropy, which is conventionally characterized by the orientational order parameter tensor $\langle P_2(n \cdot h)\rangle$ defined by Eq. (4.358). (We remind the reader that for rigid dipolar particles there is no difference between the unit vectors e and n.) As in the case of the internal order parameter S_2, [see Eq. (4.81)], one may define the set of quantities $\langle P_l(n \cdot h)\rangle$ for an arbitrary l. Of those, the first statistical moment $\langle P_1\rangle$ is proportional to the polarization (magnetization) of the medium, and the moments with $l > 2$, although not having meanings of directly observable quantities, determine those via the chain-linked set [see Eq. (4.369)].

The statistical properties of the particle assembly in equilibrium follow from the Gibbs distribution

$$W(n) = W_0 = Z_0^{-1} \exp\left[\frac{\mu(n \cdot H)}{T}\right], \qquad Z_0 = \int \exp\left[\frac{\mu(n \cdot H)}{T}\right] dn \quad (4.366)$$

and hence all the equilibrium moments $\langle P_l\rangle_0$ can be expressed analytically. For example

$$\langle P_1\rangle_0 = L(\xi_0), \qquad \langle P_2\rangle_0 = 1 - \frac{3L(\xi_0)}{\xi_0}, \qquad \xi_0 = \frac{\mu H}{T} \quad (4.367)$$

where $L(x)$ is the Langevin function.

The orientational kinetics of the dipolar suspension is described by the rotary diffusion equation presented in Section II as Eq. (4.51); its form for the electric dipoles is also well known [147,148]. The only modification one has to perform in Eq. (4.51) to make it account for the particles suspended in a fluid is to change the relaxation time from τ_D to τ_B. Defining the latter in a more general form than in Eq. (4.29), we write

$$\tau_B = \frac{C\eta V}{T} \tag{4.368}$$

noting that formfactor is minimal ($C = 3$) for a sphere and grows with the particle anisometricity. Since in a nonequilibrium situation equation (4.51) has no exact solution, the orientational order parameter cannot be expressed in any closed form. Instead we arrive at a chain-linked set

$$\frac{2\tau_B}{l(l+1)}\frac{d}{dt}\langle P_l\rangle + \langle P_l\rangle = \frac{\xi_0}{2l+1}[\langle P_{l-1}\rangle - \langle P_{l+1}\rangle] \tag{4.369}$$

where $l = 1, 2, 3\ldots$ and $\langle P_0\rangle = 1$. Note that equation (4.369) is in fact a reduced form of Eq. (4.60) setting there $\sigma = \psi = m = 0$. We remark that setting $\sigma = 0$ is neither an error nor an inconsistency. Indeed, assuming $\sigma = 0$ together with keeping $\tau = \tau_D$ implies the internal motion in a magnetically isotropic particle. Noteworthy, the description admits yet another interpretation, also completely justified. Namely, to model a mechanical motion of a magnetically hard particle suspended in an isotropic liquid, one takes $\tau = \tau_B$ [this is the case of Eq. (4.369)] and then has to set $\sigma = 0$ because in the absence of an external field the system does not have any specific direction for the particle axis.

2. Orientation Optical Anisotropy

While considering optical properties of dilute ultradisperse systems, two circumstances usually provide important simplifications: the smallness of the particle linear dimension with respect to the light wavelength, and the smallness of the mean interparticle distance compared to the cross section of the lightbeam involved in the measurement. Both scale inequalities are satisfied in a fairly wide range of stable colloidal systems and thus enable us to describe them as continuous media using the effective refractive indices. The principal values of the latter in the ensemble of anisometric particles are given by

$$v_i = v_0\left[1 + \mathscr{A}\phi + \mathscr{B}\phi(\langle n_i^2\rangle - \tfrac{1}{3})\right] \tag{4.370}$$

where v_0 is the refractive index of the dispersing liquid, ϕ is the solid-phase volume fraction of the suspension, and the coefficients $\mathscr{A}$ and $\mathscr{B}$ are determined only by the particle depolarizing factors and the relative dielectric permeability of its substance. If there are no optical transitions in the particle molecular structure, then $\mathscr{A}$ and $\mathscr{B}$ do not depend on the light frequency. Since the time period of the lightwave oscillations is by all means much smaller than that of the particle orientation relaxation, it is clear that the refractive indices v_i have no dispersion in the optical frequency range. If so, only the low-frequency ($\omega \sim 1/\tau_B$) dispersion of v_i, arising from particle orientation motion under the influence of the external field, should be taken into account. To estimate the characteristic values of τ_B, let us set $T = 300$ K, $V = 5 \times 10^{-19}$ cm^3 (colloidal particles with linear size ~ 10 nm), $\eta = 10^{-2}$ P (water), and $C = 6$. Using these parameters, we find that $\tau = C\eta V/T \sim 10^{-6}$ s.

Formula (4.370) leads to a simple expression for the macroscopic optical anisotropy. If the Oz axis of the coordinate framework coincides with the applied field direction h, then the uniaxial symmetry condition yields $\langle n_x^2 \rangle = \langle n_y^2 \rangle$ and $v_x = v_y$. Hence for the refractive index anisotropy (birefringence) we have

$$\Delta v = v_z - v_x = v_0 \phi \mathscr{B} \langle P_2 \rangle \tag{4.371}$$

3. Spectral Expansion

Depending on the scheme chosen, the birefringence experiments provide [143,144] direct measurements of either Δv or $(\Delta v)^2$. To present the theoretical results in a form suitable for comparison with the experimental data, let us consider the orientational oscillations induced in the dipolar suspension by a harmonic field $H = H_0 \cos \omega t$ and analyze the frequency dependencies of the spectra of the order parameters $\langle P_2 \rangle$ and $\langle P_2 \rangle^2$. As formula (4.371) shows, the latter quantities are directly proportional to Δv and $(\Delta v)^2$, respectively. Since the oscillations are steady, let us expand the time-dependent orientational parameters into the Fourier series

$$\langle P_2 \rangle = \sum_{m=0}^{\infty} A_{2m}(\xi_0, \omega t) \cos(2m\omega t - \beta_m)$$

$$\langle P_2 \rangle^2 = \sum_{m=0}^{\infty} B_{2m}(\xi_0, \omega t) \cos(2m\omega t - \gamma_m) \tag{4.372}$$

where $\xi_0 = \mu H_0/T$ is the dimensionless field amplitude.

For extremely low frequencies ($\omega \tau_B \ll 1$), when the phase lag between the field and orientation oscillations is negligible, the dynamics of the order parameters may be described with the aid of the equilibrium distribution function (4.366), setting there $H = H(t)$. For example, in this limiting case formula (4.367) gives

$$\langle P_2 \rangle|_{\omega\tau_B \to 0} = \int P_2 \, W_0 \, dn = 1 - \frac{3L(\xi_0 \cos \omega t)}{\xi_0 \cos \omega t} \qquad (4.373)$$

Thus, evaluation of the Fourier coefficients, although always remaining numerical, in the quasistatic regime is relatively simple.

For arbitrary $\omega\tau_B$, the properties of the steady-state orientational oscillations should be obtained from the chain-linked equations (4.369). We do this by expanding every $\langle P_l \rangle$ in the Fourier series

$$\langle P_l \rangle = \sum_{k=-\infty}^{\infty} Q_{lk}(\omega) e^{ik\omega t}$$

Following this, Eq. (4.369) transforms into a homogeneous tridiagonal vector recurrence relation

$$\left[\frac{2i\omega\tau_B}{l(l+1)} + 1 \right] Q_{lk} - \frac{\xi_0}{4l+2} \left(Q_{l-1,k-1} + Q_{l-1,k+1} - Q_{l+1,k-1} - Q_{l+1,k+1} \right) = 0$$

$$(4.374)$$

with the initial conditions $Q_{00} = 1$, $Q_{0k} = 0$ at $k \neq 0$. Solving it by the matrix sweep method, (see Section II.C.3), the set of amplitudes $\{Q_{lk}\}$ may be evaluated with any desired accuracy. The resulting numerical array was used to obtain the dependencies of A_{2m} and B_{2m}, [see Eq. (4.372)] on ξ_0 and $\omega\tau_B$.

It is worth noting that there exists one more limiting case where analytical representation of $\langle P_2 \rangle$ is possible. If the $\xi_0, \omega\tau_B \gg 1$ are fulfilled, that is, if the Brownian motion effects are weak, then we may transit from the statistical description to the dynamical one. In such a situation the particle rotation is determined simply by the balance of viscous and field-induced torques and thus is governed by the equation

$$\frac{\partial \vartheta}{\partial t} = -\frac{\xi_0}{2\tau_B} \sin \vartheta \, \cos \omega t \qquad (4.375)$$

The angular trajectory derived by integration of this equation is

$$\tan\left(\frac{\vartheta}{2}\right) = \tan\left(\frac{\vartheta_0}{2}\right) \exp\left[-\frac{\xi_0}{2\omega\tau_B} \sin \omega t \right] \qquad (4.376)$$

and it depends parametrically on the initial orientation ϑ_0 of the particle. Since the statistical averaging in the dynamical system makes no sense, it is logical, while treating the particle ensemble, to replace it by the averaging over the initial orientation distribution $f(\vartheta_0)$. Assuming that at the moment when the field was turned on, the particles were oriented randomly [i.e., $f(\vartheta_0) = 1/4\pi$], for the orientation order parameter we find

$$\langle P_2 \rangle = 1 - \frac{3zL(z)}{\sinh^2 z}, \qquad z = \frac{\xi_0}{2\omega\tau_B}\sin\omega t \qquad (4.377)$$

In this approximation, unlike the general case, the harmonic amplitudes are determined by a single self-similar argument $\xi_0/\omega\tau_B$. Note also that for the suspensions of ferromagnetic particles, where the hysteretic remagnetization process is possible, Eq. (4.377) holds only if an additional condition

$$\frac{2\omega\tau_B}{\xi_0} = \frac{2C\omega V\eta}{\mu H_0} \leq 1$$

is satisfied [138].

It is necessary to mention that Eqs. (4.375)–(4.377) are close analogs of the expressions derived by O'Konski et al. while studying transient electric birefringence in solutions. The only difference between our Eqs. (4.376) and (4.377) and those of Ref. 149 is in the representation of the self-similar argument; in Ref. 149 it is $\xi_0 t/\tau_B$, since that paper deals with a transient process.

4. Results and Discussion

Formula (4.371) proves explicitly that the suspension birefringence is an even function of the applied field amplitude. For this reason, in response to the excitation of the frequency ω the optical anisotropy Δv oscillates with the basic frequency 2ω. The higher-rank harmonics induced by the saturation behavior of $\Delta v(\xi_0)$ are the multiples of the basic one. It is also clear that besides the oscillatory contribution, the frequency spectrum of Δv contains a constant component.

The amplitude–frequency characteristics of the first four birefringence harmonics (including the zero-rank one) of a dipolar system are plotted in Figures 4.37 and 4.38. All the curves are calculated by the abovementioned numerical procedure; the asymptotics (4.373) and (4.377) were used as the auxiliary relations. The graphs show that for $m = 0, 1$ (i.e., for the zero and second overtones), the behavior of the harmonics is similar no matter which quantity, Δv or $(\Delta v)^2$, is considered. The amplitudes A_0 and B_0 increase monotonically, tending to unity as the field amplitude ξ_0 grows, and they diminish with frequency for $\xi_0 = $ const. The amplitudes A_2 and B_2 first ascend with ξ_0, then pass through a maximum [see curves 1 and 2 in Fig. 4.37(b) and

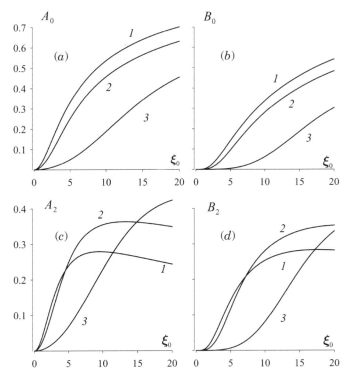

Figure 4.37. Field–amplitude dependencies of the spectral coefficients of the zeroth $(m = 0)$ and first $(m = 1)$ harmonics of $\langle P_2 \rangle$ and $\langle P_2 \rangle^2$; oscillation frequency is $\omega\tau_B = 0$ (1), 1 (2), 4 (3).

curve 1 Fig. 4.37(d) and finally decrease slowly proportional to $(\ln \xi_0 / \xi_0$. This logarithmic diminution is inherent to every harmonic amplitude except for the constant components $(m = 0)$ and is due to cessation of all the orientational vibrations at $\xi_0 \to \infty$.

The essential difference between the field dependencies of the spectral amplitudes A_{2m} and B_{2m}, namely, the experimentally measurable Fourier components of Δv and $(\Delta v)^2$, manifests itself at $m \geq 2$. Below we take $m = 2$ and $m = 3$ as an example. As Figure 4.38 shows, the coefficients A_4 and A_6 in the range $\xi_0 \leq 20$ increase monotonically, while the behavior of the amplitudes B_4 and B_6 is more complicated. In particular, each B_{2m} curve at $\omega\tau_B \to 0$ undergoes a minimum $(B_{2m} = 0)$ at a certain finite value $\xi_0^{(m)}$. These points constitute the following numerical sequence

m	2	3	4	$\cdots$
$\xi_0^{(m)}$	4.06	6.57	8.97	$\cdots$

$$(4.378)$$

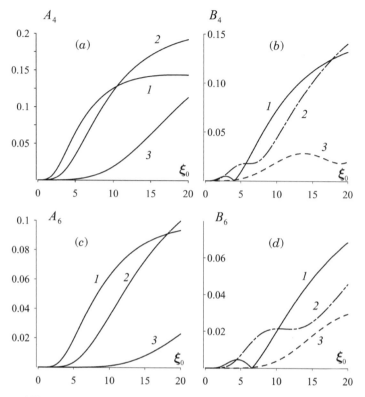

Figure 4.38. Field–amplitude dependencies of the spectral coefficients of the second ($m = 2$) and third ($m = 3$) harmonics of $\langle P_2 \rangle$ and $\langle P_2 \rangle^2$ oscillations; $\omega\tau_B = 0$ (1), 1 (2), 4 (3).

which is derived from Eq. (4.373). When ω grows, the minima at the B_{2m} curves become more shallow and finally cease to exist in the dispersion region $\omega\tau_B \sim 1$. But with further growth of frequency they again acquire the wavelike form [see curves 3 in Figs. 4.38(c) and 4.38(d)]. The surfaces plotted in Figures 4.39 and 4.40 give the global overview on the amplitude–frequency behavior of the functions $B_{4,6}(\xi_0, \omega\tau_B)$. The same features are inherent to every B_{2m} at $m > 3$ as well. At ξ_0, $\omega\tau_B$, using Eq. (4.377), one finds that the positions of the characteristic minima of B_{2m} tend to the values

$$
\begin{array}{c|ccc|c}
m & 2 & 3 & 4 & \cdots \\
\hline
(\xi_0/\omega\tau_B)^{(m)} & 4.13 & 6.54 & 8.84 & \cdots
\end{array}
\qquad (4.379)
$$

Hence, for large ξ_0 and $\omega\tau_B$ the projections of the bottoms of the valleys in Figures 4.39 and 4.40 are straight lines in the (ξ_0, $\omega\tau_B$) plane. The obvious

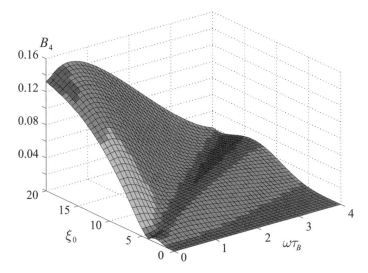

Figure 4.39. Amplitude–frequency dependencies of the spectral coefficient $B_4(\xi_0, \omega\tau_B)$ of the squared order parameter $\langle P_2 \rangle^2$.

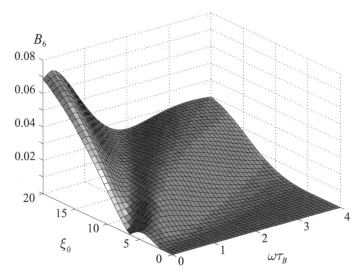

Figure 4.40. Amplitude–frequency dependencies of the spectral coefficient $B_6(\xi_0, \omega\tau_B)$ of the squared order parameter $\langle P_2 \rangle^2$.

resemblance between series (4.378) and (4.379) can hardly be occasional and seems to be due to the qualitative similarity in the behavior of functions (4.373) and (4.377).

The characteristic series (4.378) and (4.379) were already given in our papers [150,151]. In Ref. 150, while considering the nonlinear Cotton–Mouton effect in a magnetic colloid, we obtained Eq. (4.378) for the first time, although with less accuracy. There we proposed and proved the possibility of using the effect of causing a certain B_{2m} amplitude for the evaluation of the mean magnetic moment of the particles in a ferrocolloid to vanish. The parameter in question was obtained by comparing the dimensionless number $\xi_0^{(m)}$ predicted by the theory and the experimental value $H_0^{(m)}$ of the magnetic field amplitude at which in the transmitted light intensity spectrum the harmonic of the frequency $2m\omega$ vanishes. From definition of $\xi_0^{(m)}$ it follows that

$$\mu = \frac{\xi_0^{(m)} T}{H_0^{(m)}} \qquad (4.380)$$

Substituting in Eq. (4.380) the numbers $\xi_0^{(2)}$ and $\xi_0^{(3)}$ from the tabulated values in (4.378) and the values $H_0^{(2)}$ and $H_0^{(2)}$ reported in Ref. 153, we obtained values of μ that are in a satisfactory agreement with the results of other independent measurements.

It is worthwhile to make one comment here. In Ref. 150 we dealt with the orientation optical properties of a superparamagnetic ferrocolloid: the system of type 3, discussed in Section VI.B.1. Then we believed that merely the existence of the specific "spectrum of nodes" (4.378) is sufficient to distinguish a superparamagnetic system from a rigid-dipole one. However, this concept turned out to be invalid. The identity of the numerical series $\{\xi_0^{(m)}\}$ for a superparamagnetic ensemble and a system of magnetically hard (dipolar) particles is established by the fact that in both cases the field dependencies of the orientation order parameter coincide: $p \propto 1 - 3L(\xi_0)/\xi_0$ (see Ref. 66 for details). The real difference in the birefringence dynamics between the systems of types 1 and 3 manifests itself not in the shape of the $B_{2m}(\xi_0)$ curves but in the magnitude of the optical effect. Comparison of the present results with those of Ref. 149 shows that in superparamagnetic systems the "swing" of all the curves $B_{2m}(\xi_0)$ is $15/2\sigma$ times smaller than in the dipolar ones. Here $\sigma = \mu H_a/2T$ is the ratio of the particle magnetic anisotropy energy to its thermal one; for superparamagnetic colloids, σ hardly exceeds unity.

It is clear that the rotation relaxation time of the particles cannot be evaluated using only the quasistatic ($\omega \tau_B \ll 1$) measurements. However, with the aid of the series (4.378) and (4.379), one can manage to find τ_B employing a minimal amount of additional experimental data. Let us assume that we have measured by some method the field behavior of one of the optical harmonics, say, $B_4(H_0)$, at just two frequencies: one low ($\omega_1 \tau_B \ll 1$) and one high ($\omega_2 \tau_B \gg 1$). According to our results, in both cases the curves have to pass through a minimum. We denote by H_{01} and H_{02} the dimensional field amplitude values corresponding to the measured points. Extracting from Eqs. (4.378) and (4.379) the characteristic numbers $\xi_0^{(2)}$ and $(\xi_0/\omega\tau_B)^{(2)}$ and using the ξ_0 definition, we arrive at the relation

$$\tau_B = \frac{\xi_0^{(2)} H_{02}}{(\xi_0/\omega\tau_B)^{(2)} H_{01} \omega_2} \approx \frac{H_{02}}{H_{01} \omega_2} \qquad (4.381)$$

which enables us to estimate the relaxation time of the particle orientation through the observed quantities: the positions of the field dependence minima of the lightbeam intensity harmonics.

In conclusion, let us estimate the dimensional values of particle parameters and field amplitudes at which the orientational effects discussed would be most pronounced and, hence, observable. For a magnetic colloid (single-domain particles of linear size ~ 10 nm, volume $V_m \sim 5 \times 10^{-19}$ cm^3, and magnetization $I_s \sim 500$ G) at room temperature, one finds $\mu = I_s V_m \sim 2.5 \cdot 10^{-16}$ erg/G. The magnetic field amplitude that can cause a distinctive orientational nonlinearity is determined by the condition $\xi_0 \geq 1$, which yields $H_0 \geq T/\mu \sim 160$ Oe. According to Ref. 152, the actual values $H_0^{(2)}$ and $H_0^{(3)}$ were 500 and 683 Oe, respectively.

For ferroelectric particles (the same size as in a magnetic colloid) with the spontaneous polarization $P \sim 5 \times 10^{-6}$ C/cm^2, we get $\mu = PV \sim 7.5 \times 10^{-15}$ CGSE units. Hence, the electric field amplitude that satisfies the condition $\xi_0 \geq 1$ is $E_0 \geq 5$ CGSE units or 1.5 kV/cm.

For a solution of stiff-chain macromolecules (for example, PBLG), with $\mu \sim 3 \times 10^3$ Debye [153], the corresponding estimation shifts to $E_0 \geq 4$ kV/cm but still remains within the experimentally feasible range.

Acknowledgments

The authors are grateful to Yu. P. Kalmykov, W. T. Coffey, V. V. Rusakov, and P. C. Fannin for fruitful discussions and stimulating advice.

Partial financial support from the International Association for the Promotion of Cooperation with Scientists from the New Independent States of the Former Soviet Union (INTAS) under Grant 01–2341 and Award No. PE–009–0 of the U.S. Civilian Research & Development Foundation for the Independent States of the Former Soviet Union (CRDF) is gratefully acknowledged.

The package Maple V for PC used in the calculations was obtained by the Institute of Continuous Media Mechanics in the framework of the EuroMath Network and Services for the New Independent States—Phase II (EmNet/NIS/II) project funded by INTAS under grant IA-003.

References

1. R. E. Rosensweig, *Ferrohydrodynamics*, 2nd ed., Dover, New York, 1997.
2. E. Blums, A. Cebers, and M. M. Maiorov, *Magnetic Fluids,* de Gruyter, Berlin, 1997.
3. Proc. 7th Int. Conf. Magnetic Fluids, *J. Magn. Magn. Mat.* **149**(1–2) (1995).
4. Proc. 8th Int. Conf. Magnetic Fluids, *J. Magn. Magn. Mat.* **201**(1–3) (1999).
5. Proc. 9th Int. Conf. Magnetic Fluids, *J. Magn. Magn. Mat.* **252**(1–3) (2003).
6. Proc. 2nd Int. Conf. Scientific and Clinical Applications of Magnetic Carriers, *J. Magn. Magn. Mat.* **194**(1–3) (1999).
7. E. Hasmonay, E. Dubois, J.-C. Bacri, R. Perzynski, Yu. L. Raikher, and V. I. Stepanov, *Eur. Phys. J. B* **5**, 859 (1998).
8. E. Hasmonay, J. Depeyrot, M. H. Sousa, F. A. Tourinho, J.-C. Bacri, R. Perzynski, Yu. L. Raikher, and I. Rosenman, *J. Appl. Phys.* **88**, 6628 (2000).
9. A. A. Potanin, G. J. Reynolds, and R. Hirko, *J. Magn. Magn. Mat.* **211**, 277 (2000).
10. V. T. Peikov and A. M. Lane, *J. Colloid Interf. Sci.* **206**, 350 (1998).
11. V. T. Peikov, K. S. Jeon, and A. M. Lane, *J. Magn. Magn. Mat.* **193**, 307 (1999).
12. A. Moser, K. Takano, D. T. Margulies, M. Albrecht, Y. Sonobe, Y. Ikeda, S. Sun, and E. E. Fullerton, *J. Phys. D: Appl. Phys.* **35**, R157 (2002).
13. L. Néel, *C. R. Acad. Sci.* (Paris) **228**, 664 (1949); *Ann. Geophys.* **5**, 99 (1949).
14. C. P. Bean and J. D. Livingston, *J. Appl. Phys.* **30**, 120S (1959).
15. S. Chikatzumi, *Physics of Magnetism*, Wiley, New York, 1964.
16. S. H. Charap, P.-L. Lui, and Y. He, *IEEE Trans. Magn.* **33**, 978 (1997).
17. D. Weller and A. Moser, *IEEE Trans. Magn.* **35**, 4423 (1999).
18. L. Gammaitoni, P. Hänggi, P. Jung, and F. Marchesoni, *Rev. Mod. Phys.* **70**, 223 (1998).
19. R. Benzi, R. A. Sutera, and A. Vulpiani, *J. Phys. A: Gen. Phys.* **14**, L453 (1981).
20. K. Weisenfeld and F. Moss, *Nature* **373**, 33 (1994).
21. R. Bartussek, P. Hänggi, and P. Jung, *Phys. Rev. E* **49**, 3930 (1994).
22. P. Jung and P. Talkner, *Phys. Rev. E* **51**, 2640 (1995).
23. Yu. L. Raikher and V. I. Stepanov, *Fiz. Tverd. Tela* (St. Petersburg) **43**, 270 (2001) [*Phys. Solid State*, **43**, 279 (2001)].
24. Yu. L. Raikher and V. I. Stepanov, *Phys. Rev. Lett.* **86**, 1923 (2001).
25. A. N. Grigorenko, V. I. Konov, and P. I. Nikitin, *Pis'ma v Zh. Eksp. Teor. Fiz.* **52**, 1182 (1990) [*Sov. Phys. JETP Lett.* **52**, 593 (1990)].
26. E. K. Sadykov, *J. Phys. C: Condens. Mat.* **4**, 3295 (1992).
27. L. B. Kiss, Z. Gingl, Z. Márton, J. Kertész, F. Moss, G. Schmera, and A. Bulsara, *J. Stat. Phys.* **70**, 451 (1993).
28. A. N. Grigorenko, P. I. Nikitin, A. N. Slavin, and P. Y. Zhou, *J. Appl. Phys.* **76**, 6335 (1994).
29. Yu. L. Raikher and V. I. Stepanov, *J. Phys. C: Condens. Mat.* **6**, 4137 (1994).
30. Yu. L. Raikher and V. I. Stepanov, *Phys. Rev. B* **52**, 3493 (1995).
31. A. N. Grigorenko, P. I. Nikitin, and G. V. Roschepkin, *J. Appl. Phys.* **79**, 6113 (1996).
32. N. Smith and P. Arnett, *Appl. Phys. Lett.* **78**, 1448 (2001).
33. V. L. Safonov and H. N. Bertram, *J. Appl. Phys.* **87**, 5681 (2000).
34. H. N. Bertram, *J. Appl. Phys.* **90**, 5768 (2001).

35. H. Risken, *The Fokker–Planck Equation, Methods of Solutions and Applications*, Springer-Verlag, Berlin, 1989.

36. W. T. Coffey, Yu. P. Kalmykov, and J. T. Waldron, *The Langevin Equation. With applications in Physics, Chemistry and Electrical Engineering*, World Scientific, Singapore, 1996.

37. L. D. Landau and E. M. Lifshitz, *Phys. Z. Sowjetunion* **8**, 153 (1935).

38. R. Kaiser and G. Miskolczy, *J. Appl. Phys.* **41**, 1064 (1970).

39. X. Zianni, K. N. Trohidou, and J. A. Blackman, *J. Appl. Phys.* **81**, 4739 (1997).

40. R. H. Kodama, A. E. Berkowitz, E. J. McNiff, and S. Foner, *Phys. Rev. Lett.* **77**, 394 (1996).

41. C. Chen, O. Kitakami, and Y. Shimada, *J. Appl. Phys.* **84**, 2184 (1998).

42. V. P. Shilov, J. C. Bacri, F. Gazeau, F. Gendron, R. Perzynski, and Yu. L. Raikher, *J. Appl. Phys.* **85**, 6642 (1999).

43. V. P. Shilov, Yu. L. Raikher, J. C. Bacri, F. Gazeau, and R. Perzynski, *Phys. Rev. B* **60**, 11902 (1999).

44. X. Batlle and A. Labarta, *J. Phys. D: Appl. Phys.* **35**, R15 (2002).

45. T. L. Gilbert, *Phys. Rev.* **100**, 1243 (1955).

46. G. V. Skrotskii and L. V. Kurbatov, in *Ferromagnetic Resonance*, S. V. Vonsovsii, ed., Pergamon Press, Oxford, 1966.

47. W. F. Brown, Jr., *Phys. Rev.* **130**, 1677 (1963).

48. Yu. L. Raikher and M. I. Shliomis, *Adv. Chem. Phys.* **87**, 595 (1994).

49. L. D. Landau and E. M. Lifshitz, *Quantum Mechanics*, Pergamon Press, Oxford, 1959.

50. W. T. Coffey, D. A. Garanin, and D. J. McCarthy, *Adv. Chem. Phys.* **117**, 483 (2001).

51. C. W. Gardiner, *Handbook of Stochastic Methods, 2nd ed.,* Springer Series in Synergetics, **13**, H. Haken, ed., Springer, Berlin, 1985.

52. J. S. Langer, *Phys. Rev. Lett.* **21**, 973 (1968).

53. J. S. Langer, *Ann. Phys.* **54**, 258 (1969).

54. Yu. L. Raikher and M. I. Shliomis, *Zh. Eksp. Teor. Fiz.* **74**, 1060 (1974) [*Sov. Phys. JETP* **40**, 526 (1974)].

55. A. Aharoni, *Phys. Rev.* **177**, 799 (1969).

56. A. Aharoni, *Phys. Rev. B* **7**, 1103 (1976).

57. A. Aharoni, *Phys. Rev. B* **46**, 5434 (1994).

58. W. T. Coffey, D. S. F. Crothers, Yu. P. Kalmykov, E. S. Massawe, and J. T. Waldron, *Phys. Rev. E* **49**, 1869 (1994).

59. W. T. Coffey, D. S. F. Crothers, Yu. P. Kalmykov, and J. T. Waldron, *Phys. Rev. B* **51**, 15947 (1995).

60. W. T. Coffey, D. S. F. Crothers, J. L. Dormann, L. J. Geoghegan, Yu. P. Kalmykov, J. T. Waldron, and A. W. Wickstead, *Phys. Rev. B* **52**, 15951 (1995).

61. C. A. J. Fletcher, *Computational Technique for Fluid Dynamics*, Vol. 1, Springer, Berlin, 1988.

62. E. Kneller, in *Magnetism and Metallurgy*, Vol.1., A. Berkowitz and E. Kneller, eds., Academic Press, New York, 1982, Ch.8.

63. S. L. Ginzburg, *Irreversible Phenomena in Spin Glasses*, Nauka, Moscow, 1989.

64. T. Bitoh, K. Ohba, M. Takamatsu, T. Shirane, and S. Chikzawa, *J. Phys. Soc. Japan* **64**, 1311 (1995).

65. S. Chikazawa, T. Bitoh, K. Ohba, M. Takamatsu, and T. Shirane, *J. Magn. Magn. Mat.* **154**, 59 (1996).

66. D. A. Krueger, *J. Appl. Phys.* **50**, 8169 (1979).

67. Yu. L. Raikher and V. I. Stepanov, *Phys. Rev. B* **55**, 15005 (1997).

68. J. L. García-Palacios and F. J. Lázaro, *Phys. Rev. B* **55**, 1006 (1997).

69. W. Coffey, M. Evans, and P. Grigolini, *Molecular Diffusion and Spectra*, Wiley, New York, 1984.

70. Yu. L. Raikher and V. I. Stepanov, *Europhys. Lett.* **32**, 589 (1995).

71. W. T. Coffey, P. J. Cregg, D. S. F. Crothers, J. T. Waldron, and A. W. Wickstead, *J. Magn. Magn. Mat.* **131**, L301 (1994).

72. M. I. Shliomis and V. I. Stepanov, *J. Magn. Magn. Mat.* **122**, 176 (1993).

73. J. I. Gittleman, B. Abeles, and S. Bozowski, *Phys. Rev. B* **9**, 3891 (1974).

74. A. F. Pshenichnikov and A. V. Lebedev, *Sov. Phys. JETP* **68**, 498 (1989).

75. W. H. Press, B. P. Flannery, S. A. Teulolsky, and W. T. Vetterling, *Numerical Recipes. The Art of Scientific Computing*, Cambridge Univ. Press, Cambridge, U.K., 1989.

76. J. Zhang, C. Boyd, and W. Luo, *Phys. Rev. Lett.* **77**, 390 (1996).

77. L. Spinu, D. Fiorani, H. Srikanth, F. Lucari, F. D'Orazio, E. Tronc, and M. Nogués, *J. Magn. Magn. Mat.* **226**, 1927 (2001).

78. Yu. L. Raikher and V. I. Stepanov, *Phys. Rev. B* **66**, 214406 (2002).

79. D. A. Garanin, E. C. Kennedy, D. C. F. Crothers, W. T. Coffey, *Phys. Rev. E* **60**, 6499 (1999).

80. J. García-Palacios and P. Svedlidh, *Phys. Rev. Lett.* **85**, 3724 (2000).

81. V. I. Stepanov and Yu. L. Raikher, *J. Magn. Magn. Mat.* **272–276**, E1277 (2004).

82. B. A. Storonkin, *Kristallografiya* **30**, 841 (1985) [*Sov. Phys. Crystallogr.* **30**, 489 (1985)].

83. A. Aharoni, *Phys. Rev.* **135A**, 447 (1964).

84. Yu. P. Kalmykov, *Phys. Rev. E* **61**, 6320 (2000).

85. L. Bessias, L. BenJaffel, and J.-L. Dormann, *Phys. Rev. B* **45**, 7805 (1992).

86. A. Aharoni, *Phys. Rev. B* **46**, 5434 (1992).

87. P. G. Cregg, D. S. F. Crothers, and A. W. Weakstead, *J. Appl. Phys.* **76**, 4900 (1994).

88. W. F. Brown (Jr.), *Physica B* **86**, 1423 (1977).

89. W. F. Brown (Jr.), *IEEE Trans. Magn.* **15**, 1196 (1979).

90. W. T. Coffey, *Adv. Chem. Phys.* **103**, 259 (1998).

91. D. A. Garanin, V. V. Ishchenko, and L. V. Panina, *Teor. Mat. Fiz.* **82**, 242 (1990) [*Theor. Math. Phys.* **82**, 169 (1990)].

92. W. T. Coffey and D. S. F. Crothers, *Phys. Rev. E* **54**, 4768 (1996).

93. I. Klik and Y. D. Yao, *J. Magn. Magn. Mat.* **186**, 233 (1998).

94. W. F. Brown (Jr.), *J. Appl. Phys.* **30**, Suppl. 5, 130S (1959).

95. M. I. Dykman, D. G. Luchinsky, R. Manella, P. V. E. McClintock, N. D. Stein, and N. G. Stocks, *Nuovo Cimento D* **17**, 661 (1995).

96. M. I. Dykman, R. Mannella, P. V. E. McClintock, and N. G. Stocks, *Phys. Rev. Lett.* **65**, 2606 (1990).

97. M. I. Dykman, R. Mannella, P. V. E. McClintock, and N. G. Stocks, *Pis'ma v ZhETF* **52**, 281 (1990); *Sov. Phys. JETP Lett.* **52**, 141 (1990).

98. L. Gammaitoni, F. Marchesoni, M. Martinelli, L. Pardi, and S. Santucci, *Phys. Lett. A* **158**, 449 (1991).

99. M. I. Dykman, R. Mannella, P. V. E. McClintock, and N. G. Stocks, *Phys. Rev. Lett.* **68**, 2985 (1992).

100. M. Morillo and J. Gómez-Ordóñez, *Phys. Rev. A* **46**, 6738 (1992).

101. L. Gammaitoni and F. Marchesoni, *Phys. Rev. Lett.* **70**, 873 (1993).

102. M. I. Dykman, R. Mannella, P. V. E. McClintock, and N. G. Stocks, *Phys. Rev. Lett.* **70**, 874 (1993).

103. M. Morillo and J. Gómez-Ordóñez, *Phys. Rev. Lett.* **71**, 9 (1993).

104. C. P. Bean, *J. Appl. Phys.* **26**, 1381 (1955).

105. W. Wernsdorfer, E. Bonet Orozco, K. Hasselbach, A. Benoit, B. Barbara, N. Demoncy, A. Loiseau, H. Pascard, and D. Mailly, *Phys. Rev. Lett.* **78**, 1791 (1997).

106. H. Casalta, P. Schleger, C. Bellouard, M. Hennion, I. Mirebeau, G. Ehlers, B. Farago, J.-L. Dormann, M. Kelsch, M. Linde, and F. Phillipp, *Phys. Rev. Lett.* **82**, 1301 (1999).

107. T. Leiber, F. Marchesoni, and H. Risken, *Phys. Rev. A* **38**, 983 (1988).

108. T. Leiber, F. Marchesoni, and H. Risken, *Phys. Rev. Lett.* **59**, 1381 (1992).

109. A. J. Martin, G. Meier, and A. Saupe, *Symp. Faraday Soc.* **5**, 119 (1971).

110. P. Jung and P. Hanggi, *Z. Phys. B* **90**, 225 (1993).

111. P. C. Fannin and Yu. L. Raikher, *J. Phys. D: Appl. Phys.* **34**, 1612 (2001).

112. P. C. Fannin, Yu. L. Raikher, A. T. Giannitis, and S. W. Charles, *J. Magn. Magn. Mat.* **252**, 114 (2002).

113. Yu. L. Raikher, V. I. Stepanov, and P. C. Fannin, *J. Magn. Magn. Mat.* **258**, 369 (2003).

114. H. Gang, G. Nicolis, and C. Nicolis, *Phys. Rev. A* **42**, 2030 (1990).

115. M. E. Inchiosa, A. R. Bulsara, and L. Gammaitoni, *Phys. Rev. E* **55**, 4049 (1997).

116. D. A. Garanin, *Phys. Rev. E* **54**, 3250 (1996).

117. M. Sparks, *Ferromagnetic Relaxation*, McGraw-Hill, New York, 1964.

118. S. V. Vonsovskii, ed., *Ferromagnetic Resonance*, Pergamon Press, Oxford, 1966.

119. W. Wernsdorfer, E. Bonet Orozco, K. Hasselbach, A. Benoit, B. Barbara, N. Demoncy, and A. Loiseau, *Phys. Rev. Lett.* **78**, 1791 (1997).

120. Yu. L. Raikher, V. I. Stepanov, A. N. Grigorenko, and P. I. Nikitin, *Phys. Rev. E* **56**, 6400 (1997).

121. E. K. Sadykov and A. G. Isavnin, *Fiz. Tverd. Tela (St. Petersburg)* **38**, 2104 (1996) [*Phys. Solid State* **38**, 1160 (1998).

122. M. E. Inchiosa and A. R. Bulsara, *Phys. Rev. E* **58**, 115 (1998).

123. K. Loerincz, Z. Gingl, L. B. Kiss, and A. R. Bulsara, *Phys. Lett. A* **254**, 154 (1999).

124. Yu. L. Raikher and V. I. Stepanov, *Phys. Metals Metallogr.* **91** (Suppl. 1), 100 (2001).

125. Yu. L. Raikher and V. I. Stepanov, *J. Magn. Magn. Mat.* **252**, 129 (2002).

126. A. Perez-Madrid and J. M. Rubi, *Phys. Rev. E* **51**, 4159 (1995).

127. J. M. G. Vilar and J. M. Rubi, *Phys. Rev. Lett.* **77**, 2863 (1996).

128. J. M. G. Vilar, A. Perez-Madrid, and J. M. Rubi, *Phys. Rev. E* **54**, 6929 (1996).

129. Yu. P. Kalmykov and S. V. Titov, *Fiz. Tverd. Tela* (St. Petersburg) **40**, 1642 (1998) [*Phys. Solid State* **40**, 1492 (1998).

130. M. A. Martsenyuk, Yu. L. Raikher, and M. I. Shliomis, *Zh. Eksp. Teor. Fiz.* **65**, 834 (1973) [*Sov. Phys. JETP Lett.* **38**, 413 (1974)].

131. V. I. Stepanov and M. I. Shliomis, *Bull. Acad. Sci. USSR, Ser. Phys.* **55**, 1 (1991).

132. M. I. Shliomis and V. I. Stepanov, *Adv. Chem. Phys.* **87**, 1 (1994).

133. Yu. L. Raikher and V. I. Stepanov, *J. Magn. Magn. Mat.* **196**, 88 (1999).

134. Yu. L. Raikher and V. I. Stepanov, in *Proc. Moscow Int. Symp. Magnetism*, Moscow, June 20–24, 1999, MSU, Part. 1, p. 50.

135. J. C. Bacri and R. Perzynski, in *Lecture Notes in Physics: Complex Fluids*, Vol. 85, L. Garrido, ed., Springer, Berlin, 1993.

136. Yu. N. Skibin, V. V. Chekanov, and Yu. L. Raikher, *Zh. Eksp. Teor. Fiz.* **72**, 949 (1977) [*Sov. Phys. JETP* **45**, 496 (1977)].

137. Yu. L. Raikher and Yu. N. Skibin, *C. R. (Comptes Rendu) Acad. Sci. USSR* **302**, 1088 (1988) [*Sov. Phys. Doklady* **302**, 746 (1988)].

138. Yu. L. Raikher and P. C. Scholten, *J. Magn. Magn. Mat.* **74**, 275 (1988).

139. M. I. Shliomis and V. I. Stepanov, *J. Magn. Magn. Mat.* **122**, 196 (1993).

140. P. C. Scholten, *IEEE Trans. Magn.* **16**, 77 (1980).

141. P. C. Scholten, *IEEE Trans. Magn.* **16**, 221 (1980).

142. D. P. E. Dickson, N. M. K. Reid, C. Hunt, H. D. Williams, M. El-Hilo, and K. O'Grady, *J. Magn. Magn. Mat.* **125**, 345 (1993).

143. M. V. Volkenshtein, *Molecular Optics, GITTL,* Moscow, 1951.

144. H. Haus, *Waves and Fields in Optoelectronics*, Prentice-Hall, Englewood Cliffs, NJ, 1984.

145. H. Watanabe and A. Morita, *Adv. Chem. Phys.* **56**(1), 225 (1984).

146. Yu. L. Raikher, V. I. Stepanov, and S. V. Burylov, *Kolloid. Zh.* **52**, 887 (1990) [*Colloid J. USSR* **52**, 768 (1990)].

147. R. Ullman, *J. Chem. Phys.* **56**, 1869 (1972).

148. H. Watanabe and A. Morita, *J. Chem. Phys.* **78**, 5311 (1983).

149. C. T. O'Konski, K. Yoshioka, and W. H. Orttung, *J. Phys. Chem.* **63**, 1558 (1959).

150. Yu. L. Raikher, S. V. Burylov, and V. I. Stepanov, *Pis'ma Zh. Eksp. Teor. Fiz.* **47**, 273 (1988). [*Sov. Phys. JETP Lett.* **47**, 330 (1988)].

151. Yu. L. Raikher, V. I. Stepanov, and S. V. Burylov, in *Dynamic Behavior of Macromolecules, Colloids, Liquid Crystals and Biological Systems*, H. Watanabe, ed., Hirokawa Publ. Co., Tokyo, 1989, p. 23.

152. Yu. N. Skibin, in *Abstracts 4th All-Union Conf. Magnetic Fluids,* Plyos, USSR (now Russia), 1985, Vol. 2, p. 98.

153. V. N. Tzvetkov, *Stiff-Chain Polymer Molecules*, Nauka, Leningrad, 1986.

AUTHOR INDEX

Numbers in parentheses are reference numbers and indicate that the author's work is referred to although his name is not mentioned in the text. Numbers in *italic* show the pages on which the complete references are listed.

Advances in Chemical Physics, Volume 129, edited by Stuart A. Rice
ISBN 0-471-44527-4 Copyright © 2004 John Wiley & Sons, Inc.

SUBJECT INDEX

Advances in Chemical Physics, Volume 129, edited by Stuart A. Rice
ISBN 0-471-44527-4 Copyright © 2004 John Wiley & Sons, Inc.